Geophysical Monograph Series

Including

IUGG Volumes
Maurice Ewing Volumes
Mineral Physics Volumes

88 **Active Margins and Marginal Basins of the Western Pacific** *Brian Taylor and James Natland (Eds.)*

89 **Natural and Anthropogenic Influences in Fluvial Geomorphology** *John E. Costa, Andrew J. Miller, Kenneth W. Potter, and Peter R. Wilcock (Eds.)*

90 **Physics of the Magnetopause** *Paul Song, B.U.Ö. Sonnerup, and M.F. Thomsen (Eds.)*

91 **Seafloor Hydrothermal Systems: Physical, Chemical, Biological, and Geological Interactions** *Susan E. Humphris, Robert A. Zierenberg, Lauren S. Mullineaux, and Richard E. Thomson (Eds.)*

92 **Mauna Loa Revealed: Structure, Composition, History, and Hazards** *J. M. Rhodes and John P. Lockwood (Eds.)*

93 **Cross-Scale Coupling in Space Plasmas** *James L. Horowitz, Nagendra Singh, and James L. Burch (Eds.)*

94 **Double-Diffusive Convection** *Alan Brandt and H. J. S. Fernando (Eds.)*

95 **Earth Processes: Reading the Isotopic Code** *Asish Basu and Stan Hart (Eds.)*

96 **Subduction Top to Bottom** *Gray E. Bebout, David Scholl, Stephen Kirby, and John Platt (Eds.)*

97 **Radiation Belts: Models and Standards** *J. F. Lemaire, D. Heynderickx, and D. N. Baker (Eds.)*

98 **Magnetic Storms** Bruce *T. Tsurutani, Walter D. Gonzalez, Yohsuke Kamide, and John K. Arballo (Eds.)*

99 **Coronal Mass Ejections** *Nancy Crooker, Jo Ann Joselyn, and Joan Feynman (Eds.)*

100 **Large Igneous Provinces** *John J. Mahoney and Millard F. Coffin (Eds.)*

101 **Properties of Earth and Planetary Materials at High Pressure and Temperature** *Murli Manghnani and Takehiki Yagi (Eds.)*

102 **Measurement Techniques in Space Plasmas: Particles** *Robert F. Pfaff, Joseph E. Borovsky, and David T. Young (Eds.)*

103 **Measurement Techniques in Space Plasmas: Fields** *Robert F. Pfaff, Joseph E. Borovsky, and David T. Young (Eds.)*

104 **Geospace Mass and Energy Flow: Results From the International Solar-Terrestrial Physics Program** *James L. Horwitz, Dennis L. Gallagher, and William K. Peterson (Eds.)*

105 **New Perspectives on the Earth's Magnetotail** *A. Nishida, D. N. Baker, and S. W. H. Cowley (Eds.)*

106 **Faulting and Magmatism at Mid-Ocean Ridges** *W. Roger Buck, Paul T. Delaney, Jeffrey A. Karson, and Yves Lagabrielle (Eds.)*

107 **Rivers Over Rock: Fluvial Processes in Bedrock Channels** *Keith J. Tinkler and Ellen E. Wohl (Eds.)*

108 **Assessment of Non-Point Source Pollution in the Vadose Zone** *Dennis L. Corwin, Keith Loague, and Timothy R. Ellsworth (Eds.)*

109 **Sun-Earth Plasma Interactions** *J. L. Burch, R. L. Carovillano, and S. K. Antiochos (Eds.)*

110 **The Controlled Flood in Grand Canyon** *Robert H. Webb, John C. Schmidt, G. Richard Marzolf, and Richard A. Valdez (Eds.)*

111 **Magnetic Helicity in Space and Laboratory Plasmas** *Michael R. Brown, Richard C. Canfield, and Alexei A. Pevtsov (Eds.)*

112 **Mechanisms of Global Climate Change at Millennial Time Scales** *Peter U. Clark, Robert S. Webb, and Lloyd D. Keigwin (Eds.)*

113 **Faults and Subsurface Fluid Flow in the Shallow Crust** *William C. Haneberg, Peter S. Mozley, J. Casey Moore, and Laurel B. Goodwin (Eds.)*

114 **Inverse Methods in Global Biogeochemical Cycles** *Prasad Kasibhatla, Martin Heimann, Peter Rayner, Natalie Mahowald, Ronald G. Prinn, and Dana E. Hartley (Eds.)*

115 **Atlantic Rifts and Continental Margins** *Webster Mohriak and Manik Talwani (Eds.)*

116 **Remote Sensing of Active Volcanism** *Peter J. Mouginis-Mark, Joy A. Crisp, and Jonathan H. Fink (Eds.)*

117 **Earth's Deep Interior: Mineral Physics and Tomography From the Atomic to the Global Scale** *Shun-ichiro Karato, Alessandro Forte, Robert Liebermann, Guy Masters, Lars Stixrude (Eds.)*

118 **Magnetospheric Current Systems** *Shin-ichi Ohtani, Ryoichi Fujii, Michael Hesse, and Robert L. Lysak (Eds.)*

119 **Radio Astronomy at Long Wavelengths** *Robert G. Stone, Kurt W. Weiler, Melvyn L. Goldstein, and Jean-Louis Bougeret (Eds.)*

120 **GeoComplexity and the Physics of Earthquakes** *John B. Rundle, Donald L. Turcotte, and William Klein (Eds.)*

121 **The History and Dynamics of Global Plate Motions** *Mark A. Richards, Richard G. Gordon, Rob D. van der Hilst (Eds.)*

122 **Dynamics of Fluids in Fractured Rock** *Boris Faybishenko, Paul A. Witherspoon, and Sally M. Benson (Eds.)*

123 **Atmospheric Science Across the Stratopause** *David E. Siskind, Stephen D. Eckerman, and Michael E. Summers (Eds.)*

Geophysical Monograph 124

Natural Gas Hydrates
Occurrence, Distribution, and Detection

Charles K. Paull
William P. Dillon
Editors

American Geophysical Union
Washington, DC

Library of Congress Cataloging-in-Publication Data
Natural gas hydrates: occurrence, distribution, and detection / Charles K. Paull, William P. Dillon, editors
p. cm. -- (Geophysical monograph ; 124)
Includes bibliographical references.
ISBN 0-87590-982-5
1. Natural gas hydrates. I. Paull, Charles K. II. Dillon, William P. III. Series

TN884.N39.2001
553.2'85--dc21

2001022907

ISBN 0-87590-982-5
ISSN 0065-8448

2000 Florida Avenue, N.W.
Washington, DC 20009

Printed in the United States of America.

CONTENTS

Preface
Charles K. Paull and William P. Dillon . . . vii

GENERAL ISSUES

The Global Occurrence of Natural Gas Hydrates
Keith A. Kvenvolden and Thomas D. Lorenson . . . 3

Modeling the Global Carbon Cycle With a Gas Hydrate Capacitor: Significance for the Latest Paleocene Thermal Maximum
Gerald R. Dickens . . . 19

GEOCHEMISTRY OF GAS HYDRATE

Overviews and Methods of Geochemistry

Ion Exclusion Associated With Marine Gas Hydrate Deposits
William Ussler III and Charles K. Paull . . . 41

History and Significance of Gas Sampling During the DSDP and ODP
Charles K. Paull and William Ussler III. . . . 53

Gas Hydrates in Convergent Margins: Formation, Occurrence, Geochemistry and Global Significance
Miriam Kastner. . . . 67

Geochemical Regional Studies

Cascadia

Sea Floor Methane Hydrates at Hydrate Ridge, Cascadia Margin
E. Suess, M.E. Torres, G. Bohrmann, R.W. Collier, D. Rickert, C. Goldfinger, P. Linke, A. Heuser, H. Sahling, K. Heeschen, C. Jung, K. Nakamura, J. Greinert, O. Pfannkuche, A. Trehu, G.Klinkhammer, M.J. Whiticar, A. Eisenhauer, B. Teichert, and M. Elvert . . . 87

Gas Hydrate-Associated Carbonates and Methane-Venting at Hydrate Ridge: Classification, Distribution and Origin of Authigenic Lithologies
Jens Greinert, Gerhard Bohrmann, and Erwin Suess . . . 99

Carbon Isotopes of Biomarkers Derived from Methane-Oxidizing Microbes at Hydrate Ridge, Cascadia Convergent Margin
Marcus Elvert, Jens Greinert, Erwin Suess, and Michael J. Whiticar . . . 115

Gulf of Mexico

Stability of Thermogenic Gas Hydrate in the Gulf of Mexico: Constraints on Models of Climate Change
Roger Sassen, Stephen T. Sweet, Alexei V. Milkov, Debra A. DeFreitas, Mahlon C. Kennicutt, and Harry H. Roberts . . . 131

CONTENTS

Fluid and Gas Expulsion on the Northern Gulf of Mexico Continental Slope: Mud-Prone to Mineral-Prone Responses
Harry H. Roberts 145

GEOPHYSICS

Geophysical Overviews and Methods

Deep-tow Seismic Investigations of Methane Hydrates
Warren T. Wood and Joseph F. Gettrust 165

Comparison of Elastic Velocity Models for Gas-Hydrate-Bearing Sediments
Myung W. Lee and Timothy S. Collett 179

Quantitative Well-Log Analysis of In-Situ Natural Gas Hydrates
Timothy S. Collett 189

Geophysical Regional Studies

Blake Ridge

Seafloor Collapse and Methane Venting Associated with Gas Hydrate on the Blake Ridge—Causes and Implications to Seafloor Stability and Methane Release
William P. Dillon, Jeffrey W. Nealon, Michael H. Taylor, Myung W. Lee, Rebecca M. Drury, and Christopher H. Anton 211

Seismic Studies of the Blake Ridge: Implications for Hydrate Distribution,Methane Expulsion, and Free Gas Dynamics
W. Steven Holbrook 235

Peru/Middle American Trenches

Gas Hydrates Along the Peru and Middle America Trench System
Ingo A. Percher, Nina Kukowski, Cesar R. Ranero, and Roland von Huene 257

Cascadia

Geophysical Studies of Marine Gas Hydrates in Northern Cascadia
R. D. Hyndman, G. D. Spence, R. Chapman, M. Reidel, and R. N. Edwards 273

High-resolution Multibeam Survey of Hydrate Ridge, Offshore Oregon
David Clague, Norm Maher, and Charles K. Paull 297

IMPLICATIONS

Potential Influence of Gas Hydrates on Seabed Installations
Martin Hovland and Ove Tobias Gudmestad 307

PREFACE

We publish this volume at a time when there is a growing interest in gas hydrates and major expansion in international research efforts. The first recognition of natural gas hydrate on land in Arctic conditions was in the mid-1960's (by I. Makogon) and in the seabed environment only in the early 1970's, after natural seafloor gas hydrate was drilled on the Blake Ridge during Deep Sea Drilling Project Leg 11. Initial scientific investigations were slow to develop because the study of natural gas hydrates is unusually challenging. Gas hydrate exists in nature in conditions of temperature and pressure where human beings cannot survive, and if gas hydrate is transported from its region of stability to normal Earth-surface conditions, it dissociates. Thus, in contrast to most minerals, we cannot depend on drilled samples to provide accurate estimates of the amount of gas hydrate present. Even the heat and changes in chemistry (methane saturation, salinity, etc.) introduced by the drilling process affect the gas hydrate, independent of the changes brought about by moving a sample to the surface. Gas hydrate has been identified in nature generally by inference from indirect evidence in drilling data or by using remotely sensed indications, mostly from seismic data. Obviously, the established techniques of geologic analysis, which require direct observation and sampling, do not apply to gas hydrate studies, and controversy has surrounded many interpretations. Pressure/temperature conditions appropriate for the existence of gas hydrate occur over the greater part of the shallow subsurface of the Earth beneath the ocean at water depths exceeding about 500 m (shallower beneath colder Arctic seas) and on land beneath high-latitude permafrost. Gas hydrate actually will be present in such conditions, however, only where methane is present at high concentrations. In the Arctic, these methane concentrations are often associated with petroleum deposits, whereas at continental margins in the oceans, where by far the greatest amount of gas hydrate occurs, the gas is almost all microbially derived methane. The margins of the oceans are where the flux of organic carbon to the sea floor is greatest because oceanic biological productivity is highest and organic detritus from the continents also collects to some extent. Furthermore, the continental margins are where sedimentation rates are fastest, so that the rapid accumulation of sediment serves to cover and seal the organic material before it is oxidized, allowing the microorganisms in the sediments to use it as food and form the methane that becomes incorporated into gas hydrate.

Initially, gas hydrate in the natural environment was looked on as little more than a curiosity. However, estimates by various workers shortly began to point to a huge volume of methane bound up in the gas hydrate reservoir. Over the past 10-12 years, estimates have converged on a "consensus value" of about 21×10^{15} cubic meters of methane in natural gas hydrate, according to Keith Kvenvolden. Agreement on this value is within an order of magnitude despite the varying approaches that have been taken by 10 different sets of workers. The amount may represent 3,000 times the amount of methane in the present atmosphere, where it is a powerful greenhouse gas. It may also represent twice the amount of all fossil fuels on Earth (compared on the basis of the mass of carbon).

Gas hydrate is considered important for several reasons. It may dissociate by natural processes in the sediments, creating overpressures that can trigger landslides, thus controlling natural seafloor mass movements. It may be dissociated by petroleum drilling and production activity, causing sediment mobilization and representing a potential safety concern to such industrial operations. When released to the atmosphere it is a powerful greenhouse gas that may have had impact on past climates. Finally, if only a small fraction of the methane hydrate in the world occurs in extractable concentrations, it might be a significant energy resource of a hydrocarbon fuel. The use of methane as a fuel, although it would continue to increase the global greenhouse effect in the atmosphere, would be significantly less polluting than other fossil fuels; methane generates considerably less carbon dioxide than fuel oil or coal, no particulates and no sulfur compounds.

Studies of gas hydrate in the natural environment involve the skills of a diverse community of scientists and the use of a wide range of geophysical techniques, wire-line logging tools, seafloor observation and mapping systems, and geochemical sampling techniques. The editors hope that the present volume will provide insight into the diversity of approaches and the complexities of gas hydrate research through reviews of some approaches and research directions and examples of specific studies that highlight individual problems.

Charles K. Paull
MBARI Moss Landing, CA

William P. Dillon
USGS Woods Hole, MA

Section 1

GENERAL ISSUES

The Global Occurrence of Natural Gas Hydrate

Keith A. Kvenvolden and Thomas D. Lorenson

U.S. Geological Survey, Menlo Park, California 94025

Natural gas hydrate occurs worldwide in oceanic sediment of continental and insular slopes and rises of active and passive margins, in deep-water sediment of inland lakes and seas, and in polar sediment on both continents and continental shelves. In aquatic sediment, where water depths exceed about 300 m and bottom water temperatures approach 0° C, gas hydrate is found at the seafloor to sediment depths of about 1,100 m. In polar continental regions, gas hydrate can be present in sediment at depths between about 150 and 2000 m. Thus, natural gas hydrate is restricted to the shallow geosphere where its presence affects the physical and chemical properties of near-surface sediment. An updated global inventory reports on natural gas hydrate recovered from 19 places worldwide and includes 77 places where the presence of gas hydrate has been inferred from geophysical, geochemical, and geological evidence. The potential amount of methane in natural gas hydrate is enormous, with current estimates converging around about 10 teratonnes (10^{19} g) of methane carbon.

INTRODUCTION

Recognition that natural gas could form gas-hydrate deposits by Vasil'ev et al. (1970) led to the idea that natural gas hydrate should occur globally wherever the pressure and temperature conditions were favorable. Such conditions exist in sediment of polar regions as well as in sediment that covers about 90% of the ocean bottom (Makogon et al., 1971, 1972, 1973; Trofimuk et al., 1973, 1975, 1979). Although gas hydrate is found in polar regions, commonly associated with onshore and offshore permafrost, it is not found in 90% of the ocean bottom. With the exception of polar continental shelves, oceanic gas hydrate is restricted to continental slopes and rises of outer continental and insular margins around the world (summarized by Kvenvolden, 1993). Recognition of the potential for natural gas-hydrate occurrence was followed by its discovery, first in permafrost regions of northern Russia (Makogon et al., 1971, 1972). It was then observed in sediment of the Black Sea (Yefremova and Zhizhchenko, 1974) and inferred by seismic surveys to be present in sediment of the Blake Ridge, of the western Atlantic Ocean (Stoll et al., 1971; Ewing and Hollister, 1972). By the early 1980s, gas hydrate had been found in outer continental margin sediment of the Middle America Trench offshore from Mexico (Shipley and Didyk, 1982) and Guatemala (Harrison and Curiale, 1982). Since then, the rate of discovery of evidence for gas hydrate has accelerated.

Natural Gas Hydrates: Occurrence, Distribution, and Detection
Geophysical Monograph 124

Published in 2001 by the American Geophysical Union

Definition

Natural gas hydrate is a solid substance composed of water and natural gas (mainly methane). The water molecules are arranged in a rigid framework of cages (called a clathrate), many of which are occupied and stabilized by a molecule of methane. The physical chemistry of methane hydrate is well known (Sloan, 1998), and the occurrence of gas hydrate is controlled by an interrelation among factors of temperature, pressure, and compositions of gas and water (Kvenvolden, 1993). Fig. 1A and B show arbitrary examples of the depth zones of

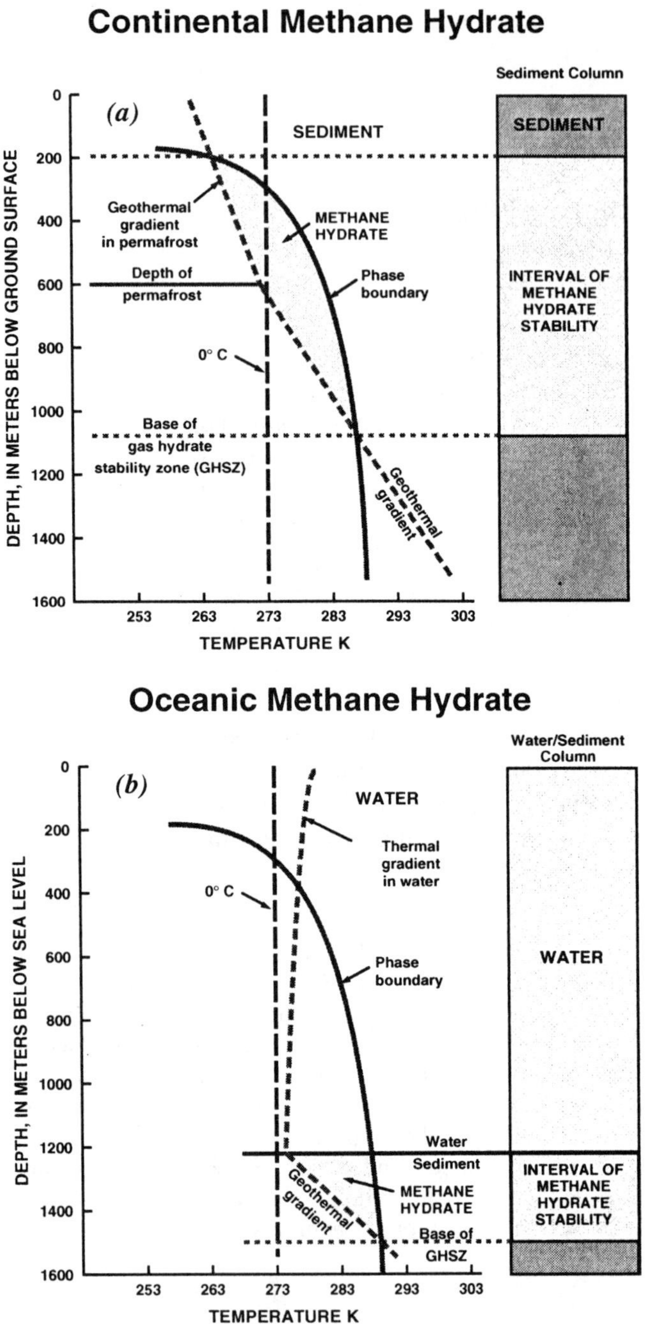

Figure 1. Arbitrary examples of different depth-temperature zones in which gas hydrate is stable in: (A) a polar continental region; and (B) an oceanic setting in an outer continental margin. The phase boundary is for a methane hydrate and pure water. Adapted from Kvenvolden and McMenamin (1980).

methane-hydrate occurrence in polar and outer continental margin settings. The upper depth limit for methane hydrate is about 150 m below the surface in continental polar regions, where surface temperatures are well below 0° C (273 K). In oceanic sediment, gas hydrate occurs where bottom-water temperatures approach 0° C, and the water depths exceed about 300 m. The maximum lower limit of methane hydrate occurrence is determined by the geothermal gradient; this lower limit is about 2000 m beneath the continental surface and the ocean floor, although the depths of occurrence are typically less than 1000 m, depending on local conditions. The occurrence of natural gas hydrate is restricted to the shallow geosphere where the amount of methane exceeds its solubility in water and thus is present in large enough quantities to form methane hydrate.

Composition

Methane comprises more than 99% of the hydrocarbon gases in most natural gas hydrate. Minor quantities of CO_2 and H_2S are often present, but because the main gas component is methane, the gas hydrate is called methane hydrate which has Structure I crystallography (Sloan, 1998). In the Gulf of Mexico (Brooks et al., 1984), Caspian Sea (Yefremova and Gritchina, 1981; Ginsburg et al., 1992), and Mackenzie Delta (Dallimore et al., 1999), gas hydrate has been recovered in which methane is accompanied by significant amounts of ethane and propane; this gas hydrate has Structure II crystallography (Sloan, 1998). Structure H gas hydrate with even higher molecular weight hydrocarbon gases has also been observed in the Gulf of Mexico (Sassen and MacDonald, 1994). A Structure I methane hydrate containing ~10% H_2S has been recovered from the Cascadia Margin offshore from Oregon (Kastner et al., 1998).

Information on the molecular composition of hydrocarbon gases in gas-hydrate samples, coupled with measurements of carbon- and deuterium-isotopic compositions ($\delta^{13}C$ and δD, respectively) of methane, provide a basis for interpreting the origin of this gas. In most of the methane hydrate samples, methane has a carbon-isotopic composition lighter than -60‰ (relative to the Peedee Belemnite standard), suggesting that the methane is mainly microbial in origin. Microbial methane in methane hydrate may have either been generated in place, migrated only short distances, or recycled during the concomitant processes of hydrate formation and dissociation accompanying sedimentation (Paull et al., 1994). This microbial methane likely results from methanogenic processes, taking place in shallow sediment, in which CO_2 from organic matter is reduced to methane. To a much lesser extent, especially in oceanic sediment, acetate from organic matter is converted to methane by fermentation reactions. The pathways of natural methane production and the range of expected carbon- and deuterium-isotopic compositions of methane have been discussed in detail by Whiticar et al. (1986) and Whiticar (1999).

Table 1. Carbon isotopic compositions and concentrations (% of hydrocarbon gases) of methane in natural gas hydrate

Regions	CH_4 (%)	$\delta^{13}C$ (‰)	Reference
Offshore SE USA, Blake Outer Ridge			
DSDP Leg 76	>99	-68.0	Brooks et al. (1983)
ODP Leg 164	>99	-65.8 to –70.7	Lorenson and Collett (1999)
Offshore Peru, Peru-Chile Trench			
ODP Leg 112	>99	-65.0, -59.6	Kvenvolden and Kastner (1990)
Offshore northern California			
Eel River Basin	>99	-69.1 to -57.6	Brooks et al. (1991)
Offshore Oregon			
ODP Leg 146	>99	-64.5 to –67.5	Kastner et al. (1998)
Hydrate Ridge	>99	-62.4 to –71.5	Suess et al. (1999a)
Black Sea	>99	-63.3, -61.8	Ginsburg et al. (1990b)
Sea of Okhotsk	>99	-64.3	Ginsburg et al. (1993)
Lake Baikal	>99	-57.6 to –68.2	Kuz'min et al. (1998)
Offshore Norway			
Haakon-Mosaby mud volcano	>99	-56.0 to –61.9	Ginsburg et al. (1999); Lein et al. (1999)
Gulf of Mexico			
DSDP Leg 96	>99	-71.3	Pflaum et al. (1986)
Garden Banks	>99	-70.4	Brooks et al. (1986)
Green Canyon	>99	-69.2, -66.5	Brooks et al. (1986)
Green Canyon	62, 74, 78	-44.6, -56.5, -43.2	Brooks et al. (1986)
Mississippi Canyon	97	-48.2	Brooks et al. (1986)
Caspian Sea	59 to 96	-44.8 to -55.7	Ginsburg et al. (1992)
Offshore Guatemala, Middle America Trench			
DSDP Leg 84	>99	-43.6 to -36.1	Kvenvolden et al. (1984)
DSDP Leg 84	>99	-46.2 to -40.7	Brooks et al. (1985)
Mackenzie Delta (Mallik 2L-38)	>99	-39.8 to -45.5	Lorenson et al. (1999)

In contrast, methane with carbon-isotopic compositions heavier than about -60‰, accompanied by significant amounts of higher-molecular-weight hydrocarbons, such as ethane, propane and larger hydrocarbon gases, is of thermal origin, according to the criteria of Schoell (1988). This gas mixture results from the thermal decomposition of organic matter at sediment depths greater than about 1000 m. The mixture has migrated long distances from deeply buried sediment in order to form gas hydrate. In the case of the Gulf of Mexico and Caspian Sea, gas hydrate has been found at or near the seafloor, whereas in the Mackenzie Delta, gas hydrate occurs to a depth of about 1100 m. Table 1 (modified from Kvenvolden, 1995) summarizes the concentrations and methane-carbon-isotopic compositions of natural-gas hydrate.

Amount of Methane

All estimates of the methane content of methane hydrate are highly speculative, but they suggest that methane quantities are very large (Kvenvolden, 1999). Cherskiy and Makogon (1970) proposed that the amount of methane in natural gas hydrate is potentially "enormous." Early estimates ranged from 1.7 to 4.1×10^3 teratonnes (teratonne = 10^{18} g) of methane carbon for oceanic sediment and from 0.0075 to 18 teratonnes of methane carbon for polar regions (summarized by the Potential Gas Committee, 1981). The SI prefix 'tera' with the word 'tonne,' which specifically indicates metric ton, is used here and in later discussion because it minimizes the need for exponents. Because oceanic gas hydrate deposits are much more voluminous than polar-region gas hydrate accumulations, the oceanic gas hydrate is emphasized in global estimates of the methane content of gas hydrate.

By the end of the 1980s, estimates of methane in gas hydrate were greatly constrained when Kvenvolden (1988) and MacDonald (1990a), working independently, estimated the global methane-carbon content of gas hydrate to be 11 and 12 teratonnes, respectively. These nearly equal estimates are coincidental, but the convergence of independent ideas made this value, expressed to one significant figure as 10 teratonnes, the consensus estimate.

In the 1990s new estimates were made using the power of general circulation models (GCM) in which pertinent variables, such as pressure-temperature-pressure phase relations, thermal gradients, and sediment porosities, are entered into a 1° x 1° global grid. Estimates by Gornitz and Fung (1994) ranged between 14 and 75 teratonnes of methane carbon, with the most likely estimate of 14 teratonnes for their *in situ* microbial-gas-generation model

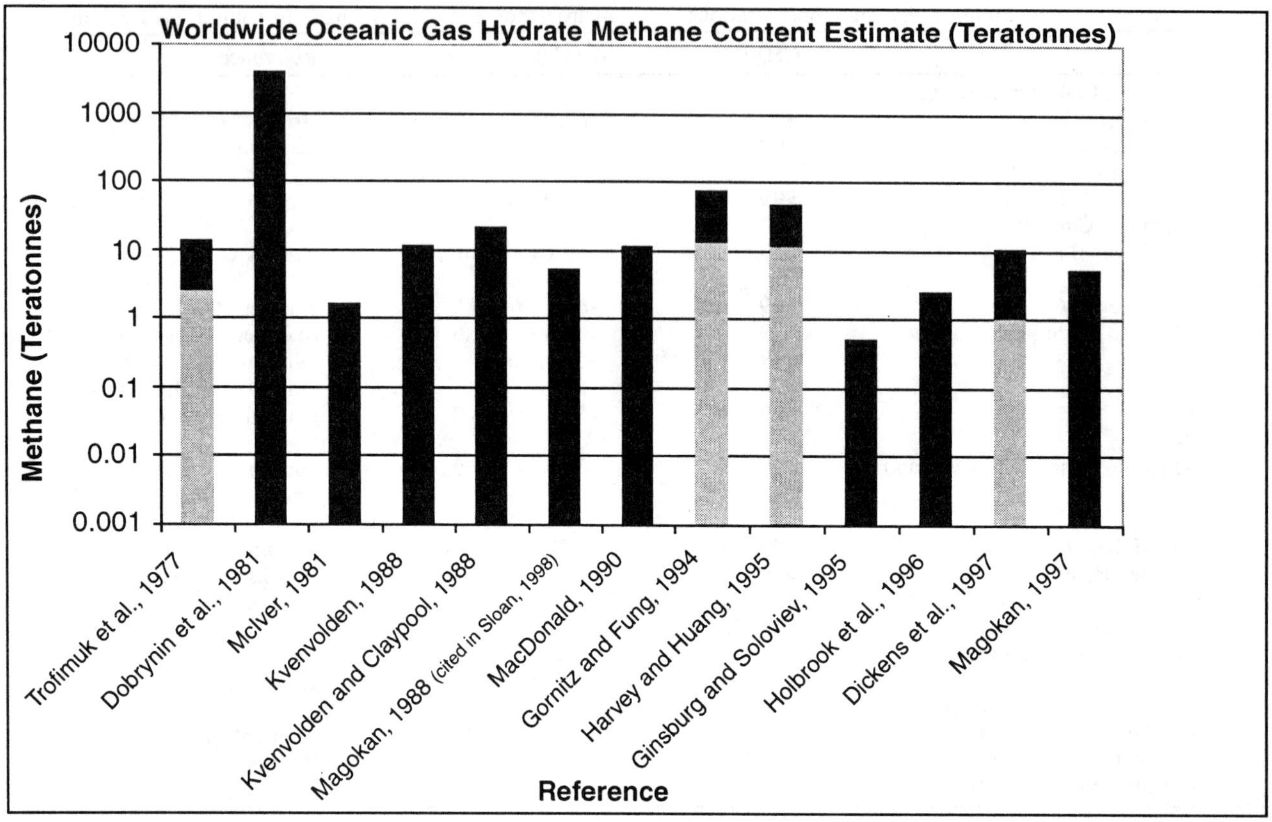

Figure 2. Histogram of estimates of the methane content (teratonnes of methane carbon) of worldwide gas hydrate deposits. In black/gray bars, black segments indicate range of estimates.

and 62 teratonnes for their pore-fluid model. Harvey and Huang (1995) calculated that the total methane-carbon content of oceanic hydrate to be 15, 24, and 49 teratonnes, depending on the assumptions regarding pore fractions occupied by gas hydrate. They selected the intermediate value of 24 teratonnes as their best estimate. Other global estimates of methane-carbon in gas hydrate, made during the same period, are 3 teratonnes (Holbrook et al., 1996), 1 to 10 teratonnes (Dickens et al., 1997), 8 teratonnes (Makogon, 1997) and 0.5 teratonnes (Ginsburg and Soloviev, 1995). Thus, the best estimates range from a low of 0.5 to a high of 24 teratonnes. The consensus value of 10 teratonnes of methane carbon in gas hydrate worldwide remains about midway between these extremes (Fig. 2).

EVIDENCE FOR GAS HYDRATE

Three kinds of evidence can be used to identify the presence of natural gas hydrate—geological, geochemical, and geophysical. Geological evidence includes sediment properties, stratigraphic relationships, gas-migration pathways, and, most importantly, the actual recovery and description of gas-hydrate samples. Pore fluid chemistry and gas compositions (molecular and isotopic) are important aspects of gas-hydrate geochemistry. Finally, geophysical evidence includes data from seismic reflection profiling, vertical seismic profiling, and various kinds of well-logging. Seismic reflection profiling is especially valuable when accessing the areal extent of gas-hydrate deposits.

Geological Evidence

The presence of gas hydrate alters the physical properties of sediment by replacing pore-space water and/or gas with a solid, thereby reducing the effective pore space and permeability, and increasing the acoustic velocity which can be assessed by seismic methods. These physical-property changes can be examined with well-logging tools. When cores containing gas hydrate are recovered during drilling, the *in situ* sediment properties are altered due to the dissociation of gas hydrate, which is unstable at ambient conditions. Sediment confined in core liners often develops gas-expansion cracks as destabilized gas hydrate releases gas. Also, because gas-hydrate dissociation is an endothermic process, there is a measurable temperature decrease when gas hydrate is present in the cores (Kastner et al., 1995). Where gas hydrate is dispersed in fine-grained sediment, recovered material is often frothy and fluidized because of gas and water released from decomposing gas hydrate. Only when gas hydrate in sediment is present in quantity as nodules, layers along bedding planes, fractures and joints, or masses at the seafloor, and when core recovery is rapid, can samples of gas hydrate be recovered for examination, analysis, and storage.

Submarine slides and slumps can provide inferential but not diagnostic evidence for gas hydrate. A possible connection between seafloor failures and gas-hydrate occurrence was first recognized by McIver (1977; 1982). Several possible examples have been described since then including the following: (1) surficial slides and slumps on the continental slope and rise of S.W. Africa (Summerhayes et al., 1979); (2) slumps on the U.S. Atlantic continental margin (Carpenter, 1981); (3) collapse structures on the Blake Ridge (Dillon et al., 1998); (4) Storegga slide offshore from Norway (Jansen et al., 1987; Bugge et al., 1987); (5) sediment blocks on seafloor of fjords in British Columbia (Bornholt and Prior 1989); and (6) massive bedding plane slides and rotational slumps on the Alaskan Beaufort Sea continental margin (Kayen and Lee, 1991).

Geochemical Evidence

Analyses of gas hydrate, gas-hydrate-bearing sediment, and sediment suspected of having contained gas hydrate provide a wealth of information which can be used to interpret the presence of gas hydrate, gas sources, and processes that took place during gas-hydrate formation and decomposition. Molecular and isotopic compositions of the gases provide clues to sources, whereas ionic and isotopic compositions of pore waters are used to identify sediment that may have contained gas hydrate. For example, during gas-hydrate formation, water molecules crystallize into a cubic lattice structure, and the hydrate crystals exclude chloride ions. During Deep Sea Drilling Project (DSDP) Leg 67, offshore from Guatemala, Hesse and Harrison (1981) first noticed that chlorinities of pore-water samples in gas-hydrate-bearing zones decreased significantly with depth from about 19 to 9‰. Now decreasing chlorinity trends with depth in gas-hydrate-bearing sediment have been observed in the Blake Ridge [DSDP Leg 76 (Kvenvolden and Barnard, 1983; Jenden and Gieskes, 1983)], [Ocean Drilling Program (ODP) Leg 164 (Paull et al., 2000)], offshore from Guatemala [DSDP Leg 84 (Kvenvolden and McDonald, 1985; Hesse et al., 1985)], offshore from Peru [ODP Leg 112 (Kvenvolden and Kastner, 1990)], the Cascadia Margin [ODP Leg 146 (Whiticar et al., 1995)], and the Mackenzie Delta [Mallik 2L-38 well (Cranston, 1999)].

These chlorinity trends are derived to a large extent from the dissociation of gas hydrate during core recovery. The chloride concentrations are measured on mixtures composed of *in situ* pore water and dissociated gas-hydrate water. Thus, a dilution artifact is produced in the pore water that can be squeezed from the sediment. Pore-water freshening is a useful signal for the presence of gas hydrate although other processes can cause pore waters to have decreasing chlorinities (Kvenvolden, 1993). Nevertheless, where pore-water freshening is observed in sediment containing large amounts of methane within the zone of gas-hydrate stability, gas hydrate is likely to have been present.

Concurrent with trends in chloride concentrations, oxygen isotopic compositions ($\delta^{18}O$) of pore waters increase with depth. Data from DSDP Leg 67 (Hesse and Harrison, 1981) show the increase in oxygen isotopic composition of pore water from near zero close to the sediment surface to compositions as large as +2.6‰ (relative to Standard Mean Ocean Water, SMOW) at depth. Oxygen isotopic fractionation of pore water apparently takes place during gas-hydrate formation, leading to an enrichment of ^{18}O with depth of pore fluids recovered after gas-hydrate dissociation. This same trend of oxygen isotopic compositions was observed in pore waters of gas-hydrate-bearing sediment recovered on ODP Leg 112 offshore from Peru (Kvenvolden and Kastner, 1990) and on Leg 164 offshore from the southeastern U.S. (Matsumoto and Borowski, 2000).

Geophysical Evidence

Most oceanic occurrences of gas hydrate are inferred, based mainly on marine seismic reflection profiles showing an anomalous bottom-simulating-reflection (BSR). This reflection coincides with the depth predicted from the phase diagram (Fig. 1) as the base of the gas-hydrate stability zone (Fig. 3). BSRs mark the interface between higher sonic velocity, hydrate-bearing sediment above and lower sonic velocity, free gas-bearing sediment below. The seismic reflection from the base of gas hydrate is generally characterized by a reflection polarity reversal resulting in a negative reflection coefficient. The reflection usually mimics seafloor topography, often increasing in sediment depth with increasing water depth (Shipley et al., 1979).

In addition to seismic data, combining geophysical information from various wire-line well logs can aid in the detection and evaluation of gas-hydrate intervals (Kvenvolden and Grantz, 1990). Well logs for gas-hydrate studies include caliper, gamma ray, spontaneous potential, resistivity, sonic velocity, and neutron porosity (Goodman, 1980). In a comprehensive survey of the Mallik 2L-38 well in the Mackenzie Delta, the downhole logging-tool strings

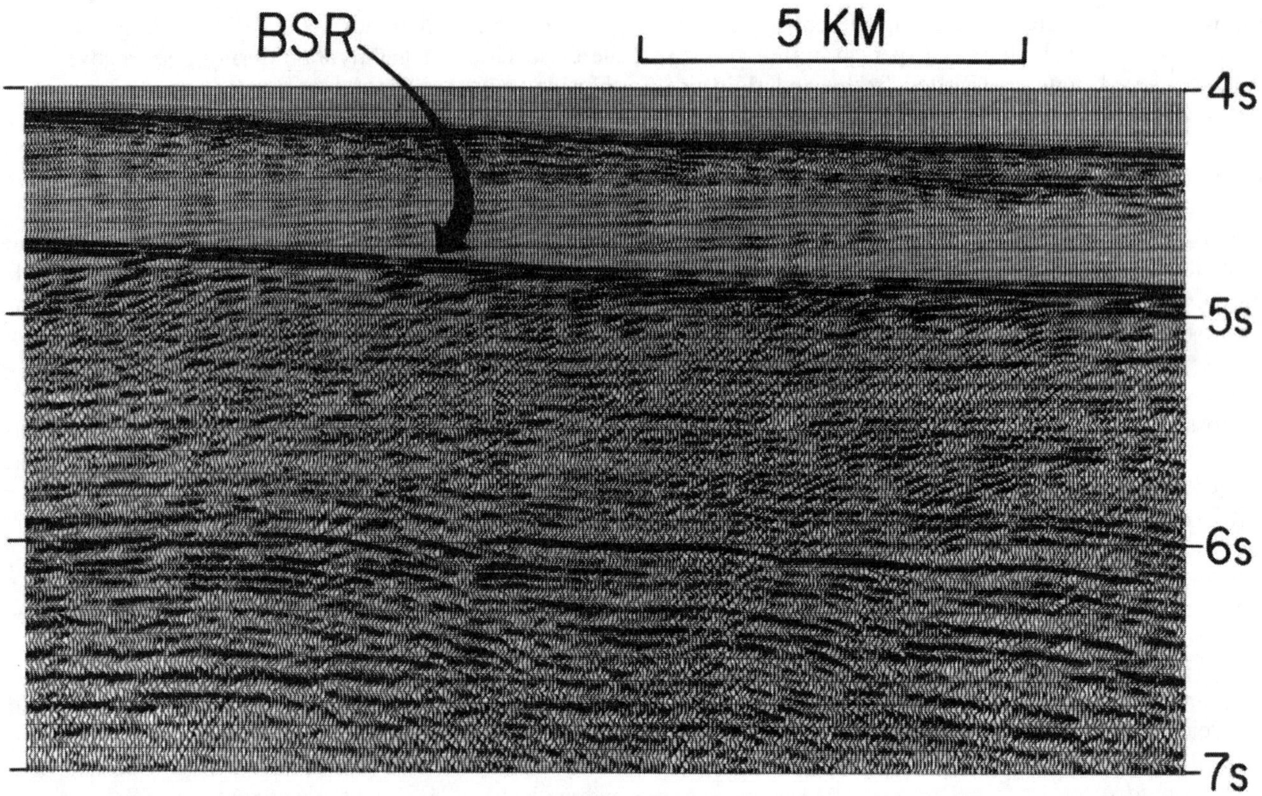

Figure 3. Multichannel seismic reflection profile from the flank of the Blake Ridge. The strong BSR is inferred to represent the base of the gas hydrate stability zone. From U.S. Geological Survey line BT1, courtesy of W.P. Dillon, USGS.

included the Schlumberger Platform Express with a high resolution laterolog, Array Induction Imager Tool, Dipole Shear Sonic Imager, and a Fullbore Formation MicroImager (Collett et al., 1999). The resistivity and sonic velocity logs have proved most useful in the study of gas hydrate on the North Slope of Alaska (Collett, 1993). Well logs provide a basis for estimating gas quantity, and in conjunction with seismic data, hold the key to future global gas-hydrate assessments.

INVENTORY OF GAS HYDRATE

A compilation of global gas-hydrate occurrences was first undertaken by Kvenvolden and McMenamin (1980). They showed 14 areas (9 oceanic and 5 continental) where geophysical and some geochemical evidence indicated gas hydrate. Much of the geophysical evidence for oceanic gas hydrate was based on the observations of BSRs on marine seismic records (Fig. 3). By 1988 the number of identified gas-hydrate regions had increased to 36 (30 oceanic including inland seas and 8 continental) (Kvenvolden, 1988). Independently, Panayev (1987) and Ginsburg et al. (1990a) identified 22 and 36 regions, respectively, where subaquatic gas hydrate occurs. Kvenvolden et al. (1993) listed 47 locations worldwide where sub-aquatic gas hydrate could be expected, and gas-hydrate samples were recovered at 14 of these locations.

International interest in gas hydrate is increasing rapidly with the recognition that the sequestered methane may be useful as an energy resource and may be a factor in global climate change (summarized in Kvenvolden, 1993). With this increasing interest it seems appropriate to update the inventory of known and inferred gas hydrate occurrences, both oceanic and continental. To organize the inventory, we have divided the gas-hydrate worldwide distribution into seven regions and have tabulated basic information in Tables 2 and 3.

The seven geographic regions of gas-hydrate occurrence are Pacific Ocean, Indian Ocean, Atlantic Ocean, North (Arctic, oceanic), South (Antarctica), Other (Inland Seas and Lakes), and Continental. At present there are 35 locations in the Pacific Ocean (P1-P35), 3 in the Indian Ocean, (I1-I3), 19 in the Atlantic Ocean (A1-A19), 3 in the North (Arctic Ocean) (N1-N3), 4 in the South (Antarctica)

Table 2. Summary of known and inferred subaquatic gas-hydrate occurrences in the Pacific and Indian Oceans.
[BSR, bottom-simulating reflection; $<Cl^-$, low chloride content of pore water; CH_4, high methane content; VAMPs, velocity amplitude pulldowns; Logs, well-log response]

Designation	Location Offshore	Evidence	References
PACIFIC OCEAN			
P1	Panama	BSR	Shipley et al. (1979)
P2	Costa Rica (Middle America Trench)	BSR	Shipley et al. (1979)
	(DSDP Leg 84)	Samples	Kvenvolden and McDonald (1985)
	(ODP Leg 170)	Samples	Shipboard Scientific Party (1997)
P3	Nicaragua (Middle America Trench)	BSR	Shipley et al. (1979)
P4	Guatemala (Middle America Trench)	BSR	Shipley et al. (1979)
	(DSDP Leg 67)	Samples	Harrison and Curiale (1982)
		$<Cl^-$	Hesse and Harrison (1981); Harrison et al. (1982)
	(DSDP Leg 84)	Samples	Kvenvolden and McDonald (1985)
		Logs	Kvenvolden and McDonald (1985)
		$<Cl^-$	Hesse et al. (1985)
P5	Mexico (Middle America Trench	BSR	Shipley et al. (1979)
	(DSDP Leg 66)	Samples	Shipley and Didyk (1982)
P6	Mexico (Gulf of California, Guaymas Basin)	BSR	Lonsdale (1985)
P7	California, USA (Eel River Basin)	BSR	Field and Kvenvolden (1985)
		Samples	Brooks et al. (1991)
P8	Oregon, USA (Cascadia Basin)	BSR	Moore et al. (1992)
	(ODP Leg 146)	Samples	Whiticar et al. (1995); Kastner et al. (1998)
	(Hydrate Ridge)	Samples	Suess et al. (1999a; (1999b)
P9	Canada (Cascadia Basin)	BSR	Davis et al. (1990); Hyndman and Spence (1992)
P10	(Fjords of British Columbia)	Slumps	Bornholt and Prior (1989)
P11	Alaska, USA (Eastern Aleutian Trench)	BSR	Kvenvolden and von Huene (1985)
P12	Alaska, USA (Middle Aleutian Trench)	BSR	McCarthy et al. (1984)
		$<Cl^-$	Hesse and Harrison (1981)
P13	Bering Sea (Alaska, USA)	VAMPs	Scholl and Cooper (1978)
P14	(USA, Bering Sea Shelf)	BSR	Hammond and Gaither (1983)
P15	(USA/Russia, Navarin Margin)	BSR	Carlson et al. (1985)
P16	(Russia, Shirshov Ridge)	BSR	Saltykova et al. (1987)
P17	Okhotsk Sea (Paramushir Island, Russia)	Samples	Zonenshayn et al. (1987)
P18	(Sahkalin Island, Russia)	Samples	Ginsburg et al. (1993)
P19	(off Abashiri, Kitami-Yamato Mount, Japan)	BSR	Matsumoto et al. (1994); Sato (1994)
P20	Japan, Japan Sea (DSDP Leg 57)	$<Cl^-$	Moore and Gieskes (1980)
	(Okushiri Ridge) (ODP Leg 127)	Sample	Shipboard Scientific Party (1990)
		BSR	Kuramoto and Okamura (1995)
P21	(Western Tsugaru Basin)	BSR	Sato (1994)
P22	(Tatar Trough)	BSR	Sato (1994)
P23	Japan, Nankai Trough (off eastern Miyazaki)	BSR	Aoki et al. (1983); Tamano et al. (1984)
P24	(off southern Shikoku)	BSR	Shipboard Scientific Party (1991)
	(ODP Leg 131)	Sample	Shipboard Scientific Party (1991)
P25	(Muroto Trough)	BSR	Sato (1994); Matsumoto et al. (1994)
P26	(Kumano-nada thru off Omaezaki Cape)	BSR	Okuda (1995); Gas Epoch (1995)
		CH_4, $<Cl^-$	JNOC press release (2000)
P27	Japan, Offshore Chiba Basin	BSR	Arato et al. (1995)
P28	Japan, Kuril Trench (off Tokachi/Hidaka)	BSR	Sato (1994)
P29	New Zealand (Hikurangi Trough)	BSR	Katz (1981)
P30	Chile (Peru-Chile Trench)	BSR	Cande et al. (1987)
		CH_4, $<Cl^-$	Froelich et al. (1995)
P31	Peru (Peru-Chile Trench)	BSR	Miller et al. (1991)
	(ODP Leg 112)	Samples	Kvenvolden and Kastner (1990)
P32	Taiwan (South China Sea)	BSR	Chi et al. (1998)
P33	Australia (Tasman Sea, Lord Howe Rise)	BSR	Exon et al. (1998)
P34	(Timor Trough)	CH_4	McKirdy and Cook (1980)
P35	Indonesia (Celebes Sea)	BSR	Neben et al. (1998)
INDIAN OCEAN			
I1	Oman (Gulf of Oman, Makran Margin)	BSR	White (1979)
I2	India (Arabian Sea)	BSR	Veerayya et al. (1998)
I3	(Bay of Bengal)	BSR	Rastogi et al. (1999)

Table 3. Summary of known and inferred subaquatic gas-hydrate occurrences in Atlantic Ocean, polar oceans, continents, and in inland seas and lakes. BSR, bottom-simulating reflection; $<Cl^{-}$, low chloride content of pore water; CH_4, high methane content; Logs, well-log response]; PCS, Pressure core sample; Geophysical, seismic evidence of past occurrence of gas hydrate

Designation	Location Offshore	Evidence	References
ATLANTIC OCEAN			
A1	Argentina (Central Argentine Basin)	BSR	Manley and Flood (1989)
A2	Brazil (Amazon Fan)	BSR	Manley and Flood (1989)
A3	(Pelotas Basin)	BSR	Fontana and Mussumeci (1994)
A4	Barbados (Barbados Ridge)	BSR	Ladd et al. (1982)
A5	Southern Caribbean Sea	BSR	Ladd et al. (1984)
A6	Panama and Colombia (Colombia Basin)	BSR	Shipley et al. (1979)
A7	Gulf of Mexico (off Mexico, western area)	BSR	Shipley et al. (1979); Hedberg (1980)
A8	(off USA, northern area)	Samples	Brooks et al. (1984, 1986); Pflaum et al. (1986); Sassen and MacDonald (1994, 1997)
A9	Southeastern USA (Blake Ridge)	BSR	Markl et al. (1970); Shipley et al. (1979)
		Slumps	Carpenter (1981); Dillon et al. (1998)
	(DSDP Leg 76)	Samples	Kvenvolden and Barnard (1983)
		$<Cl^{-}$	Jenden and Gieskes (1983)
	(ODP Leg 164)	BSR	Holbrook et al. (1996)
	(ODP Leg 164)	Samples	Paull et al. (1996, 2000); Lorenson and Collett (1999)
A10	Eastern USA (Carolina Trough)	BSR	Dillon et al. (1983)
A11	Eastern USA (Continental Rise)	BSR	Tucholke et al. (1977); Dillon et al. (1993, 1995)
A12	Newfoundland, Canada (Laborador Shelf)	BSR	Taylor et al. (1979)
A13	Norway (Storegga Slide)	BSR	Bugge et al. (1987); Mienert et al. (1998)
		Slumps	Jansen et al. (1987); Bugge et al (1987)
		$<Cl^{-}$	Hesse and Harrison (1981)
	(ODP Leg 104)	CH_4, $<Cl^{-}$	Kvenvolden et al. (1989)
A14	(Barents Sea)	BSR	Andreassen et al. (1990)
A15	(off Svalbard)	BSR	Posewang and Mienert (1999)
A16	(Haakon-Mosby mud volcano)	Samples	Ginsburg et al. (1999)
A17	Ireland (Porcupine Basin)	Geophysical	Henriet et al. (1998)
A18	Africa (S.W. Africa)	Slumps	Summerhayes et al. (1979)
A19	(Nigeria)	Samples	Brooks et al. (1994)
NORTH (Arctic)			
N1	Alaska (Beaufort Sea)	BSR	Grantz and Dinter (1980) Kvenvolden and Grantz (1990)
		Slumps	Kayen and Lee (1991)
N2	Canada (Beaufort Sea)	Logs	Weaver and Stewart (1982)
N3	Canada (Sverdrup Basin)	Logs	Judge (1982)
SOUTH (Antarctic)			
S1	Antarctica (Wilkes Land Margin)	BSR	Kvenvolden et al. (1987)
S2	(Ross Sea)	CH_4	McIver (1975)
		$<Cl^{-}$	Mann and Gieskes (1975)
S3	(Weddell Sea)	BSR	Lonsdale (1990)
S4	(South Shetland margin)	BSR	Lodolo et al. (1993)
OTHER (INLAND SEAS AND LAKES)			
O1	Black Sea, Russia	Samples	Yefremova and Zhizhchenko (1974) Ginsburg et al. (1990b)
O2	Caspian Sea, Russia		Samples Yefremova and Gritchina (1981) Ginsburg et al. (1992)
O3	Lake Baikal, Russia	BSR	Hutchinson et al. (1991)
		Samples	Kus'min et al. (1998)
O4	Mediterranean Sea (ODP Leg 160)	$<Cl^{-}$, CH_4	DeLange and Brumsack (1998)
O5	Turkey (Kula mud volcano)	Sample	Woodside et al., (1998)
CONTINENTAL			
C1	Alaska (North Slope)	Logs	Collett (1993)
		PCS	Summary by Collett (1993)
C2	Canada (Mackenzie Delta)	Logs	Bily and Dick (1974)
		Samples	Dallimore et al. (1999)
C3	(Arctic Islands)	Logs	Davidson et al. (1978); Judge (1982)
C4	Russia (Messoyakah Field)	CH_4	Makogon et al. (1972)
C5	(Timan-Pechora Province)	Interpretation	Cherskiy et al. (1985)
C6	(Western Siberian Platform)	Interpretation	Cherskiy et al. (1985)
C7	(Eastern Siberian Craton)	Interpretation	Cherskiy et al. (1985)
C8	(Northeast Siberia)	Interpretation	Cherskiy et al. (1985)

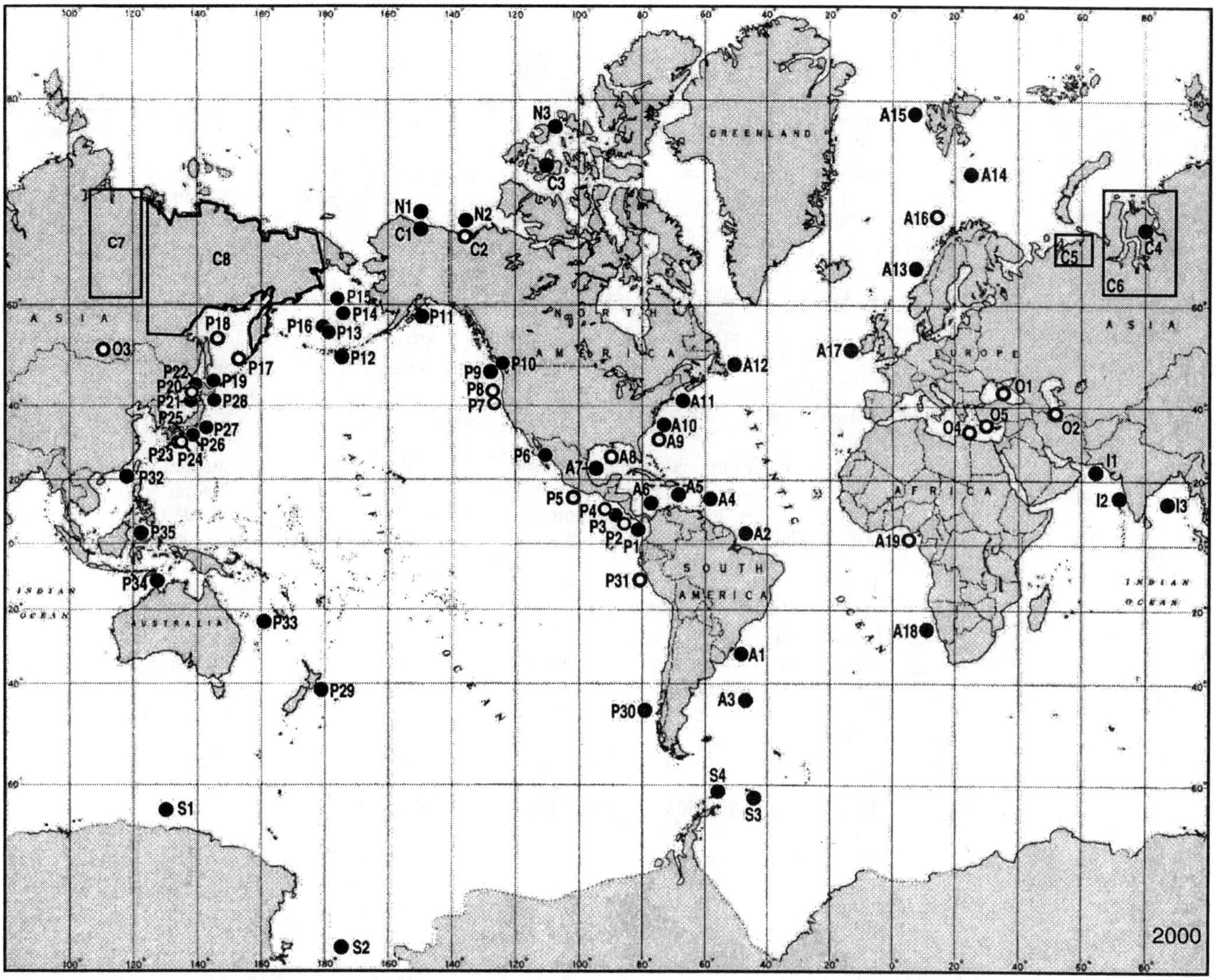

Figure 4. Map showing worldwide locations of known and inferred gas hydrate in aquatic (oceans, seas, and lakes) sediment and in polar continental sediment. Open circles indicated locations where samples of natural gas hydrate have been recovered. Filled circles indicate location of inferred presence of natural gas hydrate. Regions labeled C5 through C8 indicate areas where gas hydrate is inferred to occur.

(S1-S4), 4 in Other (O1-O4), and 8 in Continental (C1-C8). Occurrences of gas hydrate are listed for the Pacific and Indian Ocean (Table 2), the Atlantic Ocean and other locations (Table 3). Fig. 4 shows the location of known and inferred natural gas hydrate in the world.

Gas-hydrate samples have been recovered from 19 regions worldwide, and a brief description, adapted from Booth et al. (1996), is given in Table 4. These samples provide the most direct evidence for the global presence of gas hydrate. The first documented recovery of gas hydrate from sediment was obtained in 1972 using a pressure-core-barrel at the Arco-Exxon N.W. Eileen State No. 2 wildcat well on the North Slope of Alaska in the West Prudhoe Bay oil field (summarized by Collett, 1993). Pressure reduction tests on the contents of the core barrel showed that gas hydrate had been recovered, but the gas-hydrate sample was never observed. Earlier, methane was released from gas hydrate by methanol injection at the Messoyaka gas field in western Siberia, but there is no report of actual recovery and analysis of gas-hydrate samples (Makogon, et al., 1971). The first direct observation of gas hydrate apparently was made by Yefremova and Zhizhchenko (1974), who described gas-hydrate crystals found in near-surface sediment of the Black Sea.

Table 4. Descriptions of natural gas hydrate

Designation (Tables 2-4)	Geographic Location	Core/Sample Count	Water:Sediment Depths (m)	Description
P2	Middle America Trench (Costa Rica)			
	DSDP Leg 84, Site 565	2 samples	3099: 285&319	Inclusions in mud and sand.
	ODP Leg 170, Site 1041	9 samples	3306: 119-259	Disseminated and sheets in Microfractures.
P4	Middle America Trench (Guatemala)			
	DSDP Leg 67, Site 497	1 sample	2347: 368	Inclusion in sediment.
	Site 498	1 sample	5478: 307	Cement in coarse vitric sand.
	DSDP Leg 84, Site 568	1 sample	2010: 404	Inclusion in mudstone.
	Site 570	7 samples	1698: 192-338	Laminated ash; massive core at 249 m (1.05 m core).
P5	Middle America Trench (Mexico)			
	DSDP Leg 66, Site 490	4 samples	1761: 140-364	Laminated ash and mud.
	Site 491	3 samples	2883: 89-168	Inclusions in mud.
	Site 492	2 samples	1935: 141&170	Laminated ash.
P7	Eel River Basin (California)	7 cores	510-642: 0-2.8	Layers, nodules in mud.
P8	Cascadia Basin (Oregon)			
	DSDP Leg 146, Site 892	3 cores	647: 2-19	Aggregates, layers in silt.
	Hydrate Ridge	3 samples	600: ~0	Layers, massive in carbonate.
P17	Okhotsk Sea (Russia)			
	Paramushir Island	2 cores	768: 1.8	Layers in ooze.
P18	Okhotsh Sea (Russia)			
	Sahkalin Island	5 cores	710: 0.3-2	Layers in silt and clay.
P20	Japan Sea (Japan)			
	ODP Leg 127, Site 796	1 sample	2571: 68	Crystals in sand with clay.
P24	Nankai Trough (Japan)			
	ODP Leg 131, Site 808	1 sample	4684: 90-140	Fragment in wash core.
P31	Peru-Chile Trench (Peru)			
	ODP Leg 112, Site 685	2 samples	5070: 99-166	Fragments in mud.
	Site 688	1 sample	3820: 141	Grains in mud.
A8	Gulf of Mexico (Texas and Louisiana)			
	DSDP Leg 96, Site 618	2 samples	2400: 20	Nodules, crystals in mud.
	Green Canyon	14 cores	530-880: 1.2-4.8	Nodules, layers in rubble.
	Garden Banks	1 core	850: 2.8&3.8	Nodules, layers in rubble.
	Mississippi Canyon	1 core	1300: ~3.8	Pieces in coarse sediment.
	Bush Hill	1 core	540: 0	Mounds at seafloor.
A9	Blake Ridge (Southeastern USA)			
	DSDP Leg 76, Site 533	1 sample	3191: 238	Fragment in mud.
	ODP Leg 164, Site 994	2 cores	2799: 260	Fragments in clay.
	Site 996	5 cores	2170: 0-66	Nodules, veins in mud.
	Site 997	1 sample	2770: 331	Massive core (~30 cm).
A16	Haakon-Mosby Mud Volcano (Norway)	11 samples	~1255: 0-3	Inclusions and plates.
A19	Niger Delta (Nigeria)	6 cores	561-770: 0-4.6	Nodules, dispersed in clay.
O1	Black Sea (Russia)	3 cores	2052: 0.7-2.2	Veinlets in silty clay.
O2	Caspian Sea (Russia)	24 cores	475-600: 0-1.2	Laminated in clayey silt.
O3	Lake Baikal (Russia)	2 samples	1433: 121&161	Disseminated in sand, silt.
O5	Mediterranean Sea			
	Kula Mud Volcano	~2 samples?	~1700: not given	Not described.
C2	Mackenzie Delta (Canada)	6 cores	N/A: 887-920	Dispersed in sand, gravel.

Deep ocean drilling (DSDP and ODP) has obtained gas hydrate at nine locations [offshore Japan (two sites), Peru, Costa Rica, Guatemala, Mexico, and southeastern, southern (Gulf of Mexico) and western United States]. Coring with piston, gravity, and push cores, dredging, and individual site sampling have also recovered gas hydrate from near the seafloor in eight regions [Black Sea, Caspian Sea, Sea of Okhotsk (two locations), offshore from northern California and central Oregon, offshore from Norway, and the northern Gulf of Mexico].

In some cases only a single sample of gas hydrate was obtained, but more often gas hydrate samples were observed in several cores from a given area (Table 4). For example, in the Gulf of Mexico, microbial as well as thermal methane occurs in different samples of gas hydrate. Samples were recovered at a single site on DSDP Leg 96 (Pflaum et al., 1986), at 13 piston-core stations located more than 300 km away (Brooks et al., 1986), and at specific sites by grab sampling from a research submarine (MacDonald et al., 1994). Offshore from Guatemala, gas hydrate was observed in cores from a total of five drill sites on DSDP Legs 67 (Harrison and Curiale, 1982) and 84 (Kvenvolden and McDonald, 1985). At the Blake Ridge, offshore southeastern United States, gas hydrate was observed in cores from a total of five drilling sites on DSDP Leg 76 (Kvenvolden and Barnard, 1983) and ODP Leg 164 (Paull et al., 1998). On the Cascadia Margin, offshore from central Oregon, gas hydrate has been found both by drilling (ODP Leg 146, Whiticar et al., 1995; Kastner et al., 1998) and by dredging (Suess et al., 1999b). In the southern Caspian Sea, Ginsburg et al. (1992) found gas hydrate in 24 cores from two areas about 40 km apart. Surface-sediment cores at two locations in the Sea of Okhotsk (Ginsburg et al., 1993) and at the Haakon-Mosby mud volcano, offshore from Norway (Ginsburg et al., 1999) contained gas hydrate.

Up to the present time most gas-hydrate samples have been found in oceanic sediment, but now gas hydrate has been found in polar continental sediment of Canada. Two oil and gas exploratory wells, drilling in Mackenzie Delta permafrost, had penetrated shallow sand reservoirs thought to contain gas hydrate at depths of 820 to 1100 m (Bily and Dick, 1974), but no gas hydrate samples were recovered. The amount of formation gas in the drilling mud increased significantly during penetration of the sands, which were porous but had low permeability. These characteristics suggested the presence of gas hydrate. Well-log responses, such as high resistivity and acoustic velocity, also indicated gas hydrate. The presence of gas hydrate in the Mackenzie Delta region has now been proven by the recovery of gas-hydrate samples from depths between 890 and 1100 m during coring at the Mallik 2L-38 well (Dallimore et al., 1999).

Chloride anomalies (pore-water freshening) have been noted at eight oceanic locations and one continental location from which the presence of gas hydrate can be inferred. In fact the chloride concentrations have been used to estimate the amount of gas hydrate present in sediment of the Cascadia Margin (Kastner et al., 1995) and the Blake Ridge (Paull et al., 1996).

Most of the inferred locations, however, are based on the appearance of BSRs on marine seismic records. A total of 53 locations on Fig. 4 are identified because of BSR observations. Recent marine seismic studies have suggested that gas hydrate is present at nine sites around Japan (summarized by Murota, 1996); on the Lord Howe Rise, offshore from Australia (Exon et al., 1998); offshore from Ireland (Henriet et al., 1998); offshore from India (Veerayya et al., 1998; Rastogi et al., 1999); offshore from Taiwan (Chi et al., 1998); in the Celebes Sea (Neben et al., 1998); and in the Barents Sea offshore from Norway (Andreassen et al., 1990; Mienert and Posewang, 1999). These observations augment the compilations of Kvenvolden et al. (1993), Booth et al. (1996), and Ginsburg and Soloviev (1998), reinforcing the idea that submarine gas hydrates are common worldwide.

CONCLUSION

Natural gas hydrate is a global phenomenon now receiving international attention. It occurs worldwide but is restricted to two environments—polar continental and deep water (mostly on outer continental and insular margins) in sediment of the shallow geosphere (<2000 m below the solid surface). An inventory of global occurrences shows 77 places in which the presence of gas hydrate is inferred by geophysical, geochemical, and geological methods. This inventory includes 19 places where samples of the substance have actually been recovered. Most natural gas hydrate contains methane that was generated by microbial processes. Methane with a thermal origin occurs in gas hydrate only where deeper deposits of thermogenic natural gas provide a source of the methane. The potential amount of methane in global gas-hydrate occurrences is very large, with current estimates converging at about 10 teratonnes of methane carbon.

REFERENCES

Andreassen, K., K. Hogstad, and K.A. Berteussen, Gas hydrate in the southern Barents Sea, indicated by a shallow seismic anomaly, *First Break, 8*, 235-245, 1990.

Aoki, Y., T. Tamano, and S. Kato, Detailed structure of the Nankai Trough from migrated seismic sections, in Studies in Continental Margin Geology, *American Association Petroleum Geologists Memoir 34*, 309-322, 1983.

Arato, H., G. Akai, S. Uchiyama, T. Kudo, and K. Sekiguchi, BSRs recognized within the Offshore Chiba basin, Abstract, Spring Meeting, Japanese Association for Petroleum Technology, 1995.

Bily, C., and J.W.L. Dick, Naturally occurring gas hydrates in the Mackenzie Delta, N.W.T, *Bulletin of Canadian Petroleum Geology, 22*, 320-352, 1974.

Booth, J.S., M.M. Rowe, K.M. Fischer, Offshore gas hydrate sample database, *U.S. Geological Survey Open-File Report 96-272*, 1996.

Bornhold, B.D., and D.B. Prior, Sediment blocks on the sea floor in British Columbia fjords, *GeoMarine Letters, 9*, 135-144, 1989.

Brooks, J.M., A.L. Anderson, R. Sassen, I.R. MacDonald, M.C. Kennicutt II, and N.L. Guinasso, Jr., Hydrate occurrences in shallow subsurface cores from continental slope sediments, *Annals of the New York Academy of Sciences 715*, 318-391, 1994.

Brooks, J.M., L.A. Barnard, D.A. Wiesenburg, M.C. Kennicutt III, and K.A. Kvenvolden, Molecular and isotopic compositions of hydrocarbons at Site 533, Deep Sea Drilling Project Leg 76, *Initial Reports, Deep Sea Drilling Project, 76*, 377-389, 1983.

Brooks, J.M., B.H. Cox, W.R. Bryant, M.C. Kennicutt III, R.G. Mann, and T.J. McDonald, Association of gas hydrates and oil seepage in the Gulf of Mexico, *Organic Geochemistry, 10*, 221-234, 1986.

Brooks, J.M., M.E. Field, and M.C. Kennicutt, Observations of gas hydrates in marine sediments, offshore northern California, *Marine Geology, 96*, 103-109, 1991.

Brooks, J.M., A.W.A. Jeffrey, T.J. McDonald, R.C. Pflaum, and K.A. Kvenvolden, Geochemistry of hydrate gas and water from Site 570, Deep Sea Drilling Project Leg 84, *Initial Reports, Deep Sea Drilling Project, 84*, 699-703, 1985.

Brooks, J.M., M.C. Kennicutt III, R.R. Fay, T.J. McDonald, and R. Sassen, Thermogenic gas hydrates in the Gulf of Mexico: *Science, 225*, 409-411, 1984.

Bugge, T., S. Befring, R.H. Belderson, T. Eidvin, E. Jansen, N.H. Kenyon, H. Holtedahl, and H.P. Sejrup, A giant three-stage submarine slide off Norway, *Geo-Marine Letters, 7*, 191-198, 1987.

Cande, S.C., R.B. Leslie, J.C. Parra, and M. Hobart, Interaction between the Chile Ridge and Chile Trench: geophysical and geothermal evidence, *Journal Geophysical Research 92(B1)*, 495-520, 1987

Carlson, P.R., M. Golan-Bac, H.A. Karl, and K.A. Kvenvolden, Seismic and geochemical evidence for shallow gas in sediment on Navarin continental margin, Bering Sea, *American Association Petroleum Geologists Bulletin 69*, 422-436, 1985.

Carpenter, G., Coincident sediment slump/clathrate complexes on the U.S. Atlantic continental slope, *Geo-Marine Letters, 1*, 29-32, 1981.

Cherskiy, N., and Yu.. F. Makogon, Solid gas—World reserves are enormous, *Oil and Gas International, 10(8)*, 82-84, 1970.

Cherskiy, N.V., V.P. Tsarev, and S.P. Nikitin, Investigation and prediction of conditions of accumulation of gas resources in gas-hydrate pools, *Petroleum Geology, 21*, 65-89, 1985.

Chi, W-C., D.L. Reed, C-S Liu, and N. Lundberg, Distribution of the bottom-simulating reflector in the offshore Taiwan collision zone, *Terrestrial, Atmospheric and Oceanic Sciences, 9*, 779-794, 1998.

Collett, T.S., Natural gas hydrates of the Prudhoe Bay and Kuparuk River area, North Slope, Alaska, *American Association of Petroleum Geologists Bulletin 77*, 793-812, 1993.

Collett, T.S., and J. Ladd, Detection of gas hydrate with downhole logs and assessment of gas hydrate concentrations (saturations) and gas volumes on the Blake Ridge with electrical resistivity log data, *Proc. Ocean Drill. Program, Sci. Results, 164*, 179-191, 1999.

Collett, T.S., and R.F. Wendlandt, Formation evaluation of gas hydrate-bearing marine sediments on the Blake Ridge with downhole geochemical log measurements, *Proc. Ocean Drill. Program, Sci. Results, 164*, 199-215, 1999.

Collett, T.S., R.E. Lewis, S.R. Dallimore, M.W. Lee, T.H. Mroz, and T. Uchida, Detailed evaluation of gas hydrate reservoir properties using JAPEX/JNOC/GSC Mallik 2L-38 gas hydrate research well downhole woll-log desplays, *Geological Survey of Canada Bulletin 544*, 295-311, 1999.

Cranston, R.E., Pore-water geochemistry, JA:EX/JNOC/GNC Mallik 2L-38 gas hydrate research well, *Geological Survey of Canada Bulletin 544*, 165-175, 1999.

Dallimore, S.R., T. Uchida, and T.S. Collett, Scientific results from JAPEX/JNOC/GCS Mallik 2L-38 gas hydrate research well, Mackenzie Delta, Northwest Territories, Canada, *Geological Survey of Canada Bulletin 544*, 403 pp, 1999.

Davidson, D.W., M.K. El-Defrawy, M.O. Fuglem, and A.S. Judge, Natural gas hydrates in northern Canada, in National Research Council of Canada, *Proceedings 3rd International Conference on Permafrost 1*, 938-943, 1978.

Davis, E.E., R.D. Hyndman, and H. Villinger, Rates of fluid expulsion across the northern Cascadia accretionary prism: Constraints from new heat flow and multichannel seismic reflection data, *Journal Geophysical Research 95(B6)*, 8869-8889, 1990.

DeLange, G.J., and H.-J. Brumsack, The occurrence of gas hydrates in eastern Mediterranean mud dome structures as indicated by pore-water composition, in *Gas Hydrates—Relevance to World Margin Stability and Climatic Change*, edited by J.-P Henriet and J. Mienert, pp. 167-175, The Geological Society, London, Special Publication 137, 1998.

Dickens, G.R., C.K. Paull, P. Wallace, and ODP Leg 164 Scientific Party, Direct measurement of in situ methane quantities in a large gas-hydrate reservoir, *Nature (London), 385 (6615)*, 426-428, 1997.

Dillon, W.P., W.W. Danforth, D.R. Hutchinson, R.M. Drury, M.H. Taylor, and J.S. Booth, Evidence for faulting related to dissociation of gas hydrate and release of methane off the southeastern United States, in *Gas Hydrates—Relevance to World Margin Stability and Climatic Change*, edited by J.-P Henriet and J. Mienert, pp. 293-302, The Geological Society, London, Special Publication 137, 1998.

Dillon, W.P., K. Fehlhaber, D.F. Coleman, M.W. Lee, and D.R. Hutchinson, Maps showing gas-hydrate distribution off the east coast of the United States, *U.S. Geological Survey, Miscellaneous Field Studies Map, MF 2268*, 2 sheets, 1995.

Dillon, W.P., M.W. Lee, K. Fehlhaber, and D.G. Coleman, Gas hydrates on the Atlantic continental margin of the United States—controls on concentration, *The Future of Energy Gases, U.S. Geological Survey, Professional Paper 1570*, 313-330, 1993.

Dillon, W.P., P. Popenoe, J.A. Grow, K.D. Klitgord, B.A. Swift, C.K. Paull, and K.V. Cashman, Growth faulting and salt diapirism: their relationship and control in the Carolina Trough, eastern North America, in Studies in Continental Margin Geology, *American Association Petroleum Geologists Memoir 34*, 21-46, 1983.

Dobrynin, V.M., Yu. P. Korotajev, and D.V. Plyuschev, Gas hydrates—a possible energy resource, In R.G. Meyer and J.C. Olson, eds., *Long-term Energy Resources*, Pitman, Boston, 727-729, 1981.

Exon, N.F., G.R. Dickens, J-M. Auzende, Y. Lafoy, P.A. Symonds, and S. Van de Beuque, Gas hydrates and free gas on the Lord Howe Rise, Tasman Sea, *Petroleum Exploration Society of Australia Journal*, No. 26, 148-158, 1998.

Ewing, J.I., and C.H. Hollister, Regional aspects of deep sea drilling in the western North Atlantic, *Initial Reports, Deep Sea Drilling Project11*, 951-973, 1972.

Field, M.E., and K.A. Kvenvolden, Gas hydrates on the northern California continental margin, *Geology, 13*, 517-520, 1985.

Fontana, R.L., and A. Mussumeci, Hydrates offshore Brazil, *Annals of the New York Academy of Sciences 715*, 106-113, 1994.

Froelich, P.N., K.A. Kvenvolden, M.E. Torres, A. Waseda, B.M. Didyk, and T.D. Lorenson, Geochemical evidence for gas hydrate in sediment near the Chile Triple Junction, *Proc. Ocean Drill. Program, Sci. Results, 141*, 279-286, 1995.

Gas Epoch, Potential of unconventional natural gas: methane hydrate, Japan Gas Association, *Gas Epoch, 8*, 18-21, 1995.

Ginsburg, G.D., and V.A. Soloviev, Submarine gas hydrate estimation: theoretical and empirical approaches, *Offshore Technology Conference, OTC 7693, 1*, 513-518, 1995.

Ginsburg, G.D. and V.A. Soloviev, V.A. Submarine Gas Hydrates, Translated from Russian, Norma Publishers, St. Petersburg, Russia, 216 pp., 1998.

Ginsburg, G.D., R.A. Guseynov, A.A. Dadashev, G.A. Ivanova, S.A. Kazantsev, V.A. Soloviev, E.V., Telepnev, P. Ye. Askeri-Nasirov, A.A. Yesikov, V.I. Mal'tseva, G.Yu. Mashirov, and I. Yu. Shabayeva, Gas hydrates of the southern Caspian, I*nt. Geol. Rev., 43*, 765-782, 1992.

Ginsburg, G.D., I.S. Gramberg, and V.A. Soloviev, Gas hydrates of continental slopes, *Geologiya Nefti i Gaza, 11*, 43-45, 1990a.

Ginsburg, G.D., A.N. Kremlev, M.N. Grigor'ev, G.V. Larkin, A.D. Pavlenkin, and N.A. Saltykova, Filtrogenic gas hydrates in the Black Sea (Twenty-first voyage of the research vessel *Evpatoriya*), *Geologiya I Geofizika, 31(3)*, 10-19, 1990b.

Ginsburg, G.D., A.V. Milkov, V.A. Soloviev, A.V. Egorov, G.A. Cherdashev, P.R. Vogt, K. Crane, T.D. Lorenson, and M.D. Khutorskoy, Gas hydrate accumulation at the Haakon-Mosby mud volcano, *Geo-Marine Letters, 19*, 57-67, 1999.

Ginsburg, G.D., V.A. Soloviev, R.E. Cranston, T.D. Lorenson, and K.A. Kvenvolden, Gas hydrates from the continental slope, offshore Sakhalin Island, Okhotsk Sea, *Geo-Marine Letters, 13*, 41-48, 1993.

Goodman, M.A., *Insitu* gas hydrates: past experience and exploitation concepts, *Proc. 1st International Gas Research Conference*, 376-391, 1980.

Gornitz, V., and I. Fung, Potential distribution of methane hydrates in the world's ocean, *Global Biogeochemical Cycles, 8*, 335-347, 1994.

Grantz, A., and D.A. Dinter, Constraints of geologic processes on western Beaufort Sea oil developments, *Oil and Gas Journal 78(18)*, 304-319, 1980.

Hammond, R.D., and J.R. Gaither, Anomalous seismic character—Bering Sea shelf, *Geophysics, 48*, 590-605, 1983.

Harrison, W.E., and J.A. Curiale, Gas hydrates in sediments of holes 497 and 498A, Deep Sea Drilling Project Leg 67, *Initial Reports, Deep Sea Drilling Project, 67*, 591-594, 1982.

Harrison, W.E., R. Hesse, and J.M. Gieskes, Relationship between sedimentary facies and interstitial water chemistry of slope, trench, and Cocos plate sites from the Middle America Trench transect, active margin off Guatemala, Deep Sea Drilling Project Leg 67, *Initial Reports, Deep Sea Drilling Project, 67*, 603-613, 1982.

Harvey, L.D.D., and Z. Huang, Evaluation of the potential impact of methane clathrate destabilization on future global warming, *Journal of Geophysical Research, 100(D20)*, 2905-2926, 1995.

Hedberg, H.D., Methane generation and petroleum migration, in Problems of Petroleum Migration, *American Association Petroleum Geologists, Studies in Geology 10*, 179-206, 1980.

Henriet, J.P., B. De Mol, S. Pillen, M. Vanneste, D. Van Rooij, W. Versteeg, P.F. Croker, P.M. Shannon, V. Unnithan, S. Bouriak, P. Chachkine, and The Porcupine—Beligica 97 Shipboard Party, Gas hydrate crystals may help build reefs, *Nature, 391*, 648-649, 1998.

Hesse, R., and W.E. Harrison, Gas hydrates (clathrates) causing pore-water freshening and oxygen isotope fractionation in deep-water sedimentary sections of terrigenous continental margins, *Earth and Planetary Science Letters, 55*, 453-462, 1981.

Hesse, R., J. Lebel, and J.M. Gieskes, Interstitial water chemistry of gas-hydrate-bearing sections on the Middle America Trench slope, Deep-Sea Drilling Project leg 84, *Initial Reports, Deep Sea Drilling Project, 84*, 727-736, 1985.

Holbrook, W.S., H. Hoskins, W.T. Wood, R.A. Stephen, D. Lizarralde, and Leg 164 Science Party, Methane hydrate and free gas on the Blake Ridge from vertical seismic profiling, *Science, 273*, 1840-1843, 1996.

Hutchinson, D.R., A.J. Golmshtok, C.A. Scholz, T.C. Moore, M.W. Lee, and M. Kuzmin, Bottom simulating reflector in Lake Baikal, *Transactions, American Geophysical Union 72(17, Supplement)*, 307, 1991.

Hyndman, R.D., and G.D. Spence, A seismic study of methane hydrate marine bottom simulating reflectors, *Journal Geophysical Research 97(B5)*, 6683-6698, 1992.

Jansen, E., S. Befring, T. Bugge, T. Eidvin, G. Hotedahl, and H.P. Sejrup, Large submarine slides on the Norwegian continental margin: sediments, transport and timing, *Marine Geology, 78*, 77-107, 1987.

Jenden, P.D., and J.M. Gieskes, Chemical and isotopic composition of interstitial water from Deep Sea Drilling Project Sites 533 and 534, *Initial Reports Deep Sea Drilling Project, 76*, 453-461, 1983.

JNOC press release, Proof of the methane-hydrate-containing layer in the national wildcat, "Nankai Trough", based on the National Domestic Petroleum Exploration Program, Japan National Oil Corporation, January 20, 2000.

Judge, A.S., Natural gas hydrates in Canada, in Roger J.E. Brown Memorial Volume, *Proceedings of the 4th Canadian Permafrost Conference 1981*, 320-328, 1982.

Kastner, M., K.A. Kvenvolden, M.J. Whiticar, A. Camerlenghi, and T.D. Lorenson, Relation between pore fluid chemistry and gas hydrates associated with bottom-simulating reflectors at the Cascadia Margin, Sites 889 and 892, *Proc. Ocean Drill. Program, Sci. Results, 146 (Pt. 1)*, 175-187, 1995.

Kastner, M., K.A. Kvenvolden, and T.D. Lorenson, Chemistry, isotopic composition, and origin of a methane-hydrogen sulfide hydrate at the Cascadia subduction zone, *Earth and Planetary Science Letters,156*, 173-183, 1998.

Katz, H.R., Probable gas hydrate in continental slope east of the North Island, New Zealand, *Journal of Petroleum Geology, 3*, 315-324, 1981.

Kayen, R.E., and H.J. Lee, Pleistocene slope instability of gas hydrate-laden sediment on the Beaufort Sea margin, *Marine Geotechnology 10*, 125-141, 1991.

Kuramoto, S., and I. Okamura, The single-channel acoustic investigation and 3.5 kHz SBP search of the ocean area off southwestern Hokkaido, in General study on the environmental variations of the ocean area west off Hokkaido, *Geological Survey of Japan*, 28-41, 1995.

Kus'min, M.I., G.V. Kalmychkov, R.F. Geletii, V.A. Gnilusha, A.V. Goreglyad, B.N. Khakhaev, L.A. Pevzner, T. Kawai, N. Yoshida, A.D. Duchkov, V.A. Ponomarchuk, A.E. Kontorovich, N.M. Abazhin, G.A. Makhov, Yu. A. Dyadin, F.A. Kuznetsov, E.G. Larionov, A.Yu. Manakov, B.S. Smolyakov, M.M. Mandel'baum, and N.K. Zheleznyakov, The first find of gas-hydrates in the sedimentary rocks of Lake Baikal, *Doklady Akademii Nauk, 362*, 541-543, 1998, [Translation in *Doklady Earth Sciences, 362*, 1029-1031, 1998].

Kvenvolden, K.A., Methane hydrate—A major reservoir of carbon in the shallow geosphere?, *Chemical Geology, 71*, 41-51, 1988.

Kvenvolden, K.A., Gas hydrates—Geological perspective and global change, *Reviews of Geophysics, 31*, 173-187, 1993.

Kvenvolden, K.A., A review of the geochemistry of methane in natural gas hydrate, *Organic Geochemistry, 23*, 997-1008, 1995.

Kvenvolden, K.A., Potential effects of gas hydrate on human welfare, *Proceedings of the National Academy of Sciences (USA), 96*, 3420-3426, 1999.

Kvenvolden, K.A., and L.A. Barnard, Gas hydrates of the Blake Outer Ridge, Site 533, Deep Sea Drilling Project Leg 76, *Initial Reports, Deep Sea Drilling Project, 76*, 353-365, 1983.

Kvenvolden, K.A., and G.E. Claypool, Gas hydrates in oceanic sediment, *U.S. Geological Survey Open-File Report 88-216*, 50 pp.

Kvenvolden, K.A. and A. Grantz, Gas hydrates of the Arctic Ocean region, in *The Geology of North America, 50, The Arctic Ocean Region,* Geological Society of America, 539-549, 1990.

Kvenvolden, K.A., and M. Kastner, Gas hydrates of the Peruvian outer continental margin, *Proc. Ocean Drill. Program, Sci. Results, 112*, 517-526, 1990.

Kvenvolden, K.A., and T.J. McDonald, Gas hydrates of the Middle America Trench, Deep Sea Drilling Project Leg 84, *Initial Reports, Deep Sea Drilling Project, 84*, 667-682, 1985.

Kvenvolden, K.A., and M.A. McMenamin, Hydrates of natural gas: a review of their geologic occurrence, *U.S. Geological Survey Circular 825*, 11 pp, 1980.

Kvenvolden K.A., and R. von Huene, Natural gas generation in sediments of the convergent margin of the eastern Aleutian Trench area, in Tectonostratigraphic Terranes of the Circum-Pacific Region, *Circum-Pacific Council on Energy and Minerals, Earth Science Series 1*, 31-49, 1985.

Kvenvolden, K.A., G.E. Claypool, C.N. Threlkeld, and E.D. Sloan, Geochemistry of a naturally occurring massive marine gas hydrate, *Organic Geochemistry, 6*, 703-713, 1984.

Kvenvolden, K.A., G.D. Ginsburg, and V.A. Soloviev, Worldwide distribution of subaquatic gas hydrates, *Geo-Marine Letters, 13*, 32-40, 1993.

Kvenvolden, K.A., M. Golan-Bac, and J.B. Rapp, Hydrocarbon geochemistry of sediments offshore from Antarctica. I. Wilkes Land continental margin in The Antarctic Continental Margin: Geology and Geophysics of Offshore Wilkes Land and the Western Ross Sea, *Circum-Pacific Council on Energy and Mineral Resources, Earth Science Series 5A*, 205-213, 1987.

Kvenvolden, K.A., M. Golan-Bac, T.J. McDonald, R.C. Pflaum, and J.M. Brooks, Hydrocarbon gases in sediment of the Vøring Plateau, Norwegian Sea, *Proc. Ocean Drill. Program, Sci. Results, 104*, 319-326, 1989.

Ladd, J., G. Westbrook, and S. Lewis, Subduction tectonics in forearcs: Guatemala vs. Barbados, *Lamont-Doherty Geological Observatory Yearbook 1981-1982, 8*, 17-22, 1982.

Ladd, J.W., M. Truchan, M. Talwani, P.L. Stoffa, P. Buh, R. Houtz, A. Mauffret, and G. Westbrook, Seismic reflection profiles across the southern margin of the Caribbean, *The Geological Society of America Memoir 162*, 153-159, 1984.

Lee, M.W., Gas hydrates amount estimated from acoustic logs at the Blake Ridge, Sites 994, 995, and 997, *Proc. Ocean Drill. Program, Sci. Results, 164*, 193-198, 1999.

Lein, A., P. Vogt, K. Crane, A. Ergorov, and M. Ivanov, Chemical and isotopic evidence for the nature of the fluid in CH_4-containing sediments of the Haakon-Mosby mud volcano, *Geo-Marine Letters, 19*, 76-83, 1999.

Lodolo, E., A. Camerlenghi, and G. Brancolini, A bottom simulating reflector on the South Shetland margin, Antarctic Peninsula, *Antarctic Science, 5*, 207-210, 1993.

Lonsdale, M.J., The relationship between silica diagenesis, methane, and seismic reflections on the South Orkney microcontinent. *Proc. Ocean Drill. Program, Sci. Results, 113*, 27-37, 1990.

Lonsdale P., A transform continental margin rich in hydrocarbons, Gulf of California, *American Association of Petroleum Geologists Bulletin, 69*, 1160-1180, 1985.

Lorenson, T.D., and T.S. Collett, Gas content and composition of gas hydrate from sediments of the southeastern North American continental margin, *Proc. Ocean Drill. Program, Sci. Results, 164*, 37-46, 1999.

Lorenson, T.D., M.J. Whiticar, A. Waseda, S.R. Dallimore, and T.S. Collett, Gas composition and isotopic geochemistry of cuttings, core, and gas hydrate from the JAPEX/JNOC/GSC Mallik 2L-38 gas hydrate research well, *Geologic Survey of Canada, Bulletin 544*, 143-163, 1999.

MacDonald, G.T., Role of methane clathrates in past and future climates, *Clim. Change, 16*, 247-281, 1990.

MacDonald, I.R., N.L. Guinasso Jr., R. Sassen, J.M. Brooks, L. Lee, and K.T. Scott, Gas hydrate that breaches the sea floor on the continental slope of the Gulf of Mexico, *Geology, 22*, 699-702, 1994.

Makogon, Yu., F., Hydrates of Hydrocarbons, PennWell, Tulsa, Oklahoma, 482 pp., 1997.

Makogon, Yu. F., Natural gas hydrates the state of study of the USSR and perspectives for its use, oral presentation, no abstract, *Chemical Congress of North America*, Toronto, Canada, June 5-10, 1988 [cited in Sloan (1998)].

Makogon, Yu. F., F.A. Trebin, A.A. Trofimuk, V.P. Tsarev, and N.V. Cherskiy, Detection of a pool of natural gas in a solid (hydrated gas) state, *Doklady Akademii Nauk SSSR, 196*, 203-206, 1971, [Translation in *Doklady-Earth Science Section, 196*, 197-200, 1972].

Makogon, Yu. F., A.A. Trofimuk, V.P. Tsarev, and N.V. Cherskiy, Possible origin of natural gas hydrates at floors of seas and oceans, *Academiya Nauk SSSR, Sibirskoe Otdeleniye Geologiya I Geofizika, 4*, 3-6, 1973, [Translation in *International Geology Review, 16*, 553-556, 1974].

Makogon, Yu. F., V.I. Tsarev, and V.V. Cherskiy, Formation of large natural gas fields in zones of permanently low temperatures, *Doklady Akademii Nauk SSSR, 205*, 700-703, 1972, [Translation in *Doklady-Earth Science Section, 205*, 215-218, 1973].

Manley, P.L., and R.D. Flood, Anomalous sound velocities in near-surface, organic-rich gassy sediments in the central Argentine Basin, *Deep-Sea Research, 36*, 611-623, 1989.

Mann, R., and J. Gieskes, Interstitial water studies, Leg 28, *Initial Reports, Deep Sea Drilling Project 28*, 805-817, 1975.

Markl, R.G., G.M. Bryan, and J.I. Ewing, Structure of the Blake-Bahama Outer Ridge, *Journal Geophysical Research, 75*, 4539-4555, 1970.

Matsumoto, R., and W.S. Borowski, Gas hydrate estimates from newly determined oxygen isotopic fractionation (α_{GH-IW}) and $\delta^{18}O$ anomalies of the interstitial waters: Leg 164, Blake Ridge, *Proc. Ocean Drill. Program, Sci. Results, 164*, 59-66, 2000.

Matsumoto, R., Y. Okuda, and Y. Aoki, Methane hydrate, Tokoyo Nikkei Science Publishers, 1994.

McCarthy, J., A.J. Stevenson, D.W. Scholl, and T.L.Vallier, Speculations on the petroleum geology of the accretionary body: an example from the central Aleutians, *Marine and Petroleum Geology, 1*, 151-167, 1984.

McIver, R.D., Hydrocarbon gases in canned core samples from Leg 28 sites 271, 272, and 273, Ross Sea, *Initial Reports, Deep Sea Drilling Project, 28*, 815-817, 1975.

McIver, R.D., Role of natural gas—important agent in geological processes, *Geol. Soc. Am. Abst. Programs, 9*, 1089-1090, 1977.

McIver, R.D., Gas hydrates, in R.F. Meyer, and J.C. Olson, eds., *Long-term Energy Resources*, Pitman, Boston, 713-726, 1981.

McIver, R.D., Role of naturally occurring gas hydrates in sediment transport, *American Association of Petroleum Geologists Bulletin 66*, 789-792, 1982.

McKirdy, D.M., and P.J. Cook, Organic geochemistry of Pliocene-Pleistocene calcareous sediments, DSDP Site 262, Timor Trough, *American Association Petroleum Geologists Bulletin 64*, 2118-2138, 1980.

Mienert, J., and J. Posewang, Evidence of shallow- and deep-water gas hydrate destabilizations in North Atlantic polar continental margin sediments: *Geo-Marine Letters, 19*, 143-149, 1999.

Mienert, J., J. Posewang, and M. Baumann, Gas hydrates along the northeastern Atlantic margin: possible hydrate-bound margin instabilities and possible release of methane, in *Gas Hydrates—Relevance to World Margin Stability and Climatic Change*, edited by J.-P Henriet and J. Mienert, pp. 275-291, The Geological Society, London, Special Publication 137, 1998.

Miller, J.J., M.W. Lee, and R. von Huene, An analysis of a seismic reflection from the base of a gas hydrate zone, offshore Peru, *American Association Petroleum Geologist Bulletin 75*, 910-924, 1991.

Moore, G.W., and J.M. Gieskes, Interaction between sediment and interstitial water near the Japan Trench, Leg 57, Deep Sea Drilling Project, *Initial Reports, Deep Sea Drilling Project,56, 57(2)*, 1269-1275, 1980.

Moore, J.C., K.M. Brown, F. Horath, G. Cochrane, M. MacKay, and G. Moore, Plumbing accretionary prisms: effects of permeability variations. *Philosophical Transactions Royal Society of London Series A, 335*, 275-288, 1992.

Murota, T., Gas hydrate exploration: Its technology-environment interface in the world and Japan, *Hitotsubashi Journal of Economics, 37*, 21-44, 1996.

Neben, S., K. Hinz, and H. Beiersdorf, Reflection characteristics, depth and geographical distribution of bottom simulating reflectors within the accritionary wedge of Sulawesi, in *Gas Hydrates—Relevance to World Margin Stability and Climatic Change*, edited by J.-P Henriet and J. Mienert, pp. 255-265, The Geological Society, London, Special Publication 137, 1998.

Okuda, Y., Methane hydrate with special reference to potentialities as future natural gas resources, *Proceedings of Thermal Engineering Symposium of July 7, 1995; Japan Association of Mechanical Engineering*, 5-8, 1995.

Panayev, V.A., Gas hydrates in the oceans, *International Geology Review, 29*, 569-602, 1987.

Paull, C.K., W.S. Borowski, N.M. Rodriguez, and the ODP Leg 164 Shipboard Scientific Party, Marine gas hydrate inventory: preliminary results of ODP Leg 164 and implications for gas venting and slumping associated with the Blake Ridge gas hydrate field, in *Gas Hydrates—Relevance to World Margin Stability and Climatic Change*, edited by J.-P Henriet and J. Mienert, pp. 153-160, The Geological Society, London, Special Publication 137, 1998.

Paull, C.K., T.D. Lorenson, W.S. Boroworski, W. Ussler, K. Olsen, N.M. Rodriguez, and H. Wehner, Isotopic composition of CH_4, CO_2 species, and sedimentary organic matter within ODP Leg 164 sample from the Blake Ridge: Gas source implications, *Proc. Ocean Drill. Program, Sci. Results, 164*, 67-78, 2000.

Paull, C.K., R. Matsumoto, P.J. Wallace, et al., *Proc. Ocean Drill. Program, Initial Reports 164*, 623 pp., 1996.

Paull, C.K., W. Ussler, III, and W.S. Borowski, Sources of biogenic methane to form marine gas hydrates—in situ production or upward migration?, *Annals of the New York Academy of Sciences, 715*, 392-409, 1994.

Pflaum, R.C., J.M. Brooks, H.B. Cox, M.C. Kennicutt, and D-D. Sheu, Molecular and isotopic analysis of core gases and gas hydrates, Deep Sea Drilling Project Leg 96, *Initial Reports, Deep Sea Drilling Project, 96*, 781-784, 1986.

Posewang, J., and J. Mienert, High-resolution seismic studies of gas hydrates west of Svalbard, *Geo-Marine Letters, 19*, 150-156, 1999.

Potential Gas Committee, Potential Supply of Natural Gas in the United States (as of December 31, 1980), *Potential Gas Agency, Colorado School of Mines, Golden, Colorado*, 119 pp., 1981.

Rastogi, A., B. Deka, I.L. Budhiraja, and G.C. Agarfwal, Possibility of large deposits of gas hydrates in deeper waters of India, *Marine Georesources and Geotechnology, 17*, 49-63, 1999.

Saltykova, N.A., V.A. Soloviev, and A.D. Pavlenkin, Seismic indications of submarine gas hydrates in the Bering Sea,

Geological-Geophysical Investigations of the World Oceans, *Sevmorgeologia, Lenigrad*, 119-130, 1987.

Sassen, R., and I.R. MacDonald, Evidence of structure H hydrate, Gulf of Mexico continental slope, *Organic Geochemistry, 22*, 1029-1032, 1994.

Sassen, R., and I.R. MacDonald, Hydrocarbons of experimental and natural gas hydrates, Gulf of Mexico continental slope, *Organic Geochemistry, 26*, 289-293, 1997.

Sato, M., Geographical distribution of naturally occurring gas hydrates, *The Earth Monthly, 16*, 533-538, 1994.

Schoell, M., Multiple origins of methane in the Earth, *Chemical Geology, 71*, 1-10, 1988.

Scholl, D.W., and A.K. Cooper, VAMPs--Possible hydrocarbon-bearing structures in Bering Sea basin, *American Association Petroleum Geologists Bulletin 62*, 2481-2488, 1978.

Shipboard Scientific Party, Site 796, *Proc. Ocean Drill. Program, Initial Reports 127*, 247-322, 1990.

Shipboard Scientific Party, Site 808 *Proc. Ocean Drill. Program, Initial Reports 131*, 71-269, 1991.

Shipboard Scientific Party, Site 1041, *Proc. Ocean Drilling Program, Initial Reports, 170*, 153-188, 1997.

Shipley, T.H., and B.M. Didyk, Occurrence of methane hydrates offshore southern Mexico, *Initial Reports, Deep Sea Drilling Project, 66*, 547-555, 1982.

Shipley, T.H., M.H. Houston, R.T. Buffler, F.J. Shaub, K.J. McMillen, J.W. Ladd, and J.L. Worzel, J.L., Seismic reflection evidence for the widespread occurrence of possible gas-hydrate horizons on continental slopes and rises, *American Association of Petroleum Geologists Bulletin, 63*, 2204-2213, 1979.

Sloan, E.D., Clathrate Hydrates of Natural Gas, 2nd Ed., Marcel Dekker, New York, 705 pp., 1998.

Stoll, R.D., J.I. Ewing, and G.M. Bryan, Anomalous wave velocities in sediments containing gas hydrates, *Journal of Geophysical Research, 76*, 2090-2094, 1971.

Suess, E., M.E. Torres, G. Bohrmann, R.W. Collier, J. Greinert, P. Linke, G. Rehder, A. Trehu, K. Wallmann, G. Winckler, and E. Zuleger, Gas hydrate destabilization: enhanced dewatering, benthic material turnover and large methane plumes at the Cascadia convergent margin, *Earth and Planetary Science Letters, 170*, 1-15, 1999a.

Suess, E., G. Bohrman, J. Greinert, and E. Lausch, Flammable Ice, *Scientific American, 281 (5)*, 76l-83, 1999b.

Summerhayes, C.P., B.D. Bornhold, and R.W. Embley, Surficial slides and slumps on the continental slopes and rises of Southwest Africa: a reconnaissance study, *Marine Geology, 31*, 265-277, 1979.

Tamano T, T. Toba, and Y. Aoki, Development of fore-arc continental margins and their potential for hydrocarbon accumulation, *Proceedings 11th World Petroleum Congress, New York, Wiley, 2*, 135-145, 1984.

Taylor, A.E., R.J. Wetmiller, and A.S. Judge, Two risks to drilling and production off the east coast of Canada--earthquakes and gas hydrates, in W. Denner (Ed.), *Symposium on Research in the Labrador Coastal and Offshore Region*, Memorial University of Newfoundland, 91-105, 1979.

Trofimuk, A.A., N.V. Cherskiy, and V.P. Tsarev, Accumulation of natural gases in zones of hydrate formation in the hydrosphere, *Doklady Akademii Aauk SSSR, 212*, 931-934, 1973.

Trofimuk, A.A., N.V. Cherskiy, and V.P. Tsarev, The biogenic methane resources of the oceans, *Doklady Academii Nauk SSSR, 225*, 936-943, 1975.

Trofimuk, A.A., N.V. Cherskiy, and V.P. Tsarev, The role of continental glaciation and hydrate formation on petroleum occurrences, in R.F. Meyer, ed., *Future Supply of Nature-made Petroleum and Gas*, Pergamon Press, New York, 919-926, 1977.

Trofimuk, A.A., N.V. Cherskiy, and V.P. Tsarev, The gas-hydrate sources of hydrocarbons, *Priroda, 1*, 18-27, 1979.

Tucholke, B.E., G.M. Bryan, and J.I. Ewing, Gas-hydrate horizons detected in seismic-profiler data from the western North Atlantic, *American Association Petroleum Geologists Bulletin, 61*, 698-707, 1977.

Vasil'ev, V.G., Yu. F. Makogon, F.A. Trebin, A.A. Trofimuk, and N.V. Cherskiy, The property of natural gases to occur in the Earth crust in a solid state and to form gas hydrate deposits, *Otkrytiya v SSSR, 1968-1969*, 15-17, 1970.

Veerayya, M., S.M. Karisiddaiah, K.H. Vora, B.G. Wagle, and R. Almeida, in *Gas Hydrates—Relevance to World Margin Stability and Climatic Change*, edited by J.-P Henriet and J. Mienert, pp. 239-253, The Geological Society, London, Special Publication 137, 1998.

Weaver, J.S., and J.M. Stewart, In-situ hydrates under the Beaufort Sea Shelf, in Roger J.E. Brown Memorial Volume, *Proceedings 4th Canadian Permafrost Conference 1981*, 312-319, 1982.

White, R.S., Gas hydrate layers trapping free gas in the Gulf of Oman, *Earth and Planetary Science Letters, 42*, 114-120, 1979.

Whiticar, M.J., Carbon and hydrogen isotope systematics of bacterial formation and oxidation of methane, *Chemical Geology, 161*, 291-314, 1999

Whiticar, M.J., E. Faber, and M. Schoell, Biogenic methane formation in marine and freshwater environments: CO_2 reduction vs. acetate fermentation—isotopic evidence, *Geochim. Cosmochim. Acta, 50*, 693-709, 1986.

Whiticar, M.J., M. Hovland, M. Kastner, and J.C. Sample, Organic geochemistry of gases, fluids, and hydrates at the Cascadia accretionary margin, *Proc. Ocean Drill. Program, Sci. Results, 146 (Pt. 1)*, 385-397, 1995.

Woodside, J.M., M.K. Ivanov, A.F. Limonov, and Shipboard Scientists of the Anaxiprobe expeditions, Shallow gas and gas hydrates in the Anaximander Mountains region, eastern Mediterranean Sea, in *Gas Hydrates—Relevance to World Margin Stability and Climatic Change*, edited by J.-P Henriet and J. Mienert, pp. 177-193, The Geological Society, London, Special Publication 137, 1998.

Yefremova, A.G., and B.P. Zhizhchenko, Occurrence of crystal hydrates of gas in sediments of modern marine basins, *Doklady Akademii Nauk SSSR, 214*, 1179-1181, 1974.

Yefremova, A.G., and N.D. Gritchina, Gas hydrates in sediments beneath seas and the problem of their exploitation, *Geologiya Nefti i Gaza, 2*, 32-35, 1981.

Zonenshayn, L.P., I.O. Murdmaa, B.V. Baranov, A.P. Kusnetsov, V.S. Kuzin, M.I. Kuz'min, G.P. Avdeyko, P.A. Stunzhas, V.N. Lukashin, M.S. Barash, G.M. Valyashko, and L.L. Demina, An underwater gas source in the Sea of Okhotsk west of Paramushir Island, *Oceanology, 27*, 795-800, 1987.

K.A. Kvenvolden, U.S. Geological Survey, 345 Middlefield Road, MS 999, Menlo Park, CA 94025
kkvenvolden@usgs.gov

T.D. Lorenson, U.S. Geological Survey, 345 Middlefield Foad, MS 999, Menlo Park, CA 94025
tlorenson@usgs.gov

Modeling the Global Carbon Cycle With a Gas Hydrate Capacitor: Significance for the Latest Paleocene Thermal Maximum

Gerald R. Dickens[1]

School of Earth Sciences, James Cook University, Townsville, Queensland 4811, Australia

Abstract. The "Latest Paleocene Thermal Maximum" (or LPTM) ca. 55 Ma was characterized by a 4 to 6 °C rise in deep ocean water temperature and an extraordinary injection of ^{12}C-rich carbon into the exogenic carbon cycle. The best explanation for the carbon cycle perturbation is that the bottom water warming converted massive amounts of marine gas hydrate to free CH_4 gas, and this CH_4 was added to the ocean. If correct, basic models for the global carbon cycle must be reconstructed with a gas hydrate "capacitor" so that large quantities of ^{12}C-rich carbon can be stored and released over time. Although recent work has discussed CH_4 release from gas hydrate, the first operational model of the global carbon cycle including CH_4 storage is presented here both conceptually and mathematically. Using the Blake Ridge gas hydrate deposit as an example, the capacitor contains three internal reservoirs: dissolved gas, gas hydrate and free gas. Carbon enters dissolved gas through methanogenesis of organic matter. Upon saturation of pore waters, carbon is transferred to gas hydrate and then free gas at appropriate temperature and pressure conditions. Carbon leaves free gas to the exogenic carbon cycle through anaerobic CH_4 oxidation or, if there is no overlying gas hydrate, direct injection to the water column. Because the amount and location of free gas depends on temperature, the capacitor can suddenly release massive amounts of ^{12}C-rich carbon with abrupt bottom water warming. When the "gas hydrate" capacitor is connected to a Paleogene exogenic carbon cycle and subjected to a 5°C warming, the response is similar to that observed for the LPTM.

1. INTRODUCTION

Clathrate hydrates of gas ("gas hydrates") are crystalline substances composed of cages of water molecules surrounding molecules of low-molecular-weight gas. The stability of these compounds depends on pressure, temperature, activity of water, and gas concentration [*Handa*, 1990; *Dickens and Quinby-Hunt*, 1997]. Gas hydrates will dissociate to constituent water and gas phases with a prescribed increase in temperature or decrease in pressure, activity of water or gas concentration.

Gas hydrates predominantly composed of CH_4 can naturally occur in sediment pore space of under appropriate conditions of high pressure, low temperature, and adequate CH_4 concentrations. These criteria are met between the seafloor

[1]Also at Department of Geology and Geophysics, Rice University, Houston TX 77005, USA

Natural Gas Hydrates: Occurrence, Distribution, and Detection
Geophysical Monograph 124

and a relatively shallow depth on many continental margins [e.g., *Kvenvolden et al.*, 1993; *Dickens et al.*, 1997a]. Consequently, the upper few hundred meters of marine sediment in many regions contain laterally extensive zones of CH_4 hydrate. Current estimates place the total mass of carbon in present marine gas hydrate deposits between 7,500 and 15,000 Gt (Gt = 10^{15} g) [*MacDonald*, 1990; *Kvenvolden*, 1993, 1998; *Gornitz and Fung*, 1994]. Most of the CH_4 in these deposits is formed through bacterial methanogenesis and is therefore extremely enriched in ^{12}C (average $\delta^{13}C$ of –60‰ [*Kvenvolden*, 1995]).

The late early Paleogene ca. 59 to 46 million years ago (Ma) was characterized by significantly warmer climates than present-day, including relatively high temperatures (7-10 °C) for water at depths of 1000 to 3000 m in the ocean [*Miller et al.*, 1987; *Zachos et al.*, 1994]. Superimposed on this general warmth was a brief interval ca. 55 Ma [*Norris and Röhl*, 1999] when terrestrial, high-latitude marine, and deep ocean temperatures soared by at least an additional 4 to 6 °C [*Kennett and Stott*, 1991; *Thomas and Shackleton*, 1996; *Fricke et al.*, 1998; *Katz et al.*, 1999; *Thomas et al.*, 1999]. This latest Paleocene thermal maximum, or LPTM [*Zachos et al.*, 1993], coincided with profound global environmental change, including an extraordinary negative carbon isotope ($\delta^{13}C$) excursion [e.g., *Thomas and Shackleton*, 1996]. At least 25 different Paleogene stable isotope records, including those constructed from bulk and foraminiferal carbonate of all oceans, and terrestrial carbonate and organic matter from multiple continents, now show part of a –2.5 to –3 ‰ anomaly in $\delta^{13}C$ across the LPTM [e.g., *Koch et al.*, 1992; *Thomas*, 1998]. In high-resolution isotope records from open-ocean locations (Fig. 1), the $\delta^{13}C$ excursion can be described as a rapid decrease of at least –2.5 ‰ over 10 to 40 cm followed by a return to near initial values in a roughly logarithmic pattern over 100 to 400 cm, depending on sedimentation rate. The main drop and gradual return in $\delta^{13}C$ spanned less than 10,000 years and ~170,000 years, respectively [*Kennett and Stott*, 1991; *Pardo et al.*, 1997; *Bralower et al.*, 1998; *Katz et al.*, 1999; *Norris and Röhl*, 1999; *Dickens*, 2000].

On the basis of geological observations and simple carbon mass balance equations, it has been proposed that the LPTM $\delta^{13}C$ excursion reflects thermal dissociation of marine gas hydrates and release of massive quantities of ^{12}C-rich CH_4 to the ocean [*Dickens et al.*, 1995]. While support for this hypothesis has grown considerably over the past five years, there remains a major conceptual problem: no working model of the global carbon cycle, present or Paleogene, simple or complex, allows for carbon mass transfer between the ocean, atmosphere and gas hydrates over time [*Dickens*, 1999, 2000b]. This paper synthesizes recent work to construct the first operational carbon cycle that includes gas hydrates. The model is speculative and purposely simple. Nevertheless, the model nicely explains the isotope excursion across the LPTM.

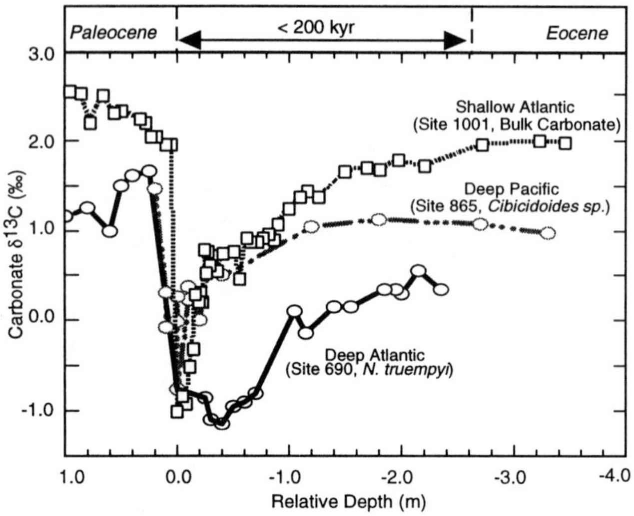

Figure 1. High-resolution carbon isotope records at three widely separated Ocean Drilling Program (ODP) sites placed on a common depth scale where the $\delta^{13}C$ minimum is at 0.0 m. Benthic foraminifera at Site 690, Maud Rise, South Atlantic [*Kennett and Stott*, 1991]. Benthic foraminifera at Site 865, Allison Guyot, Equatorial Pacific [*Bralower et al.*, 1995]. Bulk carbonate at Site 1001, Nicaraguan Rise, Caribbean [*Bralower et al.*, 1997]. Note the abrupt drop in $\delta^{13}C$ of –2.5 to –3 ‰ over 5 to 20 cm followed by a roughly logarithmic return to near initial values over 1 to 4 m. In this figure, "deep" refers to intermediate water depths, 1500-2000 m [*Thomas*, 1998].

2. THE PALEOGENE STEADY-STATE EXOGENIC CARBON CYCLE

The exogenic carbon cycle includes all carbon in the ocean, atmosphere and biomass (Fig. 2). Because carbon cycles through these internal reservoirs in less than 2,000 years at present (and presumably in the past), the mass (M_{Ex}) and isotopic composition (δ_{Ex}) of the exogenic carbon cycle can be considered a single entity over moderate time scales [*Broecker*, 1974; *Sundquist*, 1986]. Carbon is added to and removed from the exogenic carbon cycle through several external carbon inputs (rivers, weathering and volcanoes) and outputs (carbonate and organic matter) of different isotopic composition (Fig. 2). Variations in these external fluxes can thus change the mass and isotopic composition of the exogenic carbon cycle through time [e.g., *Berner et al.*, 1983; *Berger and Vincent*, 1986; *Sundquist*, 1986; *Kump and Arthur*, 1999].

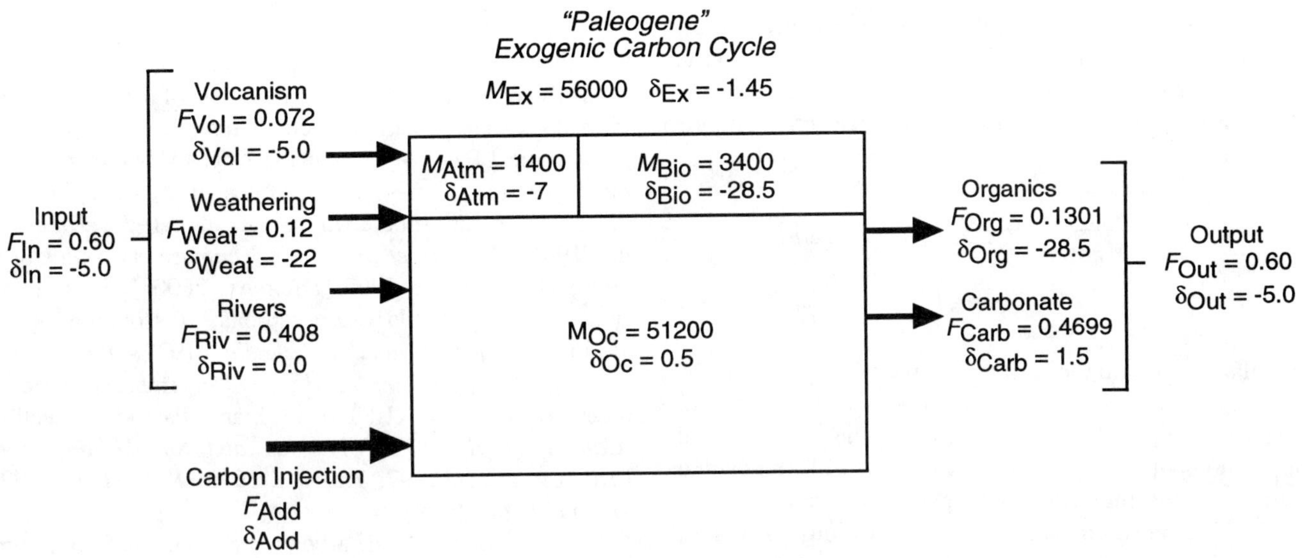

Figure 2. A Paleogene exogenic carbon cycle at steady state conditions just prior to the LPTM and sudden addition of carbon. Carbon enters through volcanism, weathering and rivers, cycles through the ocean, atmosphere and biomass, and leaves through carbonate and organic matter. Masses, fluxes and isotopic compositions for the various reservoirs have been modified slightly from *Kump and Arthur* [1999].

The simplest equations to describe carbon mass transfer between the exogenic carbon cycle and external carbon fluxes are:

$$M_{\rm Ex} = F_{\rm In}t - F_{\rm Out}t + M_{\rm Ex(o)}\,, \tag{1}$$

$$M_{\rm Ex}\delta_{\rm Ex} = F_{\rm In}\delta_{\rm In}t - F_{\rm Out}\delta_{\rm Out}t + M_{\rm Ex(o)}\delta_{\rm Ex(o)}\,, \tag{2}$$

where $M_{\rm EX(o)}$ and $\delta_{\rm EX(o)}$ are the mass and isotope composition of the exogenic carbon cycle at initial conditions, $F_{\rm In}$, $F_{\rm Out}$, $\delta_{\rm In}$ and $\delta_{\rm Out}$ are the fluxes and isotopic compositions of external carbon inputs and outputs, respectively, and t is time. Both equations can be differentiated with respect to t to render:

$$\frac{dM_{Ex}}{dt} = F_{In} - F_{Out}\,, \tag{3}$$

$$\frac{dM_{Ex}\delta_{Ex}}{dt} = F_{In}\delta_{In} - F_{Out}\delta_{Out}\,, \tag{4}$$

Combining the equations gives the fundamental expression for the change in δ^{13}C of the exogenic carbon cycle over time resulting from changes in carbon inputs and outputs (adapted from *Kump* [1991] and *Kump and Arthur* [1999]):

$$\frac{d\delta_{Ex}}{dt} = \frac{F_{In}}{M_{Ex}}\left(\delta_{In} - \delta_{Ex}\right) - \frac{F_{Out}}{M_{Ex}}\left(\delta_{Out} - \delta_{Ex}\right). \tag{5}$$

In this paper, for simplicity, conventional carbon inputs (i.e., volcanism, weathering, and rivers) are combined as a single flux. However, for reasons that shall become clear later, carbonate and organic matter outputs must be considered separately. Thus, Equation 5 needs to be expressed as (adapted from *Kump* [1991] and *Kump and Arthur* [1999]):

$$\frac{d\delta_{Ex}}{dt} = \frac{F_{In}}{M_{Ex}}\left(\delta_{In} - \delta_{Ex}\right) - \frac{F_{Carb}}{M_{Ex}}\left(\delta_{Carb} - \delta_{Ex}\right) - \frac{F_{Org}}{M_{Ex}}\left(\delta_{Org} - \delta_{Ex}\right), \tag{6}$$

where $F_{\rm Carb}$, $F_{\rm Org}$, $\delta_{\rm Carb}$, and $\delta_{\rm Org}$ are the fluxes and isotopic compositions of carbonate and organic matter, respectively. Carbonate and organic carbon are formed from carbon within the exogenic carbon cycle. Hence, $F_{\rm Carb}$, $F_{\rm Org}$, $\delta_{\rm Carb}$, and $\delta_{\rm Org}$ are functions of $M_{\rm Ex}$ and $\delta_{\rm Ex}$. For initial modeling, the assumed mass-composition relationships for these external carbon outputs are:

$$F_{\rm Carb} = k_{Carb}M_{\rm Ex}\,, \tag{7}$$

$$F_{\rm Org} = k_{Org}M_{\rm Ex}\,, \tag{8}$$

$$\delta_{\rm Carb} = \delta_{Ex} + \Delta_{\rm Carb}\,, \tag{9}$$

$$\delta_{\rm Org} = \delta_{Ex} + \Delta_{\rm Org}\,, \tag{10}$$

where k_{Carb} and k_{Org} are reciprocal e-folding (residence) times, and Δ_{Carb} and Δ_{Org} are isotopic fractionations of carbonate and organic matter, respectively. Equation 5 can include the mass-composition relationships of external carbon outputs as follows:

$$\frac{d\delta_{Ex}}{dt} = \frac{F_{In}}{M_{Ex}}(\delta_{In} - \delta_{Ex}) - k_{Carb}(\Delta_{Carb}) - k_{Org}(\Delta_{Org}) \; . \tag{11}$$

Secular changes in the carbon isotopic composition of the exogenic carbon cycle can be understood with knowledge of its initial mass and composition, the fluxes and compositions of external inputs, and the mass-composition relationships of the external outputs (Equation 11). Although there are discrepancies between various estimates, the pre-industrial Holocene exogenic carbon cycle contained ~40,000 Gt with a δ_{Ex} of about –2.0 ‰ [cf. *Broecker*, 1974; *Bolin et al.*, 1979; *Berner et al.*, 1983; *Sundquist*, 1986; *Walker and Kasting*, 1992; *Broecker and Peng*, 1993; *Siegenthaler*, 1993]. According to these workers, volcanism, weathering, and rivers supplied ~0.40 Gt/yr with δ_{In} of about -5 ‰, whereas carbonate and organic matter removed ~0.31 and ~0.09 Gt/yr with δ_{Carb} and δ_{Org} of about 1 and -25 ‰, respectively. In a simple model of the Holocene global carbon cycle, $1/k_{Carb}$, $1/k_{Org}$ are 130,000 and 433,000 years, and Δ_{Carb} and Δ_{Org} are 3 and -23 ‰.

Faster seafloor spreading rates, general warmth, and shallower carbonate compensation depths suggest that pCO_2, M_{Ex}, F_{In}, and F_{Out} were significantly greater in the late Paleocene than at present [e.g., *Berner et al.*, 1983; *Kump*, 1991; *Compton and Mallinson*, 1996; *Kump and Arthur*, 1999]. Long-term carbon isotope records also show that δ_{Carb} and δ_{Org} were ~1.5 and ~-28.5 ‰, respectively, in the late Paleocene [*Shackleton and Hall*, 1984; *Lewan*, 1986]. *Kump and Arthur* [1999] have presented masses, fluxes and isotopic compositions for a steady-state exogenic carbon cycle that is compatible with these observations. After modifications to include biomass, and a 1 ‰ fractionation between marine inorganic carbon and carbonate, their carbon cycle (Fig. 2) is considered a good approximation for the exogenic carbon cycle prior to the LPTM. In this carbon cycle, $1/k_{Carb}$, $1/k_{Org}$ are 119,200 and 430,500 years, and Δ_{Carb} and Δ_{Org} are 2.95 and –27.05 ‰.

3. THE LATEST PALEOCENE CARBON ISOTOPE EXCURSION: THE PROBLEM

The magnitude, timing, and global nature of the $\delta^{13}C$ excursion across the LPTM (Fig. 1) indicates an abrupt increase in the $^{12}C/^{13}C$ ratio of all carbon in the exogenic carbon cycle (Fig. 2). There is no conceivable mechanism to suddenly remove enormous quantities of ^{13}C-rich carbon from the exogenic carbon cycle. Thus, all recent papers regarding the LPTM agree that the $\delta^{13}C$ excursion reflects a rapid injection of massive quantities of ^{12}C-rich carbon to the ocean or atmosphere [*Lu and Keller*, 1993; *Zachos et al.*, 1993; *Dickens et al.*, 1995; *Thomas and Shackleton*, 1996; *Bains et al.*, 1999; *Dickens*, 2000a]. Such an increase in M_{EX} should cause carbonate dissolution on the seafloor [*Broecker and Takahashi*, 1977; *Walker and Kasting*, 1992; *Dickens et al.*, 1997b]. This dissolution is observed across the LPTM in many Paleogene sections [*Canudo et al.*, 1995; *Thomas and Shackleton*, 1996; *Bralower et al.*, 1997; *Thomas*, 1998; *Katz et al.*, 1999; *Thomas et al.*, 1999].

The effect of a sudden carbon injection upon δ_{Ex} over time can be evaluated by modifying Equation 11 [*Berger and Vincent*, 1986; *Dickens*, 2000b]:

$$\frac{d\delta_{Ex}}{dt} = \frac{F_{Add}}{M_{Ex}}(\delta_{Add} - \delta_{Ex}) + \frac{F_{In}}{M_{Ex}}(\delta_{In} - \delta_{Ex}) - k_{Carb}(\Delta_{Carb}) - k_{Org}(\Delta_{Org}) \; , \tag{12}$$

where F_{Add} and δ_{Add} are the flux and isotopic composition of the carbon input (Fig. 2). The first term accounts for the change in $\delta^{13}C$ caused by carbon addition whereas the latter terms account for the change in $\delta^{13}C$ caused by external carbon "throughput" [*Berger and Vincent*, 1986].

For a given $\delta^{13}C$ excursion, potential carbon injections of flux F_{Add} and isotopic composition δ_{Add} can be determined from Equation 12 by stepwise integration. Figure 3 shows possible combinations of F_{Add} and δ_{Add} that can explain a –2.5 ‰ shift in the $\delta^{13}C$ of the Paleogene exogenic carbon cycle (Fig. 2) given a simple one-step release of carbon over 10,000 years. As seen in Figure 3, the amount of carbon required to cause a global negative $\delta^{13}C$ excursion is (1) directly related to the magnitude of the $\delta^{13}C$ excursion, but (2) inversely related to the absolute value of its isotopic composition [*Dickens*, 2000b]. Moreover, if δ_{Ex} is negative, disproportionately large masses of carbon with less enriched in ^{12}C are needed to cause a given negative excursion. Thus, a relatively small carbon input with a $\delta^{13}C$ of –60 ‰ can have the same effect on the isotopic composition of the ocean, atmosphere and biomass as an immensely large carbon input with a $\delta^{13}C$ of –5 ‰ [*Dickens et al.*, 1995; *Dickens*, 2000b].

Importantly, most theoretical carbon inputs are implausible causes for a –2.5 ‰ excursion in δ_{Ex} over 10,000 years.

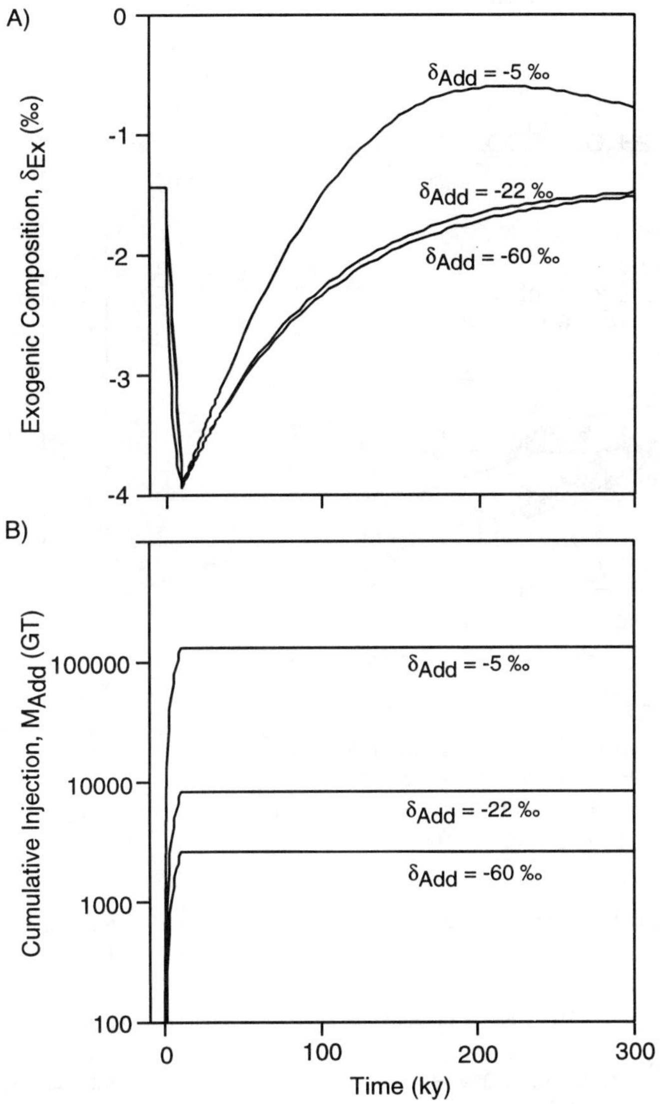

Figure 3. Mathematical solutions to Equation 12 using parameters for the Paleogene exogenic carbon cycle (Fig. 2). (A) Carbon isotopic composition of the entire exogenic carbon cycle over time after a carbon injection. (B) Required mass of the carbon injection in order to cause a given $\delta^{13}C$ excursion. Note the exponential scaling on the axis for mass.

First, there are only a limited number of significant external carbon reservoirs, each with a specific $\delta^{13}C$ (Fig. 2). Second, most of these reservoirs cannot supply a sufficient mass of carbon over short time scales [*Dickens et al.*, 1995; *Thomas and Shackleton*, 1996]. For example, global volcanism (F_{Vol}) would have to increase at least 100 times to explain a –2.5 ‰ excursion in the Paleocene exogenic carbon cycle (Figs. 2, 3). There is no evidence (or mechanism) for such prodigious volcanism [*Larson*, 1991], in spite of considerable early Paleogene basalt production in the North Atlantic [e.g., *Eldholm and Grue,* 1994]. Indeed, given the large mass of the Paleogene exogenic carbon cycle, a rapid global $\delta^{13}C$ excursion of –2.5 ‰ is impossible to explain without invoking a carbon source outside of widely accepted models of the global carbon cycle [cf. *Sundquist*, 1986; *Dickens et al.*, 1995]. This problem remains if one considers potential Paleogene carbon cycles with a wide range in M_{Ex} and δ_{Ex} [*Dickens*, 2000b].

4. THE GAS HYDRATE DISSOCIATION HYPOTHESIS

A scenario involving thermohaline circulation and release of CH_4 from marine gas hydrates (Fig. 4) provides a general explanation for the $\delta^{13}C$ excursion across the LPTM [*Dickens et al.*, 1995, 1997b; *Katz et al.*, 1999]. Similar to present day, vast quantities of CH_4 greatly enriched in ^{12}C were stored as gas hydrate in the upper few hundred meters of sediment on continental margins. However, long-term global warming during the late Paleocene [*Shackleton and Hall*, 1984; *Zachos et al.*, 1994] pushed the ocean circulation system past a critical threshold [e.g., *Stocker and Schmittner*, 1997], causing warm surface waters to sink and temperatures to rise at intermediate water depths. This bottom water warming on continental margins induced a change in sediment geotherms so that temperatures increased significantly at depth, converting once solid gas hydrates into free gas (Fig. 4). The resulting increase in pore pressure at depth led to injection of CH_4 into the ocean. The CH_4 reacted with dissolved O_2 to produce CO_2 (perhaps through bacteria), which then dissolved carbonate on the seafloor. Once oxidized in the ocean, the excess ^{12}C-rich carbon propagated throughout the exogenic carbon cycle.

Release and oxidation of CH_4 from gas hydrates is an appealing mechanism to explain carbon mass transfer during the LPTM because it is a direct and plausible response of the abrupt bottom water warming [*Dickens et al.*, 1995]. Faunal assemblages and oxygen isotopes of benthic foraminifera indicate a sudden warming of intermediate to deep waters in the ocean by 4-6 °C [*Kennett and Stott, 1991*; *Zachos et al.*, 1993; *Thomas and Shackleton*, 1996; *Katz et al.*, 1999]. Moreover, in at least one section from the Atlantic, evidence for bottom water warming appears to precedes the $\delta^{13}C$ excursion [*Kennett and Stott*, 1991], suggesting that a temperature rise caused massive carbon injection. Several researchers [e.g., *Kennett and Stott*, 1991; *Kaiho et al.*, 1996; *Bralower et al.*, 1997; *Pardo et al.*, 1997; *Katz et al.*, 1999] also have inferred from laminated sediments or benthic foraminifera assemblages that dissolved O_2 decreased at intermediate water depths during and immediately after massive carbon injection. This drop in

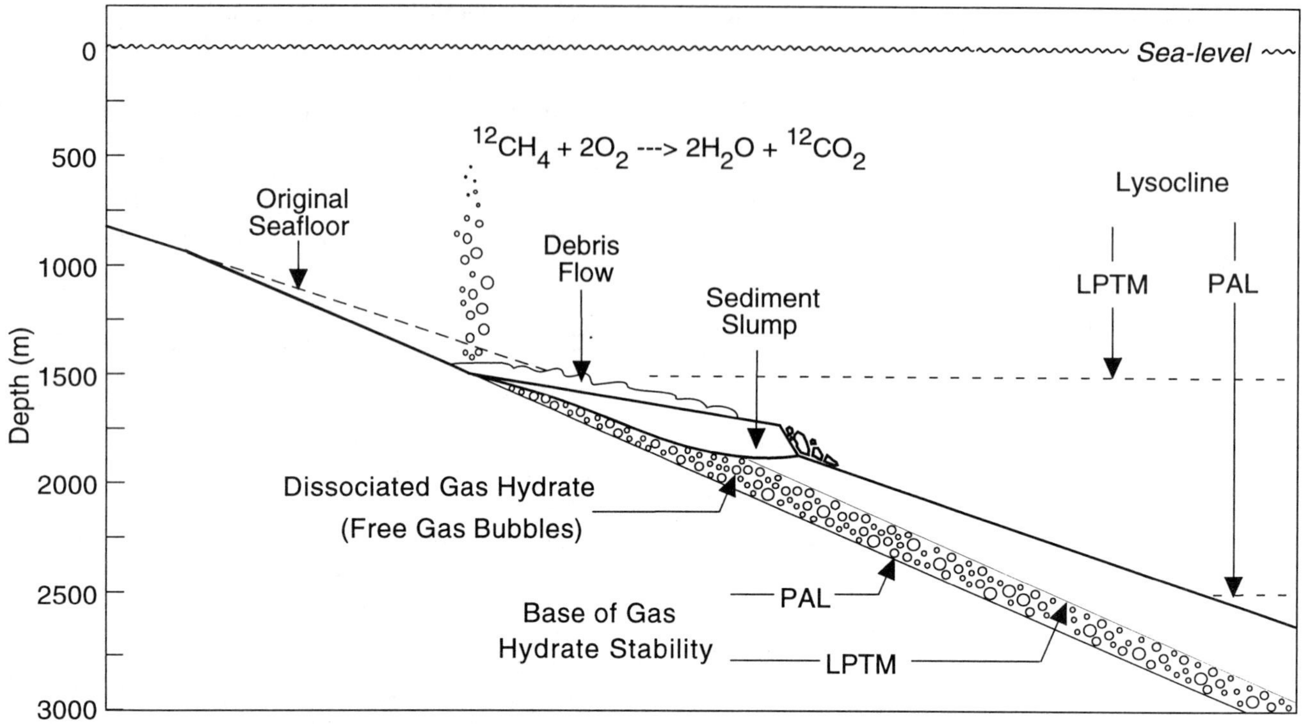

Figure 4. Schematic of the LPTM hydrate dissociation hypothesis [from *Katz et al.*, 1999]. Massive quantities of ^{12}C-rich CH_4 are released to the water column via sediment failure after conversion of gas hydrate to free gas in the sediment column. Three expected effects of gas release are removal of dissolved oxygen, production of ^{12}C-rich CO_2, and a shoaling of the lysocline (depth of significant carbonate dissolution).

dissolved O_2 appears to exceed that caused by the warming of water, perhaps signifying an additional O_2 demand in the water column, such as from oxidation of CH_4 [*Thomas*, 1998; *Dickens*, 2000a].

Marine gas hydrates appear to be the only reservoir that can supply an adequate flux of ^{12}C-rich carbon to account for the LPTM $\delta^{13}C$ excursion [*Dickens et al.*, 1995; *Bralower et al.*, 1997; *Bains et al.*, 1999; *Dickens*, 1999, 2000b *Norris and Röhl*, 1999]. According to Equation 12, release of ~2600 Gt of carbon with a $\delta^{13}C$ of –60 ‰ over 10,000 years would cause a –2.5 ‰ $\delta^{13}C$ excursion in the Paleogene exogenic carbon cycle (Fig. 3). This mass exceeds the <1800 Gt estimated previously [*Dickens et al.*, 1995, 1997; *Dickens*, 2000a, 2000b] because past work considered injection into a smaller Holocene carbon cycle. Such carbon transfer immediately raises a profound conceptual problem. Widely cited models for the global carbon cycle [e.g., *Bolin et al.*, 1979; *Berner et al.*, 1983; *Sundquist*, 1986; *Walker and Kasting*, 1992] are portrayed as reasonable representations of how carbon is distributed and moved on Earth. However, no published carbon cycle model contains gas hydrates and exchange fluxes to and from this carbon pool. Models of the global carbon cycle have to be modified at a basic level to understand rapid injection of CH_4 during the LPTM [*Dickens*, 1999b, 2000a, 2000b].

5. OBSERVATIONS FROM THE BLAKE RIDGE

The Blake Ridge is a sediment drift off the southeast coast of the United States characterized by gas-charged sediment and a regionally extensive bottom-simulating reflector (BSR) on seismic profiles. The BSR (Fig. 5) constitutes an acoustic phase reversal between upper sediment with gas hydrate and lower sediment with free gas [*Shipley et al.*, 1979; *Dillon and Paull*, 1983; *Katzman et al.*, 1994]. In 1995, Leg 164 of the Ocean Drilling Program (ODP) drilled sites 994, 995 and 997 on the crest of the Blake Ridge to understand *in situ* characteristics and amounts of gas stored in this region [*Paull et al.*, 1996a]. The Blake Ridge is currently the best-understood marine gas hydrate deposit. Although it may not be representative of such deposits, the Blake Ridge provides an essential framework for incorporating gas hydrates and associated fluxes into models of the global carbon cycle.

Figure 5 shows a seismic profile across the Blake Ridge [*Paull et al.*, 1996a] with downcore profiles of pore water

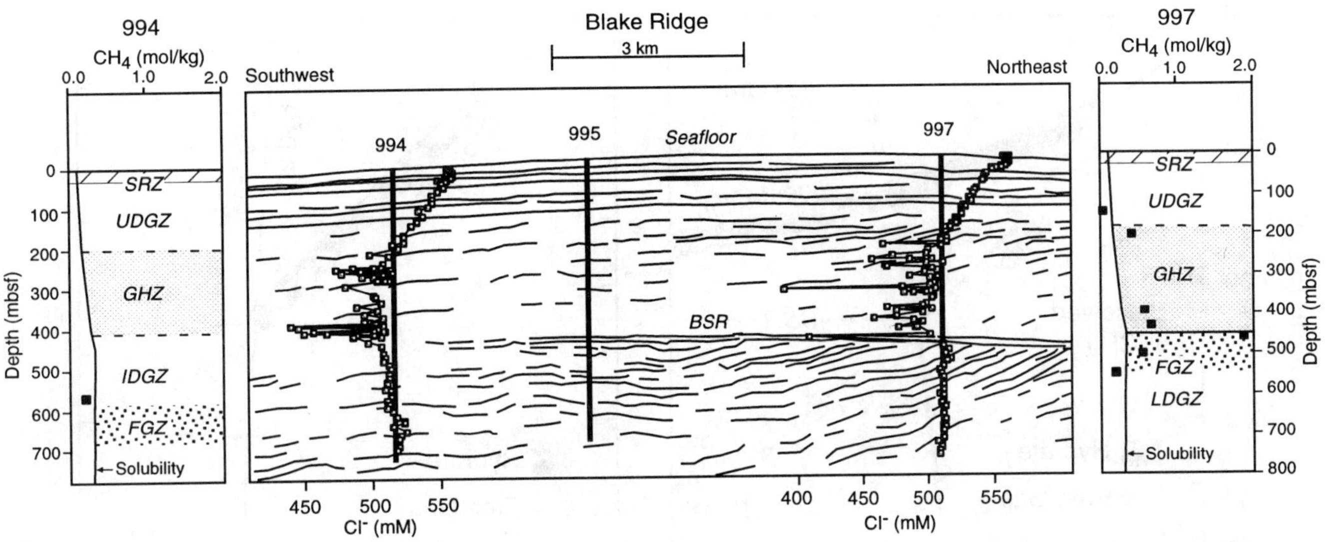

Figure 5. Seismic profile across the crest of the Blake Ridge (adapted from *Paull et al.* [1996]) showing the BSR and ODP Sites drilled by Leg 164. Also shown are pore water Cl^- concentrations squeezed from sediment, *in situ* CH_4 concentrations, and appropriate CH_4 solubility curves for sites 994 and 997 [*Paull et al.*, 1996; *Dickens et al.*, 1997a, 2000; *Egeberg and Dickens*, 1999]. Chemical and physical observations are explained best if gas is located in laterally extensive gas zones according to solubility. SRZ = sulfate reduction zone; UDGZ = upper dissolved gas zone; GHZ = gas hydrate zone; IDGZ = intermediate dissolved gas zone; FGZ = free gas zone; LDGZ = lower dissolved gas zone.

Cl^- squeezed from sediment [*Paull et al.*, 1996a; *Egeberg and Dickens*, 1999] and *in situ* CH_4 concentrations recovered from the pressure core sampler or PCS [*Dickens et al.*, 1997a, 2000]. Many workers have examined these combined data [e.g., *Holbrook et al.*, 1996; *Paull et al.*, 1996a; *Dickens et al.*, 1997a; *Borowski et al.*, 1999; *Egeberg and Dickens*, 1999; *Dickens*, 2001] and generally agree on four key points. (1) The BSR represents an interface between significant amounts of overlying gas hydrate and underlying free gas; the BSR is present at site 997 but not at site 994. (2) Anomalous pore water freshening (low Cl^-) mostly signifies dissociation of hydrate during core retrieval; sites 994 and 997 contain thick (>200 m) intervals of anomalous freshening, although it is thicker and fresher at site 997. (3) Methane concentrations average about 0.55 mol/kg at site 997 and generally decrease toward the seafloor; although not evident in Figure 5, no CH_4 exists in the top 20 m at sites 994, 995 and 997 [*Paull et al.*, 1996a; *Borowski et al.*, 1999]. Gas hydrate occupies between 0-9% of porosity.

Information from ODP Leg 164 suggests that significant CH_4 exists in gas horizons on the Blake Ridge [*Dickens et al.*, 1997a; *Xu and Ruppel*, 1999]. From top to bottom, these zones include (Fig. 5): a SO_4^{2-} reduction zone (SRZ) without CH_4, an upper dissolved gas zone (UDGZ) without gas hydrate, a gas hydrate zone (GHZ), an intermediate dissolved gas zone (IDGZ) without gas hydrate or free gas, a free gas zone (FGZ), and a lower dissolved gas zone (LDGZ). The origin of these six gas horizons is explained and expected from considerations of CH_4 solubility.

Methane solubility in water at high pressure and low temperature is governed by two equilibrium curves (Fig. 6): a CH_4 hydrate-pore water "partial saturation curve" [*Handa*, 1990; *Zatsepina and Buffett*, 1997], and a free CH_4 gas-pore water saturation curve [*Duan et al.*, 1992]. In marine sequences with geotherms, equilibrium CH_4 concentrations increase with depth on the first curve but decrease with depth on the second curve. Thus, the CH_4 hydrate-pore water curve applies to the upper part of a deep marine sediment column, the free CH_4 gas-pore water curve applies to a lower part of the sediment column, and the two curves intersect at some depth (Fig. 6). This depth is where pressure and temperature on the geotherm equal those on the three-phase free CH_4 gas-CH_4 hydrate-pore water equilibrium curve [*Dickens and Quinby-Hunt*, 1997]. Depending on pressure, temperature and total CH_4 concentration, CH_4 in pore space can be dissolved in water, in solid gas hydrate, or as free gas bubbles. Multiple gas horizons naturally arise on the Blake Ridge because of variable CH_4 saturation with depth (Figs. 5, 6).

Gas horizons are crucial to understanding gas distribution on the Blake Ridge and important to building a satisfactory global carbon cycle model with gas hydrates. First, large

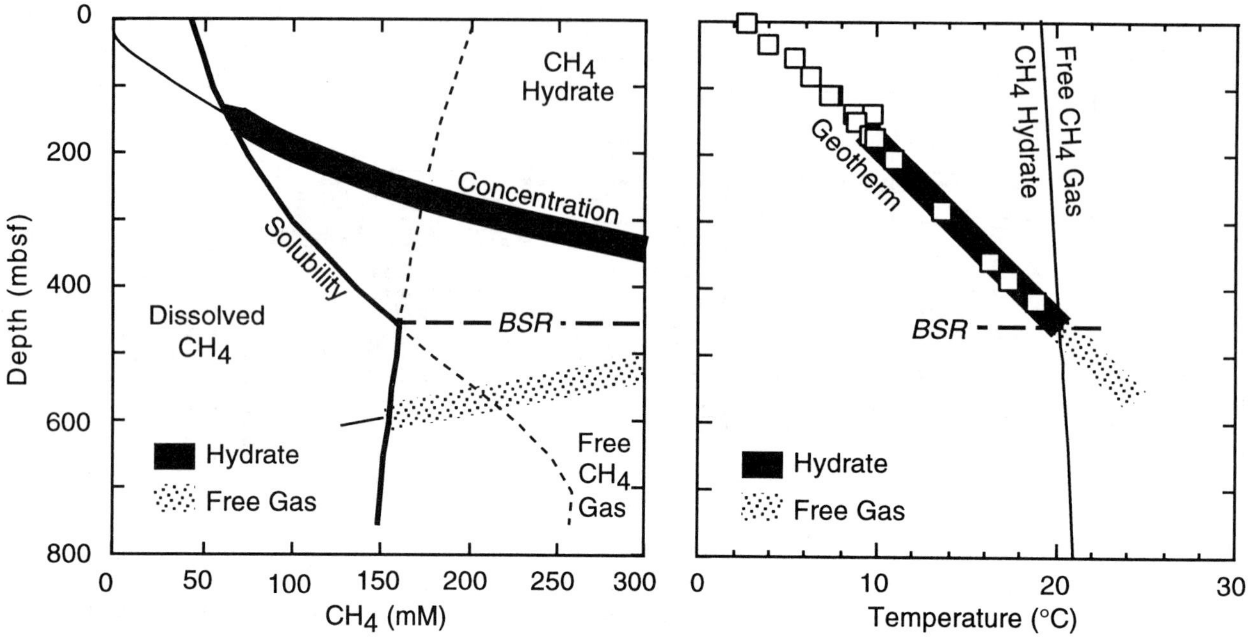

Figure 6. Gas hydrate and free gas occurrence at site 997. (A) The solubility curves for pore water with an activity of 0.956 [*Egeberg and Dickens*, 1999] and assumed *in situ* gas concentration profiles [see *Dickens et al.*, 1997a] in concentration-depth space. (B) The geotherm [*Paull et al.*, 1996] and the three phase equilibrium curve for pore water with an activity of 0.956 [*Dickens and Quinby-Hunt*, 1997].

quantities of CH_4 are stored in three different phases: dissolved gas, gas hydrate, and free gas bubbles. Second, for a given CH_4 concentration profile, the proportion of free gas to gas hydrate in the sediment column depends on the position of the CH_4 saturation curves, which in turn, depend on the geotherm (Fig. 6). A steeper geotherm will increase the amount of free gas unless there is a change in the CH_4 concentration profile. Third, once pore waters are saturated with CH_4, the amount of dissolved gas remains relatively constant with increasing CH_4 concentration or a change in the position of CH_4 saturation curves. Fourth, concentrations of CH_4 in the GHZ, while high for marine sediment, are not a maximum. The Blake Ridge could store significantly more CH_4 than is found at present-day.

Geochemical, microbiological, and geophysical investigations together indicate multiple sources and sinks of CH_4 that have variable magnitudes across the Blake Ridge [*Borowski et al.*, 1996, 1999; *Wellsbury et al.*, 1997; *Egeberg and Dickens*, 1999; *Xu and Ruppel*, 1999; *Paull et al.*, 2000]. This dynamic CH_4 cycling probably involves several bacterial pathways [e.g., *Wellsbury et al.*, 1997] and large internal fluxes such as the burial of solid gas hydrate and the advection of free gas [e.g., *Paull et al.*, 1994; *Egeberg and Dickens*, 1999; *Xu and Ruppel*, 1999]. For this paper, two CH_4 fluxes on the Blake Ridge are of special interest: the input and output of gas.

Hydrocarbon gas on the Blake Ridge is almost entirely CH_4 with a $\delta^{13}C$ <-63 ‰ [*Claypool and Threlkeld*, 1983; *Paull et al.*, 1996a, 2000; *Dickens et al.*, 1997a; *Wallace et al.*, 2000]. These molecular and isotopic characteristics suggest the vast majority of gas on the Blake Ridge was generated through bacterial methanogenesis of organic matter at shallow depth [*Claypool and Threlkeld*, 1983; *Paull et al.*, 1999]. This view is consistent with parallel depth profiles of the $\delta^{13}C$ of CH_4 and dissolved ΣCO_2 [*Claypool and Threlkeld*, 1983; *Kvenvolden*, 1993, 1998; *Paull et al.*, 2000]. Importantly, the isotopic composition of gas is similar for samples collected above and below the BSR, and at ambient and *in situ* pressure [*Paull et al.*, 2000; *Wallace et al.*, 2000]. This suggests that there is limited fractionation of carbon isotopes as carbon passes between dissolved gas, gas hydrate, and free gas on the Blake Ridge.

Pore water SO_4^{2-} profiles across the SRZ (Fig. 7) decrease linearly from a seawater concentration at the seafloor to zero concentration at the base of the SRZ [*Borowski et al.*, 1996, 1999; *Paull et al.*, 1996a]. The linearity of these profiles suggest that SO_4^{2-} depletion is largely driven by an upward flux of CH_4 and anaerobic CH_4 oxidation (CH_4 +

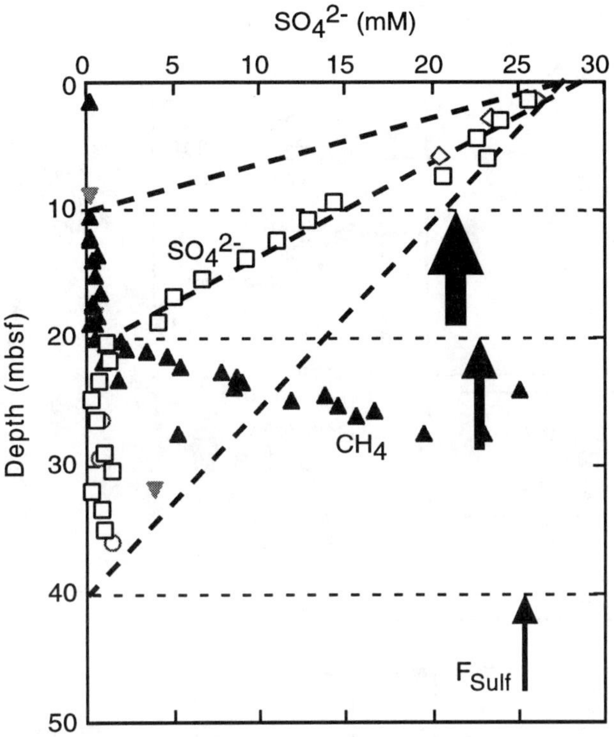

Figure 7. The relationship between linear pore water SO_4^{2-} profiles and upward CH_4 fluxes (F_{Sulf}) in a steady-state system (adapted from *Borowski et al.* [1996]). Note that a doubling of F_{out} halves the depth of SO_4^{2-} depletion. Concentrations of SO_4^{2-} and CH_4 shown are from site 994 [*Paull et al.,* 1996].

SO_4^{2-} ---> HS^- + HCO_3^- + H_2O) at a SO_4^{2-}/CH_4 interface [*Borowski et al.*, 1996; *Niewöhner et al.*, 1998; *Dickens*, 2001]. Such an interpretation is strongly supported on the Blake Ridge through analyses of various phases in shallow sediment. This includes carbon isotopes of dissolved ΣCO_2 and CH_4, sulfur isotopes of total solid sulfur, and concentrations of dissolved NH_4^+ and headspace CH_4 [*Claypool and Threlkeld*, 1983; *Borowski et al.*, 1996, 2000; *Hoehler et al.*, 2000). Tracer based studies using ^{35}S also indicate significantly elevated SO_4^{2-} reduction rates (SRRs) at the SO_4^{2-}/CH_4 interface [*Hoehler et al.*, 2000]. Anaerobic CH_4 oxidation (Fig. 7) provides the major pathway for removing CH_4 from the Blake Ridge [*Borowski et al.*, 1996, 1999, 2000; *Egeberg and Dickens*, 1999; *Dickens*, 2001]. However, in at least one location on the Blake Ridge, CH_4 moves upward along a fault and discharges directly into the water column [*Paull et al.*, 1995, 1996a].

For reasons of reaction stoichiometry, downward SO_4^{2-} fluxes, and hence depths of pore water SO_4^{2-} depletion, should be directly related to upward CH_4 fluxes [*Borowski et al.*, 1996; *Niewöhner et al.*, 1998]. The depth of zero pore water SO_4^{2-} occurs between 10-155 m below seafloor across the Blake Ridge [*Borowski et al.*, 1996, 1999], suggesting that outward CH_4 fluxes vary by at least a factor of 16. Importantly, regions on the Blake Ridge with the shallowest depths of pore water SO_4^{2-} depletion, and by inference maximum outward CH_4 flux, correspond to areas with BSRs [*Borowski et al.*, 1999]. Evidently, there is a direct link between the presence of free gas and the outward flux of CH_4 (Fig. 7). Although the actual link is not specified in this paper, it probably represents upward fluid flow [*Egeberg and Dickens*, 1999; *Xu and Ruppel*, 1999].

6. A GLOBAL GAS HYDRATE CAPACITOR

A capacitor is a device in an electrical circuit that stores and releases electrical charge depending on external forcing (electrical supply). In a simple circuit with a capacitor, charge slowly builds in the device over time. If the capacitor reaches steady-state conditions, the device is full and inputs and outputs are equal. With a change in external forcing, such as the closing of a switch, a sudden discharge from the capacitor will occur.

Isotopic records across the LPTM (Fig. 1) suggest that a large gas hydrate "capacitor" is connected to the exogenic carbon cycle (Fig. 8). Similar to other reservoirs, there would be a relatively constant input of carbon into gas hydrates over time. At steady-state conditions, there would also be a relatively constant output. However, an increase in bottom water temperature would steepen sediment geotherms and dissociate gas hydrate to free CH_4 gas. Free gas could then escape to the ocean or atmosphere so that the output of carbon could vary significantly over time.

Results from the Blake Ridge provide constraints for incorporating a gas hydrate capacitor into the global carbon cycle. Consider a global gas hydrate capacitor of carbon mass M_{GH} and carbon isotopic composition δ_{GH} with three internal reservoirs (Fig. 8): dissolved gas of mass M_{Diss} and composition δ_{Diss}, gas hydrate of mass M_{Hyd} and composition δ_{Hyd}, and free gas of mass M_{Free} and composition δ_{Free}. The total mass of M_{GH} at any time is:

$$M_{GH} = M_{Diss} + M_{Hyd} + M_{Free} \, , \tag{13}$$

Gas analyses from various different depths and pressures on the Blake Ridge indicate that the carbon isotopic composition of dissolved gas, gas hydrate and free gas are similar [*Paull et al.*, 2000; *Wallace et al.*, 2000]. Hence, for this paper, $\delta_{GH} = \delta_{Diss} = \delta_{Hyd} = \delta_{Free}$ = -60 ‰, and only expressions for carbon mass transfer are developed.

The simplest equation to describe the mass of dissolved gas over time is:

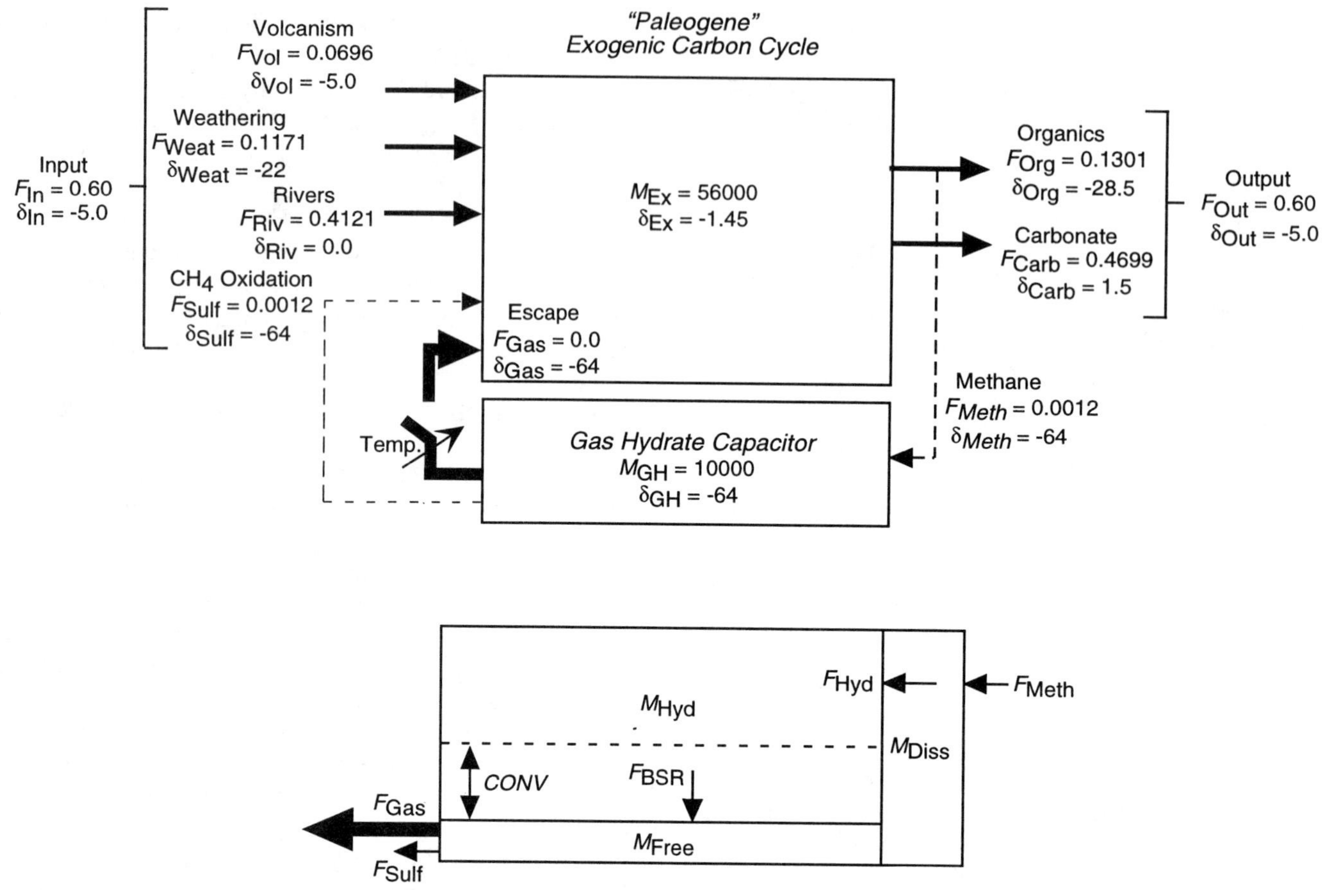

Figure 8. A Paleogene exogenic carbon cycle at steady state conditions just prior to the LPTM (Fig. 2) with a gas hydrate capacitor. Carbon enters the capacitor as ^{12}C-rich CH_4 through methanogenesis of organic matter. The CH_4 saturates pore waters to form gas hydrate, which then is buried past gas hydrate stability (Fig. 6). Carbon leaves the capacitor through anaerobic CH_4 oxidation (Fig. 7) of free gas, or in special cases, direct injection of free gas. Because the proportion of gas hydrate and free gas depends on temperature (here shown by a function CONV, Fig. 10), the overall mass and output of the capacitor depend on temperature.

$$M_{Diss} = F_{Meth}t - F_{Hyd}t + M_{Diss(o)}, \tag{14}$$

where $M_{Diss(o)}$ is the mass of dissolved gas at initial conditions, F_{Meth} is an input of carbon from methanogenesis, and F_{Hyd} is an output of carbon to gas hydrate with two stipulations relating to saturation. When the dissolved gas reservoir is undersaturated, F_{Hyd} is zero and the reservoir increases in mass. Alternatively, when the dissolved gas reservoir is saturated, F_{Hyd} is equal to F_{Meth} and the reservoir is at constant mass.

The simplest equation to describe the mass of gas hydrate over time is:

$$M_{Hyd} = F_{Hyd}t - F_{BSR}t - F_{Conv}t + M_{Hyd(o)}, \tag{15}$$

where $M_{Hyd(o)}$ is the mass of gas hydrate at initial conditions, F_{Hyd} is the input of carbon from dissolved gas as above, F_{BSR} is the net output of carbon as solid gas hydrate is buried past stability conditions, and F_{Conv} is the output of carbon as gas hydrate converts to free gas with a change in temperature (Figs. 6, 8). The net downward flux of carbon from gas hydrate to free gas on the Blake Ridge at constant bottom water temperature probably depends on the sedimentation rate (SR) and the amount of gas hydrate in pore space [*Egeberg and Dickens*, 1999]. For convenience, global SRs are considered constant over time. Thus, F_{BSR} is proportional to the mass of gas hydrate:

$$F_{BSR} = k_{Hyd}M_{Hyd}, \tag{16}$$

where k_{Hyd} is a constant.

The simplest equation to describe the mass of free gas over time is:

$$M_{Free} = F_{BSR}t + F_{Conv}t - F_{Sulf}t - F_{Gas}t + M_{Free(o)}\ , \quad (17)$$

where $M_{Free(o)}$ is the mass of free gas at initial conditions, F_{BSR} and F_{Conv} are the same as above, F_{Sulf} is the output of carbon through anaerobic CH_4 oxidation (Fig. 7), and F_{Gas} is direct loss of free gas to the exogenic carbon cycle. This last output is needed to discharge carbon suddenly.

The relationship between bottom water temperature and the distribution of gas in marine sediment is complex because temperature propagation through a sediment column is a nonlinear function when the temperature change causes a chemical phase change. The simplest approach is to ignore the phase change, an assumption that should be fairly accurate on thousand year scales [*MacDonald*, 1990]. Figure 9 shows the evolution of the CH_4 concentration profile at site 997 after a hypothetical 5 °C increase in bottom water temperature over 10,000 years without a change in water depth [see *Dickens et al.*, 1995]. As exemplified by this figure, an expected effect at any location saturated with gas is for a large fraction of original gas hydrate to dissociate to free gas bubbles at depth. However, the fraction of converted gas hydrate depends on the gas concentration profile and initial geotherm (as well as secondary factors such as thermal conductivity).

For this paper, three assumptions are made [*Dickens et al.*, 1995]: the geotherm is the same at all ocean locations, the gas hydrate concentration is the same at all depths, and gas hydrate is found only between two water depths and the base of gas hydrate stability. These assumptions result in a gas hydrate capacitor where carbon mass is evenly distributed across an area in temperature-pressure space (Fig. 10). Given such conditions, a change in bottom water temperature will shift all geotherms similarly and F_{Conv} is directly related to a change in "area":

$$F_{Conv} = \frac{CONV}{AREA}\left(M_{Hyd} + M_{Free}\right)\ , \quad (18)$$

where *AREA* is the total area with gas hydrate and free gas under the chosen three phase equilibrium curve (in this case, for seawater), and *CONV* is the area where gas hydrate is converted to free gas for a given change in temperature. Note that CONV can be positive or negative depending on whether temperature is increasing or decreasing. Both regions can be solved by integration over fixed limits in temperature-pressure space.

Perhaps the most crucial observation in Figure 10 is that bottom water warming will dissociate gas hydrate completely in relatively shallow sediment [*Dickens et al.*, 1995]. Presumably, gas in these upper sediment columns would be over-pressured and, without an overlying zone to reprecipitate gas hydrate, could escape directly to the water column via faults or slumps [*McIver*, 1982; *Kayen and Lee*, 1991; *Paull et al.*, 1996b; *Dillon et al.*, 1998]. The output of carbon from this "exposed" zone F_{Gas} (above) can be expressed as:

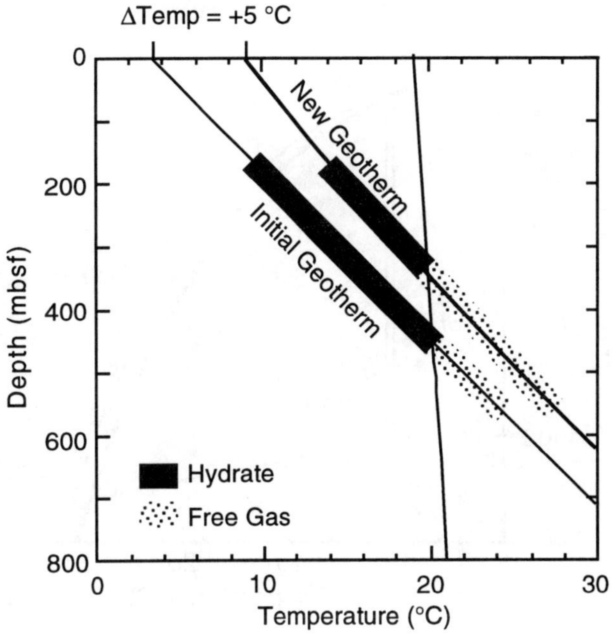

Figure 9. The expected effect of a 5°C increase in bottom water temperature over 10,000 years on the geotherm and gas hydrate stability.

$$F_{Gas} = k_{Gas}\frac{EXP}{AREA}\left(M_{Hyd} + M_{Free}\right)\ , \quad (19)$$

where *EXP* is the area where free gas is entirely exposed for a given change in temperature (Fig. 10), and k_{Gas} is a constant representing the fraction of carbon released from the exposed region. Unlike CONV, EXP can only be positive because once carbon has been released from sediment, it cannot be returned directly to free gas with a decrease in temperature. The region EXP also can be solved by integration over fixed limits in temperature-pressure space.

Changes in carbon mass within the global gas hydrate capacitor over time can be calculated from the above equations (Equations 13-19) with knowledge of the initial mass distribution, the carbon input (methanogenesis), constants regarding internal mass-flux relationships, and the thermal history of bottom water. Current estimates from somewhat independent calculations suggest that marine gas hydrates

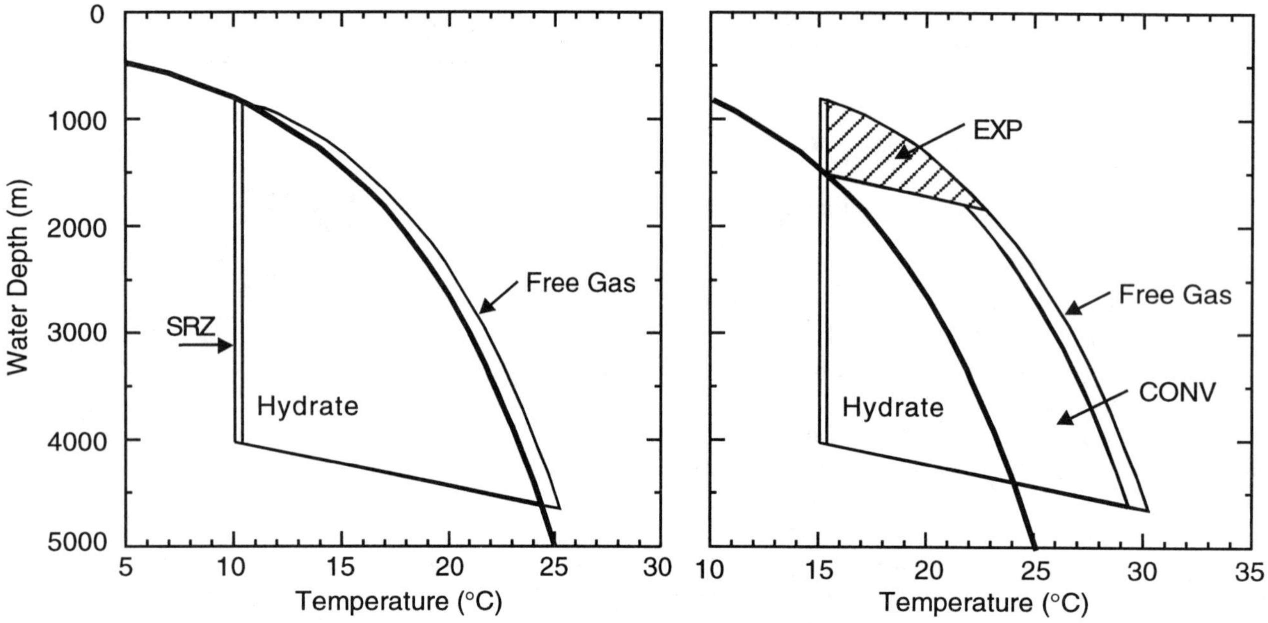

Figure 10. The change in the overall distribution of gas hydrates and free gas with an increase in bottom water temperature by 5 °C. CONV = the area where gas hydrate has been converted to free gas. EXP = the area where there is no gas hydrate remaining. The diagram assumes CH_4 saturation of pore waters wherever gas hydrate and free gas are found.

contain 7,500-15,000 Gt of carbon as CH_4 [e.g., *MacDonald*, 1990; *Kvenvolden*, 1993; *Gornitz and Fung*, 1994]. These estimates are problematic [*Kvenvolden*, 1998]. First, no global estimate has distinguished between dissolved gas, gas hydrate and free gas. Second, the overall distribution of gas hydrates at present-day has not been mapped in sufficient detail. Certainly the commonly discussed link between gas hydrate abundance and BSR presence is fundamentally flawed. Gas hydrates and free gas can exist without a BSR when CH_4 concentrations are sufficiently low at the depth where saturation curves intersect [*Xu and Ruppel*, 1999], for example at site 994 on the Blake Ridge (Fig. 5). A "speculative and uncertain" mass of 10,000 Gt for the gas hydrate capacitor [*Kvenvolden*, 1998] is accepted as plausible for the present-day. In reaching this estimate, various authors have assumed that gas hydrate fills on average 2 to 10 % of pore space [see *Kvenvolden*, 1998]. This means that ~1,000 Gt of carbon are stored as dissolved gas considering typical solubility curves (Fig. 6). There must also be a nominal amount of free gas because solid hydrate will dissociate to free gas when it is buried past the intersection of the solubility curves [*Paull et al.*, 1994; *Egeberg and Dickens*, 1999]. Indeed, in the model presented here, free gas is necessary to drive anearobic CH_4 oxidation and return carbon to the exogenic carbon cycle. A free gas mass of 500 Gt is assumed for the initial capacitor, leaving 8,500 Gt of carbon in gas hydrates.

No operational model for the Holocene global carbon cycle contains gas hydrates [e.g., *Broecker*, 1974; *Bolin et al.*, 1979; *Sundquist*, 1986; *Walker and Kasting*, 1992]. Yet these models are widely cited and appear to be reasonable representations for basic carbon cycling in the late Pleistocene and Holocene. The best explanation is that the present gas hydrate capacitor is at steady-state conditions and fluxes out of the capacitor are small compared to other external fluxes. Although this view has only limited support in the literature [*Dickens*, 2001], gas hydrates and free gas are assumed to be in steady-state conditions when bottom water temperature is constant.

A crude estimate for the current global flux F_{Sulf} can be made from considerations of SO_4^{2-} profiles on the Blake Ridge (Fig. 7). Pore water SO_4^{2-} profiles indicate a regional upward CH_4 flux of ~12 mol/m^2-kyr to the SRZ [*Borowski et al.*, 1996, 1999; *Dickens*, 2001]. Gas concentration profiles combined with geophysical mapping suggest the Blake Ridge contains ~35 Gt of CH_4 across about 25,000 km^2 [*Dillon and Paull*, 1983; *Dickens et al.*, 1997a]. Extrapolating these numbers to all marine gas hydrates gives a F_{Sulf} of ~0.0012 Gt/yr. If anaerobic CH_4 oxidation is the primary mechanism to remove carbon from marine gas hydrates during constant bottom water temperature, carbon output from gas hydrates is indeed small compared to other carbon fluxes of the global carbon cycle (Fig. 8).

The primary carbon input to gas hydrates is through bacterial methanogenesis of organic matter, a process that significantly fractionates carbon isotopes and preferentially places ^{12}C into CH_4 [*Kvenvolden*, 1995, 1998; *Paull et al.*, 2000]. The simplest equations to describe this carbon input to gas hydrates are:

$$F_{Meth} = k_{Meth} F_{Org} , \quad (20)$$

$$\delta_{Meth} = \delta_{Org} + \Delta_{Meth} , \quad (21)$$

where F_{Meth} and δ_{Meth} are the flux and composition of CH_4 entering gas hydrates, k_{Meth} is the fraction of organic carbon converted to CH_4, and Δ_{Meth} is the isotopic fractionation between organic matter and CH_4. When the gas hydrate capacitor is at steady-state conditions, $F_{Meth} = F_{Sulf}$.

7. A BASIC SIMULATION OF THE LPTM

Marine gas hydrates as simplified above can be connected to the Paleogene exogenic carbon cycle through three fluxes (Fig. 8). Carbon is added to M_{Diss} through methanogenesis (F_{Meth}) and removed from M_{Free} by anaerobic CH_4 oxidation (F_{Sulf}) and direct methane injection (F_{Gas}).

For convenience, the hypothetical present-day gas hydrate capacitor has been connected to the hypothetical Paleogene carbon cycle. The development of a Paleogene gas hydrate capacitor cannot be justified given the poor constraints on present masses and fluxes for marine gas hydrates. The connection can be made at steady-state conditions with three modifications. First, individual carbon inputs from rivers, weathering and volcanoes are adjusted slightly to account for the small additional carbon input from anaerobic CH_4 oxidation (Fig. 8). Second, the isotopic composition of gas hydrates is decreased to –64 ‰ to account for the lower $\delta^{13}C$ of Paleogene marine organic matter [*Lewan*, 1986]. Third, the fraction of organic matter entering gas hydrates (k_{Meth}) is lowered to account for the greater output of organic carbon during the Paleogene (F_{Org}). The last of these adjustments has an important implication noted later.

The overall model for the global carbon cycle constructed here (Fig. 8) transfers ^{12}C-rich CH_4 between the Paleogene exogenic carbon cycle and gas hydrates. At a constant bottom water temperature of 10 °C, and for a steady-state exogenic carbon cycle, 0.0012 Gt/yr of CH_4 with a $\delta^{13}C$ of –64 % enter and leave the exogenic carbon cycle by moving through marine sediments as dissolved gas, gas hydrate and free gas. Gas hydrates are stable everywhere below a shallow SRZ where temperatures on a 0.035 °C/m geotherm are less than temperatures on the three-phase equilibrium curve for seawater, and the top of the sediment column is between 900 and 4000 m below sealevel. A change in temperature, carbon inputs, or carbon outputs will cause variations in mass and composition within this coupled circuit.

Figure 11 shows variations in the gas hydrate capacitor after a steady 5 °C increase to all geotherms over 10,000 years. In this scenario, there is no cooling and carbon only leaves marine gas hydrates through anaerobic CH_4 oxidation. Free gas, although exposed at shallow depth (Fig. 10), does not escape to the water column directly ($k_{Gas} = 0$). During the initial warming, approximately 5,000 Gt of gas hydrate is converted to free gas. The increased abundance of free gas drives elevated CH_4 oxidation rates at the SRZ, causing the gas hydrate capacitor to lose carbon to the exogenic carbon cycle. Once the warming has stopped, however, the capacitor still loses carbon at a significant rate for hundreds of thousands of years because an enormous mass of free gas remains. Thus, ^{12}C-rich carbon is bled into the exogenic carbon cycle over a considerable length of time, resulting in a small and protracted $\delta^{13}C$ excursion that is not similar to the isotope anomaly during the LPTM.

Figure 12 shows variations in the gas hydrate capacitor after a steady 5 °C increase to all geotherms over 10,000 years as above. The difference between this scenario and the last is the gradual 5 °C cooling over 50,000 years immediately after the abrupt warming, much like during the LPTM. The drop in temperature returns the potential region of gas hydrates to its original size, converting large amounts of free gas generated during warming back to gas hydrate. At some point in time (in this case, 55,000 years after initial warming), the abundance of free gas is depleted below its initial mass. This condition makes CH_4 oxidation rates at the SRZ lower than CH_4 production rates by methanogenesis. The gas hydrate capacitor then begins to recharge. In this scenario, therefore, ^{12}C-rich carbon is bled into the exogenic carbon cycle over some fraction of the thermal perturbation, and then preferentially removed from the exogenic carbon cycle slightly faster than expected from external carbon throughput. The result is a small and protracted $\delta^{13}C$ excursion that is similar in shape but not magnitude to the isotope anomaly during the LPTM.

Figure 13 shows variations in the gas hydrate capacitor after a steady 5 °C increase to all geotherms over 10,000 followed by gradual 5 °C cooling over 50,000 years as above. The difference between this scenario and the last is that free gas is vented directly into the water column ($k_{Gas} = 1$) as all gas hydrate is dissociated in sediment where the seafloor lies between 900-2,000 m below sea level (Fig.

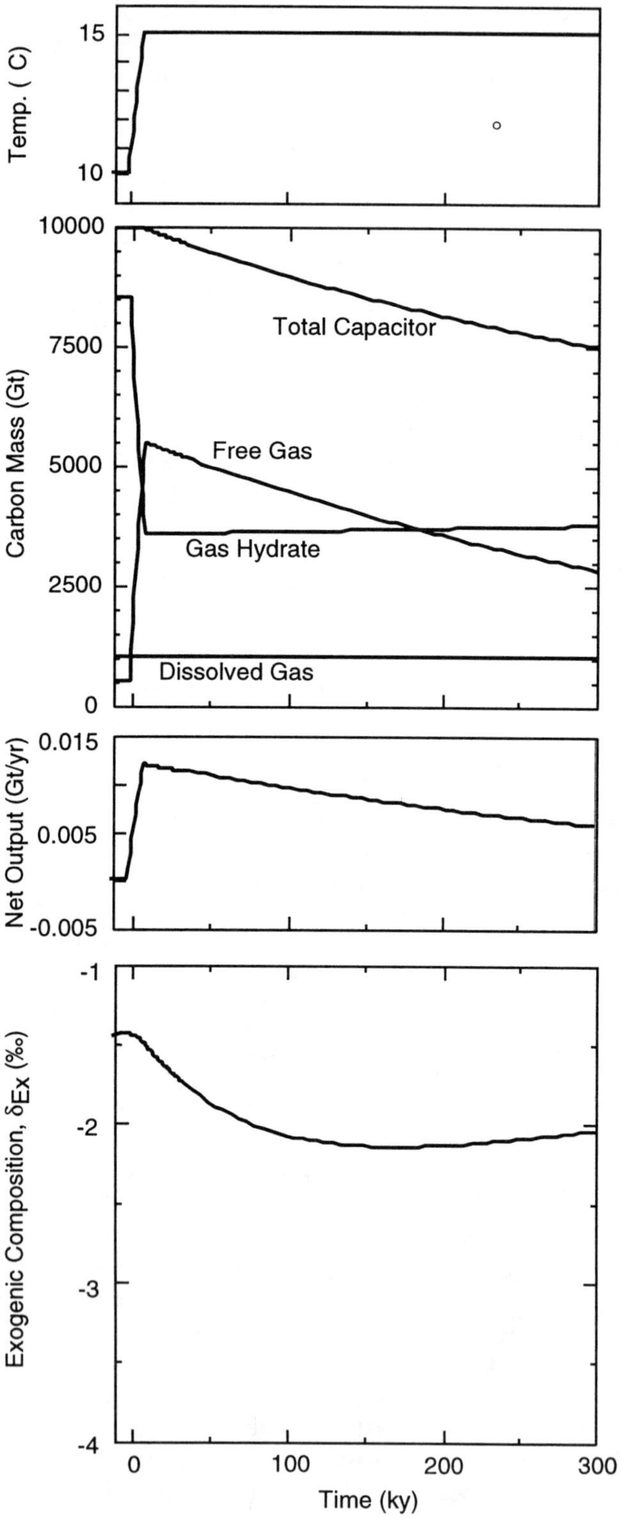

Figure 11. The predicted change in the mass of the gas hydrate capacitor and isotopic composition of the exogenic carbon cycle after a 5°C increase in bottom water temperature over 10,000 years. In this case, no exposed free gas is released.

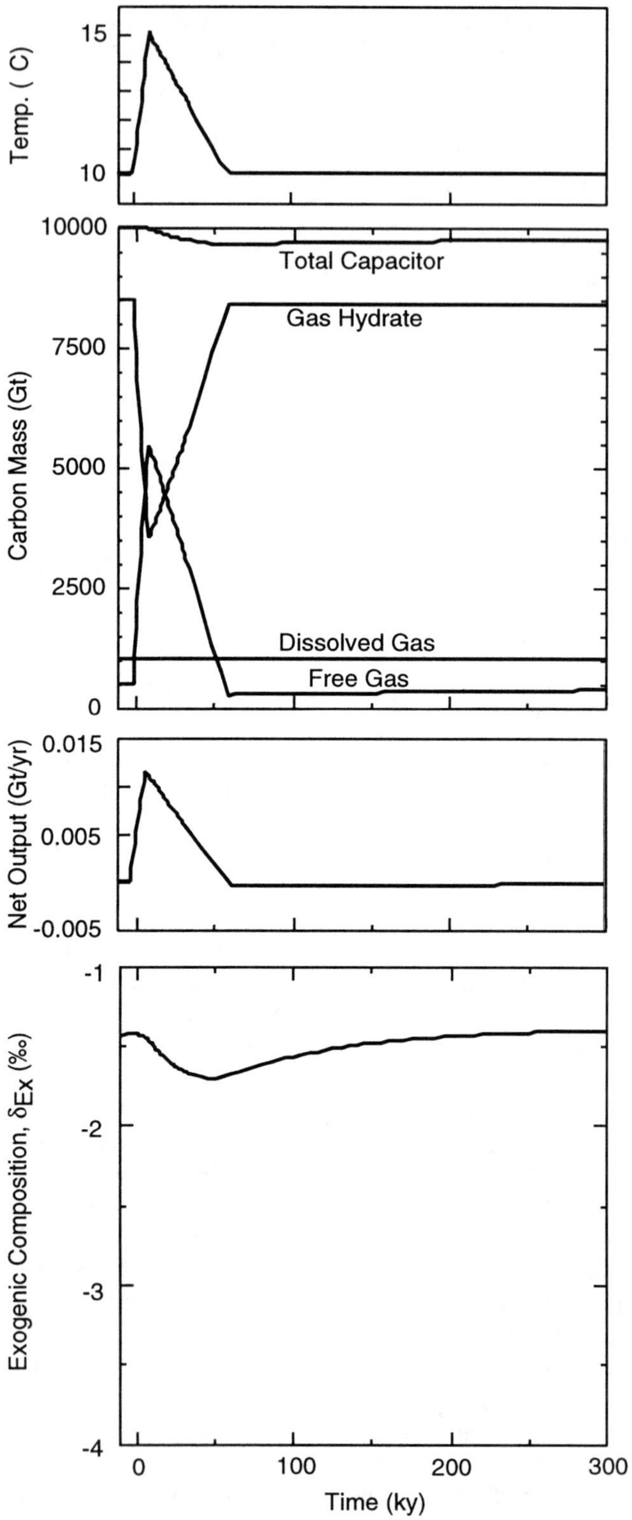

Figure 12. The predicted change in the mass of the gas hydrate capacitor and isotopic composition of the exogenic carbon cycle after a 5°C increase in bottom water temperature over 10,000 years followed by a 5°C decrease in bottom water temperature over 50,000 years. In this case, no exposed free gas is released.

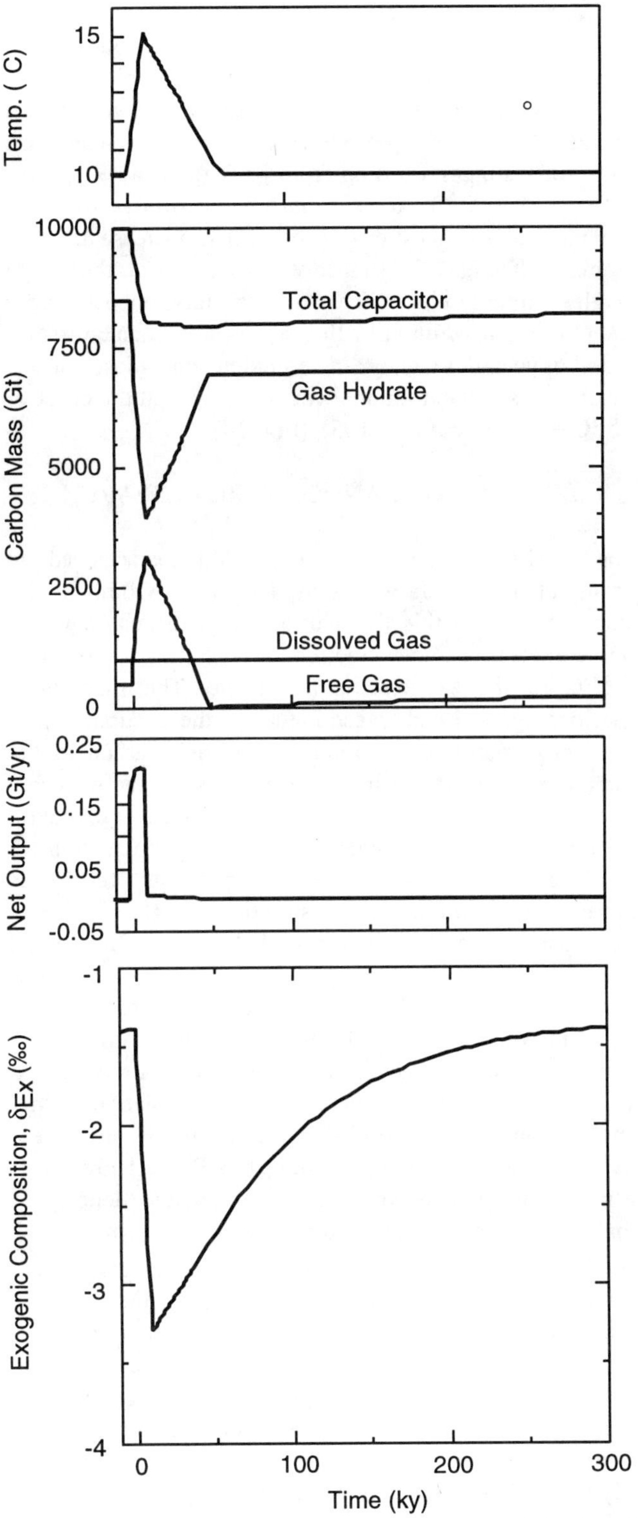

Figure 13. The predicted change in the mass of the gas hydrate capacitor and isotopic composition of the exogenic carbon cycle after a 5°C increase in bottom water temperature over 10,000 years followed by a 5°C decrease in bottom water temperature over 50,000 years. In this case, exposed free gas is released to the water column from shallow depths (Fig. 10). Note the change in scale for the net carbon output.

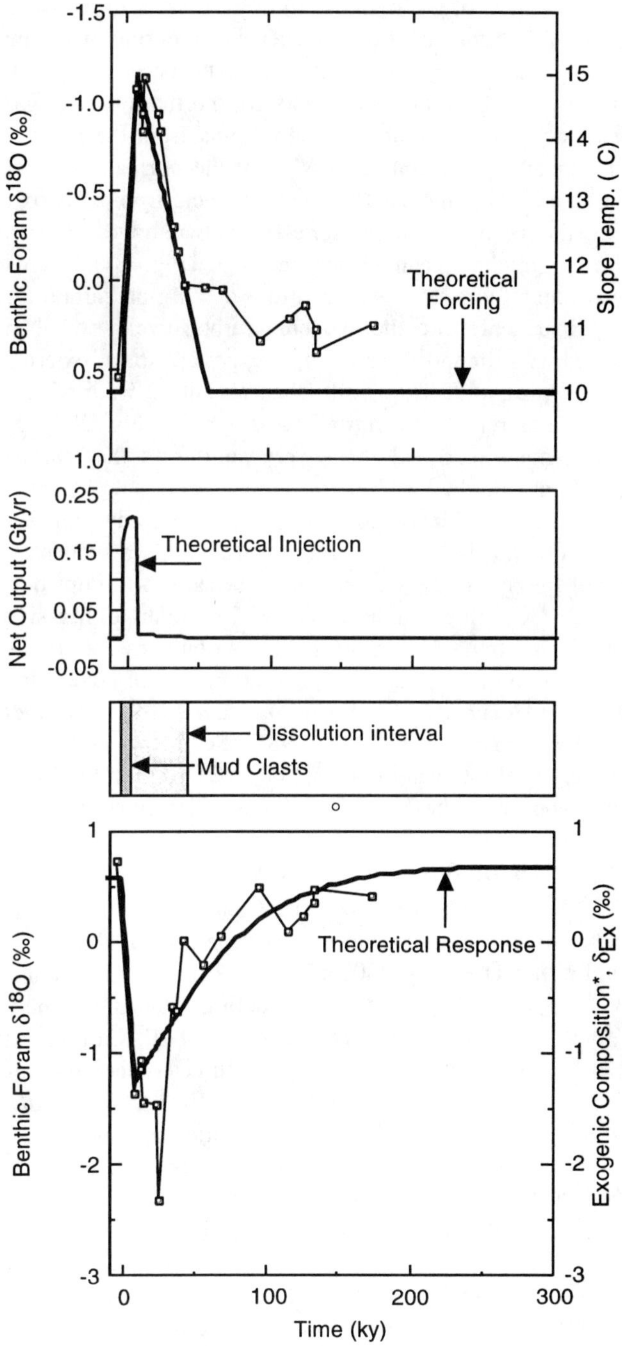

Figure 14. The predicted change in the mass of the gas hydrate capacitor and isotopic composition of exogenic carbon cycle as in Figure 13 compared to oxygen and carbon isotope records of single species benthic foraminifera at ODP Site 1051 [*Katz et al.*, 1999]. Note the presence of a mud clast interval precisely at the onset of the LPTM $\delta^{13}C$ excursion.

10). Bottom water warming thus causes a sudden release of 1,950 Gt of ^{12}C-rich carbon to the exogenic carbon cycle, followed by an additional slow venting of 100 Gt through CH_4 oxidation. Again, the drop in temperature returns the potential region of gas hydrates to its original size, converting large amounts of free gas generated during warming back to gas hydrate. In this case, however, because a prodigious amount of carbon was removed, the mass of free gas quickly drops to zero, and carbon outputs from the gas hydrate capacitor cease entirely. Without the normal input of ^{12}C-rich carbon to the exogenic carbon cycle from CH_4 oxidation, the gas hydrate capacitor effectively removes ^{12}C-rich carbon from the ocean during recharge (Fig. 8). In this scenario, therefore, massive quantities of ^{12}C-rich carbon are released suddenly into the exogenic carbon cycle, and then removed significantly faster than expected from external carbon throughput. The result is an abrupt –2 ‰ $\delta^{13}C$ excursion that returns to initial conditions within 200,000 years, a magnitude and shape very similar to the isotope anomaly during the LPTM (Fig. 1).

With the model developed here (Fig. 8), the only explanation for chemical changes during the LPTM is a sudden release of free gas to the water column because of abrupt bottom water warming and dissociation of gas hydrate. Massive quantities of free CH_4 gas probably can only escape to the water column with faulting or slumping of overlying sediment [*McIver*, 1982; *Kayen and Lee*, 1991; *Paull et al.*, 1996b; *Dillon et al.*, 1998]. Sediment failure at relatively shallow water depths, 900 to 2,000 m in the Paleogene, appears to be a requisite for a satisfactory explanation of the LPTM.

Leg 171B of the ODP recovered sediment across the LPTM at site 1051 in the western Atlantic. This section is special because it is expanded, was deposited on the continental slope at ~1500-2000 m water depth, and contains obvious sediment cycles for constructing detailed age models [*Katz et al.*, 1999; *Norris and Röhl*, 1999]. Carbon and oxygen isotope compositions have been determined on single species benthic foraminifera at Site 1051 [*Katz et al.*, 1999]. The combination of a high-resolution age model, detailed isotope records and a slope location provides the first opportunity to test critical aspects of the overall model.

Oxygen and carbon isotope records of *Oridorsalis* spp. display -1.5 ‰ and –3.0 ‰ excursions, respectively, across the LPTM at Site 1051 (Fig. 14). The drop in $\delta^{18}O$ occurs over about 7,000 yrs and suggests a >5°C increase in water temperature on the continental slope. This amount, duration, and location of bottom water warming is sufficient to dissociate large quantities of gas hydrate in middle to lower slope sediments (Fig. 10). The drop in $\delta^{13}C$ is similar in shape and magnitude to those in other LPTM sections (Fig. 1), and represents a sudden increase in the ^{12}C composition of the exogenic carbon cycle. Considering isotopic fractionations between the exogenic carbon cycle, the ocean, and carbonate (Figs. 2, 8), the observed excursion in carbonate at Site 1051 suggests a drop from –1.3 to –3.4 ‰ in δ_{Ex} (Fig. 14). This $\delta^{13}C$ excursion is broadly similar in shape and magnitude to that predicted by a 5°C increase in temperature if free gas is suddenly released from the slope through sediment failure (Fig. 13). As highlighted by *Katz et al.* [1999], in addition to the expected carbonate dissolution and oxygen deficiency, an anomalous mud clast interval indicative of sediment failure occurs precisely at the onset of the $\delta^{13}C$ excursion at Site 1051 (Fig. 14).

8. CONCLUSIONS AND FUTURE RESEARCH

The LPTM was a interval ca. 55 Ma characterized by warming of deep ocean waters by 4 to 6 °C within 10,000 years. Coeval with this thermal anomaly, the cause which remains unclear, is an extraordinary –2.5 ‰ excursion in the $\delta^{13}C$ of the exogenic carbon cycle. The magnitude, shape, timing, and widespread nature of the excursion indicate a sudden and massive input of ^{12}C-rich carbon to the ocean and atmosphere. Using basic models for the exogenic carbon cycle, it becomes obvious that release and oxidation of CH4 from marine gas hydrates is the best explanation for the $\delta^{13}C$ excursion. However, such a hypothesis necessitates a partial but significant reconstruction of global carbon cycle models. Specifically, the global carbon cycle must contain a large mass of CH_4 stored as gas hydrates and free gas that is sensitive to changes in ocean temperature.

The first operational model for the global carbon cycle that includes gas hydrates is developed here both conceptually and mathematically. The model is speculative, somewhat deterministic and purposely simple, but is consistent with basic interpretations regarding the Blake Ridge, currently the best understood marine gas hydrate deposit. A global gas hydrate capacitor stores a large mass of ^{12}C-rich carbon in three internal reservoirs: dissolved gas, gas hydrate and free gas. Carbon enters the capacitor from the exogenic carbon cycle when organic matter is converted to CH_4. Upon saturation of pore waters, carbon is transferred to gas hydrate and free gas, the proportion depending on temperature and pressure. Carbon leaves the capacitor as free gas through anaerobic CH_4 oxidation or, in cases without overlying hydrate, direct injection to the water column. Massive amounts of ^{12}C-rich carbon can suddenly escape the capacitor with abrupt bottom water warming because the amount and location of free gas depend on temperature.

Given a Paleogene exogenic carbon cycle, and a gas hydrate capacitor similar in mass to that inferred for the present, the following explanation is offered for the LPTM. The 5 °C warming of bottom waters increased temperatures in pore space at depth, converting approximately 5,000 Gt of gas hydrate to free gas. The increased free gas drove elevated CH_4 oxidation at the SRZ. More importantly, about 1,900 Gt of free gas were directly injected into the water column with sediment failure between 900-2,000 m below sea level. The immense quantity of released gas was oxidized in the water column to CO_2, which dissolved carbonate on the seafloor. A drop in temperature following gas injection returned the potential region of gas hydrates to its original size, converting large amounts of free gas generated during warming back to gas hydrate. Because such an enormous amount of carbon was removed, the mass of free gas dropped to zero during cooling, and carbon outputs from the gas hydrate capacitor stopped completely. Without the normal input of ^{12}C-rich carbon to the ocean through CH_4 oxidation, ^{12}C-rich carbon was effectively removed from the exogenic carbon cycle faster than expected from considerations of carbon throughput by external fluxes. The consequences would be coeval $\delta^{18}O$ and $\delta^{13}C$ excursions coincident with evidence for carbonate dissolution, oxygen deficiency and, on some margins, sediment failure. The record at ODP Site 1051 provides a compelling argument that this scenario actually occurred.

The growing evidence suggesting massive CH_4 release during the LPTM forces several intriguing and debatable issues. Perhaps the most important of these is the mass of marine gas hydrates in the Paleogene prior to the LPTM. Deep waters in the ocean were about 10 °C warmer in the late Paleocene than at present [e.g., *Miller et al.*, 1987], making the overall dimensions of the gas hydrate capacitor smaller in temperature-pressure space (Fig. 10). However, it is entirely possible that the overall mass was larger in the late Paleocene because the external flux of organic matter (F_{Org}), and hence rates of methanogenesis, were probably greater (Fig. 8). The average pore space filled by gas hydrate and free gas may have been greater in the late Paleocene than at present.

The model presented here can be improved significantly. Immediate effects of CH_4 release, carbonate dissolution and oxygen consumption, have not been incorporated. Some parameters, such as isotopic fractionation between the carbon cycle and organic matter (Δ_{Org}), are not constants but functions of carbon mass [e.g., *Kump and Arthur*, 1999]. Moreover, a series of potential feedbacks have not been appreciated. For example, higher atmospheric pCO_2 after CH_4 release [*Dickens*, 2000a] might have caused increased weathering, greater delivery of nutrients to the ocean, and higher marine biological productivity [e.g., *Walker et al.*, 1981; *Berner et al.*, 1983; *Zachos and Dickens*, 2000]. In turn, enhanced productivity could lead to greater organic matter output, higher rates of methanogenesis, and faster recharge of the gas hydrate capacitor (Fig. 8). Indeed, the entire concept of a gas hydrate capacitor with multiple internal reservoirs is itself open to debate. The real importance of this paper is that it leads to issues, problems and insights that have been overlooked previously.

Acknowledgments. The ideas in this paper, and the paper itself, have grown and lost many appendages since their hatching in Charly's bar with Jim O'Neil in 1994. I am greatly indebted to many colleagues who have helped my efforts at understanding gas hydrates and the LPTM. For this particular work, I especially thank Mimi Katz and Dotti Pak for sharing their isotope analyses, Carolyn Ruppel and Tim Bralower for constructive reviews, and Lee Kump for an exceptional review on a different paper that immediately sparked the one here.

REFERENCES

Bains, S., R.M. Corfield, and R.D. Norris, Mechanisms of climate warming at the end of the Paleocene, *Science*, *285*, 724-727, 1999.

Berger, W.H., and E. Vincent, Deep-sea carbonates: Reading the carbon-isotope signal, *Geol. Rundsch.*, 75, 249-269, 1986.

Berner, R.A., A.C. Lasaga, and R.M. Garrels, The carbonate-silicate geochemical cycle and its effect on atmospheric carbon dioxide over the past 100 million years, *Am. J. Science*, *283*, 641-683, 1983.

Bralower, T.J., J.C. Zachos, E. Thomas, M. Parrow, C.K. Paull, D.C. Kelly, I. Premoli Silva, W.V. Sliter, and K.C. Lohmann, Late Paleocene to Eocene paleoceanography of the equatorial Pacific Ocean: Stable isotopes recorded at Ocean Drilling Program Site 865, Allison Guyot, *Paleoceanography*, *10*, 841-865, 1995.

Bralower, T.J., D.J. Thomas, J.C. Zachos, M.M. Hirschmann, U. Röhl, H. Sigurdsson, E. Thomas, and D.L. Whitney, High-resolution records of the late Palaeocene thermal maximum and circum-Caribbean volcanism: Is there a causal link?, *Geology*, *25*, 963-966, 1997.

Bralower, T.J., D.J. Thomas, E. Thomas, and J.C. Zachos, High-resolution records of the late Palaeocene thermal maximum and circum-Caribbean volcanism: Is there a causal link?: Reply, *Geology*, *26*, 671, 1998.

Bolin, B., E.T. Degens, P. Duvigneaud, and S. Kempe, Carbon cycle modeling, in *The Global Carbon Cycle*, edited by B. Bolin, E.T. Degens, S. Kempe and P. Ketner, *SCOPE Report 13*, Wiley, New York, USA, 1-28, 1979.

Borowski, W.S., T.M. Hoehler, M.J. Alperin, M.J., N.M. Rodriguez, and C.K. Paull, Significance of anaerobic methane oxidation in methane-rich sediments overlying the Blake Ridge gas hydrates, Proc. ODP, Sci. Res., 164, 87-99, 2000.

Borowski, W.S., C.K. Paull, and W. Ussler III, Marine pore-water sulfate profiles indicate in situ methane flux from underlying gas hydrate, *Geology*, *24*, 655-658, 1996.

Borowski, W.S., C.K. Paull, and W. Ussler III, Global and local variations of interstitial sulfate gradients in deep-water, continental margin sediments: Sensitivity to underlying methane and gas hydrates, *Mar. Geol.*, *159*, 131-154, 1999.

Broecker, W.S., *Chemical Oceanography*, 214 pp., Harcourt, Brace, Jovanovich, Inc., New York, 1974.

Broecker, W.S., and T.-H. Peng, What caused the glacial to interglacial CO_2 change? in *The Global Carbon Cycle*, edited by M. Heimann, *NATO ASI Series I, 15*, 95-115, 1993.

Broecker, W.S., and T. Takahashi, Neutralization of fossil fuel CO_2 by marine calcium carbonate, in *Fate of Fossil Fuel CO_2 in the Oceans*, edited by N. Andersen and A. Malahof, Plenum, New York, USA, 213-241, 1977.

Canudo, J.I., G. Keller, E. Molina, and N. Ortiz, Planktic foraminiferal turnover and $\delta^{13}C$ isotopes across the Paleocene-Eocene transition at Caravaca and Zumaya, Spain, *Palaeogeogr. Palaeoclimatol. Palaeoecol.*, *114*, 75-100, 1995.

Claypool, G.E., and C.N. Threlkeld, Anoxic diagenesis and methane generation in sediments of the Blake Outer Ridge, Deep Sea Drilling Project Site 533, Leg 76, *Init. Repts, DSDP*, *76*, 391-402, 1983.

Compton, J.S., and D.J. Mallinson, Geochemical consequences of increased late Cenozoic weathering rates and the global CO_2 balance since 100 Ma, *Paleoceanography*, *11*, 431-446, 1996.

Dickens, G.R., The blast in the past, *Nature*, *401*, 752-755, 1999.

Dickens, G.R., Methane oxidation during the Late Palaeocene Thermal Maximum, *Bull. Soc. Géol. France*, *171*, 37-49, 2000a.

Dickens, G.R., Carbon addition and removal during the Late Palaeocene Thermal Maximum: Basic theory with a preliminary treatment of the isotope record at Ocean Drilling Program Site 1051, Blake Nose, in *Western North Atlantic Paleogene and Cretaceous Paleoceanography*, edited by D. Kroon, R. Norris, and A. Klaus, *Geol. Soc. London Spec. Pub.*, 183, 293-305, 2000b.

Dickens, G.R., Sulfate profiles and barium fronts in sediment on the Blake Ridge: Present and past methane fluxes through a large gas hydrate reservoir, *Geochim. Cosmochim. Acta*, 2001 (in press).

Dickens, G.R., M.M. Castillo, and J.C.G. Walker, A blast of gas in the latest Paleocene: Simulating first-order effects of massive dissociation of oceanic gas hydrate, *Geology*, *25*, 259-262, 1997b.

Dickens, G.R., J.R. O'Neil, D.K. Rea, and R.M. Owen, Dissociation of oceanic methane hydrate as a cause of the carbon isotope excursion at the end of the Paleocene, *Paleoceanography*, *10*, 965-971, 1995.

Dickens, G.R., C.K. Paull, P. Wallace, and ODP Leg 164 Scientific Party, Direct measurement of in situ gas volumes in a large gas hydrate reservoir, *Nature*, *385*, 426-428, 1997a.

Dickens, G.R., and M.S. Quinby-Hunt, Methane hydrate stability in pore water: A simple theoretical approach for geophysical applications, *J. Geophys. Res.*, *102*, 773-783, 1997.

Dickens, G.R., P. Wallace, C.K. Paull, and W.S. Borowski, Detection of methane gas hydrate in the pressure core sampler (PCS): Volume-pressure-time relations during controlled degassing experiments, *Proc. ODP, Sci. Res.*, *164*, 113-128, 2000.

Dillon, W.P., W.W. Danforth, D.R. Hutchinson, R.M. Drury, M.H. Taylor, and J.S. Booth, Evidence for faulting related to dissociation of gas hydrate and release of methane off the southeastern United States, in *Gas Hydrates: Relevance to world margin stability and climatic change*, edited by J.-P. Henriet and J. Mienert, *Geol. Soc. London Spec. Pub.*, *137*, 293-302, 1998.

Dillon, W.P., and C.K. Paull, Marine gas hydrates, II. Geophysical evidence, in *Natural Gas Hydrates: Properties, Occurrences and Recovery*, edited by J.L Cox, Butterworth, London, pp. 73-90, 1983.

Duan, Z., N. Møller, J. Greenberg, and J.H. Weare, The prediction of methane solubility in natural waters to high ionic strengths from 0° to 250 °C and from 0 to 1600 bar, *Geochim. Cosmochim. Acta*, *56*, 1451-1460, 1992.

Egeberg, P.K., and G.R. Dickens, Thermodynamic and pore water halogen constraints on gas hydrate distribution at ODP Site 997 (Blake Ridge), *Chem. Geol.*, *153*, 53-79, 1999.

Eldholm, O., and K. Grue, North Atlantic volcanic margins: Dimensions and production rates, *J. Geophys. Res.*, *99*, 2955-2968, 1994.

Fricke, H.C., W.C. Clyde, J.R. O'Neil, and P.D. Gingerich, Evidence for rapid climate change in North America during the Latest Palaeocene thermal maximum: Oxygen isotope compositions of biogenic phosphate from the Bighorn Basin (Wyoming), *Earth Planet. Sci. Lett.*, *160*, 193-208, 1998.

Gornitz, V., and I. Fung, Potential distribution of methane hydrates in the world's oceans, *Glob. Biogeochem. Cycles*, *8*, 335-347, 1994.

Handa, Y.P., Effect of hydrostatic pressure and salinity on the stability of gas hydrates, *J. Phys. Chem.*, *94*, 2652-2657, 1990.

Hoehler, T.M., W.S. Borowski, M.J. Alperin, N.M. Rodriguez, and C.K. Paull, Model, stable isotope, and radiotracer characterization of anaerobic methane oxidation in gas hydrate-bearing sediments of the Blake Ridge, *Proc. ODP, Sci. Res.*, *164*, 79-85, 2000.

Holbrook, W.S., H. Hoskins, W.T. Wood, R.A. Stephen, D. Lizarralde, and Leg 164 Science Party, Methane hydrate and free gas on the Blake Ridge from vertical seismic profiling, *Science*, *273*, 1840-1843, 1996.

Kaiho, K., T. Arinobu, R. Ishiwatari, H.E.G. Morgans, H. Okada, N. Takeda, K. Tazaki, G. Zhou, Y. Kajiwara, R. Matsumoto, A. Hirai, N. Niitsuma, and H. Wada, Latest Paleocene benthic foramiferal extinction and environmental changes at Tawanui, New Zealand, *Paleoceanography*, *11*, 447-465, 1996.

Katz, M.E., D.K. Pak, G.R. Dickens, and K.G. Miller, The source and fate of massive carbon input during the Latest Paleocene Thermal Maximum, *Science*, *286*, 1531-1533, 1999.

Katzman, R., W.S. Holbrook, and C.K. Paull, Combined vertical-incidence and wide-angle seismic study of a gas hydrate zone, Blake Ridge, *J. Geophys. Res.*, *99*, 17,975-17,995, 1994.

Kayen, R.E., and H. Lee, 1991, Pleistocene slope instability of gas hydrate-laden sediment on the Beaufort Sea Margin, *Mar. Geotech.*, *10*, 125-141, 1991.

Kennett, J.P., and L.D. Stott, Abrupt deep sea warming, paleoceanographic changes and benthic extinctions at the end of the Palaeocene, *Nature*, *353*, 319-322, 1991.

Koch, P.L., J.C. Zachos, and P. Gingerich, Correlation between isotope records in marine and continental carbon reservoirs near the Palaeocene/Eocene boundary, *Nature*, *358*, 319-322, 1992.

Kump, L.R., Interpreting carbon isotope excursions: Strangelove oceans, *Geology*, *19*, 299-302, 1991.

Kump, L.R., and M.A. Arthur, Interpreting carbon-isotope excursions: Carbonates and organic matter, *Chem. Geol.*, *161*, 181-198, 1999.

Kvenvolden, K.A., Gas hydrates: Geological perspective and global change, *Rev. Geophys.*, *31*, 173-187, 1993.

Kvenvolden, K.A., A review of the geochemistry of methane in natural gas hydrate, *Org. Geochem.*, *23*, 997-1008, 1995.

Kvenvolden, K.A., A primer on the geological occurrence of gas hydrate, in *Gas Hydrates: Relevance to world margin stability and climatic change*, edited by J.-P. Henriet and J. Mienert, *Geol. Soc. London Spec. Pub.*, *137*, 9-30, 1998.

Larson, R., Geological consequences of superplumes, *Geology*, *19*, 963-966, 1991.

Lewan, M.D., Stable carbon isotopes of amorphous kerogen from Phanerozoic sedimentary rocks, *Geochim. Cosmochim. Acta*, *50*, 1583-1591, 1986.

Lu, G., and G. Keller, Climatic and oceanographic events across the Paleocene-Eocene transition in the Antarctic Indian Ocean: Inference from planktic foraminifera, *Mar. Micropaleo.*, *21*, 101-142, 1993.

MacDonald, G., Role of methane clathrates in past and future climates, *Clim. Change*, *16*, 247-281, 1990.

McIver, R.D., Role of naturally occurring gas hydrates in sediment transport, *AAPG Bull.*, *66*, 789-792, 1982.

Miller, K.G., R.G. Fairbanks, and G.S. Mountain, Tertiary oxygen isotope synthesis, sea level history, and continental margin erosion, *Paleoceanography*, *2*, 741-761, 1987.

Niewöhner, C., C. Hensen, S. Kasten, M. Zabel, and H.D. Schulz, Deep sulfate reduction completely mediated by anaerobic methane oxidation in sediments of the upwelling area off Namibia, *Geochim. Cosmochim. Acta*, *62*, 455-464, 1998.

Norris, R.D., and U. Röhl, Carbon cycling and chronology of climate warming during the Palaeocene/Eocene transition, *Nature*, *401*, 775-778.

Pardo, A., G. Keller, E. Molina, and J.I. Canudo, Planktic foraminiferal turnover across the Palaeocene-Eocene transition at DSDP Site 401, Bay of Biscay, North Atlantic, *Mar. Micropaleo.*, *29*, 129-15, 1997.

Paull, C.K., W.J. Buelow, W. Ussler III, and W.S. Borowski, Increased continental-margin slumping frequency during lowstands above gas hydrate-bearing sediments, *Geology*, *24*, 143-146, 1996b.

Paull, C.K., R. Matsumoto, P.J. Wallace, et al., *Proc. ODP, Init. Repts.*, *164*, Ocean Drilling Program, TX, 623 pp, 1996a.

Paull, C.K., T.D. Lorenson, W.S. Borowski, W. Ussler III, K. Olsen, and N.M. Rodriguez, N.M., Isotopic composition of CH_4, CO_2 species, and sedimentary organic matter within samples from the Blake Ridge: Gas source implications, *Proc. ODP, Sci. Res.*, *164*, 229-236, 2000.

Paull, C.K., W. Ussler III, and W.S. Borowski, Sources of biogenic methane to form marine gas hydrates. *Ann. New York Acad. Sci.*, *715*, 392-409, 1994.

Paull, C.K., W. Ussler III, W.S. Borowski, and F.N. Speiss, Methane-rich plumes on the Carolina continental rise: Associations with gas hydrates, *Geology*, *23*, 89-92, 1995.

Shackleton, N.J. and M.A. Hall, Carbon isotope data from Leg 74 sediments, *Init. Repts. DSDP*, *74*, 613-620, 1984.

Shipley, T.H., M.H. Houston, R.T. Buffler, F.J. Shaub, K.J. McMillen, J.W. Ladd, and J.L. Worzel, Seismic reflection evidence for the widespread occurrence of possible gas-hydrate horizons on continental slopes and rises, *AAPG Bull.*, *63*, 2204-2213, 1979.

Siegenthaler, U., Modeling the present-day oceanic carbon cycle, in *The Global Carbon Cycle*, edited by M. Heimann, *NATO ASI Series I*, *15*, 367-395, 1993.

Stocker, T.F. and A. Schmittner, Influence of CO_2 emission rates on the stability of the thermohaline circulation, *Nature*, *388*, 862-865. 1997.

Sundquist, E.T., Geologic analogs: Their value and limitations in carbon dioxide research, in *The Changing Carbon Cycle: A Global Analysis*, edited by J.R. Trabalka and D.E. Reichle, Springer-Verlag, New York, USA, 371-402, 1986.

Thomas, D.J., T.J. Bralower, and J.C. Zachos, New evidence for subtropical warming during the late Paleocene thermal maximum: Stable isotopes from Deep Sea Drilling Project Site 527, Walvis Ridge, *Paleoceanography*, *14*, 561-570, 1999.

Thomas, E., Biogeography of the late Palaeocene benthic foraminiferal extinction, in *Late Palaeocene-Early Eocene Climatic and Biotic Events*, edited by M.-P. Aubry, S. Lucas and W.A. Berggren, Columbia University Press, New York, USA, 214-243, 1998.

Thomas, E., and N.J. Shackleton, The Paleocene-Eocene benthic foraminiferal extinction and stable isotope anomalies, in *Correlations of the early Paleogene in Northwest Europe*, edited by R.O'B. Knox, R. Corfield, and R.E. Dunay, *Geol. Soc. London Spec. Pub.*, *101*, 401-411, 1996.

Walker, J.C.G., P.B. Hays, and J.F. Kasting, A negative feedback mechanism for the long-term stabilization of Earth's surface temperture, *J. Geophys. Res.*, *86*, 9776-9782, 1981.

Walker, J.C.G., and J.F. Kasting, Effects of fuel and forest conservation on future levels of atmospheric carbon dioxide, *Palaeogeogr. Palaeoclimatol., Palaeoecol.*, *97*, 151-189, 1992.

Wallace, P.J., G.R. Dickens, C.K. Paull, and W. Ussler, III, Carbon isotopic composition of methane in gas hydrates and free-gas-bearing sediments from the Blake Ridge: Effects of degassing on isotopic compositions of gas released from pressurized core samples, *Proc. ODP, Sci. Res.*, *164*, 101-112, 2000.

Wellsbury, P., K. Goodman, T. Barth, B. Cragg, S. Barnes, and R.J. Parkes, A deep bacterial biosphere fuelled by low-temperature acetate generation during burial, *Nature*, *388*, 573-576, 1997.

Xu, W., and C. Ruppel, Predicting the occurrence, distribution, and evolution of methane gas hydrate in porous marine sediments, *J. Geophys. Res.*, *104*, 5081-5095, 1999.

Zatsepina, O.Y., and B.A. Buffett, Phase equilibrium of gas hydrate: Implications for the formation of hydrate in the deep sea floor, *Geophys. Res. Lett.*, *24*, 1567-1570, 1997.

Zachos, J.C., and G.R. Dickens, An assessment of the biogeochemical feedback response to the climatic and chemical perturbations of the LPTM, GFF, 122, 188-189, 2000.

Zachos, J.C., K.C. Lohmann, J.C.G. Walker, and S.W. Wise, Abrupt climate change and transient climates during the Paleogene: A marine perspective, *J. Geology*, *101*, 191-213, 1993.

Zachos, J.C., L.D. Stott, and K.C. Lohmann, Evolution of early Cenozoic temperatures, *Paleoceanography*, *9*, 353-387, 1994.

G.R. Dickens, Department of Geology and Geophysics, Rice University, Houston, TX 77005, USA (email: Jerry@geophysics.rice.edu)

Section 2

GEOCHEMISTRY OF GAS HYDRATE

Ion Exclusion Associated With Marine Gas Hydrate Deposits

William Ussler III and Charles K. Paull

Research and Development Division, Monterey Bay Aquarium Research Institute, Moss Landing, CA

In marine sediments gas hydrate growth removes water and gas molecules from the pore water while dissolved ions are excluded from the clathrate cage. As a consequence, the surrounding pore waters become more saline. Conversely, when gas hydrates decompose, water is released, causing pore water to become fresher. Ion exclusion produces distinctive geochemical signatures that are used to identify the presence of gas hydrate and to calculate the amount of gas hydrate in marine sediment samples. These calculations, however, require knowledge of the dissolved ion concentration of the *in situ* pore water prior to gas hydrate decomposition. Because arguably no useable pore water samples have been collected *in situ* from gas-hydrate-bearing sediments, the calculated amounts of gas hydrate are very dependent on the assumed *in situ* dissolved ion concentrations. In this paper, the principles behind using measured pore water chloride concentrations to estimate gas hydrate amounts are reviewed, along with data from selected sites around the world. These data clearly define local pore water anomalies in the sediment zones that are believed to currently contain gas hydrate, but also document the occurrence of thick zones of low chloride concentrations immediately below the base of gas hydrate stability. The potential causes for these low chloride anomalies below the base of gas hydrate stability are discussed.

REVIEW OF ION EXCLUSION

Background

When gas hydrate forms, water molecules form a cage-like structure that has the capacity to trap low molecular weight gases [*Sloan*, 1990]. During gas hydrate growth in sediments, dissolved ions are excluded from the gas hydrate structure as water molecules are incorporated into gas hydrate, increasing the salinity of the surrounding pore waters (Figure 1). This process is termed "ion exclusion". Increases in interstitial ion concentrations will be proportional to the amount of gas hydrate that is formed. Chloride concentrations are commonly used to monitor the effects of ion exclusion because chloride is generally a non-reactive species in shallow sediments [*McDuff et al.*, 1978].

Natural Gas Hydrates: Occurrence, Distribution, and Detection
Geophysical Monograph 124

Models for Pore Fluid Compositional Changes in Sedimentary Pore Spaces

Simple mass balance considerations can be used to compute the effect of gas hydrate formation and decomposition on pore water chloride concentrations [*Ussler and Paull*, 1995]. During gas hydrate formation in a geochemically closed system, the resulting concentration of a non-reactive species, such as chloride can be computed from:

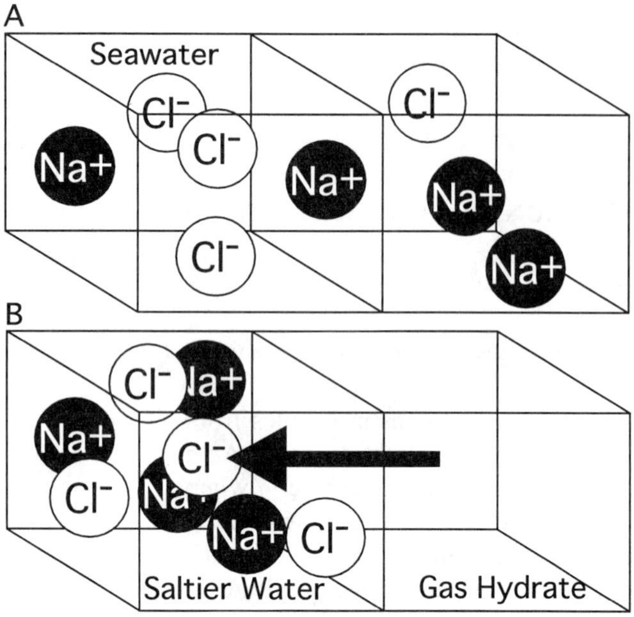

Figure 1. Cartoon illustration of how gas hydrate formation increases the salinity of the adjacent interstitial pore fluid. Dissolved ions, represented here as Na^+ and Cl^-, are initially distributed throughout the pore space (Part A). When gas hydrates form (Part B), dissolved ions are excluded from the growing hydrate structure, creating a more concentrated residual pore fluid.

$$C = \frac{C_i}{1 - \left[\frac{V_h}{(w - (w-1)V_h)}\right]} \tag{1}$$

where C_i is the initial ion concentration of the pore fluid, and C is the ion concentration after filling a volume fraction of the pore volume, V_h, with gas hydrate. This calculation does not depend on porosity because the amounts of gas hydrate and residual pore water are computed on a proportional basis. Because gas hydrates have variable amounts of guest occupancy, the term w has been defined to represent the occupancy-density characteristics of the gas hydrate formed, and is calculated from:

$$w = \frac{\rho_{fw} M_h}{\rho_h M_{fw} m_{fw}} \tag{2}$$

where ρ_{fw} and ρ_h are the densities of freshwater and gas hydrate respectively, M_h and M_{fw} are the molecular weights of freshwater and gas hydrate respectively, and m_{fw} is the number of moles of freshwater contained in 1 mole of gas hydrate. The molecular weight of gas hydrate and the number of moles of water contained within the gas hydrate structure are a function of the degree of occupancy of the gas hydrate structure. When the methane gas hydrate structure is fully occupied, 5.9 moles of water are contained in 1 mole of gas hydrate, the density is 910 kg m^{-3} (Davidson 1981), and the molecular weight is 122.2 g M^{-1}. Model results for computing the effect of gas hydrate growth on pore water chloride concentration and total salinity are shown in Figure 2A.

As long as the sediment pore spaces remain geochemically closed, the result of gas hydrate formation and subsequent decomposition will be to return the system to the original interstitial fluid composition. However, the more saline interstitial waters formed by gas hydrate growth are subject to modification by fluid exchange or solute loss by diffusion over long periods of time within the sediments. Thus, the interstitial fluid composition produced by gas hydrate decomposition will depend on the composition of the interstitial fluid at the time decomposition commenced as well as the amount of gas hydrate that decomposed (Figure 2B). The concentration of a non-reactive species (e.g., chloride ions) after gas hydrate decomposition can be computed using a simple mixing equation:

$$C_d = C_{pw}\left(1 - \frac{V_h}{(w - (w-1)V_h)}\right) + C_h\left(\frac{V_h}{(w - (w-1)V_h)}\right) \tag{3}$$

where C_d is the concentration after decomposition, C_{pw} is the pore water concentration prior to decomposition, V_h is the volume fraction of hydrate filling the pore space, C_h is the concentration in the gas hydrate, which is assumed to be zero, and w is computed using equation 2. C_{pw} will not be the same as C in equation 1 if diffusive or advective loss or gain within the dissolved ion pool has occurred after gas hydrate formation.

Equation 3 can be applied to estimating the amount of gas hydrate formerly occupying sedimentary pore space, if C_d is measured and if the value of C_{pw} can be estimated with reasonable certainty. The value selected for C_{pw} will have a significant effect on the estimate for gas hydrate amount, and the issues associated with selecting this "baseline" value will be discussed later in this paper.

Implications for Sediment Pore Fluid Geochemistry

Increased dissolved ion concentrations associated with gas hydrate formation in sediment pore fluids have the potential to establish distinctive chemical gradients between gas-hydrate-bearing and adjacent, unaffected sediments. The magnitude of these gradients will depend on the distance to gas-hydrate-bearing sediments, the amount of gas hydrate formed, paleoceanographic salinity changes,

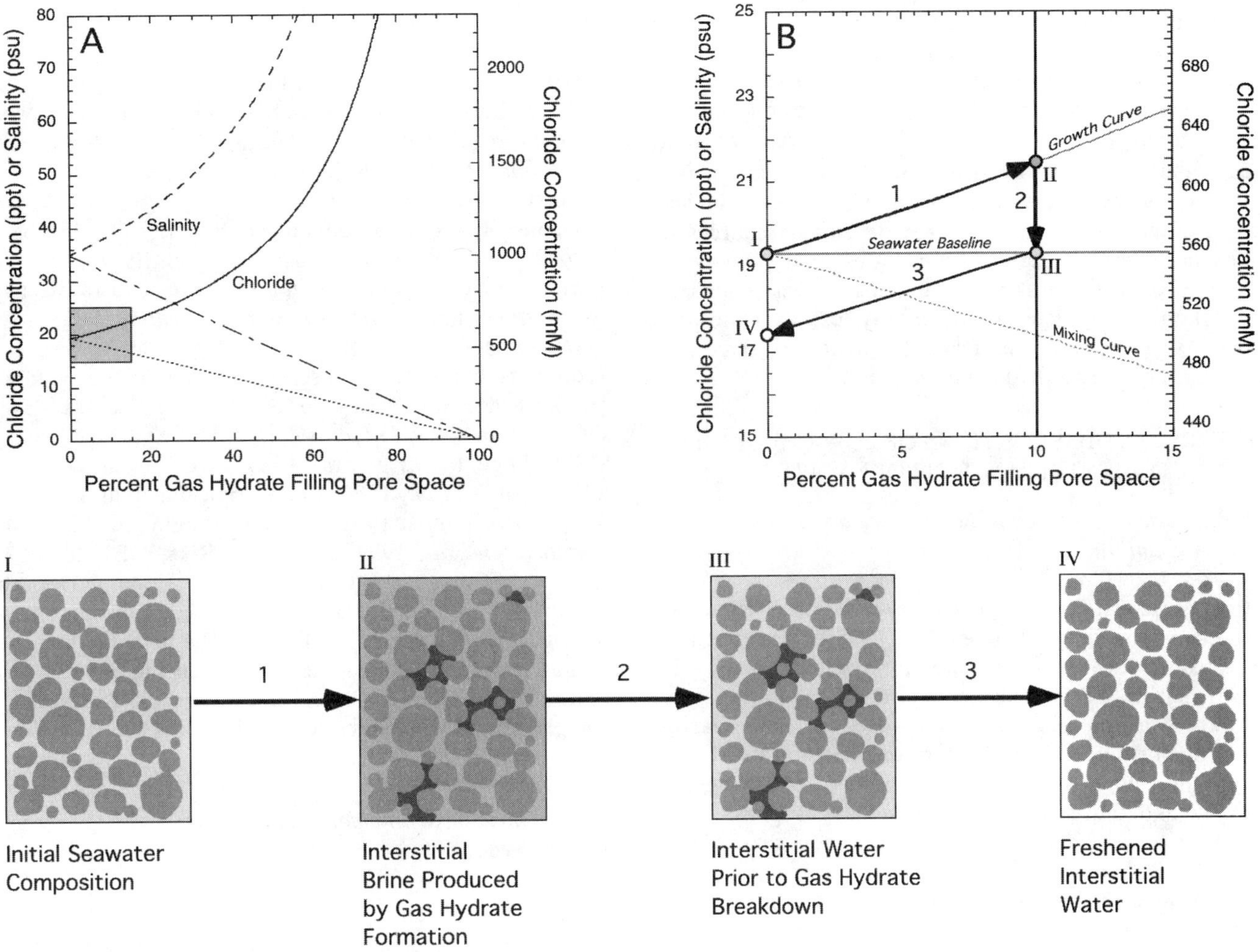

Figure 2. The effects of gas hydrate formation and decomposition on interstitial pore fluid composition with the original ionic composition of seawater are illustrated. Increasing percentages of gas hydrate filling pore space results in elevated chloride concentrations (solid curve) and salinity (dashed curve) (Part A) of the residual interstitial pore water in a closed system. Maintenance of a chemically closed system will result in the return to the original pore fluid composition after gas hydrate decomposition. However, when the concentration of dissolved ions in the locally saltier interstitial fluid decreases by diffusive loss or advective fluid exchange, then lower than seawater chloride concentrations (dotted line) and salinities (dot-dash line) may result after gas hydrate decomposition. The composition of the resulting fluid mixture is computed from the relative volumes of fresh water released by the decomposing gas hydrate and the seawater in the interstitial pore space using a linear mixing model (Equation 3). A seawater chloride concentration of 560 mM chloride and a salinity of 35 were used for the initial pore fluid composition. Full occupancy of methane gas hydrate was assumed (5.9 moles methane in 1 mole gas hydrate; gas hydrate density of 910 kg m^{-3}). The gray box in Part A is expanded in Part B to illustrate the effect of a chemically open system. Accompanying this panel is a series of four illustrations keyed to the graph. Beginning at Point I, the growth of gas hydrate causes the pore water chloride concentration to follow Trajectory 1 (labeled Growth Curve). Point II indicates the concentration of chloride in the pore water after gas hydrate fills 10% of the pore spaces. Trajectory 2 shows how chloride concentration will change as chloride is lost from the system over time by either diffusion or advection. In this example, the composition is allowed to reach that of the original starting value (e.g., seawater) at Point III. Gas hydrate decomposition will release fresh water to the pore spaces, diluting the interstitial pore fluid. The path followed during gas hydrate decomposition is indicated by Trajectory 3 and ends at Point IV. Point IV corresponds to the chloride concentration value at the intersection of the mixing curve (dotted line) and the vertical solid line representing 10% gas hydrate formation. The value of Point IV was computed from Equation 3.

and the time interval over which the gradient develops. In the simplest case, these chemical gradients will redistribute chemical constituents by diffusion, and there will be a net loss of dissolved ions from gas hydrate bearing sediments proportional to the amount of gas hydrate present. Redistribution of dissolved ions by fluid advection is a more complicated case, and a variety of net gain or loss scenarios can be constructed that depend on sources of more saline or less saline fluids. Measured pore water chloride concentrations from sediments known or inferred to contain gas hydrate are commonly less than those of modern seawater, implying that a net loss of dissolved ions has occurred after gas hydrate formation.

EMPIRICAL OBSERVATIONS OF LOW CHLORIDE CONCENTRATIONS IN PORE WATER

Correspondence Between BSRs and the Base of Gas Hydrate Stability

Remote detection of gas hydrate is based on observations of bottom simulating reflectors (BSR). BSRs are reflections in seismic profiles that are interpreted to be the manifestation of the phase boundary separating gas-hydrate-bearing sediments above from gas-bearing sediments below. A BSR is generated if there is a sufficient impedance contrast between sediments containing gas hydrate and those below containing a gas phase. This interface is believed to be located between gas-hydrate-bearing sediments and gas-bearing sediment and to correspond with the base of gas hydrate stability (BGHS) that defines the equilibrium between gas hydrate above with sediments containing free gas and gas-saturated pore water below [*Holbrook et al.*, 1996]. Thus, the BGHS is the interface responsible for the BSR.

Pore Water Geochemistry Databases

The Deep Sea Drilling Project (DSDP) and the Ocean Drilling Program (ODP) have provided large geochemical databases for pore water compositions within marine sediments. These databases were generated by the somewhat systematic sub-sampling, extraction of the pore water from sediments, and shipboard measurement of the chemical constituents [*Gieskes et al.*, 1991]. With few exceptions (e.g., Legs 11, 141, 146, and 164), safety concerns have prevented DSDP/ODP drilling in areas with BSRs at depth, which is often associated with the occurrence of gas hydrate in the overlying sediments. As more data becomes available we are beginning to understand the spatial distribution of gas hydrate from the characteristics of pore fluid chemistry. This is a technique that is completely independent of BSR distribution.

Low Chloride Anomalies and Gas Hydrate Occurrence

The first chloride concentration data for sediment pore water believed to be associated with gas hydrate were obtained on DSDP Legs 66 and 67 [*Gieskes et al.*, 1985; *Harrison et al.*, 1982]. These legs documented the occurrence of fresher pore waters from sedimentary horizons observed to contain gas hydrate. On this basis chloride anomalies were used to identify the former presence of gas hydrate in adjacent cores and to estimate gas hydrate amount [*Harrison and Curiale*, 1982; *Hesse and Harrison*, 1981]. Subsequent legs that have drilled sediments believed or observed to contain gas hydrate [Blake Ridge DSDP Leg 76; *Jenden and Gieskes*, 1983; ODP Leg 164, *Paull et al.*, 1996; Middle American Trench, DSDP Leg 84; *Hesse et al.*, 1985; Peru Trench, ODP Leg 112, *Kastner et al.*, 1990; Chile Triple Junction, ODP Leg 141, *Froelich et al.*, 1995; Cascadia Margin, ODP Leg 146, *Westbrook et al.*, 1994; and Costa Rica, ODP Leg 170, *Kimura et al.*, 1997] have further documented the occurrence of low chloride concentrations in association with gas-hydrate-bearing sediments. The potential presence of gas hydrates has been argued for on the basis of low chloride anomalies at several other borehole and seafloor sampling sites, where there generally is less supporting data [see *Ginsburg and Soloviev*, 1998].

Survey of Chloride Anomalies and Gas Hydrate Occurrence

Chloride concentration data from DSDP/ODP sites inferred to be associated with gas hydrate are summarized in Table 1 and plotted in Figure 3. These data are arranged into three groups (Figures 3B, 3C, and 3D) based on the relationship of the seismic reflection data connecting these sites with the presence of gas hydrate. Figure 3B summarizes sites that have penetrated the inferred depth of the BGHS at a site where a BSR is present. Figure 3C shows sites where a BSR does not exist directly beneath the drill site, but occurs nearby. In these cases the BGHS can be projected laterally from where the BSR is observed to the drill hole. Figure 3D shows a group of shallow sites that have not penetrated the inferred depth of the BGHS/BSR, but where seismic reflection data indicate a BSR occurs.

Four characteristics typify chloride profiles in regions where seismic data suggest the presence of gas hydrate (Figure 3): (1) Chloride concentrations decrease smoothly with depth from the seafloor. The magnitude of the chloride decrease from seafloor values ranges between 50 and 300 mM; (2) Increased scatter and erratic lower chloride values occur in a zone that is up to 200 meters thick above the inferred depth of the BGHS/BSR; (3) Distinctly smoother and near vertical chloride profiles occur beneath the

Table 1: Summary of Drill Sites in Gas-Hydrate-Bearing Sections That Have Penetrated or are Proximal to a BSR

Drill Site	Hole Depth (mbsf)	BSR[a]	BSR Depth (mbsf)	Gas Hydrates Observed	DSDP/ODP Leg	Location	References
102	661	P	620	--	11	Blake Outer Ridge	[*Hollister et al.*, 1972]
490	588.5	L	402	Yes	66	Middle America Trench - Mexico	[*Watkins et al.*, 1981]
491	542.0	P	525	Yes	66		
492	279.0	D	394	Yes	66		
496	378	D	508	Yes	67	Middle America Trench - Guatemala	[*Aubouin et al.*, 1982]
497	396.5	L	~500	--	67		
533	399	D	550	Yes	76	Blake Outer Ridge	[*Sheridan et al.*, 1983]
568	417.7	D	455	Yes	84	Middle America Trench - Costa Rica	[*von Huene and Aubouin*, 1985]
683	488	D	529	--	112	Peru Trench	[*Suess et al.*, 1988; *Froelich et al.*, 1995]
685	468.6	L	612	Yes	112		
688	779	L	473	Yes	112		
859	476.1	P	97	--	141	Chile Triple Junction	[*Behrmann et al.*, 1992; *Froelich et al.*, 1995]
860	617.8	P	200	--	141		
889	386.5	P	225	--	146	Cascadia Margin	[*Westbrook et al.*, 1994]
892	178.5	P	73	Yes	146		
994	703.5	L	450	Yes	164	Blake Ridge Gas Hydrates	[*Paull et al.*, 1996]
995	704.6	P	450	--	164		
997	750.7	P	450	Yes	164		

[a] P = penetrated BSR at site; L = penetrated lateral extension of BSR; D = BSR deeper at site.

BGHS/BSR; and (4) Mean pore water chloride values beneath the BGHS/BSR are distinctly fresher than seawater.

Drill sites that penetrated the inferred depth of the BGHS at a site with a BSR show profile characteristics 1, 2, 3, and 4 (Figure 3B). Sites that have a BSR nearby show the same pattern (Figure 3C). Where there is a BSR, but the hole did not penetrate the inferred depth of the BGHS, the existing chloride profiles also show profile characteristics 1 and 2, but these holes do not extend deep enough to display characteristics 3 and 4.

The initial freshening of measured pore waters with depth (profile characteristic 1) can be explained by either regular increases in gas hydrate amount with depth or a progressive shift toward lower *in situ* pore water salinity with depth (e.g., a change in "baseline"). The scatter seen in chloride profiles as the BGHS is approached (profile characteristic 2) is interpreted to reflect changes in the amount of gas hydrates between adjacent samples.

Profile characteristics 3 and 4 indicate that a large reservoir containing water with uniformly low chloride concentrations may exist beneath the BGHS/BSR. Apparently, low chloride zones are a characteristic geochemical feature of sediments associated with gas-hydrate-bearing sediment sequences.

CALCULATING GAS HYDRATE AMOUNT FROM BASELINE PORE WATER CONCENTRATIONS

In principle, low chloride anomalies caused by breakdown of gas hydrate during core recovery and dilution of the interstitial pore water allow the amount of gas hydrate initially present *in situ* to be calculated (Equation 3). However, the chloride concentration of the undiluted pore water prior to gas hydrate decomposition must be known.

Attempts to obtain pristine pore waters using *in situ* samplers, such as the ODP Water Sampling Temperature Probe (WSTP) [*Westbrook et al.*, 1994] and the Fisseler Tool [*Paull et al.*, 1996] have had little success. With few exceptions, these tools do not work well, especially in fine-grained sediments. Contamination with drilling fluid has been a consistent problem and has compromised the value of these *in situ* pore water measurements. Robust methods for quantifying the amount of contamination have not been identified, and the errors associated with back-calculating *in situ* values are large and hard to assess.

In the absence of *in situ* measurements, efforts have been focused on ways to estimate *in situ* pore water concentrations. Various approaches to estimate *in situ* values have been proposed. Three of the simplest and most common approaches include: (1) Assuming that the "baseline" is similar to modern seawater (Figure 4); (2) Constructing a "baseline" profile using measured chloride concentration values that are believed not to have been diluted by gas hydrate decomposition during core recovery (Figure 4); and (3) Constructing a "baseline" profile using measured *in situ* values.

A Modern Seawater Baseline

A modern seawater baseline assumption has been applied to chloride concentration data from DSDP/ODP Legs [e.g.,

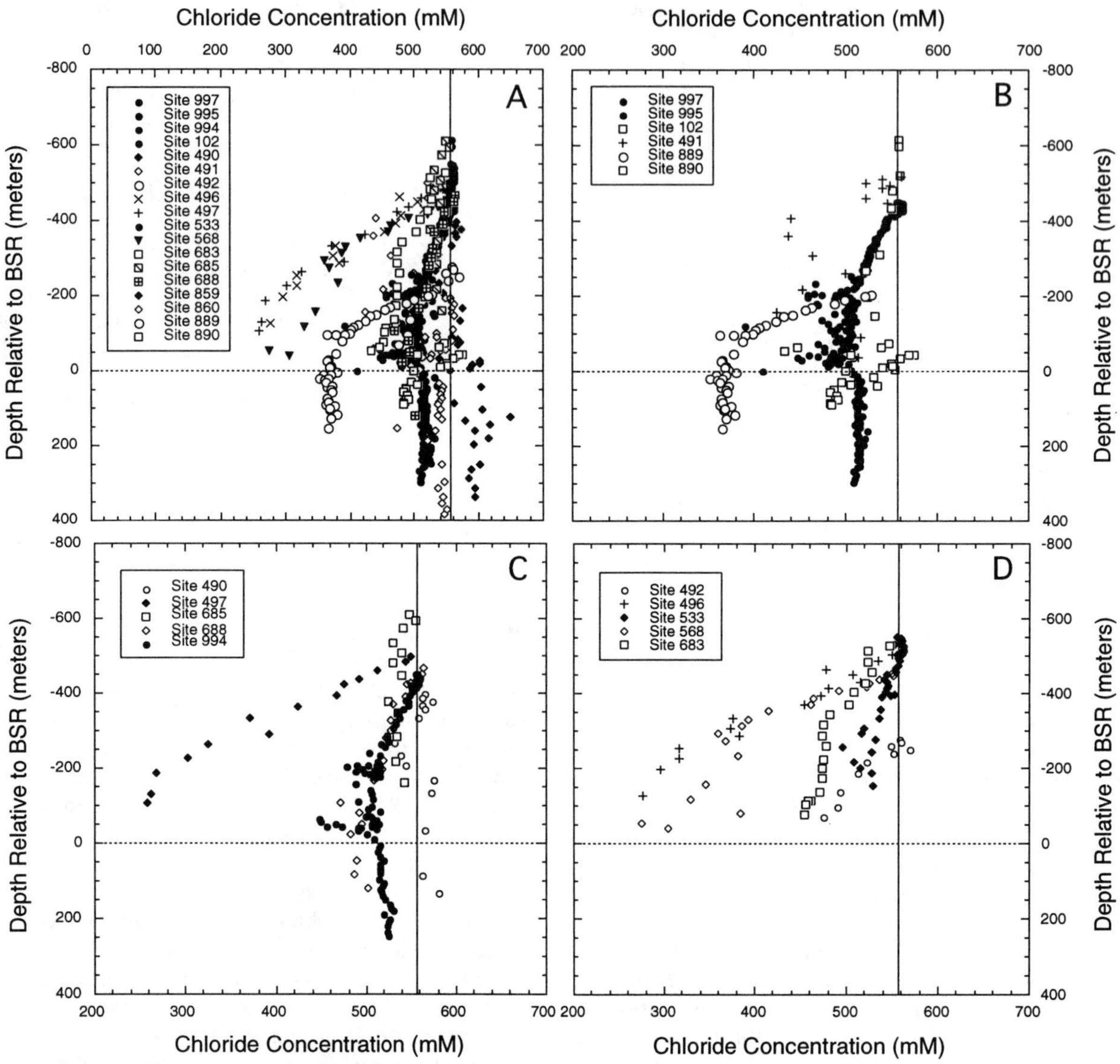

Figure 3. Chloride concentration profiles from DSDP and ODP drill sites in putative gas-hydrate-bearing sediments are plotted as a function of depth relative to the local BSR or BGHS. Negative values of relative depth indicate samples obtained from above the BSR, positive values are from depths greater than the BSR. The actual depth of the BSR for each drill site is based on the available geophysical and well-log data. Chloride data from all 17 drill sites where the presence of gas hydrates has been inferred on the basis of independent observations are shown in Part A. Chloride concentrations originally reported as parts per thousand (ppt or ‰) have been converted to mM. Six of the eight DSDP/ODP drill sites that have penetrated a BSR are illustrated in Part B. Sites 859 and 860 have been excluded from Part B because additional sources of salt are inferred to exist at these sites. Part C shows data from drill sites where a BSR was identified in close lateral proximity to the borehole. Cores from all but Site 497 recovered gas hydrate. Part D shows data from drill sites where a BSR was clearly identified beneath the site but the hole did not penetrate the BGHS. A vertical solid line in each panel indicates typical open marine chloride concentration (559 mM). Data sources: Site 102 [*Sayles et al.*, 1972]; Sites 490, 491, 492 [*Gieskes et al.*, 1985]; Sites 496 and 497 [*Harrison and Curiale*, 1982]; Site 533 [*Jenden and Gieskes*, 1983]; Site 568 [*Hesse et al.*, 1985]; Sites 683, 685, and 688 [*Kastner et al.*, 1990]; Sites 859 and 860 [*Shipboard Scientific Party*, 1992]; Sites 889 and 890 [*Shipboard Scientific Party*, 1994]; Sites 994, 995, and 997 [*Paull et al.*, 1996].

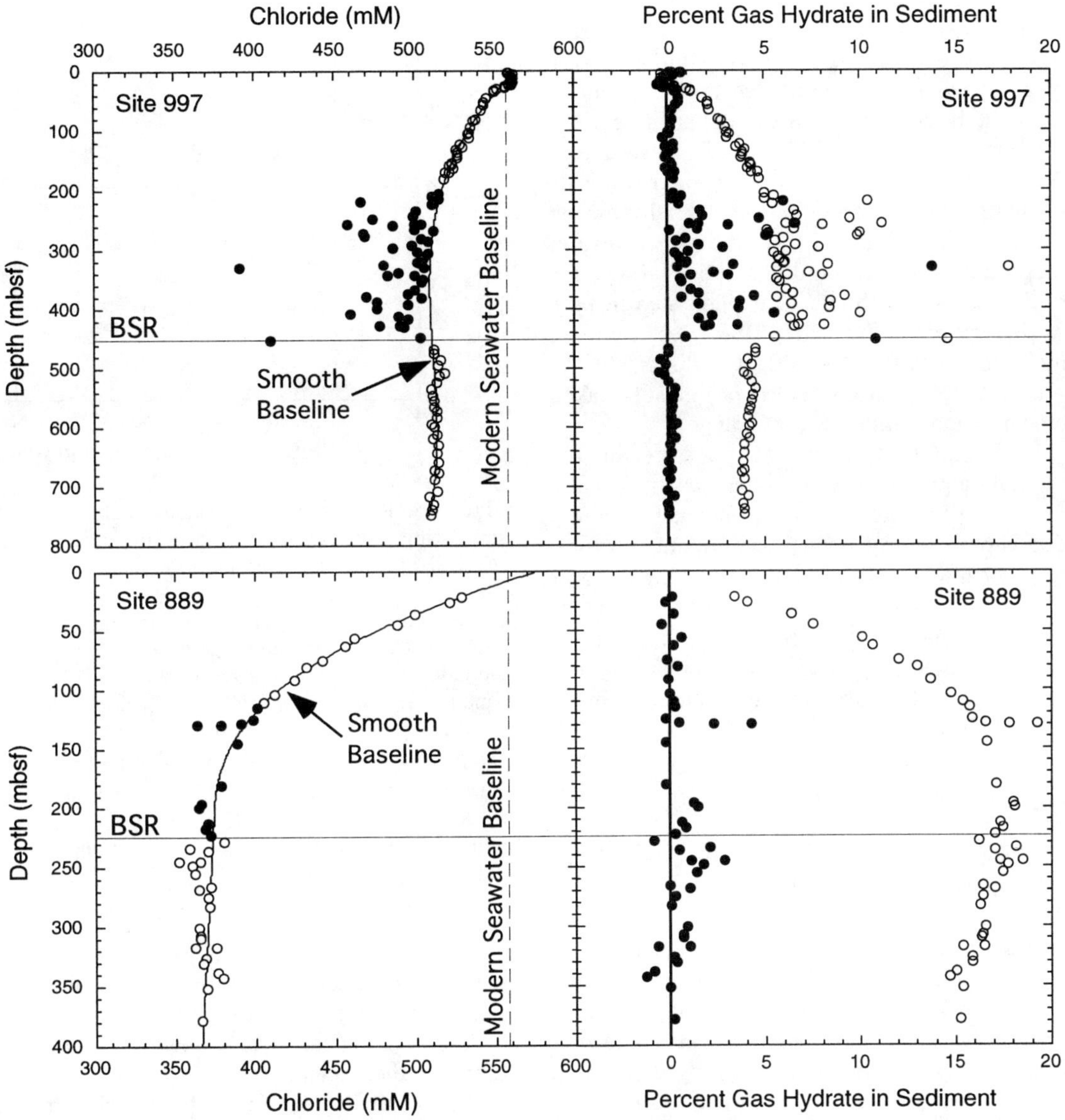

Figure 4. A comparison of two common approaches for estimating *in situ* "baseline" chloride concentration values and the corresponding estimated amount of gas hydrate is illustrated. Data from Site 997 (ODP Leg 164) is illustrated in the upper left panel and Site 889 (ODP Leg 146) in the lower left panel. In both examples, the smooth baseline was computed using a best-fit 3rd order polynomial to the values plotted as circles, except that the portion below the BSR for Site 889 was fit with a linear function. Measured pore water chloride concentrations for sediments believed to have contained gas hydrate are plotted using solid circles. Gas hydrate amount calculated using both baselines and corrected for measured sediment porosity using a linear fit to porosity-depth data is illustrated in the corresponding right-hand panels. Solid circles indicate amounts calculated using the smooth baseline, and open circles indicate amounts calculated using a modern seawater baseline. Note that the modern seawater baseline predicts the occurrence of substantial amounts of gas hydrate below the BSR and that a large geochemical discontinuity must be maintained at the BGHS/BSR. The smoothed baseline produces an estimate of gas hydrate amount more consistent with remote-sensing techniques, and with the phase change that occurs at the BGHS/BSR.

Leg 146, *Westbrook et al.*, 1994] Legs 76 and 146, [*Yuan et al.*, 1996]. However, there are two major shortcomings to using a modern seawater value for the *in situ* chloride concentration: (1) Estimated gas hydrate amounts are substantially greater than those predicted by independent, but also inadequately constrained methods, such as well-logs, vertical seismic profiling (VSP), and pressure core sampling (PCS); and (2) Measured pore water chloride values below the BGHS/BSR are often substantially fresher than modern seawater (Figure 3). If modern seawater salinity is assumed to represent the baseline above the BGHS/BSR, then a large, difficult to explain step in pore water salinity is required at the BGHS/BSR. Because there are no gas hydrates below the BGHS/BSR to dissociate, the measured salinity (or chloride concentration) must be taken to represent the baseline within that depth range.

A model simulating the diffusion across a step in the baseline like that required by a seawater baseline at the BGHS was run to illustrate how long such an anomaly could persist. For example, a chloride concentration discontinuity of 47 mM (~1.7 ppt) spreads out rapidly and is detectable 34 meters away from the interface after 20,000 years (Figure 5). If small, sharp pore water concentration discontinuities are detected in pore water profiles, then active processes must be sufficient to maintain them, because they would be erased by diffusion.

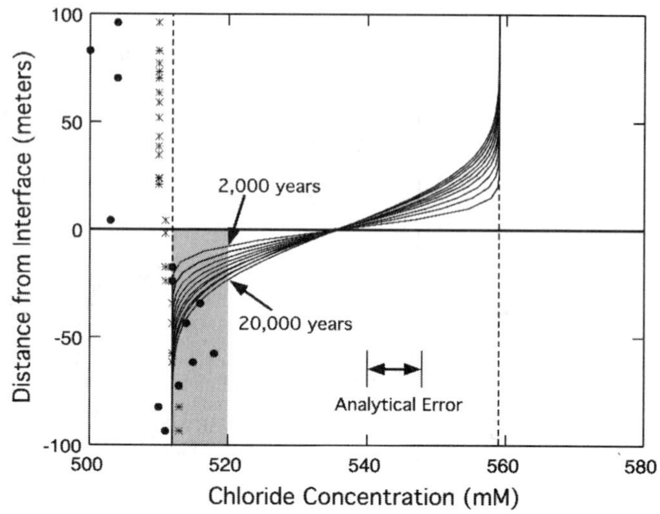

Figure 5. Results for a 1-dimensional 2-box diffusion model simulating a chloride concentration discontinuity across the BGHS/BSR and measured chloride data from ODP Site 994 are illustrated. Filled circles are measured chloride pore water concentrations and asterisks are points along a 3rd order polynomial baseline. The shaded band represents the analytical uncertainty of the chloride concentration measurement. This model is based on equation 3.13 of *Crank* [1975]. The initial chloride concentration in the upper compartment was 559 mM, and the lower compartment was 512 mM. Sediment porosity (ϕ) was 50%, and the free solution diffusion coefficient used for chloride ion was $D_o = 1.86 \times 10^{-5}\ cm^2\ s^{-1}$ at 25 °C [*Li and Gregory*, 1974]. An effective sediment diffusion coefficient (D) was computed from $D = D_o\ \phi^2$ [*Lerman*, 1979]. The analytical error typically associated with ODP chloride concentration measurements is ~0.25 ppt.

An Empirically Determined Baseline

A second approach is to define the "baseline" by either measuring chloride concentration values in close proximity to a gas hydrate occurrence [e.g., Leg 112, *Kvenvolden and Kastner*, 1990; *Kastner et al.*, 1990] or fitting a smoothing function to measured pore water chloride values that have had low chloride "spikes" removed [Figure 4; ODP Leg 164, *Paull et al.*, 1996]. The key assumption in this second approach is that gas hydrates are not uniformly distributed throughout a sedimentary section and that a smoothed envelope of maximum chloride values closely approximates the *in situ* pore water concentration prior to gas hydrate decomposition during core recovery. Fitting a smoothed curved to measured pore water chloride values is a conservative approach compared to using a seawater baseline, and the gas hydrate amounts estimated this way are minimum values.

Variable amounts of gas hydrate formation with depth will initially create local increases in pore water chloride concentration. However, these local variations in chloride and other dissolved ions will diffuse away given sufficient time. Figure 6 illustrates model results for the diffusive attenuation of a chloride anomaly created by formation of ~9% gas hydrate filling the pore space. An anomaly of 56 mM (~2 ppt) created by ion exclusion will not be detectable with current analytical methods after about 40,000 years. Thus, we would expect to see a smooth baseline within gas-hydrate-bearing sediment sequences unless the gas hydrates formed rather recently.

The physical means by which a chloride concentration "baseline" is generated *in situ* is not particularly clear. A seawater-like "baseline" could be maintained by diffusive exchange with overlying seawater and the constant removal of solutes concentrated during gas hydrate formation (Figure 4). *Kvenvolden and Kastner* [1990] have argued for upward diffusive loss of chloride excluded from underlying gas-hydrate-bearing sections. Evidence for this was provided by subtle chloride concentration increases above modern seawater values seen in the upper 100 mbsf. However, diffusion alone cannot produce pore waters less chloride-rich than average seawater. Alternatively, upward movement of chloride-depleted pore water from a large, low-chloride reservoir below the BGHS/BSR would produce substantially depleted *in situ* pore waters within gas-hydrate-bearing sediments.

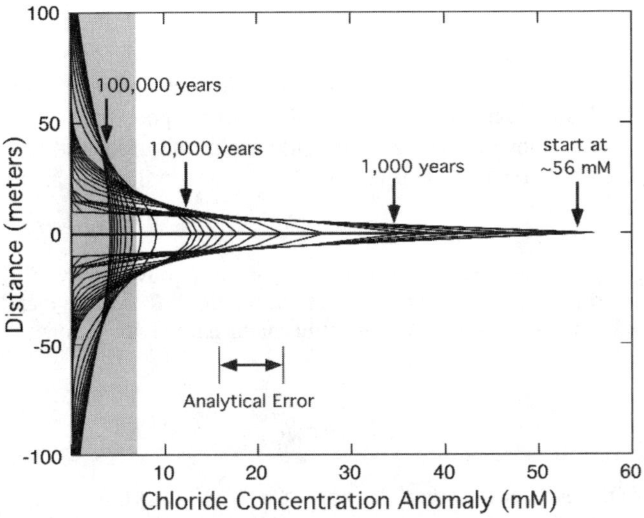

Figure 6. Results for a 1-dimensional decay model for a 2 ppt chloride anomaly computed using equation 2.15 of *Crank* [1975] are illustrated. The initial chloride concentration was 56 mM (equivalent to a ~2 ppt anomaly), and the half-width of the concentration spike was 5 meters. Sediment porosity, the effective sediment diffusion coefficient (D), and analytical errors were the same as those used in the model shown in Figure 5. The results from this model can be applied to either positive or negative concentration anomalies. The shaded band represents the analytical uncertainty of the chloride concentration measurement.

An Interpolated Baseline

A third, more sophisticated approach represents a compromise between using a seawater baseline and the conservative baseline fit to the chloride concentration data [*Egeberg and Dickens*, 1999]. This approach is strongly dependent on the occurrence of a subtle shallow chloride maximum and requires corrections for glacial/interglacial ocean water salinity changes. Their approach requires that the baseline passes approximately through *in situ* chloride measurements obtained with the WSTP, and merges with measured chloride concentrations below the BGHS/BSR. The depth-integrated amount of gas hydrate calculated with their model is slightly lower than that obtained using an empirical baseline fit, however, it indicates gas hydrate occurrence at much shallower depths than the empirical baseline fit. While this approach tries to accommodate numerous complexities, similar data to assess gas hydrate distribution does not exist elsewhere.

While gas hydrate assessment using chloride profiles has unresolved baseline issues, it is capable of producing accurate quantitative data on gas hydrate amounts. In contrast, BSRs have proven useful for remote detection of potentially gas-hydrate-bearing marine sediments, however, they provide little information about gas hydrate amount and its vertical distribution. Thus, gas hydrate distribution will not be adequately resolved until more sites with densely-spaced chloride measurements have been obtained along with *in situ* chloride sampling.

CHLORIDE ANOMALIES BELOW THE BASE OF GAS HYDRATE STABILITY

The occurrence of measured chloride concentrations that are significantly less than modern seawater chloride values at most DSDP and ODP drilling sites associated with BSRs (Figure 3) suggests that there is an inherent association between gas hydrates and low chloride pore waters beneath the BGHS/BSR. So far, only site specific explanations for these anomalies have been offered. The vertical and horizontal extent of this low-chloride reservoir is unknown anywhere, however at ODP Sites 994, 995, and 997 it spans over 300 meters of depth and across more than 10 km of width with no indication of its lower or lateral limits. Thus, these chloride profiles indicate that a large reservoir of uniformly chloride-depleted water exists beneath the BGHS/BSR.

Possible Causes of Pore Water Freshening Beneath the BGHS/BSR

Various scenarios that explain the pore water freshening beneath the BGHS/BSR can be constructed. These include: (1) Migration of chloride-depleted water not related to gas hydrate decomposition; and (2) Formation of chloride-depleted water as a consequence of gas hydrate decomposition.

The upward or lateral migration of homogeneous chloride-depleted water from an unidentified source has the potential to create pore water profiles like those observed. Although this model does not require a causal relationship between gas hydrates and chloride-depleted pore water, it is curious that gas-hydrate-bearing sediments are most often associated with thick clastic sequences primarily in accretionary prisms and sediment drifts [*Hyndman and Davis*, 1992]. These thick sequences of organic-rich clays are potential sources of low chloride pore waters and methane gas. Clay mineral dehydration reactions (particularly the smectite to illite transition at ~1.5 km depth; [*Perry and Hower*, 1972], differential-stress-induced smectite dehydration [Fitts and Brown, 1999], and the opal-A to opal-CT transformation [*Kastner et al.*, 1977]) have the potential to contribute fresh water (up to ~14 weight % from interlayer water in clays, [*Deer et al.*, 1966]; and up to ~20 weight % from opal, [*Kastner and Siever*, 1983]). These dehydration reactions could dilute pore water and result in a 90 to 160 mM reduction in pore water chloride concentration in a closed system. Release of water by mineral dehydration reactions has been suggested as a reasonable cause for low chloride pore waters below the BGHS/BSR along the Peruvian continental margin [*Kastner et al.*, 1990] and the Cascadia margin [*Westbrook et al.*, 1994].

Methane gas produced by microbial and possibly thermogenic decomposition of sedimentary organic matter has the potential to rise and accumulate within the shallow subsurface of thick clastic sequences to form gas hydrate. Fluid expulsion [e.g., *Hyndman and Davis*, 1992] of low chloride waters produced by mineral dehydration may accompany this upward flux of methane. Thus, the gas to form gas hydrates and low chloride pore waters could have a common sedimentary source. Regardless of how low chloride waters are produced beneath the BGHS/BSR, sediment compaction, increased buoyancy associated with chloride depletion, and presumably increasing temperatures, should stimulate vertical advection of these waters into the colder, more saline gas-hydrate-bearing units above.

A second scenario can be constructed in which decomposition of gas hydrate is the primary source for chloride-depleted water that invades overlying gas-hydrate-bearing sediments. Chloride profiles beneath the BGHS/BSR show very little scatter (Figure 4) suggesting that pore water chemical anomalies produced by non-uniform distribution of gas hydrate have been erased by diffusion (Figure 6).

A more radical possibility for producing chloride-depleted water below the BGHS/BSR involves the extraction of water from the interlayers of clay minerals by gas hydrate during growth. If gas hydrate growth becomes water limited [e.g., *Clennell et al.*, 1999], interlayer water in clay minerals is a potential source of water to allow continued gas hydrate formation. This irreversible loss of water by clay minerals would result in a net decrease in the salinity of the pore water without the loss of salts by diffusion, when the gas hydrates decompose below the BGHS/BSR. This low chloride reservoir beneath the BGHS/BSR may very well be part of a larger hydrologic system that is more complex than what we can easily infer from 1-dimensional analysis.

CONCLUSIONS

Within gas-hydrate-bearing sedimentary sections, low chloride anomalies are a useful proxy for indicating the presence of gas hydrate, and for calculating gas hydrate amount, if the *in situ* chloride concentration is known. However, until *in situ* measurements can be made routinely and reliably, use of chloride data as a proxy for calculating gas hydrate amount will be somewhat speculative and site specific. Simple 1-dimensional numerical modeling shows that chloride anomalies created by either gas hydrate formation or decomposition will rapidly dissipate by diffusion. Unless gas hydrates have formed or decomposed recently, relatively smooth *in situ* chloride concentration profiles are expected. DSDP/ODP sediment pore water data show that lower than seawater chloride concentrations are a characteristic geochemical feature of gas-hydrate-bearing sediments.

Acknowledgements. Financial support was provided by The David and Lucile Packard Foundation. We thank Ryo Matsumoto and Joris Gieskes for their thoughtful comments and suggestions.

REFERENCES

Aubouin, J., et al., *Initial Reports of the Deep Sea Drilling Project*, vol. 67, U.S. Government Printing Office, Washington, 1982.

Behrmann, J.H., et al., *Proceedings of the Ocean Drilling Program, Initial Reports*, vol. 141, Ocean Drilling Program, College Station, TX, 1992.

Clennell, M.B., M. Hovland, J.S. Booth, P. Henry, and W.J. Winters, Formation of natural gas hydrates in marine sediments 1. Conceptual model of gas hydrate growth conditioned by host sediment properties, *J. Geophys. Res.*, *104*, 22985-23003, 1999.

Davidson, D.W., Gas hydrates as clathrate ices, in *Natural Gas Hydrates: Properties, Occurrence, and Recovery*, edited by J.L. Cox, pp. 1-15, Butterworth, Boston, 1981.

Deer, W.A., R.A. Howie, and J. Zussman, *An Introduction to the Rock-forming Minerals*, John Wiley and Sons, Inc, New York, 1966.

Egeberg, P.K., and G.R. Dickens, Thermodynamic and pore water halogen constraints on gas hydrate distribution at ODP Site 997 (Blake Ridge), *Chem. Geol.*, *153*, 53-79, 1999.

Fitts, T.G, and K.M. Brown, Stress-induced smectite dehydration; ramifications for patterns of freshening and fluid expulsion in the North Barbados accretionary wedge, *Earth Planet. Sci. Lett.*, *172*, 179-197, 1999.

Froelich, P.N., K.A. Kvenvolden, M.E. Torres, A. Waseda, B.M. Didyk, and T.D. Lorenson, Geochemical evidence for gas hydrate in sediment near the Chile Triple Junction, in *Proceedings of the Ocean Drilling Program, Scientific Results*, edited by S.D. Lewis, et al., vol. 141, pp. 279-286, Ocean Drilling Program, College Station, TX, 1995.

Gieskes, J.M., T. Gamo, and H. Brumsack, *Chemical methods for interstitial water analysis aboard JOIDES Resolution*, Texas A & M University, College Station, TX, 1991.

Gieskes, J.M., K. Johnston, and M. Boehm, Appendix. Interstitial water studies, Leg 66, in *Initial Reports of the Deep Sea Drilling Project*, edited by R. von Huene, et. al., vol. 66, pp. 961-967, U.S. Government Printing Office, Washington, D.C., 1985.

Ginsburg, G.D., and V.A. Soloviev, *Submarine Gas Hydrates*, VNIIOkeanogeologia, St. Petersburg, 1998.

Harrison, W.E., and J.A. Curiale, Gas hydrates in sediments of holes 497 and 498A, Deep Sea Drilling Project Leg 67, in *Initial Reports of the Deep Sea Drilling Project*, edited by J. Auboin, et. al., vol. 67, pp. 591-599, U.S. Government Printing Office, Washington, D. C., 1982.

Harrison, W.E., R. Hesse, and J.M. Gieskes, Relationship between sedimentary facies and interstitial water chemistry of slope, trench, and Cocos Plate sites from the Middle America Trench transect, active margin off Guatemala, Deep Sea Drilling Project Leg 67, in *Initial Reports of the Deep Sea Drilling Project*, edited by J. Aubouin, et al., vol. 67, pp. 603-613, U.S. Government Printing Office, Washington, D. C., 1982.

Hesse, R., and W.E. Harrison, Gas hydrates (clathrates) causing pore-water freshening and oxygen isotope fractionation in deep-water sedimentary sections of terrigenous continental margins, *Earth Planet. Sci. Lett.*, *55*, 453-462, 1981.

Hesse, R., J. Lebel, and J.M. Gieskes, Interstitial water chemistry of gas-hydrate-bearing sections on the Middle America Trench Slope, Deep-Sea Drilling Project Leg 84, in *Initial Reports of the Deep Sea Drilling Project*, edited by R. von Huene, and J. Aubouin, vol. 84, pp. 727-737, U.S. Government Printing Office, Washington, 1985.

Holbrook, W.S., H. Hoskins, W.T. Wood, R.A. Stephen, and D. Lizarralde, Methane hydrate and free gas on the Blake Ridge from vertical seismic profiling, *Science*, *273*, 1840-1843, 1996.

Hollister, C.D., et al., *Initial Reports of the Deep Sea Drilling Project*, vol. 11, U.S. Government Printing Office, Washington, 1972.

Hyndman, R.D., and E.E. Davis, A mechanism for the formation of methane hydrate and seafloor bottom-simulating reflectors by vertical fluid expulsion, *J. Geophys. Res.*, *97*, 7025-7041, 1992.

Jenden, P.D., and J.M. Gieskes, Chemical and isotopic composition of interstitial water from Deep Sea Drilling Project Sites 533 and 534, in *Initial Reports of the Deep Sea Drilling Project, 76*, edited by R.E. Sheridan, et al., vol. 76, pp. 453-461, U.S. Government Printing Office, Washington, DC, 1983.

Kastner, M., H. Elderfield, J.B. Martin, E. Suess, K.A. Kvenvolden, and R.E. Garrison, Diagenesis and interstitial-water chemistry at the Peruvian continental margin--major constituents and strontium isotopes, in *Proceedings of the Ocean Drilling Program, Scientific Results*, edited by E. Suess, et al., vol. 112, pp. 413-440, Ocean Drilling Program, College Station, TX, 1990.

Kastner, M., J.B. Keene, and J.M. Gieskes, Diagenesis of siliceous oozes--I. Chemical controls on the rate of opal-A to opal-CT transformation--an experimental study, *Geochim. Cosmochim. Acta*, *41*, 1041-1059, 1977.

Kastner, M., and R. Siever, Siliceous sediments of the Guaymas Basin: The Effect of high thermal gradients on diagenesis, *J. Geol.*, *91*, 629-641, 1983.

Kimura, G., et al., Proceedings of Ocean Drilling Project Initial Reports, vol. 170, Ocean Drilling Program, College Station, TX, 1997.

Kvenvolden, K.A., and M. Kastner, Gas hydrates of the Peruvian Outer Continental Margin, in *Proceedings of the Ocean Drilling Program, Scientific Results*, edited by S.E. Suess, and R. von Huene, vol. 112, pp. 517-526, Ocean Drilling Program, College Station, TX, 1990.

Lerman, A., *Geochemical Processes: Water and Sediment Environments*, John Wiley, New York, 1979.

Li, Y., and S. Gregory, Diffusion of ions in sea water and in deep-sea sediments, *Geochim. Cosmochim. Acta*, *38*, 703-714, 1974.

McDuff, R.E., J. Gieskes, and J.R. Lawrence, Interstitial water studies, Leg 42A, in *Initial Reports of the Deep Sea Drilling Project*, edited by K.J. Hsu, et al., vol. 42, pp. 561-568, U.S. Government Printing Office, Washington, D. C., 1978.

Paull, C.K., R. Matsumoto, and P.J. Wallace, and et al., *Proceedings of the Ocean Drilling Program, Initial Reports*, vol. 164, Ocean Drilling Program, College Station, TX, 1996.

Perry, E.A., and J. Hower, Late-stage dehydration in deeply buried pelitic sediments, *Amer. Assoc. Petrol. Geol. Bull.*, *56*, 2013-2021, 1972.

Sayles, F.L., F.T. Manheim, and L.S. Waterman, Interstitial water studies on small core samples, Leg 11, in *Initial Reports of the Deep Sea Drilling Project, 11*, edited by C.D. Hollister, et al., vol. 11, pp. 997-1008, U.S. Government Printing Office, Washington, DC, 1972.

Sheridan, R.E., F.M. Gradstein, and et al., *Initial Reports of the Deep Sea Drilling Project*, vol. 76, U.S. Government Printing Office, Washington, 1983.

Shipboard Scientific Party, Site Chapters, in *Proceedings of the Ocean Drilling Program, Initial Reports*, edited by J.H. Behrmann, et al., vol. 141, pp. 1-611, Ocean Drilling Program, College Station, TX, 1992.

Shipboard Scientific Party, Site Chapters, in *Proceedings of the Ocean Drilling Program, Initial Reports*, edited by G.K. Westbrook, et al., vol. 146, pp. 55-376, Ocean Drilling Program, College Station, TX, 1994.

Shipley, T.H., M.H. Houston, R.T. Buffler, F.J. Shaub, K.J. McMillan, J.W. Ladd, and J.L. Worzel, Seismic evidence for widespread possible gas hydrate horizons on continental slopes and rises, *Amer. Assoc. Petrol. Geol. Bull.*, *63*, 2204-2213, 1979.

Sloan, E.D., *Clathrate Hydrates of Natural Gases*, Marcel Decker, Inc., New York, 1990.

Suess, E., et al., *Proceedings of the Ocean Drilling Program, Initial Reports*, vol. 112, Ocean Drilling Program, College Station, TX, 1988.

Ussler, W. III and C.K. Paull, Effects of ion exclusion and isotopic fractionation on pore water geochemistry during gas hydrate formation and decomposition, *Geo-Marine Lett.*, *15*, 37-44, 1995.

von Huene, R., and J. Aubouin, *Initial Reports of the Deep Sea Drilling Project*, vol., 84, U.S. Government Printing Office, Washington, 1985.

Watkins, J.S., et al., *Initial Reports of the Deep Sea Drilling Project*, vol., 66, U.S. Government Printing Office, Washington, 1981.

Westbrook, G.K., et al., *Proceedings of the Ocean Drilling Program, Initial Reports*, vol. 146, Ocean Drilling Program, College Station, TX, 1994.

Yuan, T., R.D. Hyndman, G.D. Spence, and B. Desmons, Seismic velocity increase and deep-sea gas hydrate concentrations above a bottom simulating reflector on the northern Cascadia continental slope, *J. Geophys. Res.*, *101*, 13655-13671, 1996.

Paull, C.K. and W. Ussler III, Monterey Bay Aquarium Research Institute, 7700 Sandholdt Road, Moss Landing, CA 95039-9644.

History and Significance of Gas Sampling During DSDP and ODP Drilling Associated with Gas Hydrates

Charles K. Paull and William Ussler III

Research and Development Division, Monterey Bay Aquarium Research Institute, Moss Landing, California

Gas hydrate formation requires adequate concentrations of gas (usually methane) to be present in the sediments. Thus, methane-bearing gas hydrate occurrence is closely coupled with the distribution of methane in marine sediments. We review the history of gas sampling from bore holes in gas-rich continental margin sediment sequences that may be associated with gas hydrate. Gas measurements have been routinely made for both drilling safety and scientific investigations during the Deep Sea Drilling Project (DSDP) and the Ocean Drilling Program (ODP). Until recently, gas samples have come only from core voids and residual gas extracted from the cored sediments. However, recent successes with down-hole tools designed to maintain cores at their original pressure during core recovery have obtained independent data on *in situ* gas amounts, gas composition, and the process of core degassing. These data show that the residual core gas concentration represents only a tiny fraction of the total *in situ* gas inventory. Because the majority of the original gas escapes from the core during core recovery, caution must be used when trying to apply traditional core gas measurements to potentially gas hydrate-bearing sediment sequences.

1. INTRODUCTION

Understanding the distribution of methane-bearing gas hydrate in marine sediment sequences requires an appreciation of the origin, distribution, and amount of methane within the host sediment. Moreover, because gas hydrate is unstable under normal surface conditions, most of the gas hydrate in sediment cores decomposes during recovery, releasing its gas. Unfortunately, knowledge of gas concentrations within recovered core samples and within the continental margin sediment sequences at *in situ* conditions is rather rudimentary because gases are extremely difficult to capture quantitatively.

The most comprehensive data set for gas measurements in marine sediments comes from the efforts of the Deep Sea Drilling Project (DSDP) and the Ocean Drilling Program (ODP). This data set has grown progressively in size and sophistication since DSDP operations started in 1969 and continues to expand with the ongoing efforts of the ODP. Over 1,040 sites have been drilled into the ocean floor by the DSDP and ODP. While gas hydrates have been visually observed in only 19 of the 1,042 sites drilled during the DSDP and ODP (through Leg 170), attempts to collect gas have been made at every site where gassy sediments were encountered. The collective experience has provided an extensive data set that is relevant to the occurrence of methane and other gases necessary for gas hydrate formation.

In this paper we review the development of residual core gas sampling techniques during DSDP and ODP with emphasis on continental margins where gas hydrates have been encountered. Comparisons made between the residual core gas data and the limited pressure core sample measurements help to establish the utility of these data and to indicate our present state of knowledge with respect to gas sampling.

Natural Gas Hydrates: Occurrence, Distribution, and Detection
Geophysical Monograph 124

2. HISTORICAL PERSPECTIVE

2.1. Core Degassing

Observations of gas expansion from freshly recovered cores have been made throughout the DSDP and ODP. Visual observation of bubbles moving along the interior surface of core liners and the occurrence of gas voids within core liners are the most common gas expansion phenomena. The presence of bubbles indicates that gas saturation was exceeded within a core during its ascent to the surface.

While never quantified, there are a number of observations that indicate that gas expansion generates significant pressures within freshly recovered cores. These observations include extrusion of core material out the ends of the liners and from holes drilled into the core liners after recovery specifically to allow gas to vent (Figure 1A). In some cases mud is forced through the check valves on the top of the down-hole coring assembly. At times gas pressure has been high enough to forcibly shoot sediments out the ends of the core liners [e.g., Site 147 – *Shipboard Scientific Party*, 1973; Sites 360 & 361 – *Bolli et al.*, 1978; and Sites 379, 380, and 381 – *Hunt and Whelan*, 1978]. In more recent years, some potentially hazardous core degassing events have occurred. On Leg 160 Site 971E a section of core liner burst, presumably because of gas pressure [*Emeis et al.*, 1996]. On Leg 164 an entire section spontaneously ejected from the core barrel and shot out ~10 meters through the air before impacting a bulkhead [*Paull et al.*, 1996].

When cores are split open for description, many of the initially gas-charged cores exhibit sediment textures that indicate the cores were disturbed during recovery. The most commonly observed features are voids that occur either alone or in clusters (Figure 1B). Usually the parting surfaces are aligned perpendicular to the long axis of the core and suggest that the core expanded as a result of degassing. In many cases it is obvious how intact pieces of sediments on both sides of a void fit back together. However, in other cases the sediments themselves look very disturbed. One common type of disruption is referred to as "mousse-like" texture (Figures 1C and D), and it is believed to be produced by decomposition of gas hydrates and/or gas expansion during core recovery [*Westbrook et al.*, 1994]. Mousse-like zones are frequently flanked by "soft and soupy" sediment that appears to have been recovered in a fluid state (Figure 1D). Cores containing zones with mousse-like textures are typically associated with large gas-voids and low core-recovery.

The possibility that some cases of core expansion are related to gas hydrate decomposition was suggested in the early days of DSDP [e.g., Leg 10 –*Worzel*, 1973; Leg 11 – *Lancelot and Ewing*, 1972; Leg 18 – *McIver*, 1973]. However, techniques to evaluate these assertions have been slow to develop.

2.2. DSDP and ODP Efforts to Avoid Pressurized Hydrocarbon Accumulations

Neither the *R/V Glomar Challenger* (used during DSDP legs 1-96) nor the *R/V JOIDES Resolution* (used by ODP beginning with Leg 101) have been equipped with blowout control systems. Thus, procedures were developed early in the DSDP to avoid encountering significant hydrocarbon accumulations [*Claypool*, 1975]. Before drilling each site, the Pollution Prevention and Safety Panel (PPSP) reviewed the geologic setting to minimize the chances of encountering pressurized hydrocarbon reservoirs. This effort clearly has biased the DSDP and ODP data sets by excluding the drilling of obvious thermogenic-gas-bearing sites.

Sites associated with a bottom simulating reflector (BSR) in pre-drilling site-surveys were also avoided throughout much of the DSDP and ODP because of concern for uncontrolled degassing. Sites with BSRs were intentionally avoided between DSDP Leg 11 and ODP Leg 112. To achieve the scientific objectives of ODP Leg 112 drilling in a region known to contain BSRs was required. Site 688 was located within a break in the regional BSR and the well conditions were monitored especially carefully during the drilling [*Suess et al.*, 1988; *Hovland et al.*, 1999]. Experience gained during Leg 112 provided more confidence that safe drilling could occur in gas-hydrate-bearing regions. However, it was not until a BSR-free site could not be found to achieve the scientific objects of Leg 141 [*Shipboard Scientific Party*, 1992] that the PPSP approved drilling through the base of gas hydrate stability at a site where a BSR was known to occur in the subsurface. At almost the same time, gas hydrates were becoming an active research topic, rather than just a drilling safety problem to avoid.

3. DEVELOPMENT OF GAS SAMPLING TECHNIQUES AND SAFETY MONITORING

3.1. Core Voids

Much of the basic gas sampling technology that is still routinely used today was developed well before the first gas hydrate sample was recovered on DSDP Leg 66. The first gas sample was taken on DSDP Leg 1 [*Erdman et al.*, 1968]. In the DSDP Leg 7 Initial Reports *Gealy and Dubois*, [1971] outlined a technique to sample gases from core voids that remains in use today. Gases in core voids are sampled by puncturing the plastic core liner with a needle that is connected through a valve to an evacuated container (referred to as a vacutainer). When the valve is opened, a sample of the gas in the void is collected and the vacutainer is subsequently sealed. Gealy and Dubois, and other early workers [e.g., *Claypool et al.*, 1973, *Doose et al.*, 1977] took their samples ashore for analysis.

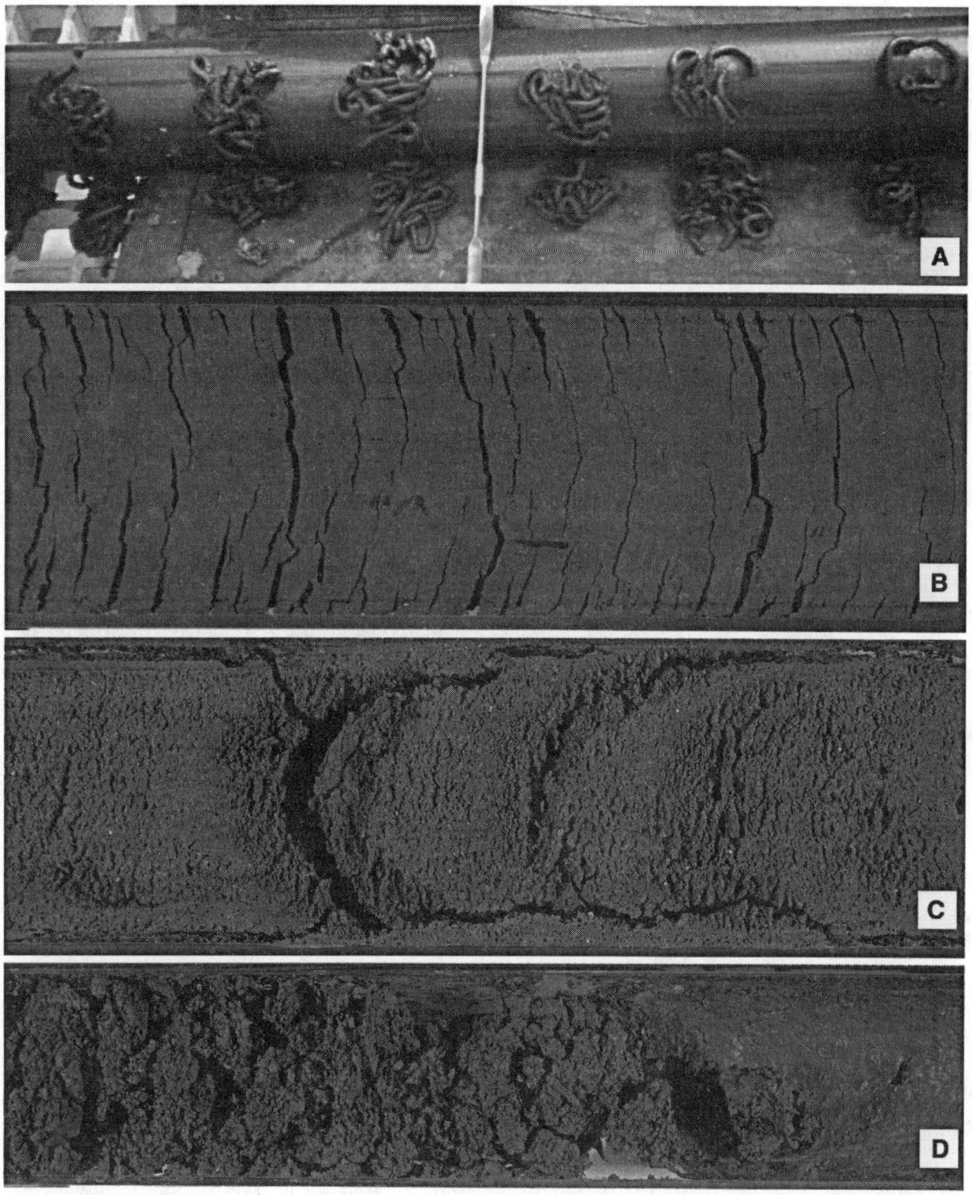

Figure 1. Photographs showing the effects of degassing on ODP cores. Part A illustrates a section of core liner (2.68 cm O.D.) that is covered with small accumulations of sediment that have been extruded through small holes drilled into the liner to release pressure and to stop material from extruding out of the ends of the liner [photograph by Bill Normark; *Flood et al.*, 1995]. These little piles of extruded sediment are called "mud worms [*Flood et al.*, 1995]. Parts B, C, and D show the surface of freshly split ODP cores. In part B, the core material has numerous voids that suggest gas expansion perpendicular to the axis of the core (Leg 146, Site 890B, Core 2H-5, 65-85 cm). Part C shows a larger gas expansion void (Leg 160, Site 971D, Core 2H-2, 55-75 cm) that is surrounded by "mousse-like" sediment texture [*Emeis et al.*, 1996]. "Mousse-like" textures are believed to be associated with gas expansion and perhaps represent voids previously occupied by gas hydrate. Note that the cracks connect with open channels along the side of the core liner. Part D shows an extreme case of mousse-like sediment texture (indicating core disruption) on the left and voids filled with soupy sediment on the right. This section of core (Leg 146, Site 892A, Core 2X-1, 20-50 cm) was interpreted to have contained gas hydrate [*Westbrook et al.*, 1994].

Shipboard measurements of the core gases became possible when a portable gas chromatograph was installed on the *Glomar Challenger* and the first results were reported for DSDP Sites 88 and 89 [*Worzel et al., 1973*]. However, because of the differences in retention times and the common disparity in the relative gas abundance, it was not practical to measure both methane and higher hydrocarbons with this system. Dual gas chromatography systems capable of measuring both methane and C_2 to C_5 alkanes were added to the *Glomar Challenger* on DSDP Leg 47 [*Whelan*, 1979]. Subsequently, regular shipboard gas sampling and measurement became a routine part of the hydrocarbon monitoring effort.

3.2. First In Situ Concentration Estimates

On DSDP Leg 15 special efforts were made to sample core gases and to measure all the gas-sensitive properties on cores from the Cariaco Trench [*Hammond et al.*, 1973]. *Horowitz et al.* [1973] collected gas from 12 gas voids using a device similar to *Gealy and Dubois's* [1971], but with a pressure gauge. Pressure in core gas pockets was as high as 2.21 atmospheres, with a median value of 1.33 atmospheres. They attributed the variation in pressure to differences in the maximum confining pressure of the soft sediment plugs, rather than indications of the *in situ* gas content of the sediment. Visual estimates of the gas void volume with respect to sediment volume and porosity measurements allowed the amount of gas to be compared to the pore water volume for the first time. Using these observations [*Hammond et al.*, 1973], the *in situ* methane concentrations in the pore water were at least 32 mM, suggesting that some of the pore waters would have become supersaturated and a significant proportion of the initial gas would have been lost during core recovery.

3.3. Headspace Techniques

Attempts to collect better gas samples from cores by correcting for the sediment volumes from which the gas evolved have been made intermittently throughout the DSDP and ODP. These techniques depend on sealing sediment subsamples into gas tight containers of known volume as soon as possible after recovery so that the residual gases that continue to evolve from the sample are captured. These techniques are called "head space" or "gas can" techniques, and use containers fitted with a rubber septum or some other device that allows the residual gases to be sub-sampled for analysis.

Gas can sampling began on DSDP Leg 18 [*McIver*, 1973] and was used sporadically throughout the DSDP and ODP [*Kvenvolden et al.*, 1990]. Gas can techniques typically involve placing a whole round section of core, ~5 cm long and ~170 cm^3 in volume, into a metal can shaped like a "paint can" with a septum in the lid. These samples were preserved within the cans for shore-based analysis. Shortly before analysis, the cans were shaken to release the remaining gas into the headspace within the gas can.

On ODP Leg 107 another headspace sampling technique was introduced [*Shipboard Scientific Party*, 1987] which has became a standard ODP gas sampling method. This headspace technique uses a #4 cork borer to collect a defined volume of sediment from the exposed end of a freshly cut core section, before the section is capped. The ~10 cm^3 samples are placed in 20 cm^3 glass vials, sealed and heated to 70° C to evolve absorbed gas. These standardized ODP headspace measurements have the advantage that they use less sediment, provide gas measurements in the absence of core voids, and are corrected for sediment volume.

3.4. Gas Collection Chambers

Gas collection chambers (GCC) were used to collect the gases that evolved from core sections as the cores warmed inside the shipboard laboratory during ODP Leg 164. Cores were cut into standard 1.5-meter sections and placed in 1.54-meter tubes that were slightly larger in diameter than the core liners. The chambers were constructed from standard PVC pipe with o-ring-fitted end caps that could be quickly attached to make a gas tight seal. Evolved gas flowed out through a port in the sealed end caps of the GCC into tygon tubing. The tubing passed into inverted 1-liter graduated cylinders that floated in baths filled with NaCl-saturated water. The cylinders filled with gas as the gas exited the tubing. Because the inverted graduated cylinders were initially water-filled, the volume of water displaced by the gas could be measured directly [*Paull et al.*, 1996].

3.5 Comparisons Among Core Gas Measurement Techniques

DSDP Leg 76 was the first leg on which a careful comparison between the core void and headspace samples was made. Both techniques provide profiles with similar shapes. While the average methane to ethane value of the headspace samples was slightly less than those in the core voids, both techniques produced similar ranges of values and depth profiles [*Gradstein and Sheridan*, 1983].

Kvenvolden et al. [1990] compared the headspace and gas can samples collected on Legs 76, 84 and 112. While there was considerable scatter in these data, both techniques produced generally comparable trends. A more detailed comparison of these techniques was subsequently conducted with the gases collected on ODP Leg 151 [*Stein et al.*, 1995]. Stein et al. showed that the vacutainer gas sample concentrations are generally higher than those from the headspace samples, although both techniques produce similar depth trends.

On ODP Leg 164, gas volumes measured with the gas collection chambers were compared to headspace data. While the gas collection chambers indicated slightly higher residual gas concentrations, both the techniques produced essentially similar stratigraphic profiles [*Paull et al.*, 1996].

3.6. Development of Pressure Core Samplers

Efforts to collect samples at *in situ* pressures using wire line coring tools were initiated early in the DSDP. The objective was to avoid the pressure-related gas loss associated with the core recovery process. Such tools are especially relevant to gas hydrate studies.

During DSDP a wire line pressure-coring tool called the Pressure Core Barrel (PCB) was developed. The PCB was designed to seal and recover a core at pressure in a chamber 5.8-meter long and 5.8-centimeter in diameter. The PCB was first tested on Leg 18 [*Creager et al.*, 1973]. Further, experimental deployments on Legs 42B, 44, 62, 64, 66, 72, and 74 were largely unsuccessful, sometimes the tool maintained down-hole pressures, at other times it recovered core, but rarely did it do both simultaneously.

The first significant scientific success of the PCB tool occurred on Leg 76 when five pressurized cores were recovered from the Blake Ridge. These cores were degassed under controlled conditions so that the pressures changes could be monitored and the gases sub-sampled throughout this process [*Kvenvolden et al.*, 1983]. Pressure within the core barrel dropped as a valve was slightly cracked to allow some gas to vent (Figure 2). However, the pressure in the core barrel would rise over the next few minutes toward the previous pressure. This pattern of pressure rise was repeated several times before ambient pressures were sustained. Unfortunately, the actual volume of gas released by the sample was not measured.

The PCB was also deployed three times at Site 568 on Leg 84. Two samples contained sediment at *in situ* pressure. The volume of gas released from one of these samples was measured to be 7.4 L from a 1.61-m long core. This was the first successful measurement of *in situ* gas volume and indicated that the volume of gas relative to volume of pore water in the sample was 0.7 at *in situ* conditions for a sample that came from between 185.7 and 195.4 mbsf [*Kvenvolden and McDonald*, 1985]. This was the only measurement of gas volume that was made using the PCB.

The DSDP PCB tool became obsolete when the ODP began because it was not compatible with the newer bottom hole assemblies. The PCB design was also limited to ~ 35 MPa sample pressures, which might severely hinder its ability to retrieve methane gas hydrate samples. There was no way to access the pressurized core, and the tool was so large that controlling the temperature after recovery was

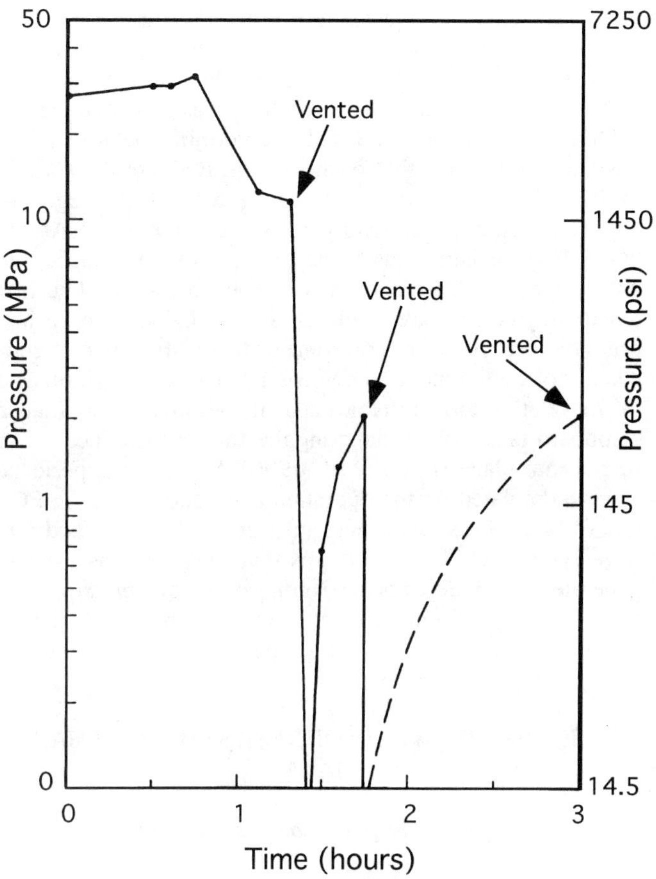

Figure 2. Plot showing the pressure change associated with stepwise degassing of a pressurized core sample while the PCB was in a bath at 25.7°C (PCB-1 from DSDP Leg 76 Site 533; from [*Kvenvolden et al.*, 1983]. The pressure drops are associated with opening a valve. Note that pressure increases occurred after each degassing event. This stepwise degassing experiment closely mimics what probably occurs within most gas-rich cores during recovery.

difficult. Thus, in 1986 the ODP decided to build a fundamentally new tool rather than adapt the existing PCB tool. A new free-fall deployed, hydraulically-actuated, wireline-retrievable tool, called the pressure core sampler (PCS) was developed to continue the effort for recovering cores at *in situ* pressure. The PCS was tested three times on Leg 124E. On the third test it was credited to have had 100% core recovery at 100% hydrostatic pressure. On this basis it was declared to be an operational tool. However, subsequent deployments (2 on Leg 131; 5 on Leg 139; 11 on Leg 141; and 10 on Leg 146) were largely unsuccessful [*Pettigrew*, 1992]. Explanations for the failures were usually attributed to deployment-specific problems. Other competing commitments for the shipboard personal and rig time prevented adequate tests on any individual leg. Shipboard staff changed between legs and sediment characteristics differed between sites, thus an understanding of how to operate the tool successfully was slow to develop.

ODP Leg 164 was dedicated to determine the amounts of gas and gas hydrate that occur *in situ*. Because the success of the PCS was critical for the success of the leg, two engineers sailed specifically to service the tool. At the outset, the decision was made that the tool would be run until it was either functioning or declared flawed. The first ten deployments were made at Site 994. While core and pressure were recovered on some of these first deployments, basic tool deployment techniques needed to be established to make the tool reliable and it became clear that a shipboard protocol for handling the samples also needed to be put into place [*Paull et al.*, 1996]. Sufficient experience was accumulated on the operation and handling of the PCS tool at Site 994, so that most attempts at Sites 995 and 997 were successful. The result was that the PCS was used to generate an independent stratigraphy of *in situ* gas concentrations (Figure 3) and compositions at 21 depths between 73.0 to 589.20 mbsf at Sites 995 and 997 [*Dickens et al.*, 1997].

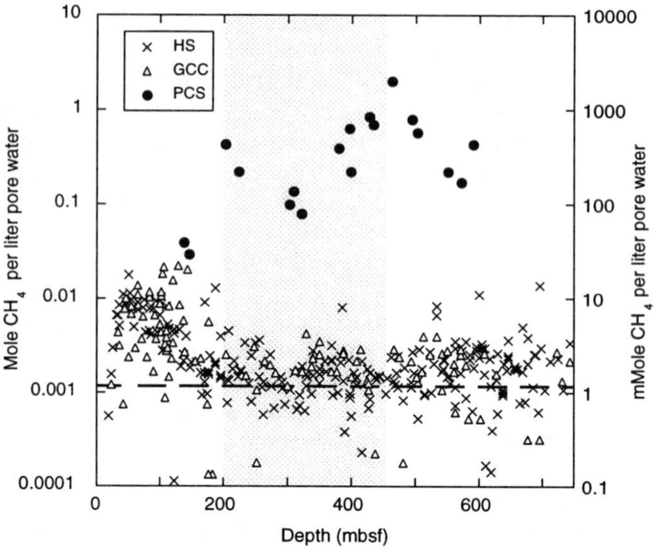

Figure 3. Plot showing methane concentration data versus depth in both mole CH_4 and mM CH_4 per liter of pore water from ODP Sites 994, 995, and 997 measured with headspace (HS), gas collection chamber (GCC), and pressure core sampler (PCS) techniques. Dashed line indicates CH_4 saturation under surface laboratory conditions (1 atm., 25° C, and 35 ‰ salinity; [*Yamamoto et al.*, 1976]. Methane is the sedimentary gas most likely to be saturated under surface conditions. Shipboard core gas sampling techniques (HS and GCC) grossly underestimate *in situ* gas concentrations as indicated by PCS data. The main gas-hydrate-bearing zone is shaded.

4. INTERPRETATION OF RESIDUAL CORE GAS DATA

4.1 Gas Void and Headspace Composition Data

Gas chromatographic analysis of the gases in the core voids provides information on the types and relative percentages of the gases that are present. Many samples contained significant proportions of oxygen and nitrogen, which are believed to indicate atmospheric contamination. Methane is typically the dominant gas in relatively uncontaminated sediment gas samples from below the base of sulfate reduction, with methane usually comprising 80-99.9% of the gas volume. Generally, carbon dioxide is the secondary core gas, typically representing less than a few percent by volume. With increasing depth, samples commonly show detectable levels of ethane and some of the higher hydrocarbons (propane, butane, and pentane). While gas compositions from core voids show a great deal of scatter, the relative proportions of the original constituents remain nearly constant after the effects of contamination with air are removed. Thus, the ratio of methane to ethane (or higher hydrocarbons) is the property that has been monitored most closely, especially from a drilling safety standpoint.

The DSDP and ODP data sets include depth profiles of methane to ethane ratios of core gases at numerous sites. Although there is considerable scatter in these data, a characteristic pattern emerges. Methane to ethane ratios typically decrease by as much as two orders of magnitude down-core (Figure 4). The prevailing paradigm is that shallow microbially-generated hydrocarbon gases are predominately composed of methane while thermally-generated gases contain significant quantities of ethane and higher hydrocarbons [*Bernard et al.*, 1978]. Thus, decreasing methane to ethane ratios reflect an increase in relative amounts of thermogenic gases (methane and higher hydrocarbons). The thermogentic gases may be produced *in situ* or may have migrated upward from greater depth and mixed with microbial gases produced at shallow depths.

The DSDP and ODP PPSP have drilling safety criteria that take into account local conditions, such as sediment temperature, organic carbon content, and stratigraphic and structural settings [*Claypool*, 1992]. However, shipboard operation personnel tend to rely on methane to ethane ratios as a simple practical drilling safety guideline. When the ratio drops below 1000, holes may be abandoned if there is much of a potential for encountering over-pressured thermogenic hydrocarbons. If the geologic conditions made the chances of encountering a hydrocarbon reservoir very unlikely, hole penetration has been allowed to continue on a core by core basis until the methane to ethane ratio falls below 100.

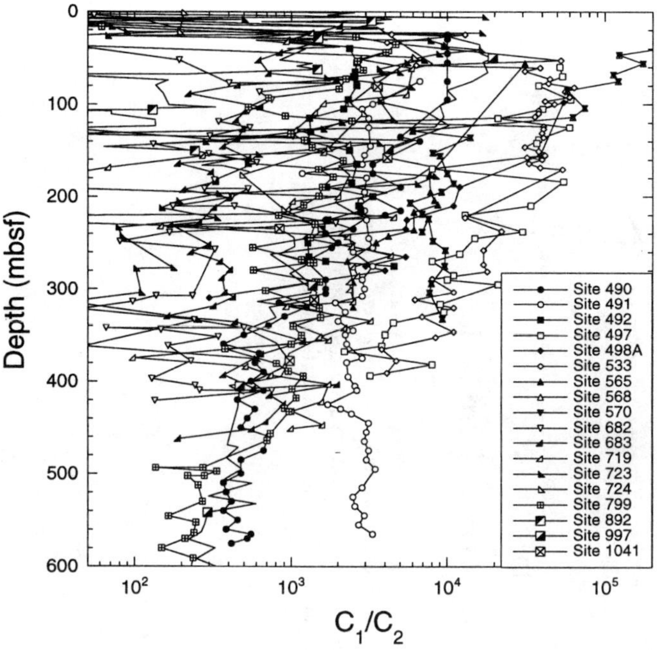

Figure 4. Plot showing methane to ethane ratios in gas samples from core voids versus depth at DSDP/ODP sites where gas hydrates were observed in cores (DSDP 490, 491, 492, 497, 498A, 533, 565, 568, 570) and from representative gas-rich sites (ODP 682, 683, 719, 723, 724, 799, 892, 997, 1041). The plot illustrates the typical wide range of scatter among cores but the methane to ethane ratios consistently trend toward lower values with depth. All these sites were intentionally located away from areas where thermogenic hydrocarbons were anticipated.

Both gas void and headspace gas measurements have a characteristic stratigraphic pattern (Figure 5). Typically the methane concentrations are very low near the top of the cores. There are numerous observations that show that the onset of an increase in methane is coincident with the level of sulfate depletion. These data are consistent with microbial methane production being inhibited by the presence of sulfate within the sulfate reduction zone [*Claypool and Kaplan,* 1974; *Martens and Berner*, 1974].

Below the sulfate reduction zone, headspace gas concentrations typically rise quickly to a subsurface maximum, and then commonly show significant scatter around the same concentration or decrease slowly with increasing depth (Figures 3 and 5A). Patterns seen in core gas data commonly vary with lithologic changes and have influenced interpretations about *in situ* gas concentrations in many of the DSDP and ODP Initial Report volumes. For example, there are numerous descriptions of the profile shapes with depth and the levels at which higher concentrations were measured which imply that the data were reflecting *in situ* conditions and sometimes these subsurface peaks have been referred to as the zone of maximum methane generation.

4.2. Gas Phase Changes During Sample Recovery

Determining the original *in situ* gas concentrations and the gas distributions between phases in bore hole samples is an inherently difficult task because extreme changes in the pressure and temperature conditions occur during core recovery. Under *in situ* conditions, methane (and small amounts of the other gases) may occur dissolved in pore water, as free gas bubbles, and/or trapped in gas hydrates. A pressure decrease will cause a considerable shift in the partitioning between the dissolved and gaseous phases (Figure 6). Under surface conditions (1 atm and 25 °C) methane and carbon dioxide saturation in pore waters of normal salinity (35 ‰) is 1.14 mM [*Yamamoto et al.*, 1976] and 29.17 mM [*Weiss*, 1974], respectively. Saturation concentration increases significantly with pressure [*Duan et al.*, 1992]. For example, at the hydrostatic pressures associated with 3 km water depths, approximately 190 mM of methane are required to saturate the pore waters at 20 °C [*Duan et al.*, 1992; Figure 6]. Thus, there are more than two orders of magnitude change in the methane saturation concentration between the surface and the seafloor.

The predominance of methane in the residual core gas samples indicates that methane saturation commonly occurs in ODP cores during recovery. Because carbon dioxide concentrations rarely approach saturation under surface conditions, gas expansion is primarily driven by methane evolution, although other gases will be partially stripped from the pore waters with the degassing methane. In gas-rich sediment cores, most of the methane that was originally dissolved in pore waters or stored in gas hydrate will be transferred into a free gas phase and escape from the core during depressurization that occurs during core recovery. Thus, how measurements of the residual core gas concentrations relate to the original *in situ* gas concentration is unclear.

4.3 Headspace Data Compared with Methane Solubility

Headspace gas concentration data can be compared with the methane saturation concentration of the pore waters in cores under surface conditions, when headspace gas measurements are corrected to reflect sediment porosity [*Paull et al.*, 2000]. The characteristic subsurface maximum in headspace data occurs at a level that is modestly in excess of saturation under surface conditions. With increasing depth headspace gas concentrations either remain approximately at the same value or decrease slightly (Figure 5B). This suggests that the peak values that can be achieved in the headspace data are limited by methane solubility at the surface and have little relationship with *in situ* methane concentrations.

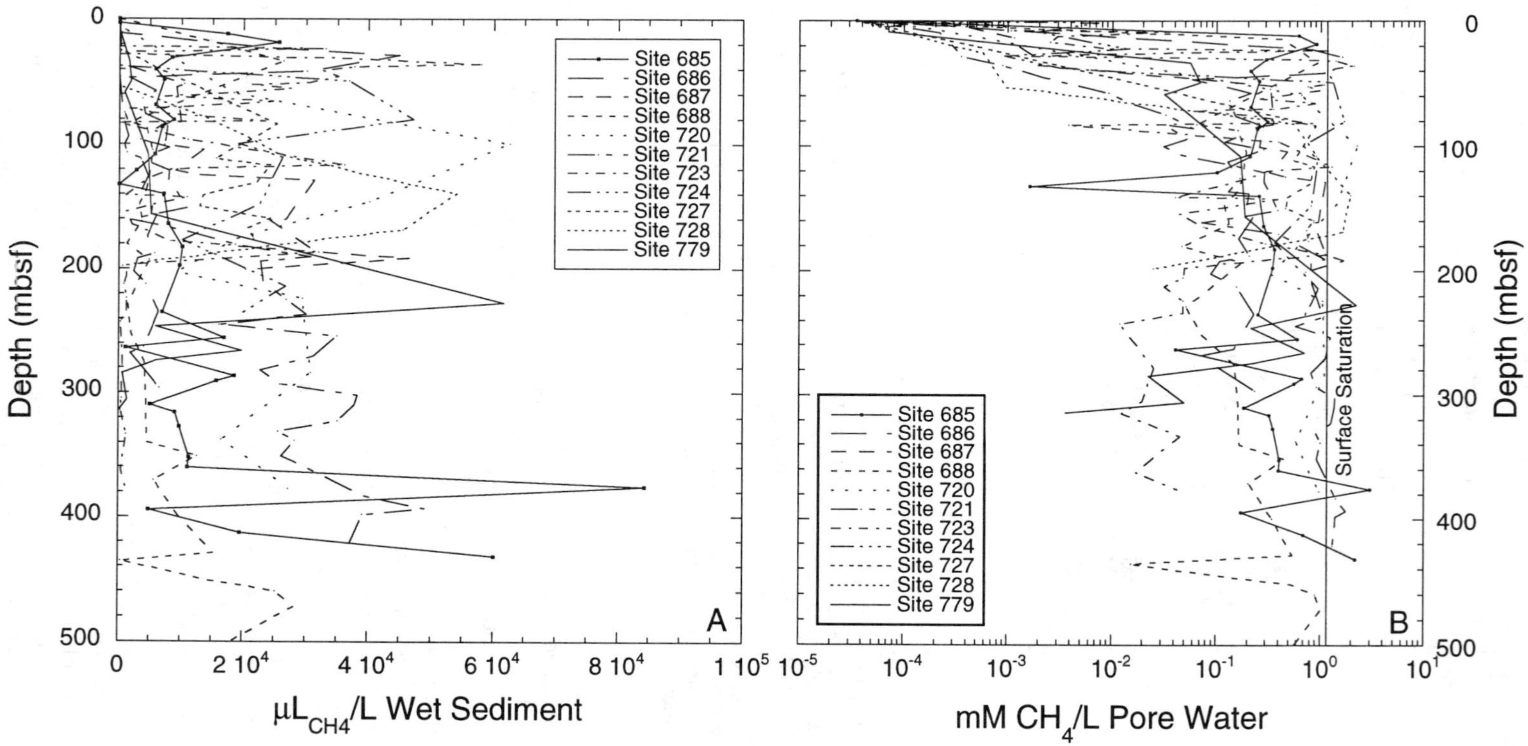

Figure 5. Plots of headspace gas concentrations from selected ODP sites illustrating the initial rise in measured values followed by a plateau of similar or even decreasing values with increasing depth. In part A, methane is plotted as μL_{CH4}/L wet sediment versus depth as reported in Initial Reports Volumes. In part B the methane concentrations are shown on a log scale with units of μL_{CH4}/L of pore water. These values were calculated assuming 50% porosity for all samples. Methane saturation under surface conditions is approximately $1 mM_{CH4}$/L of pore water.

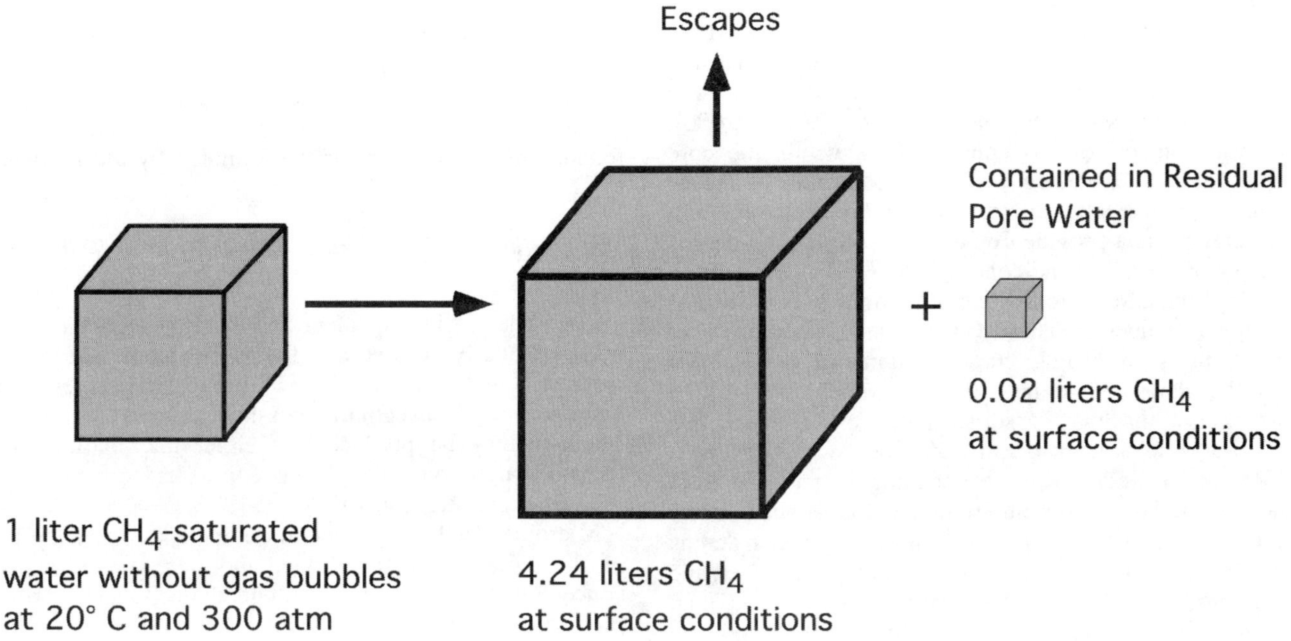

Figure 6. Cartoon showing the volume of methane that will result when a liter of methane saturated water is transferred from 3-km depth to surface conditions. For comparison, at the Earth's surface conditions ~0.02 L of methane will saturate the water.

5. INTERPRETATION OF PRESSURE CORE SAMPLE DATA

5.1. Gas Exsolution and Pressure Regeneration

The successful deployments of the PCB on DSDP Leg 76 [*Kvenvolden et al.*, 1983] and the PCS on ODP Leg 164 [*Dickens et al.*, 1997; *Dickens et al.*, 2000] both show that internal pressures within the enclosed barrel will rise after the chamber is partially vented and resealed (Figure 2). This pattern was initially interpreted as an indication of gas hydrate in the PCB sample, because pressure regeneration after each venting step would occur when gas hydrate decomposed, releasing gas and increasing the pressure in the core barrel as a new equilibrium was approached. However, subsequent observations of the pressure-time responses of pressure core samples have suggested that pressure regeneration continues to occur at higher temperatures and lower pressures than those associated with gas hydrate stability. Thus, pressure increases may also be generated by gas exsolution from super-saturated pore water [*Dickens et al.*, 2000].

The pressure increases that have been observed within the PCB and PCS tools during controlled degassing events provide some insight into the pressure generating potential associated with gas exsolution during the recovery of any gas-rich sediment core. For example, the pressure versus time plot illustrated in Figure 2 indicates that over 2 MPa of headspace pressure is regenerated twice during the controlled degassing experiment. These pressures can easily drive the observed vigorous degassing of normal sediment cores.

5.2 Comparison of Core Gas Measurements and Pressure Core Sample Measurements

Because successfully collected pressure core samples do not degas during core recovery, all the original gas can be recovered during depressurization under controlled shipboard conditions. Unfortunately, the PCB used during DSDP produced only one sample (Site 568 on Leg 84) with which headspace gas measurement can be compared [*Kvenvolden and McDonald*, 1985]. Converting this data into the units used for headspace gases indicates that this sample evolved 10 L CH_4 per L wet sediment, about 1,000 times the amount of gas that is commonly seen in headspace samples.

The availability of a larger data set provided by PCS measurement on Leg 164 allows the gas concentrations and composition data produced by routine shipboard gas sampling techniques to be evaluated [*Paull et al.*, 2000]. The measurements of gas concentrations within the core samples during ODP Leg 164 using conventional means produced dramatically different results from the *in situ* concentration data provided by the PCS. Residual core gas measurements of methane concentrations indicate an initial downward increase from low concentrations toward surface saturation values. These gradients suggest that concentrations exceeding surface saturation (~1.1 mM) would be achieved within ~30 mbsf, which is about 10 meters below the base of sulfate reduction. Thus, in this case core samples from below ~30 mbsf evolved gaseous methane during core recovery resulting in methane loss. Headspace methane concentrations peak at about 8 times surface saturation between 80-140 mbsf. At greater depths (>200 mbsf) the mean residual core gas methane concentration data decrease to values that are less than twice surface saturation. In contrast, the PCS data indicate increasing methane concentration with depth and the average methane amount below 450 mbsf was more than 400 times surface saturation values (Figure 3). More than 99% of the original methane in these samples has escaped before shipboard core sampling. Thus, headspace gas profiles do not reflect *in situ* gas abundance.

Peaks in methane concentration observed in core gas data (Figure 5) clearly have little to do with *in situ* gas volumes [*Kvenvolden and Lorenson*, 2000; *Paull et al.*, 2000]. During core recovery the amount of interstitial free gas increases until sediment failure occurs as a consequence of the internal pressure generated by evolving and expanding interstitial gas bubbles. Stratigraphic variations in the head space measurements probably reflect differences in sediment strength which may vary with lithology. The only proper use of headspace methane concentration data is to establish the rate of methane accumulation between the base of sulfate depletion and the depth at which surface saturation concentrations are achieved. However, this measurement has to be done soon after coring because methane, even in under-saturated cores, escapes quickly.

The existing PCS data allow comparisons to be made with the gas compositions measured with other traditional techniques. The methane to carbon dioxide ratios are considerably higher in the PCS samples than in all the other core gas measurements which experience depressurization prior to sampling (Figure 7). This discrepancy may result from both preferential methane loss and carbonate precipitation during depressurization.

The methane to ethane values of the gas void samples are similar to the methane to ethane values of the PCS samples (Figure 7). However, the methane to ethane values of the headspace samples are less than those of the PCS samples and gas void samples from a similar depth. The observed discrepancies are consistent with the preferential escape of methane, which will decrease the methane to ethane and methane to carbon dioxide values in the residual gas that remains in the sediment and is sampled by the headspace technique.

5.3. Origin and Geochemical Finger Print of Sedimentary Methane

Gas collected in vacutainer and headspace containers has been regularly preserved for subsequent shore-based isotopic analysis, which is particularly useful for distinguishing between the possible sources of methane. Methane may be produced by either microbiological or thermocatalytic processes from sedimentary organic matter [*Claypool and Kaplan*, 1974].

Microbial methane production commences near the base of the sulfate reduction zone, and pore water methane concentrations will increase with depth due to local methane production [*Claypool and Kaplan et al.*, 1974; Whelan et al., 1984]. Microbial methane production utilizes either carbon dioxide or acetate as the carbon source and is associated with significant isotopic fractionation. However, ethane is not a bi-product of methanogenesis, and only trace amounts of ethane are produced via fermentation reactions, which may be occurring within the same zones where microbial methane production occurs.

Thermogenic methane and other low molecular weight hydrocarbon gases are produced as a consequence of thermocatalytic degradation of kerogen. This process is optimized at temperatures near ~120°C [*Hunt,* 1996], and with normal geothermal gradients, requires burial depths significantly greater than 1 km. However, small amounts are produced at much shallower depths. Because little isotopic fractionation occurs during thermal cracking of kerogen [*Schoell*, 1980; 1983] the carbon isotopic composition of methane produced from thermal cracking of kerogen generally reflects its parent sedimentary organic matter carbon isotopic composition. Thus, if methane and ethane are transported upward from greater depths either by diffusion or advection or produced locally, thermogenic methane and ethane may dilute the microbially-produced methane.

The current paradigm for distinguishing between microbial and thermogenic sources of methane is one that uses the relative amounts of low molecular weight hydrocarbon gases and $\delta^{13}C_{CH4}$ values [*Bernard et al.*, 1978]. Microbial methane typically has $\delta^{13}C_{CH4}$ values that range from -90 to -55 °/oo (PDB) and methane to ethane ratios of >1,000, while thermogenic methane typically has $\delta^{13}C_{CH4}$ values more positive than -55 °/oo and methane to ethane ratios <100 [*Bernard et al.*, 1978]. *Bernard et al.*

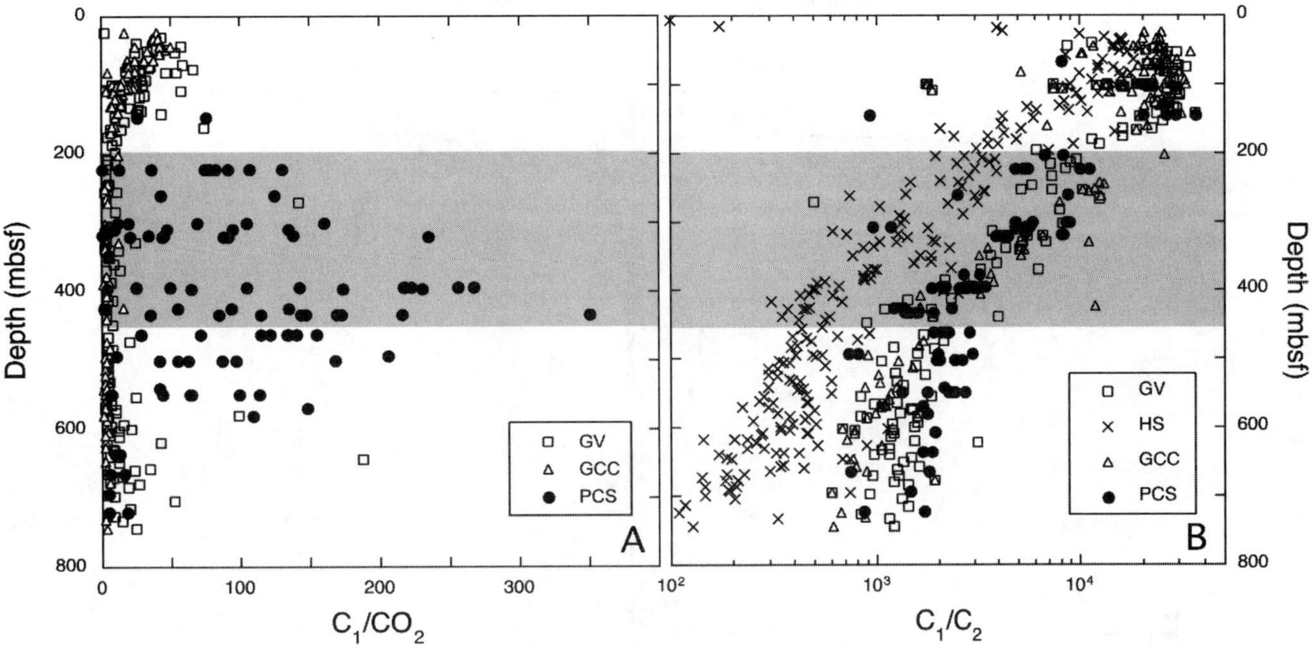

Figure 7. Plot comparing values of methane volume to carbon dioxide volume ratios (C_1 / CO_2) measured on vacutainer sampled gas void (GV), gas collection chamber (GCC), and pressure core sampler (PCS) samples from ODP Sites 994, 995, and 997 is shown in part A. Plot comparing values of methane to ethane volume ratios (C_1/ C_2) measured GV, headspace (HS), GCC, and PCS samples from ODP Sites 994, 995, and 997 is shown in part B. The main gas-hydrate-bearing zone is shaded.

[1978] established these relationships using 5-m long piston core samples from the Gulf of Mexico.

Shore-based isotopic analyses indicate that gases of both microbial and thermogenic origins have been sampled by DSDP and ODP [e.g., *Kvenvolden et al.*, 1993; *Ginsburg and Soloviev*, 1998]. Two isotopic patterns have been commonly observed where detailed gas sampling and isotopic measurements have been made. Methane samples from some sites have $\delta^{13}C_{CH4}$ values more negative than -65 °/oo in the shallowest gas sample, indicating a microbial origin but progressively shift to values more positive than -40 °/oo in the deepest samples, suggesting a thermogenic origin. However, other sites (Figure 8) contain methane that never gets more positive than -60 °/oo, suggesting that this gas is primarily microbial in origin. Methane to ethane ratios of core gases increase downward in both types of samples regardless of whether the isotopic signature suggests a thermogenic or microbial gas source.

5.4 Comparison of Isotopic Compositions of Methane from Sediment Cores and PCS Samples

Wallace et al. [2000] compared the carbon isotopic values from methane samples that were obtained from the PCS during ODP Leg 164 with the isotopic values of methane samples that were collected from adjacent sediment cores [*Paull et al.*, 2000]. Although individual methane samples collected from the progressive degassing of the PCS cores showed considerably more scatter in the carbon isotopic values than the core gas samples, the mean values of these samples were essentially the same as that produced by nearby headspace or free gas samples. Although the relative proportions of methane to other gases sampled with headspace techniques are altered by degassing during core recovery, the isotopic composition of headspace samples does not appear to be fractionated and may accurately reflect the isotopic composition of the *in situ* methane.

6. SUMMARY AND CONCLUSIONS

The data legacy left by the DSDP and ODP to date provides numerous observations of vigorous degassing from freshly recovered cores which requires gross super-saturation under Earth surface conditions. Core sampling techniques currently in use (e.g., core voids, headspace, gas can, and gas collection chambers) are capable of indicating presence or absence of dissolved gases and methane is the major residual gas. However, none of these routine core gas-sampling techniques provides accurate data on the concentration profiles with increasing depth. The common stratigraphic pattern of a shallow subsurface maximum seen

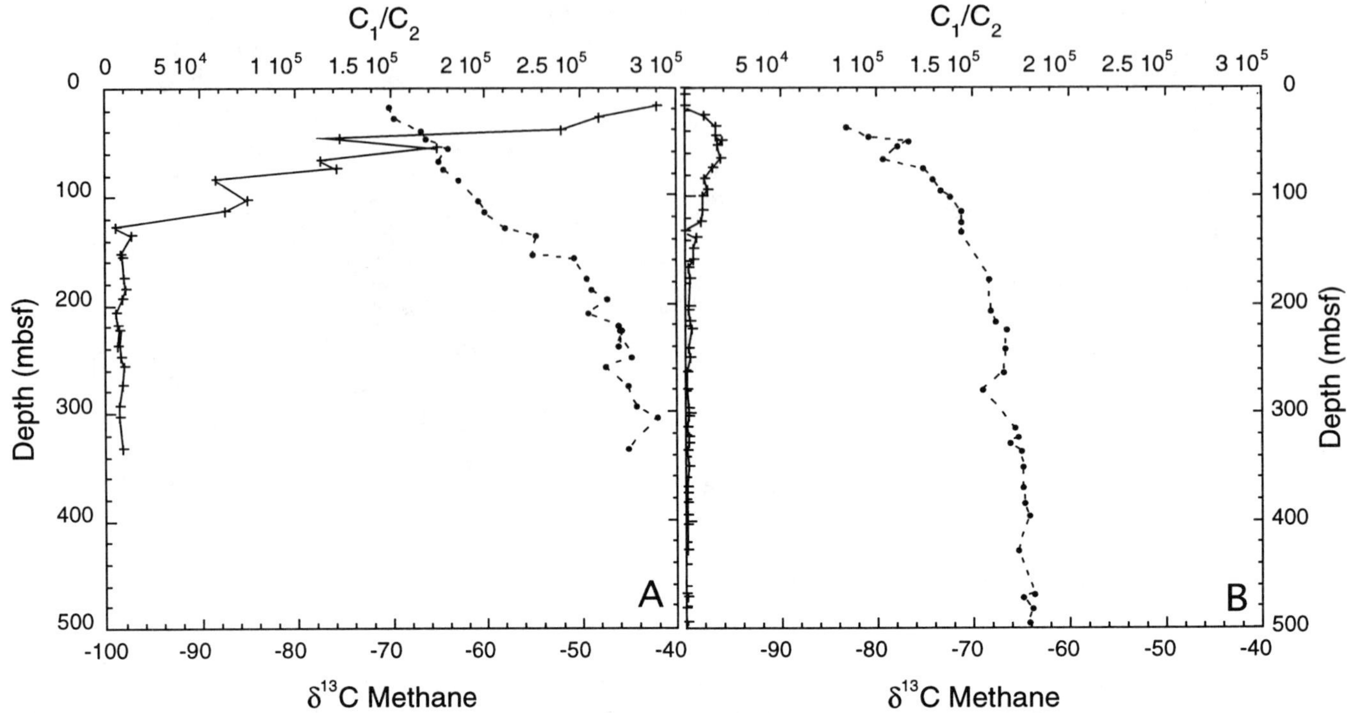

Figure 8. Plots of both the isotopic compositions of methane and the methane to ethane ratios as a function of depth from two core sites. Part A shows data from DSDP Site 568 [*Claypool et al.*, 1985], which indicate a depth trend consistent with addition of thermogenic gas with depth. Part B shows data from DSDP Site 533, which does not suggest increasing amounts of thermogenic methane with depth. However, the methane to ethane ratios at both sites trend toward to lower values. While trends toward decreasing methane to ethane values are seen at essentially all gas-rich sites that have been sampled by DSDP and ODP (Figure 4), only some of the isotopic data suggest the addition of thermogenic gas.

in headspace gas samples appears to be only an artifact of the core degassing process. Thus, except for the few successful deployments of pressure core sampling tools, there are no reliable data on the gas concentrations that occur in gas-hydrate-bearing sediments. This information gap is a major impediment toward developing our understanding of natural gas hydrate.

Acknowledgements. The David and Lucile Packard Foundation supported this work. We thank Drs. George Claypool and Jean Whelan for their comments.

REFERENCES

Bernard, B.B., J.M. Brooks, and W.M. Sackett, Light hydrocarbons in recent Texas continental shelf and slope sediments, *J. Geophy. Res.*, *83*, 4053-4061, 1978.

Bolli, H.M., W.B.F. Ryan, J.B. Foresman, W.E. Hottman, H. Kagami, J.F. Longoria, B.K. McKnight, M. Melguen, J.H. Natland, F. Proto--Decima, and W.G. Siesser, Angola continental margin; Sites 364 and 365, *Initial Rep. Deep Sea Drill. Proj.*, *40*, 357-455, 1978.

Claypool, G.E., Manual on Pollution-Prevention and Safety, 3rd Edition, *JOIDES J.*, C1-C11, 1975.

Claypool, G.E., Ocean Drilling Program Guidelines for Pollution Prevention and Safety, *JOIDES J.*, 1-23, 1992.

Claypool, G.E., and I.R. Kaplan, The origin and distribution of methane in marine sediments, in *Natural Gases in Marine Sediments*, edited by I.R. Kaplan, pp. 99-139, Plenum, New York, 1974.

Claypool, G.E., B.E. Presley, and I.R. Kaplan, Gas analysis in sediments from Legs 10, 11, 134, 14, 15, 18 and 19, *Initial Rep. Deep Sea Drill. Proj.*, *19*, 879-884, 1973.

Claypool, G.E., C.N. Threlkeld, P.N. Mankiewicz, M.A. Arthur, and T.F. Anderson, Isotopic composition of interstitial fluids and origins of methane in slope sediments of the Middle America Trench, Deep Sea Drilling Project Leg 84, *Initial Rep. Deep Sea Drill. Proj.*, *84*, 683-691, 1985.

Creager, J.S., D.W. Scholl, and Scientific Shipboard Party, Site 185, *Initial Rep. Deep Sea Drill. Proj.*, 169-216, 1973.

Dickens, G.R., W.S. Borowski, H. Wehner, and C.K. Paull, Additional shipboard information for the pressure core sampler (PCS) at sites 994, 995, 996 and 997, *Proc. Ocean Drill. Program Sci. Results*, *164*, 113-128, 2000.

Dickens, G.R., C.K. Paull, P.J. Wallace, and the ODP Leg 164 Scientific Party, Direct measurement of in situ methane quantities in a large gas-hydrate reservoir, *Nature*, *385*, 426-428, 1997.

Doose, P.R., M.W. Sandstrom, R.Z. Jodele, and I.R. Kaplan, Intersitial gas analysis of sediment samples from Site 368 and

Hole 369A, *Initial Rep. Deep Sea Drill. Proj.*, 41, 861-863, 1977.

Duan, Z., N. Moller, J. Greenberg, and J.H. Weare, The prediction of methane solubility in natural waters to high ionic strength from 0 to 250C and from 0 to 1600 bar, *Geochim. Cosmochim. Acta, 56*, 1451-1460, 1992.

Shipboard Scientific Party, Site 185, *Initial Rep. Deep Sea Drill. Proj., 19*, 169-216, 1973.

Emeis, K.C., A.H.F. Robertson, C. Richter, et al., *Proc. Ocean Drill. Program Initial Rep.*, 972 pp., 1996.

Erdman, J., R.L. Bosrt, W.J., Hines, and R.S. Scalan, Composition of gas sample 1 (core5) by component, *Initial Rep. Deep Sea Drill. Proj., 1*, 461-467, 1968.

Flood, R.D., D.J.W. Piper, A. Klaus, et al. , *Proc. Ocean Drill. Program Initial Rep., 155*, 55, 1995.

Gealy, E.L., and R. Dubois, Shipboard geochemical analysis, Leg 7, Glomar Challenger, *Initial Rep. Deep Sea Drill. Proj., 7*, 863-869, 1971.

Ginsburg, G.D., and V.A. Soloviev, *Submarine Gas Hydrates*, 215 pp., VNIIOkeanogeologia, St. Petersburg, 1998.

Gradstein, F.M., and R.E. Sheridan, Introduction, Deep Sea Drilling Project Site 533, Leg 76, *Initial Rep. Deep Sea Drill. Proj., 76*, 5-18, 1983.

Hammond, D.E., R.M. Horowitz, W.S. Broeker, and R. Bopp, Intersitial water studies, Leg 15 - Dissolved gases at Site 147, *Initial Rep. Deep Sea Drill. Proj., 20,* 765-771, 1973.

Horowitz, R.M., L.S. Waterman, and W.S. Broecker, Interstitial water studies, Leg 15, - New procedures and equipment, *Initial Rep. Deep Sea Drill. Proj., 20,* 757-763, 1973.

Hovland, M., T.J.G. Francis, G.E. Claypool, and M.M. Ball, Strategy for scientific drilling of marine gas hydrates, *JOIDES J., 25*, 20-24, 1999.

Hunt, J.M., *Petroleum Geochemistry and Geology*, 743 pp. New York, W.H. Freeman, 1996.

Hunt, J.M., and J.K. Whelan, Dissolved gases in Black Sea sediments, *Initial Rep. Deep Sea Drill. Proj., 42,* 661-665, 1978.

Kvenvolden, K., L.A. Barnard, and D.H. Cameron, Pressure core barrel; Application to the study of gas hydrates, Deep Sea Drilling Project Site 533, Leg 76, *Initial Rep. Deep Sea Drill. Proj., 76*, 367-375, 1983.

Kvenvolden, K., and T. Lorenson, Methane and other hydrocarbon gases in sediments from the southeastern North American Continental Margin, *Proc. Ocean Drill. Program Sci. Results 164*, 29-36, 2000.

Kvenvolden, K., and T.J. McDonald, Gas hydrates of the Middle America Trench, *Initial Rep. Deep Sea Drill. Proj., 84,* 667-682, 1985.

Kvenvolden, K.A., T.J. Frank, and M. Golan-Bac, Hydrocarbon gases in Tertiary and Quaternary Sediments offshore Peru - Results and Comparisons, *Proc. Ocean Drill. Program Sci. Results, 112,* 667-682, 1990.

Kvenvolden, K.A., G.D. Ginsburg, and V.A. Soloview, Worldwide distribution of subaquatic gas hydrates, *Geo-Marine Letters, 13*, 32-40, 1993.

Lancelot, Y., and J.R. Ewing, Correlation of natural gas zonation and carbonate diagenesis in Tertiary sediments for the North-West Atlantic, *Initial Rep. Deep Sea Drill. Proj., 11,* 791-799, 1972.

Martens, C.S., and R.A. Berner, Methane production in the interstitial waters of sulfate-depleted marine sediments, *Science, 185*, 1167-1169, 1974.

McIver, R.D., Hydrocarbon gases from canned core sample sites 174A, 176 and 180, *Initial Rep. Deep Sea Drill. Proj., 18,* 1013-1014, 1973.

Paull, C.K., R. Matsumoto, P. Wallace, et al., *Proc. Ocean Drill. Program Initial Rep., 164,* 620 pp., 1996.

Paull, C.K., T. Lorenson, W.S. Borowski, W.I. Ussler, K. Olsen, N.M. Rodriguez, and H. Wehner, Isotopic composition of CH4, CO2 species and sedimentary organic matter within ODP Leg 164 samples for the Blake Ridge: Gas source implications, *Proc. Ocean Drill. Program Sci. Results, 164,* 101-112, 2000a.

Paull, C.K., T.D. Lorenson, G. Dickens, W.S. Borowski, W. Ussler III, and K. Kvenvolden, Comparison of in situ and core gas measurements in ODP Leg 164 bore holes, *Annals of the New York Academy of Sciences, 912*, 23-31, 2000b.

Pettigrew, T.L., Design and operation of a wireline pressure core sampler (PCS), *Ocean Drill. Prog. Tech. Note, 17*, 1992.

Schoell, M., The hydrogen and carbon isotopic composition of methane from natural gases of various origins, *Geochim. Cosmochim. Acta, 44*, 649-661, 1980.

Schoell, M., Genetic Characterization of Natural Gases, *Am. Ass. Petrol. Geol. Bull., 67*, 2225-2238, 1983.

Shipboard Scientific Party, Explanatory Notes, *Proc. Ocean Drill. Program Initial Rep., 107*, 82, 1987.

Shipboard Scientific Party, Site Chapters, *Proc. Ocean Drill. Program Initial Rep., 141*, 1-611, 1992.

Stein, R., G. Brass, D. Graham, and A. Pimmel, Hydrocarbon measurements at Arctic Gateways Sites, *Proc. Ocean Drill. Program Initial Rep., 151*, 385-393, 1995.

Suess, E., R. von Huene, et al., *Proc. Ocean Drill. Program Initial Rep., 112,* 1015 pp., 1988.

Wallace, P.J., G.D. Dickens, C.K. Paull, and W. Ussler III, Effects of core retrieval and degasssing on the carbon isotope composition of methane in gas hydrate - and free gas-bearing sediments from the Blake Ridge, *Proc. Ocean Drill. Program Sci. Results, 164,* 101-112, 2000.

Weiss, R.F., Carbon dioxide in water and seawater: The solubility of a non-ideal gas, *Marine Chemistry, 2*, 203-215, 1974.

Westbrook, G.K., B. Carson, R.J. Musgrave, et al., *Proc. Ocean Drill. Program Initial Rep., 146*, 611 pp., 1994.

Whelan, J.K., C1 to C5 hydrocarbons from IPOD Holes 397 and 397A, *Initial Rep. Deep Sea Drill. Proj., 47*, 531-539, 1979.

Whelan, J.K., and S.S. Sato, C1 to C5 hydrocarbons from core gas pockets, Deep Sea Drilling Project Legs 56 and 57, Japan Trench Transect, *Initial Rep. Deep Sea Drill. Proj., 57,* 767-748, 1979.

Whiticar, M.J., E. Faber, and M. Schoell, Biogenic methane formation in marine and freshwater environment; CO2 reduction versus acetate fermentation, Isotopic evidence, *Geochim. et Cosomoshim. Acts, 50*, 693-709, 1986.

Worzel, J.R., W. Bryant, et al., Site 88, *Initial Rep. Deep Sea Drill. Proj., 10,* 55-61, 1973.

Worzel, J.L., Regional aspects of Deep Sea Drilling in the Gulf of Mexico, Leg 10 *Initial Rep. Deep Sea Drill. Proj., 10,* 737-748, 1973.

Yamamoto, S., J.B. Alcauskas, and T.E. Croxier, Solubility of methane in distilled water and seawater, *J. Chem. Engin. Data, 21*, 1, 78-80, 1976.

Paull, C.K. and W. Ussler III, Monterey Bay Aquarium Research Institute, 7700 Sandholdt Road, Moss Landing, CA 95039-0628

Gas Hydrates in Convergent Margins: Formation, Occurrence, Geochemistry, and Global Significance

Miriam Kastner

Scripps Institution of Oceanography, La Jolla, California

Methane is the dominant gas (>99%) occupying marine gas hydrate cages, except for a high CO_2 (23%) occurrence in the Middle America Trench and a high H_2S occurrence (~10%) at the Cascadia Margin. Methane C isotope ratios indicate that most is biogenically derived; the amount of admixed thermogenic methane is as yet undetermined. Convergent margins provide a highly favorable environment for methane hydrate formation. The physical-chemical constraints on methane hydrate stability, together with the distinct tectonic, sedimentologic, and hydrologic characteristics of convergent margins, are responsible (1) for higher rates and amounts of formation of both disseminated and massive methane hydrate. Estimates of pore volume occupancies range from a few percent to >25%; (2) for increased methane hydrate concentrations at and near the rather strong amplitude BSRs, that are often discontinuous and patchy, and along fluid conduits, resulting in a spatially irregular distribution throughout the stability field; and (3) for a more widespread occurrence of near seafloor methane hydrate, intimately associated with massive authigenic carbonates, indicating intense upward migration of methane-rich fluids. The BSR may occur above or below the base of the equilibrium stability field. Occurrence above the predicted BSR depth suggests limited supply of methane, and occurrence below it indicates ample supply of methane that causes transient time-intervals of overpressure. Based on global areal distribution, tectonics, sedimentology, hydrology, and organic matter content and type, thus potential of methane yield, it is estimated that, exclusive of shelf and permafrost occurrences, about two-thirds (60-65%) of the vast marine gas hydrate reservoir is situated in convergent margins. This tectonic environment is, therefore, of particular importance for considerations of the role of methane hydrate in the oceanic C cycle, and of the potential responses of methane hydrate to environmental stressers. The methane hydrates near the seafloor and at the BSR are most vulnerable to dissociation upon environmental stresses. Catastrophic events of methane release from hydrate, for example by major submarine landslides, may profoundly influence seawater chemistry, i.e. cause oceanic anoxia and somewhat reduce the ocean's capacity of CO_2 incorporation. The non-oxidized methane will be released to the atmosphere and enhance global warming.

INTRODUCTION

Background

Gas hydrates (gas clathrates) are crystalline substances in which an isometric ice (H_20) lattice (structure I or II) forms cages that incorporate small (3.5 - 7.5 Å) guest gas molecules. They are stable at moderate to high pressure and low temperatures, are exponentially dependent on temperature and linerly on pressure, and are only stable when the cages contain a gas molecule [*Sloan*, 1998]. The stability also depends on the composition of the gas molecule(s) and on the pore fluid chemistry [*Englesoz and Bishnoi*, 1988; *Handa*, 1990; *Dickens and Quinby-Hunt*, 1997; *Zatsepina and Buffet*, 1998]. The physical properties of the hexagonal structure H, with larger cavities that can accommodate common molecules in gasoline, first encountered in the Gulf Coast [*Sassen and MacDonald*, 1994], are being characterized [i.e. *Metha and Sloan*, 1994].

Natural Gas Hydrates: Occurrence, Distribution, and Detection
Geophysical Monograph 124

The pressure and temperature constraints restrict methane hydrate to two main environments on Earth, (1) in the modern ocean temperature regime primarily to continental margins with thick accumulations of sediment with =0.5% organic C content, where water depths exceed ~520 m, except at high latitude regions at water depths of >150-200 m, (Fig. 1) and (2) in polar regions in association with permafrost. In the oceans the amount of gas hydrates is at least an order of magnitude higher than in permafrost [*Kvenvolden*, 1988]. At ~520 m water depth the methane clathrate structure is stable up to ~ 5°C increasing to ~30°C at water depths of >6 km (Fig. 1). Methane (CH_4) hydrate (structure I) is the principal natural gas hydrate on Earth [*Kvenvolden and McMenamin*, 1980; *Sloan*, 1998] and should be important on the planets Uranus and Neptune as well as present on some of the icy satellites of Jupiter and Saturn [*Miller*, 1985]. At standard conditions, in a fully saturated structure I hydrate, with a molecular ratio of CH_4:H_2O = 1:5.75, one cm^3 of methane hydrate when dissociated contains 164 cm^3 of methane gas. Thus, availability of methane from *in situ* production and/or transport by fluids is the limiting factor of methane hydrate formation in most regions of

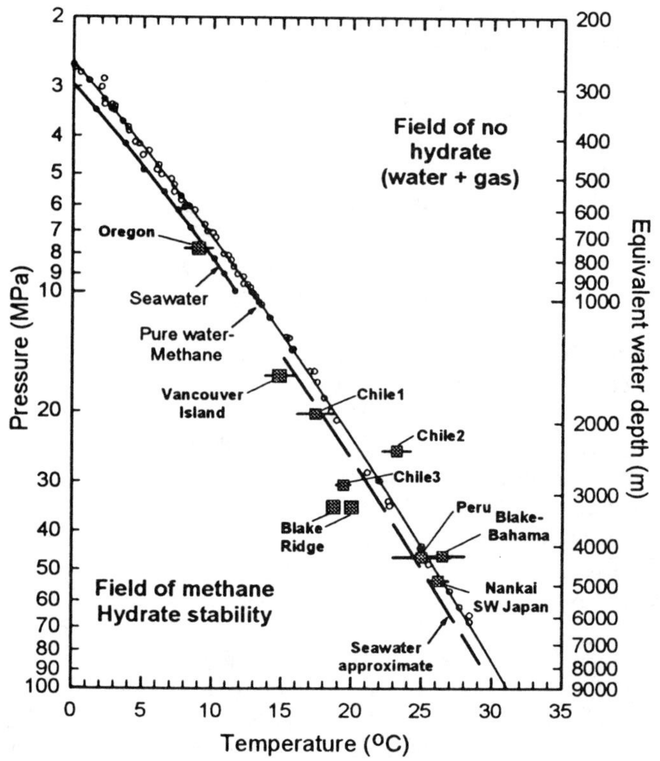

Figure 1. Phase diagram for methane - water (solid curve) and methane - seawater (dashed curve) as a function of temperature and pressure (and equivalent depth). Also shown are BSR temperatures at convergent margin DSDP - ODP gas hydrate sites, as well as at the Bahama Ridge gas hydrate field [Modified from *Hyndman*, personal communication].

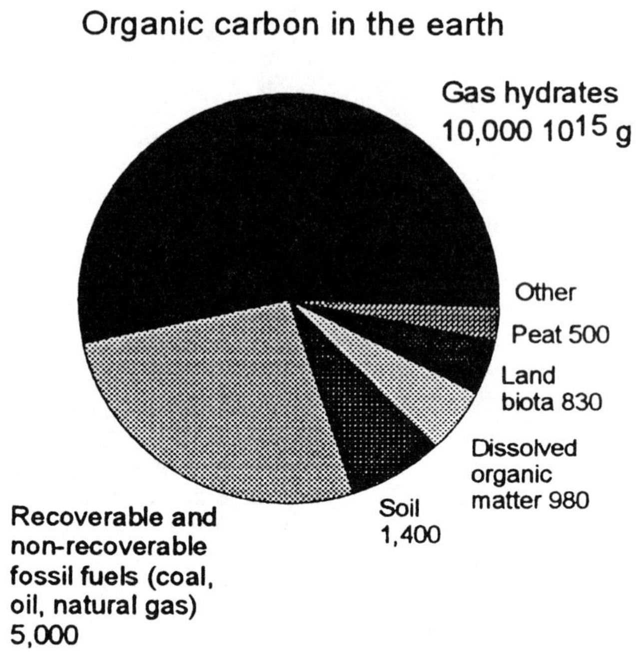

Figure 2. Distribution of organic carbon in the main Earth reservoirs, in units of 10^{15}g C. Estimates of the total global methane in gas hydrates exceeds the sum of all the other known combustible hydrocarbon reserves [Adapted from *Kvenvolden*, 1988].

the modern ocean. The estimated total amount of methane in marine gas hydrate is enormous, and exceeds the other known combustible hydrocarbon reservoirs; i.e. ~10,000 Gt (10^{19} grams) or 8 x10^{17} moles of methane carbon is stored in them (Fig. 2) [e.g., *Kvenvolden*, 1988, 1993; *MacDonald*, 1990). These hydrates store 0.5 x 10^{19} moles $H_2$0 (~10^{17} liters).

A comparison between this and the other fresh water reservoirs on Earth indicates that methane hydrate stores about 8 times the amount of fresh water in the world rivers.

Water Reservoir	Volume (liters x 10^{16})
Gas Hydrates	~9
World Lakes	12
World Rivers	1.2

Assuming seawater average salinity, their formation excluded ~3 x 10^{15} kg salt.

Geochemical studies of the methane, especially its very negative carbon isotope values (<-60‰ d^{13}C) indicate that the methane is largely biogenic in origin. The amount of admixed thermogenic methane is unknown; Mixing calculations suggests up to 20-30% thermogenic methane may be present in convergent margin gas hydrates that have d$^{13}C_{methane}$ values of -60 to -70‰ (PDB). In some regions, however, for example, in the Gulf of Mexico and Caspian Sea, the methane origin is mostly thermogenic. CO_2 and ethane are common minor constituents and traces of propane have been reported. The first documented mixed-hydrate of CH_4 - H_2S (with up to 10% H_2S and minor amounts of CO_2 and ethane) was recovered from close to the seafloor (2-19 mbsf) at the Cascadia Margin, ODP site 892 B [*Whiticar et al.*, 1995; *Kastner et al.*, 1998].

Because of the strong dependence on temperature and pressure, the phase boundry below which water and gas, instead of gas hydrate, are stable mimics the bathymetry of the seafloor, and on marine seismic records this boundry is referred to as a bottom simulating reflection (BSR). BSRs appear on these kinds of records from both convergent and rifted and sheared margins and result from the acoustic impedence contrast between inferred gas-hydrate-bearing sediment lying above sediment that is devoid of gas hydrate. In most occurrences the underlying hydrate free sediments contain some free gas [*Tucholke et al.*, 1977; *Shipley and Didyk*, 1982]. The polarity of the BSR is negative, marking a boundry between high velocity hydrate-rich sediment above and/or low velocity gas charged sediment below.

The mode of occurrence and chemistry of *in situ* oceanic gas hydrates are known from data collected by ocean drilling over the past >30 years, and from conventional coring of shallow gas hydrate occurences. Solid gas hydrate samples have been recovered from oceanic sediments by ocean drilling at continental margin sites throughout the world, at sites with and without a BSR [e.g. *Kvenvolden et al.*, 1993; *Booth et al.*, 1996]. All recoveries of gas hydrate samples have been primarily methane hydrates from sediment depths exceeding ~100 m below the seafloor (mbsf) to ~400 mbsf with three notable exceptions, at 20 - 40 mbsf in the Orca Basin Gulf of Mexico [*Pflaum et al.*, 1996], on the Blake Ridge Diapir, Southeastern U.S. from the seafloor to ~ 60 mbsf [*Paull et al.*, 1996], and at a convergent margin setting in the Cascada margin accretionary prism, at 2-19 mbsf [*Kastner et al.*, 1998; *Whiticar et al.*, 1998]. Shallow gas-hydrate samples have been recovered within 10 m of the seafloor by conventional gravity and piston coring at several locations, principally in the northern California convergent margin in the Eel River basin [e.g. *Brooks et al.*, 1991], at the Cascadia convergent margin offshore central Oregon [*Suess et al.*, 1999] and recently from offshore Vancouver Island [*Spence at al.*, 2000]; as well as in non-convergent margins, in the Gulf of Mexico [*MacDonald et al.*, 1994], the Okhotsk Sea, Black and Caspian Seas [*Kvenvolden et al.*, 1993]. As yet only little is known about the *in situ* mode of occurrence and chemistry of oceanic gas hydrates because of existing core recovery operations which are not suited for preserving gas hydrates unless at least moderately abundant or massive. Based on the recoveries of gas hydrates and inferences from pore fluid chemistry, they are distributed irregularly in ocean sediments, particularly in convergent margins. The occurrence of methane hydrate near or at the seafloor indicates active upward advection of methane-rich fluids or of methane gas [*MacDonald et al.*, 1994; *Kastner et al.*, 1998; *Suess et al.*, 1999].

Recent interest in natural gas hydrates, most of which are methane hydrates, has resulted from the realization that global warming may destabilize the enormous quantities of methane hydrate in the shallow geosphere, and that the environmental impact of release of large quantities of methane into the ocean and atmosphere could have important societal consequences [*Revelle*, 1983; *MacDonald*, 1990; *Nisbet*, 1990; *Brewer*, 2000; *Levitus et al.*, 2000; and references therein]. In the ocean, fairly rapid microbial oxidation of methane will occur [*Ward et al.*, 1987; *de Angelis et al.*, 1993; *Valentine et al.*,"in press"] and reduce the amount of methane released into the atmosphere, except during catastrophic release of methane into the ocean. If large quantities are released it will result in extensive oxygen consumption and a somewhat reduced capacity of the ocean to incorporate atmospheric fossil fuel CO_2. These chemical effects on the ocean would be harmful to microbiology. In addition to the potential oceanic and environmental impacts of gas hydrate dissociation, continental slopes may become destabilized. Sea-level changes, global warming, tectonic uplift, or rapid sedimentation episodes, would first impact gas hydrate at the 3-phase boundaries, i.e. at the BSR and in some settings at and near the seafloor. At the BSR the release of gas and water will destabilize, the sediment that will become unconsolidated, buoyant, and overpressured, and submarine landslides may be triggered [e.g. *Nisbet et al.*, 1998; *McIver*, 1982; *Dillon et al.*, 1998]. Large landslides cause local tsunamis and if the release of gas hydrate is catastrophic, some methane will be released to the atmosphere.

Methane is a very strong greenhouse gas, is second to CO_2 in importance as a contributor to the atmospheric radiation balance. Any flux of methane into the atmosphere in addition to the present annual growth rate of =1% per year would enhance global warming. New data that support links between global warming and methane hydrate dissociation, in the Santa Barbara Basin over the past 60,000 years [*Behl and Kennett*, 1996; *Kennett et al*,. 2000] and in the deep ocean ~55.6 million years ago [*Dickens et al.*, 1997] have been forwarded. The former is being related to episodes of warmer climate as seen in the temperature record of Greenland ice cores (Dansgaard Oeschger events) and the latter was a major global thermal maximum.

The fossil fuel resource potential of the estimated enormous quantities of marine methane hydrates is being debated [*Kvenvolden* 1983, 1999] and is not considered in this paper.

GAS HYDRATES IN CONVERGENT MARGNS

Summary of Previous Work

Although all margins must be considered for a global perspective of gas hydrate concentration and distribution, oceanic C cycling, and potential response to the environmental stresses discussed above, convergent margins are the focus of this paper because of their distinct characteristics and global significance. The controls on gas hydrate formation, occurrence, spatial distribution, and potential for destabilization and dissociation are considered and evaluated.

Convergent margins are the most dynamic regions on Earth; they are characterized by deep ocean trenches. In the modern ocean the total length of trenches is 43,500 km [*Von Huene and Scholl*, 1991; *Rea and Ruff*, 1996]. About 19% of convergent margins (~8,200 km) have large accretionary prisms, ~37% (~16,300 km) have small to moderate accretionary prisms, and 44% (~19,000 km) little net accretion occurs, all incoming sediments are being subducted [*von Huene and Scholl*, 1991]. About 80% of the non-accretionary convergent margins occur where the oceanic plate is subducted beneath an island arc crust, with low terrigenous organic debris, where the potential is low for gas hydrate formation. Consequently, the more significant gas hydrate findings are concentrated in ~2/3 of the convergent margins, associated with ~29,000 km of trench length where higher sedimentation rates prevail and/or higher organic C content accumulate. The estimated yearly mass flux of sedimentary material delivered to the world's trenches is 2.34 x 10^{15} g (± 10 to 20%) of which 39% is pore fluid [*Rea and Ruff*, 1996]. The terrigenous sediment input of 1.08 x 10^{15} g equals 76% of the total sediment input. The remaining 24% consist of about 2/3 carbonate sediments, 0.22 x 10^{15} g that are equivalent to ~15% of the total input, and of 1/3 biogenic opal, 0.13 x 10^{15} g equivalent to ~9% of the total flux [*Rea and Ruff*, 1996]. (These estimates exclude the turbidites in the trenches.) The average organic matter content ranges between =0.5 to ~1.5 weight % [e.g. *Revelle*, 1983; *MacDonald*, 1990; *Premuzic et al.*, 1982; *Gornitz and Fung*, 1994]. Based on tectonics, sedimentology, and organic C distribution in marine sediments, gas hydrate bearing active convergent margins are characterized by the largest sediment and organic C mass fluxes (marine and continental). Convergence rates generally range from 30 - 100 km/my (at Tonga trench the rate is exceptionally fast, 170 km/my).

Distinct Characteristic of Convergent Margins

1. The dynamics of accretion and subduction is responsible for a continuous, non-uniform, flux of a mixture of terrigenous, hemipelagic, and pelagic sediments and pore fluids, and much of the material is recycled in the subduction zone at various time scales. Accretionary prisms can, however, reach a critical size and then stop growing as along South and Middle America. In contrast, at most rifted and sheared margins the sediment flux may be very high in the earlier stages of evolution. Subsequently, however, sedimentation decreases exponentially (a notable exception is, for example, the Bengal fan) and sediments are not recycled.
2. Tectonic activity is intense as manifested in tectonic uplift, folding, faulting, and fracturing, and is linked with pervasive upward and lateral fluid flow induced by accretion, and sediment loading, subduction tectonics, and mineral reactions.
3. The active hydrology enhances fluid transport, hence fluid-rock and biochemical reactions. Thus methane-bearing fluids, generated *in situ* via both anaerobic microbial oxidation and/or thermogenically, migrate and are expelled upward. Reliable quantitative data on rates of these *in situ* processes and fluid migration are yet not available, except with respect to the upper plate venting for Oregon, the Alaska Margin, and perhaps Nankai Trough. The dynamic instability of convergent margins and the active hydrology of methane-bearing fluid transport, together with the physical chemical constraints on methane hydrate stability, produce a conducive environment for methane hydrate formation, (a) particularly at the sensitive phase equilibrium boundaries, at the BSR and in some settings at or near the seafloor, and (b) within the two-phase stability field. This results in higher concentrations and in irregular distribution of both disseminated and massive methane hydrate in convergent versus in rifted and sheared margins.
4. At both convergent and rifted and sheared margins the input of terrigenous sediment containing continental organic matter dominates; it is diluted by hemipelagic-pelagic sediments with marine organic matter from the overlying upwelling zones. The additional continuous flux of hemipelagic-pelagic sediments including marine organic matter by the subducting plate is unique at convergent margins. Consequently, although the reported average concentrations of organic matter of >0.5 to <1.5 weight percent are similar at both margin types (average accumulation rates are more difficult to calculate because often the stratigraphic controls are poor), the ratio of marine to continental organic matter is higher at convergent margins (for references see those cited in Table 1 for convergent margins, and other DSDP and ODP volumes for rifted and sheared margins). Marine organic matter enhances microbial processes responsible for sufate reduction and methane generation, hence the yield of methane per unit organic C weight should be higher at convergent margins.
5. The shelves are generally narrower and the slopes steeper at convergent margins. Although this topography reduces the slope and rise total area, it promotes (a) strong upwelling, therefore higher *in situ* input of marine organic matter from the overlying water, and (b) slumping, hence rapid sediment deposition and burial that enhance the supply and preservation of organic matter. The type and rate of supply of organic matter are important controls of methane production rate and amount.
6. At convergent margins, the average (over-time) higher sediment flux induces faster recycling of methane at the BSR than at rifted and sheared margins.

In view of these distinctive characteristics of convergent margins, the mode of occurrence, especially the inventory, the concentration, and distribution of methane hydrate in this tectonic setting and the response to environmental stresses, such as global warming, is of particular global importance and discussed below.

Gas Hydrate Occurrence, Distribution, and Concentration in Convergent Margins

Much of our knowledge about gas hydrate occurrence, distribution, and geochemistry in convergent margins is from recovered samples and geochemical data obtained through the Deep Sea Drilling Project (DSDP) and Ocean Drilling Program (ODP), summarized in Table 1.

All of the successful recoveries were from Circum-Pacific convergent margins from about 100 to =400 mbsf, primarily situated in the 2-phase region of the stability field, and most sites chosen for drilling until ODP Leg 141 at the Chile triple junction did not overlie a visible BSR. At the Cascadia Margin ODP Site 892 (Leg 146), a BSR was intersected at ~72 mbsf and gas hydrate was, however, recovered at 2-19 mbsf. Geophysical data, especially on the presence of BSRs, their depth, regional distribution, and amplitude provided important essential information about potential presence and mode of occurrence of gas hydrates and guided the choice (or avoidance) of sites for the ocean drilling in convergent margins.

The direct observations and inferences obtained from geophysical and pore fluid geochemical data indicate that gas hydrates are irregularly dissemated in turbidite sediments and in silty mudstones with increasing abundance near and at the BSR. They cement coarser sediment horizons of higher permeability such as sandy silts and ash layers. Nodular or massive gas hydrates occur in fault zones, fractures, and along lithologic boundaries; for example, off Guatemala, ODP Leg 184, a 1.05 m long massive hydrate sample was recovered from a fracture zone [*Kvenvolden and MacDonald,* 1985].

In clay rich sediments of low permeability and high anisotropy, higher surface energies are required for nucleation and growth of gas hydrate in the fine pores [*Handa et al.*, 1992; *Clennell et al.*, 1999]. Therefore, in such sediments (for example at Costa Rica, ODP Site 1041) methane-bearing fluids nucleate and crystallize in thin plates parallel to the fissility or scaly fabric of the clay-rich sediments, or occur in microfractures and in thin veins (hydrofractures?). Gas hydrate layers between sediment types parallel to bedding in inhomogeneous sediment sections are also observed. Most of the gas hydrate in such fine grained sediments is concentrated in fault and fracture zones and as cement of the interbedded coarser grained volcanic ash-rich sediments. More detailed summaries of the gas hydrate recovered through the Ocean Drilling Program are provided in *Kvenvolden* [1993, 1995], and Figure 3 shows its global distribution in margins. This dated Figure is based on BSR

Table 1. Occurrence, distribution, and characteristics of recovered gas hydrates by drilling in convergent margins.

Convergent Margin	Location (DSDP/DDP leg and site)	Sedimentology	Organic C (wt %)	Average Geothermal Gradient (°C/km)
Peru-Chile Trench	Leg 112, Sites 685, 688	Silty mudstone.	1-2 %	43-49°
Middle America Trench Mexico, Guatemala, Nicaragua, and Costa Rica	DSDP Leg 66, Sites 490, 491, 492 Leg 67, Site 497, 498 Leg 84, Sites 565, 568, 570 ODP Leg 170, Site 1041	Mostly high porosity silty sediment and abundant medium to coarser grained volcanic ash; off Costa Rica Site 1041 silty clay-rich sediment.	~1%	31° off Mexico, 32° off Guatemala, 35° off Nicaragua, 44° off Costa Rica, except at Site 1041, ~20°
Cascadia Prism	ODP Leg 146, Site 892	Terrigenous silty-clay with interlayered silty sand.	0.5 - 1.5%	51°
Nankai Trough, Japan	Leg 131, Site 808	Mostly terrigenous turbidites.	≤0.5%	~100°

Table 1. (continued) Occurrence, distribution, and characteristics of recovered gas hydrates by drilling in convergent margins.

Water depth (m) and BSR depth (mbsf)	Description of Occurrence	Pore Space Occupancy (%)	Geochemistry	References
3000-5000 water depth. No BSR was imaged at the sites drilled. Nearby BSRs at 350-570.	Cementing silty-mud.	100% in cemented sediment.	>99% methane hydrate. $\delta^{13}C$ of -59.6‰ suggests primarily biogenic origin, and possibly some mixing with thermogenic methane.	*Kvenvolden and Kastner* [1990]
1700-5500 water depth. On Legs 66, 67, 84 drilled where BSR was not imaged. On Leg 170, the calculated BSR is in the oceanic basement because of low geothermal gradient. Nearby BSR at 190-570. Off Mexico seaward the BSR extends to ~950.	Cementing coarse-grained sediment and volcanic ash; as ice-like inclusions; off Guatemala occupying fractures in mudstones. A 1.05 m massive hydrate in a fracture zone at site 570.	100% in cemented coarse-sediment and volcanic ash , but no occupancy estimates for the other modes of occurrences.	Mostly >99% methane hydrate, except for one sample high in CO_2 (23%) at Site 490. $\delta^{13}C$ of methane range from -83 to -41‰. $\delta^{13}C$ of CO_2 indicates that even the very high $\delta^{13}C$ is of a mostly biogenic origin.	*Shipley and Didyk* [1982] *Harrison and Curiale*[1982] *Kvenvolden et al.* [1984] *Kimura et al.* [1998] *Kastner* [this volume] *Claypool et al.* [1985] *Shipley et al.* [1979] *Yamano et al.* [1982]
~675 water depth BSR at ~ 72.	Mostly platy crystal aggregates, irregularly disseminated, or in thin layers and micro fractures. One piece of massive hydrate parallel to bedding, recovered from shallow depth ~18 mbsf.	Irregular, from ~5% to >30%.	The first mixed CH_4 - H_2S hydrate recovered, mostly methane with up to 10% H_2S, some CO_2 and traces of ethane. $\delta^{13}C$ of methane suggests a mostly biogenic origin.	*Kastner et al.* [1995; 1998] *Whiticar et al* [1995]
Drilled where no BSR was present; a few km westward BSR at ~200.	One piece of cemented sand in a wash core.	100% in cemented sand.	Not done, dissociated too rapidly.	*Shipboard Scientific Party* [1991]

identifications, and sites of actual gas hydrate findings. It most likely underestimates the actual global distribution of gas hydrate in margins because (1) reliable identification of BSRs may include aspects related to record quality; (2) published and yet unpublished modern seismic reflection data have shown a broader gas hydrate distribution based on BSR identification than shown in the figure [*von Huene*, personal communication].

In addition to the deeper subsurface gas hydrate occurrences in convergent margins, near seafloor gas hydrates were recovered in the Eel River basin, northern California [i.e. *Brooks et al.*, 1991], Cascadia accretionary prism, off-

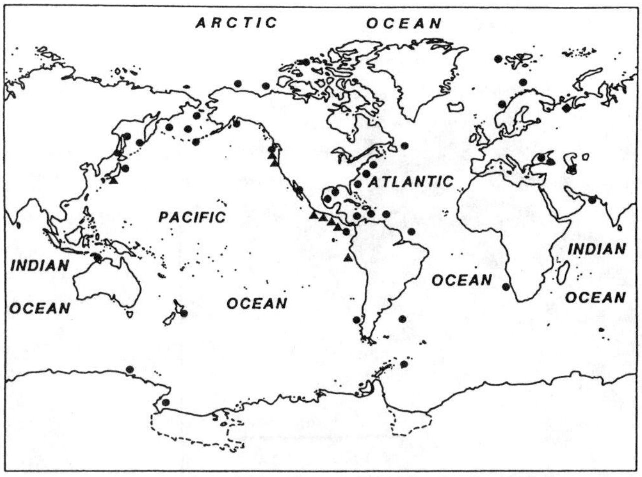

Figure 3. Distribution of the known and inferred continental margin sites with gas hydrate. Locations where gas hydrate has been recovered in convergent margins are marked with an s [Modified from *Kvenvolden*, 1993].

shore central Oregon [i.e. *Kastner et al.,* 1998; *Suess et al.*, 1999] and Vancouver Island [*Spence at al*., 2000], and Southern Barbados accretionary prism in a mud volcano field that covers ~1600 km^2 at ~1 m below the seafloor [e.g. *Westbrook and Smith*, 1983; *Le Pichon et al,* 1990; *Martin et al*., 1996]. At all these locations the shallow occurrences of gas hydrates are mostly massive to nodular. The most detailed description of the occurrence of surficial gas hydrates is from the hydrate ridge, Cascadia, in *Suess et al.* [1999]. Authigenic carbonates (aragonite, low- and high-Mg-calcite, and dolomite) with very negative d^{13}C values, indicating an origin from methane oxidation, are widespread and intimately associated with each of these near-seafloor occurrences of gas hydrates. The authigenic carbonates and gas hydrates are especially abundant and well developed in the vicinity of fluid-active thrust faults and fractures that breach the seafloor [e.g. *Kulm et al*., 1986; *Ritger et al*., 1987; *Brooks et al*., 1991; *Martin at al*., 1996; *Suess et al*., 1999]. *Carson et al.* [1994] using GLORIA imaging estimated that 2.2 x 10^7 metric tons of surficial authigenic carbonate precipitated from the venting of methane-rich fluids associated with the shallow massive gas hydrates in a 17 km^2 of the Cascadia prism, at and near the site where the hydrologically active out-of-sequence fault, shown in Figure 4, breaches the seafloor (near ODP Site 892). Preliminary observations on the relationship between the carbonate mineralogy spatial distribution and gas hydrate occurrence suggest that at the most active sites of near surface methane venting across the seafloor and gas hydrate formation, aragonite is the dominant authigenic carbonate.

Thus, at convergent margins where methane migrates into the sulfate reduction zone and reaches the seafloor, mostly through faults and fractures, some of the methane generated is sequestered in gas hydrates, some in authigenic carbonates. Some of the remaining fraction is bacterially oxidized in the oceans, and the residual fraction will reach the atmosphere. The amount sequestered in gas hydrates versus in authigenic carbonates is as yet unknown. Documenting it is important for oceanic C cycle estimates and for evaluations of the environmental impacts of methane-rich fluids venting through the seafloor on ocean chemistry and atmosphere at convergent margins in the geologic past and in the future.

Methane Occurrence and Implications for Gas Hydrate Formation and Distribution

Methane accounts for ~99% of the hydrocarbon gases analyzed in both convergent and rifted and sheared margin sediment pore fluids and gas hydrates. However, because of the existing core recovery operations there are no reliable quantitative data on *in situ* gas concentrations in convergent margin sediments. *In situ* methane concentrations are only available from the Blake Ridge, Western Atlantic Ocean (*Paull et al.,* 1996; *Dickens et al.,* 1996). (Gas concentrations can be constrained from seismic properties and from analysis of both P- and S-waves. However, conventional seismic reflection records do not provide true-amplitude seismic data required for the analysis, *von Huene*, personal communication). Thermogenic methane formation is important only at temperatures =80-90°C, therefore its contribution to gas hydrate formation is possible in regions of upward fluid migration, which is pervasive at convergent margins, as proposed by i.e. *Hyndman and Davis* [1992] and *Ginsburg and Soloviev* [1997]. Most obvious manifestations of fluid expulsion in convergent margins are the frequent association of surficial gas hydrate occurrences with chemosynthetic benthic communities, mud volcanoes, and authigenic carbonates. The presence of small amounts of thermogenic hydrocarbons, from propane to hexane, in the zone of low to moderate temperatures of diagenesis, especially near major faults that are acting as fluid conduits, indicates upward fluid migration from greater depths, from below the BSR [e.g. *Ritger et al.,* 1987; *Kulm and Suess,* 1986; *Moore et al*., 1990; *Brooks et al.*, 1991; *Whiticar et al*, 1995; *Suess et al*, 1999; *Lücke et al.,* 2000; and references therein]. A displaced BSR by a methane-bearing warm fluid migrating along a thrust fault in the Cascadia margin, associated with gas hydrate, authigenic carbonate, and prolific benthic communities, is shown in Fig. 4.

At equilibrium, methane solubility in seawater decreases from the base of the methane hydrate stability zone to the seafloor; hence a methane undersaturated fluid that is

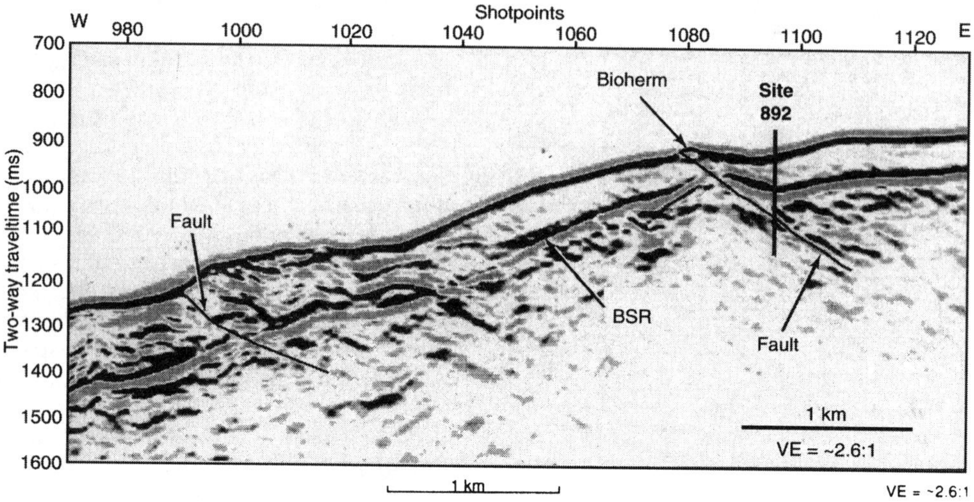

Figure 4. The strong BSR in the vicinity of ODP Site 892, Cascadia margin, is destabilized and displaced upward by a hydrologically active landward-dipping thrust fault, transporting a methane-bearing warm fluid; the fault intersects the BSR and breaches the seafloor where gas hydrate and authigenic carbonates form.

migrating upward into the hydrate stability zone may become methane hydrate saturated or oversaturated within it [*Buffet and Zatsepina*, 1997]. Accordingly, pressure-temperature changes caused by tectonic uplifting may promote gas hydrate formation. Also, at constant temperature when pressure increases methane hydrate solubility slightly decreases; at pressures >100 bars the saturation methane solubility in seawater is greater than the methane concentration required to form and stabilize methane hydrate [*Miller,*1976; *Handa,*1990]. Dissolved salt content and sediment physical and electrical properties also effect gas hydrate stability; higher salt concentrations decrease the size of the stability field in temperature-pressure space [e.g. *Englesos and Bishnoi*, 1988; *Dickens and Quinby-Hunt*, 1997], and smaller pore sizes may also decrease it [*Clennell et al.*, 1999], because of the higher surface energy required for nuclei stabilization [*Handa et al.*, 1992]. Accordingly, in a physically confined system at constant dissolved methane concentration, methane hydrate will crystallize at a lower temperature. At equilibrium, only at the base of the stability zone, at the 3-phase equilibrium, free gas must be present. Above it in most convergent margin settings methane hydrate is stable at lower dissolved methane concentrations than the methane solubility concentration [*Yamano et al.*, 1976; *Handa,* 1990]. The % cage occupancy also varies with pressure and temperature. These physical-chemical constraints require methane to be supplied continuously to preserve methane hydrate, and no free gas should be present as long as water is present in excess. As long as pore fluid is present methane gas concentrations cannot be very high at the base of the stability zone, and any methane gas entering the stability field should be captured to form gas hydrate.

The two prevailing models of oceanic gas hydrate formation are the pore fluid expulsion and migration model, [i.e., *Hyndman and Davis*, 1992, and references therein] and the *in situ* generation of methane model [i.e., *Kvenvolden and Barnard*, 1983, and references therein]. The pore fluid migration model suggested for convergent margins accounts for the observations of widespread venting of methane-rich fluids at the seafloor and the geochemical and biological consequence, discussed above, and satisfies theoretical considerations and observations of the dynamic-tectonic implications of accretion, subduction tectonics, and sediment loading. The fluid migration model depends on upward transport of biogenic and/or thermogenic dissolved methane and in some instances also of methane gas. Assuming the methane content of the migrating fluid is sufficient or exceeds the concentration necessary to form and stabilize methane hydrate within the stability field, much of the methane will be trapped in gas hydrate at the base of the stability field. Depending on the methane concentration, rate of fluid migration and gas hydrate formation, the remaining upward migrating methane may form gas hydrate within the stability field, because at lower temperatures (and pressures) the solubility of the methane hydrate decreases [*Handa*, 1990; *Zatspina and Buffet*, 1997]. Disseminated crystals of hydrate may therefore form and occur almost anywhere within the stability field, but will be more concentrated at or close to the base of the stability field and along fluid conduits, thus methane hydrate should be spatially inhomogeneously distributed, as indeed observed or inferred from pore fluid Cl concentrations. At the base of the stability field a negative polarity bottom simulating reflector (BSR) often develops, caused by the different physical properties of sediments with and without gas hydrate and/or methane gas at the base of the stability zone. The nature, distribution, and amplitude of the BSRs have been modeled for example, by *Hyndman and Spence* [1992], *Westbrook et al.* [1994], *Mackay et al.* [1994], *von Huene and Pecher* [1999], and references therein. Strong BSRs are predicted to develop where fluid expulsion is high. Recent models involving upward advection (and diffusion) of methane by *Xu and Ruppel* [1998] and *Rempel and Buffet* [1998] on the spatial distribution of methane hydrate (that do not consider the rates and sites of methane production but assume that methane is being supplied) suggest that at such advective systems the base of the field where methane hydrate actually forms and accumulates, the BSR, may occur above the base of the stability field. In contrast, if methane gas is trapped beneath the 3-phase stability boundary that is acting as a seal for gas migration, i.e. from intense or complete pore filling by methane hydrate, and gas pressures above hydrostatic evolve, the 3-phase stability boundary (BSR), would be lowered - deepened.

Utilizing satellite data on productivity, which are translated into organic matter distribution in margin sediments, gas hydrate spatial distribution predicted by the pore fluid migration models is more consistent with available observations [*Gornitz and Fung*, 1994] than the distribution predicted by the *in situ* formation model. Although most gas hydrate occurrences are associated with underlying BSRs, the upward fluid migration model also explains (1) why BSRs are often patchy in occurrence and amplitude, (2) why as observed methane hydrate occurs in regions with no underlying BSR, for example in the Peru margin [*Kvenvolden and Kastner,* 1990], and (3) why more frequent occurrences of shallow, near-seafloor gas hydrates are observed in convergent than in rifted and sheared margins [e.g. *Brooks et al.*, 1991; *Kastner et al.,* 1998; *Suess et al.*, 1999].

Although aspects related to record quality influence BSR imaging, at convergent margins BSRs are often discontinuous and patchy. Where present the amplitude is generally rather strong; low amplitude and gradational variability of BRSs also exist. At rifted and sheared margins BSRs generally are highly continuous, span over large regions, are less distinct, and have lower amplitudes. The differences in the spatial distribution and intensity of the BSRs at the two margin types arise from their distinct tectonic and hydrologic regimes responsible for methane supply; in particular the active vertical tectonics and abundance of faults and fractures that are conduits for upward migration of methane-rich fluids. At times warm fluids give rise to local discontinuous formation and dissociation of methane hydrate, and enhance the irregular distribution of methane hydrate (i.e. Fig. 4). At the 3-phase equilibrium boundary a gas phase must exist, the percent of gas hydrate pore volume occupancy at the BSR and the gas content beneath it are important variables that influence the observed variability in amplitude. At convergent margins, in addition to recycled methane from immediately beneath the BSR, primarily derived by sediment loading and continuous burial of gas hydrate-bearing sediments, as at rifted and sheared margins, the advective supply of methane through fluid expulsion and upward migration from greater depths is also first captured to form methane hydrate at the BSR, as suggested by *Hyndman and Davis* [1992]. High pore volume occupancies by methane hydrate of >20-25% evolve at convergent margins, as for example in Cascadia at ODP Site 889 [*Kastner et al.,* 1995]. The BSRs are thus enhanced, especially in high permeability sediments and in regions of diffusive advection. BSRs are discontinuous or patchy near faults and fractures, in tectonic active areas, and where warm fluids dissociate methane hydrate at the BSR and redistribute the methane by upward transport in a diffusive or focused mode. In some instances of intense fluid migration through faults the methane-rich fluids breach the seafloor, as shown in Figure 4, at the Cascadia margin.

The *in situ* biogenic methane production and supply model [*Kvenvolden and Barnard,* 1983] predicts disseminated methane hydrate formation anywhere within the stability field, assuming sufficient methane is generated. Continuous sediment loading and compaction induces some fluid advection, and sedimentation containing some organic C continuously supplies a source for *in situ* methane production. Below the base of the BSR where the methane is released from gas hydrate dissociation, caused by continuous sediment burial, the dissolved and/or released methane gas will be re-incorporated in methane hydrate at the BSR. As a result of this recycling of methane an uneven vertical distribution of gas hydrate, more at the BSR and less above, evolves. Higher methane hydrate concentrations thus must occur at the BSR also in the *in situ* formation model, as in the fluid migration model, and is supported by geochemical proxy data, for example by the depth profiles of chloride concentrations. However, the maximum pore volume occupancy by methane hydrate without vertical methane-rich fluid migration from greater depths has been modeled to be low <5-10% [i.e. *Rempel and Buffet,* 1997; 1998] A strong argument used in favor of the *in situ* production of methane are the observed depth-profiles of $\delta^{13}C$ of methane and CO_2, which almost parallel each other at several sites analyzed [*Kvenvolden and Barnard*, 1983)], for example in the Middle America Trench [*Claypool et al.*, 1985]. This indicates that CO_2 reduction is the main process by which methane is derived *in situ*.

The above two end-member models act in concert at both margin types. *In situ* methane generation, methane recycling at the BSR, and disseminated methane hydrate formation prevail at both margin types, as indicated by the Cl concentration-depth profiles (i.e. at the Cascadia, Middle America Trench, Peru, and Japan convergent margins, and the Blake Ridge [e.g. *Hesse and Harrison*, 1981, *Kastner et al.*, 1991, 1995; *Paull et al.*, 1996; and references therein]. Fluid migration, particularly along conduits of higher permeability sediment horizons, faults, and fractures is, however, rather intense and pervasive in both accreting and non-accreting convergent margins, supplying additional dissolved methane from deeper sources and leading to higher concentrations and a larger spectrum of gas hydrate distribution and mode of occurrences, than at rifted and sheared margin settings. In response to sediment loading and burial, fluid migration at most rifted and sheared margins should be more important at the earlier stages of evolution than when sedimentation has slowed down exponentially.

Preliminary modeling by *Gornitz and Fung* [1994] on methane availability for gas hydrate formation at 0.5 and 1.0% organic C contents suggests that up to 2 to 3 times more methane is probably available for gas hydrate formation in systems with extensive fluid migration as compared with systems more dominated by *in situ* formation of methane. This is supported by observed and/or inferred pore volume occupancies by gas hydrate in convergent (i.e. Cascadia) versus rifted and sheared (i.e. Blake Ridge) margins.

GEOCHEMICAL EVIDENCE FOR THE NATURE AND SEDIMENT PORE VOLUME OCCUPANCY BY GAS HYDRATE

Gas Hydrate Chemical and Isotopic Compositions

Recovered methane hydrates have been analyzed geochemically for:

1. Gas composition [e.g. *Claypool et al.*, 1985; *Kvenvolden et al.*, 1983, 1988, 1995]. Methane is the dominant gas (mostly ≥99%), minor amounts of CO_2 and ethane and traces of propane are often present, and a mixed CH_4 - H_2S hydrate was recovered from the Cascadia prism [*Kastner et al.*, 1988; Suess *et al.*, 1999]
2. C and H isotope values of the methane, for the origin of methane, as discussed below.
3. Oxygen and hydrogen isotopic values of the residual hydrate H_2O, for the mode of formation, in an open vs. closed system, and/or for the isotopic composition of the pore water at the time of formation [e.g. *Hesse and Harrison*, 1981; *Kvenvolden and Kastner*, 1990; *Kastner et al.*, 1995, 1998]. At both convergent and rifted and sheared margins, depending on the background pore water $\delta^{18}O$ and δD values, the $\delta^{18}O$ data span a small range of values, from slightly negative to about +3.2‰ (SMOW), indicating formation of gas hydrates at low temperatures in an open system with a methane hydrate-water oxygen isotope fractionation factor similar to the experimental fractionation factor obtained for sea ice of 1.0027 to 1.0031 [e.g. *O'Neil*, 1968; *Craig and Hom* 1968; *Davidson et al.* 1978]. The δD value ranges between +18 to +22‰ (SMOW) [*Craig and Hom*, 1968; *Kastner et al.*, 1998].
4. Chloride concentrations, to determine amounts of admixed sea water, hence, the porosity of the gas hydrate [*Kastner et al.*, 1995, 1998]. The majority of the existing data are, however, of contaminated gas hydrate samples with sediment pore fluid Cl, therefore cannot be used for porosity estimates of the gas hydrates themselves; and
5. CH_4/H_20 ratio in the hydrate for % cage occupancy by gas molecules, mostly methane [e.g. *Kvenvolden*, 1983, 1988, 1993, and references therein]. All the published data have considerably lower ratios, of 90 to 140, than that calculated for 90% cage occupancy, of 170 (at ~2km water depth). Because of the recovery operation of gas hydrate, some methane may escape before the hydrate is analyzed; this possibility needs to be tested experimentally. The discrepancy between the empirical data and theoretical value, however, most likely results from analyzing impure gas hydrate samples, having some admixed sediment with pore fluid. Unfortunately, no good data on the important property of CH_4/H_2O ratios in convergent margin gas hydrates exists as yet.

Origin of Methane and Yields for Methane Hydrate Formation

Sediment methane/ethane (C_1/C_2) ratios and carbon isotope values of the methane in hydrates recovered at convergent margins indicate mostly a biogenic origin for the methane. The majority of the published $\delta^{13}C$ values from convergent margin methane hydrates range between -60 to -70‰ (PDB) [e.g. data in *Kvenvolden and MacDonald*, 1985; *Claypool and Kvenvolden*, 1983, 1985; *Kastner and Kvenvolden*, 1995; *Kastner et al.*, 1998; and references therein]. Table 2 shows new data for Costa Rican hydrates at ODP Site 1041.

Table 2. Carbon and Hydrogen isotopic compositions of methane in gas hydrate samples from Costa Rica apron sediments, ODP Site 1041, and C_1/C_2 ratios of the associated gas samples. The C_1/C_2 ratios are from Kimura, Silver, Blum, et al. (1997).

Site	Sample core-section	Depth (mbsf) (mbsf)	$\delta^{13}C_{methane}$ ‰ (PDB)	$\delta D_{methane}$ ‰ (PDB)	C_1/C_2
1041A	15X-3	117.0	-67.4	-207.4	7400
1041A	15X-4	118.0	-66.5 -66.9*	-204.5	7400
1041A	16X-4	130.3	-66.9	-200.1	---
1041B	4R-5	190.1	-65.0	-192.0	2400
1041B	6R-3	205.1	-64.5	-197.1	2200
1041B	9R-2	233.7	-62.3	---	900
1041B	11R-5	256.5	-63.9	-191.3	1000

* Duplicate analysis

Thermogenic methane which forms at >80-90°C is characterized by higher, less negative $\delta^{13}C$ values than about -50‰, whereas, the $\delta^{13}C$ from anaerobic microbial reduction of CO_2 to methane in sulfate-free marine environments typically have values of -60 to -90‰, and up to -110‰ (PDB); the δD values of -170 to -250‰ (SMOW) are as well typical of biogenically produced methane. δD values for thermogenically derived methane of -150 to -300‰ largely overlap with the microbially derived methane values. [e.g. *Claypool and Kaplan*, 1974; *Schoell*, 1980; *Claypool and Kvenvolden*, 1983; *Whiticar et al.*, 1986; and references therein].

Interestingly, the majority of the observed methane hydrate C isotope values at convergent margins fall within a fairly narrow range of values of -60 to -70‰, and this range is at the lower-end (less negative) boundary of the typically microbially derived methane $\delta^{13}C$ values. It most likely suggests a mixed origin for some of the methane in convergent margin gas hydrates, a mixture of primarily microbially derived methane with some upward advected thermogenically derived methane. Preferential migration of methane relative to the higher molecular weight hydrocarbon molecules may also effect (increase) the *in situ* C_1/C_2 ratios, and thus mask the mixed origin.

Two purely microbial diagenetic processes, however, could also shift the original microbially derived methane C isotope values toward the field of thermogenic methane C-isotope values. The first is anaerobic microbial oxidation of a fraction of the methane transported into the sulfate reduction zone by advecting fluids, pervasive in convergent margins. The net result is methane depleted in ^{12}C, having less negative $\delta^{13}C$ values. Evidence for methane oxidation within the sulfate reduction zone is obtained from $\delta^{13}C$ values of the dissolved inorganic carbon (DIC). The $\delta^{13}C$ values of this DIC are more negative than -20‰, reaching values of -40‰ (PDB) and are indicative of methane oxidation [e.g. *Whiticar et al.*, 1995; *Suess et al.*, 1999; and references therein].

The second process by which ^{13}C-enriched microbial methane originates has been well documented, for example by *Claypool et al.* [1985] in slope sediments of the Middle America Trench, off Guatemala, DSDP Sites 568 and 570. At these sites the microbially derived methane $\delta^{13}C$ values range from typical biogenic values of -60‰ to -83‰ (PDB) in the uppermost ~170 m of the sediment section to -45 and even -41‰ at greater burial depths, of >350 m at Site 568 and >250 m at Site 570. The latter two are characteristic values for thermogenic methane. The associated C_1/C_2 values are supporting a biogenic origin, however, as mentioned above, because of possible preferential migration of C_1 this ratio alone is insufficient to exclude the possibility of a thermogenic origin of such ^{13}C-enriched methane. The main evidence against a thermogenic origin of methane is provided by the extremely high $\delta^{13}C$ values of the associated DIC, of up to +36 to +38‰ (PDB): the difference between the $\delta^{13}C$ methane of -41‰ and $\delta^{13}C_{DIC}$ of +38‰, which equals 77‰, is similar to the difference between biogenically produced methane $\delta^{13}C$ of about -70‰ and marine $\delta^{13}C_{DIC}$ of about +2‰. This indicates that CO_2 reduction is the main process by which CH_4 was derived at these sites during early biogeochemical diagenesis of organic C

[*Claypool and Kvenvolden*, 1983]. Bacterial reduction of CO_2 to CH_4 preferentially utilizes ^{12}C, depleting the residual CO_2 of it. The extreme kinetic enrichment in ^{13}C observed at these sites can be obtained by the open-system Raleigh distillation process, as shown for example by *Claypool et al.* [1985].

The methane $\delta^{13}C$ and δD data of the gas hydrate samples from ODP Site 1041 from offshore Costa Rica (ODP Leg 170) (Table 2) similarly show an overall systematic enrichment in both ^{13}C and D with burial depth between ~120 to ~260 mbsf. The C_1/C_2 ratios are, however, much lower than at similar burial depths observed at most other convergent margins DSDP and ODP sites, suggesting that unlike at offshore Guatemala at this site mixing with upward advecting thermogenic methane is the most likely process responsible for the observed depth gradients in the C and H isotopes. Diffusion alone could not explain the difference in the observed C and H isotope gradients in Table 2. The C isotopic analyses of the DIC, in progress, will allow us to identify the process responsible for these C and H isotopic gradients.

Methane Yields for Gas Hydrate Formation

The amount of methane that can be generated biogenically or thermogenically from 1 wt % of organic C remains somewhat conjectural. Only a fraction of the C in organic matter is available to the microbes for methane generation in the sediment and this varies greatly, for example, with the original type of organic matter, water depth, and burial rate. As mentioned previously, at convergent margins the ratio of marine to continental organic matter is higher than at rifted and sheared margins. Estimates of methane yields from organic matter on the order of a few tens of percent were suggested by *Claypool and Kvenvolden* [1983] and *Paull et al.* [1994]. An estimate based on methanogenesis rate relative to net productivity for marine organic matter [e.g. *Reeburgh et al.*, 1993] suggests a very low efficiency for methane generation of 2-3% from the residual fraction.

In convergent margins sediments with about 1% organic C and high sedimentation rates, pore fluid methane concentrations below the sulfate reduction zone range from about 20 to ~100 mM. 0.1% of marine organic C yields by biological remineralization ~200 mM methane [*Claypool*, personal communication]. Accordingly, assuming no methane loss or sequestration in hydrate, in a closed system of 50% porosity the commonly observed methane concentrations of 20 to 100 mM are obtained from 0.01 to 0.05% of the organic matter, hence the observed yield is low, 1-5%, similar to the low yield of 2-3% derived above based on *Reeburgh et al.* [1993] methane oxidation rate data.

The observed pore water concentrations either suggest that the fraction of the buried organic matter available for *in situ* biogenic remineralization to methane is very low, therefore insufficient for methane hydrate formation by this process; or that much of the methane produced has either diffused, was lost by advection, or has been stored in methane hydrate and the pore fluid concentrations approximate the methane hydrate solubility values.

Because of a longer history of oxidation the admixed continental organic matter provides an even lower yield of methane. Estimates for the thermogenic methane yields are based on pyrolysis data. Pyrolyses of marine organic matter (with H index 500) yields 50 mg methane per 1 g organic C, a 5% yield, whereas pyrolyses of continental organic matter (H index 200) yields only 20 mg methane per 1 g of organic C, a yield of 2% [*Claypool*, personal communication].

In convergent margin sediments with 1% organic C content the maximum available organic C for biogenic remineralization to methane, assuming all of it is marine organic C, is ~20% of the total organic C, less than the 50% suggested by *Paull et al.* [1994] and similar to the suggested values by *Claypool and Kvenvolden* [1983]. This is derived as follows: a maximum of ~ 50% of the organic matter is biogenically remineralizable, and at least 50% of the remineralizable fraction is utilized in the sulfate reduction zone [*Claypool and Kaplan*, 1974]. The remaining ~20%, 0.2 weight % marine organic C, could thus generate ~400 mM methane, equivalent to 2.5 to 4 times the methane hydrate solubility at about 2 km water depth (hydrostatic pressure). At this pressure the solubility concentration of methane hydrate ranges from ~90 mM at seafloor temperature to ~170 mM at 23°C [*Zatsepina and Buffet*, 1998]. However, from the 400 mM methane generated 100 to 150 mM must remain in the pore fluid to stabilize the gas hydrate, hence only 300 to 250 mM will be available for methane hydrate formation. These are maximum values, less methane will be generated because in convergent margins 20 to >50% of the organic C is of continental origin having lower biogenic and thermogenic methane yields, and loss of methane by diffusion was not accounted for. Hence, under optimal conditions, in a 50% porosity sediment, only minor amounts, a maximum of 2 - 4% of the pore volume, could be filled with methane hydrate by *in situ* biogenic methane generation. The higher concentrations observed in convergent margins with up to >20% pore space filling in a disseminated mode based on Cl concentrations, can only be obtained with addition of methane from methane-rich upward migrating fluids. Similarly in the Blake Ridge rifted and sheared margin setting, the ~9% pore space gas hydrate occupancy reported by *Dickens et al* [1997] cannot be accounted for by only *in situ* biogenic methane generation. Migration of methane-rich fluids induced by compaction and methane recycling at the BSR are probably responsible for this observation. In convergent margins fluid migration is more pervasive and aggressive than in rifted and sheared margins, and on average in convergent margins the ratio of marine to continental organic matter is higher than in rifted and sheared margins. Therefore, convergent margins provide a much more favorable environment for methane hydrate formation.

Based on the above yield analysis and the areal distributions of convergent and rifted and sheared margins within the methane hydrate stability field [i.e. *Mendard and Smith*, 1966; *Southam and Hay*, 1981; *von Huene and Scholl,* 1991; *Rea and Ruff,* 1996], exclusive of shelf and permafrost areas, it is estimated that about two-thirds, 60-65% of the vast marine gas hydrate reservoir is situated in convergent margins.

Pore Fluids Geochemistry and Estimates of Hydrate Abundances

In addition to data from actual recoveries of core samples, estimates of gas hydrate occurrence, concentration, and distribution can be achieved by:

1. Remote geophysical analyses or borehole logging. As discussed earlier, the presence of a BSR with a negative polarity indicates that some methane hydrate is present, but not how much of the amplitude is caused by hydrate and/or free gas. As yet, no other geophysical method that provides accurate yields is available. Empirical and experimental work is in progress on a large range of methods. For summaries on the geophysics of gas hydrate see for example *Hyndman and Davis* [1992], *MacKay et al.* [1994]; *von Huene and Pecher* [1999]; and references therein.
2. Measuring temperature of sediment cores containing gas hydrate. Methane hydrate dissociation is a strong endothermic reaction, [*Sloan* 1998]. Therefore during core recovery methane hydrate dissociation decreases the core temperature below the *in situ* core temperature. Lowest core temperatures measured are approximately -2° C, the freezing temperature of the pore water. The very low core temperatures, produced by dissociation of small to moderate amounts of gas hydrate "stabilize" the residual methane hydrate by forming sea ice that "protects" it, slows the dissociation rate of the remaining gas hydrate. The protective ice coating is visually distinguishable from the methane hydrate. Therefore, as suggested by *Kastner et al.* [1998] the lowering of the core temperature although an excellent indicator that gas hydrate has dissociated cannot be used to estimate reliably the amount lost, hence the pore-space occupancy by the hydrate.
3. Pore fluid geochemical analyses. Lower than seawater Cl concentrations are ubiquitous in convergent margin sediments from the seafloor to the BSR depth. Seawater Cl dilutions of 20 to 75% have been reported. The lowest observed Cl values are summarized in Table 3. The most distinct characteristics of the Cl concentration-depth profiles are:
    - The overall progressive increase in dilution with depth, with lowest concentrations near the BSR. Site 892 (Fig. 5) is an exception, as at this site near seafloor disseminated and massive hydrate occurs [*Kastner et al.*, 1988; *Suess et al.*, 1999]; the more extensive dissociation of the gas hydrate that occurred during recovery produced a stronger Cl dilution in the pore fluids of the shallowest core sediments than at the BSR.
    - Minima in Cl concentrations superimposed on the overall decrease in concentration. For example at Site 892 (Fig. 5).

The progressive dilution with depth of the pore fluids has been attributed largely to sampling artifact that causes gas hydrate partial or complete dissociation [*Hesse and Harrison*, 1981]. Indeed at every convergent margin site with pore fluid Cl dilution of >20% gas hydrate has been recovered or has been inferred to have been present but dissociated. The inference is based on the coincidence of the pore fluids low Cl concentrations with very high dissolved methane concentrations and higher than seawater $\delta^{18}O$ values. However, low Cl fluids also exist at hundreds of meters below the BSR depths and at convergent margins with no gas hydrates (e.g. Marianas). Therefore, as suggested by *Kastner et al.*, [1995, 1998] the % Cl dilution depth-profiles cannot be translated directly into percent pore fluid occupancy by gas hydrate that has decomposed, they have to be corrected for other Cl-diluting fluid-sediment processes.

The Cl dilution depth profiles also indicate the mode of the *in situ* occurrence of the hydrate. Figure 5 shows two distinct Cl dilution profiles from the Cascadia margin, ODP sites 889 and 892. At Site 889 disseminated gas hydrate dominates in the section with increasing Cl dilution toward the BSR. Whereas, at Site 892 superimposed on the disseminated hydrate profile are sediment horizons with higher concentrations of gas hydrate, probably because of higher permeabilities in select turbidite horizons at this section. The maximum seawater dilution at Site 889 is 36%, and at Site 892 is 15%, which would correspond to 39% and 16% pore volume

Table 3. Minimum Cl values recorded in convergent margin pore fluids.

Location	Max. dilution of seawater (%)	Suggested Origin	References
WESTERN PACIFIC			
Nankai Trench	20	Deep-seated hydration.	*Kastner et al.* [1992]
Japan Trench			
Deep sites	<5	?	
Shallow sites	50	Fossil aquifer.	*Moore & Gieskes* [1980]
New Hebrides	25	Deep-seated dehydration ± meteoric water.	*Martin et al.* [1992]
Mariana			
Conical Seamount	57	Deep-seated dehydration.	*Mottl* [1992]
EASTERN PACIFIC			
Peru Margin			
Lower slope & accretion complex	20	Deep-seated dehydration and gas hydrate dissociation.	*Kastner et al.* [1990]
Upper slope shallow site	40	Meteoric water.	*Kastner et al.* [1990]
Middle American Trench			
Off Guatemala & Mexico	54	Gas-hydrate dissociation.	*Hesse & Harrison* [1981]; *Gieskes et al.* [1985]
Off Costa Rica	37 to 75	Deep-seated dehydration ± gas hydrate dissociation.	*Kastner*, in preparation
Cascadia			
Off Vancouver & Central Oregon	38	Gas hydrate dissociation ± deep-seated dehydration.	*Kastner et al.* [1995]
ATLANTIC			
Barbados			
Drill Sites	30	Deep-seated dehydration.	*Gieskes et al.* [1990]; *Kastner et al.* [1995]
Mud Volcanoes	55 to 64	Deep-seated dehydration and gas hydrate dissociation.	*Martin et al.* [1993]; *Martin* [1994]

occupancy by methane hydrate, assuming hydrate dissociation has caused all the observed dilution. However, other supporting geochemical proxies, particularly Sr, Li, and Cl isotope ratios [*Kastner*, 1995, 1998; *Ransom et al.*, 1995] and the presence of traces of thermogenic hydrocarbons [*Whiticar*, 1995], indicate that a deep-seated fluid diluted by hydrous silicate transformation reactions is migrating upward in the region and is mixing with the *in situ* pore fluid. The Cl dilution profiles, when corrected for mixing with the deep-seated low-Cl fluid, suggest pore-volume methane hydrate occupancy in a disseminated mode of a minimum of 15% to ≥20% at ODP Site 889 and a maximum of ~10% occupancy at the BSR at ODP Site 892. Based on extensive geophysical data analyses *Hyndman et al*, [1992] have suggested 25 to 30% pore space occupancy by gas hydrate near the BSR at Site 889. The geochemical and geophysical estimates for ODP Site 889 are remarkably similar.

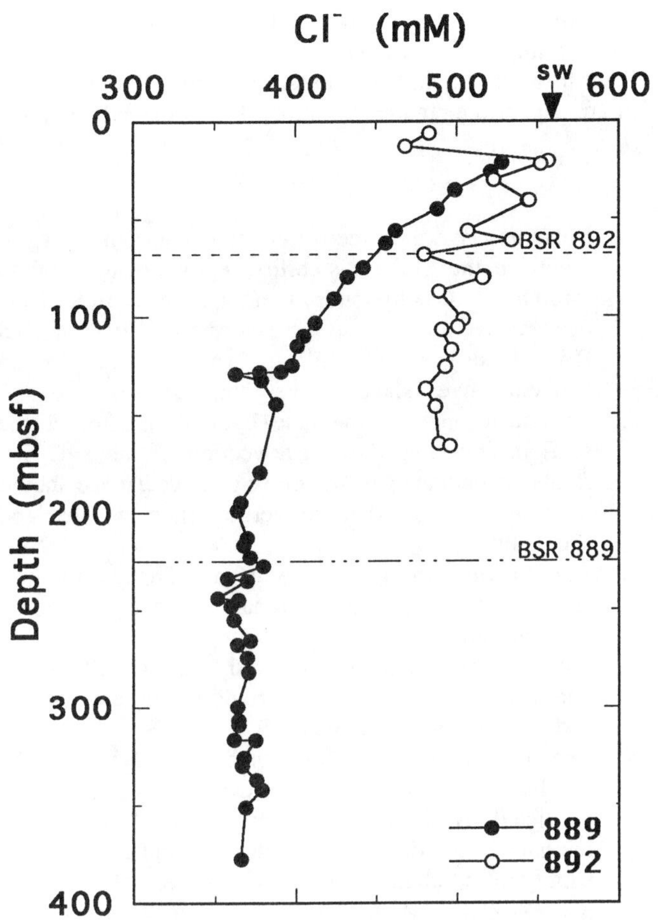

Figure 5. Depth profiles for Cl concentrations in pore fluids at ODP Sites 889 and 892 in the Cascadia convergent margin. The depths of the distinct BSRs at the two sites are marked for Site 889, offshore Victoria, by a dashed line, and for Site 892, offshore central Oregon, by a solid line. At both sites, upon recovery, dissociation of disseminated methane hydrate is responsible for most of the observed Cl dilution. The Cl minima at site 892 mark enriched in gas hydrate sediment horizons.

RESPONSES OF THE CONVERGENT MARGINS GAS HYDRATE RESERVOIR TO ENVIRONMENTAL STRESSES

The interest in the role of gas hydrates in global change and other environmental stresses stems from both the oceanic and atmospheric connections.

Most of the methane released from the seafloor will be oxidized in the ocean, some of it may result in seafloor carbonate formation, and the non-oxidized residual fraction will escape to the atmosphere. The ratio between the oxidized and residual fractions depends mostly on the methane flux through the seafloor, the water depth, and rate of microbial oxidation of the methane. The chemical effects of these reactions in the ocean and atmosphere have important societal implications. Methane flux into the ocean will lower the local seawater oxygen content via microbial oxidation of methane as indicated in equation (I); if the methane flux is significant, because of the very low ^{13}C values of methane, the oxidized methane will effect the $\delta^{13}C$ value of the dissolved inorganic carbon (DIC); this may be recorded in the $\delta^{13}C$ of select benthic foraminifers. The CO_2 produced would enhance carbonate dissolution, as shown in equation (II). Carbonate precipitation occurs according to equation (III) following dissolved oxygen consumption; anaerobic methane oxidation is coupled with sulfate reduction.

$$CH_4 + 2O_2 = CO_2 + 2H_2O \qquad \text{(I)}$$
$$CaCO_3 + CO_2 + H_2O = 2HCO_3^- + Ca^{2+} \qquad \text{(II)}$$
$$CH_4 + SO_4^{2-} + Ca^{2+} = CaCO_3 + H_2S + H_2O \qquad \text{(III)}$$

The sequestration or C in authigenic carbonates in margins, formation versus dissolution, is thus dependent on the intensity of methane flux through the seafloor in space and time. A flux of methane to the atmosphere from a seafloor source of just 0.5% of the annual atmospheric methane flux would significantly accelerate global warming. The role of methane released in convergent margins on these environmental processes is thus evaluated below.

The most direct way to reconstruct the geologic history of atmospheric methane concentration is to analyze ice core gases. The combined record of gas data from ice cores from Greenland and Antarctica over the past 160 kyrs shows (1) that at the post glacial-interglacial climatic transition, atmospheric CH_4 concentrations nearly doubled from ~350 ppbv to ~650 to 700 ppbv; (2) in the Holocene a rather constant concentration of 650 - 700 ppbv is observed up to ~200 years ago: (3) over the past 200 years CH_4 concentrations more than doubled, increased to the present level of ~1800 ppbv, but especially over the past 25 years the increase has accelerated to about 0.8% per year; the recent ~1% imbalance of the CH_4 budget, which amounts to ~18 ppbv has been attributed to human activities; [*Graedel and Crutzen*, 1993; *Raynaud and Chappellaz*, 1993]; (4) the changes in CH_4 concentrations over the past 160 kyrs closely track the atmospheric CO_2 concentration and SST records over the same time span [*Stauffer et al.*, 1988; *Raynaud and Chappellaz*, 1993]. However, recent high resolution studies of ice cores indicate that deglaciation in Antarctica started several millennia before the observed rapid rise in methane concentrations [*Blunier et al.*, 1998]. Similarly, *Severinghaus and Brook* [1999] have documented that in Greenland ice methane increase lags the rapid large temperature increase by a few decades. The timing of the methane change suggests that methane from gas hydrate did not play a role in triggering climate change at the last deglaciation. Although, the processes and rates leading to methane hydrate dissociation, and the duration of feedback scenarios of methane release and enhanced global warming are not known [e.g. *McIver*, 1981; *Nisbet et al.*, 1992]. A major catastrophic dissociation event is required to support the rapid warming in a few decades recorded in the ice cores [*Severinghaus*, 1999], and for which as yet there is no evidence.

The seafloor hydrates are more vulnerable to dissociation in response to global warming than the deeper buried ones. *Broecker and Peng* [1993] observed penetration depth of bomb ^{14}C and tritium to 360 m in 12 years. Similarly, *Brewer et al.* [1998] have observed chlorofluorocarbon (CFC) penetration to ~550m water depth in the Eel River Basin in ~30 years, and *Kastner et al.* [1998] reported CFC penetration to ~750 m water depth at Hydrate Ridge, Cascadia Margin, at a similar rate, in ~35 years. Mixing in the ocean may be approximated by eddy diffusivity, and diffusive penetration is approximated the square root of time. Eddy diffusivity of a temperature change should be the same as for tracers like ^{14}C or tritium. Accordingly, temperature change in the surface ocean will penetrate to 550 m in $12(550/360)^2 = 28$ years and to 750 m in 51 years; these calculations provide penetration times that are very similar to the above values obtained for CFC. A warming pulse may, however, take centuries to millennia to reach the deeper hydrates at the BSR [*MacDonald*, 1990].

Decomposition of the deeper gas hydrate could be enhanced by massive slumping triggered during glacials by pressure decrease from sea level drop [e.g. *McIver*, 1981; *Nisbet*, 1992], and particularly by tectonic activity in convergent margins because of the steep slopes. Even if the release of methane is gradual, the large reservoir in convergent (and rifted and sheared) margins may have a positive feedback role in global warming. Marine methane hydrates contain $\geq 10^{19}$ g C. On the basis of the above analysis of the spatial distribution of methane hydrate, assuming that 60-65% of the reservoir is stored in convergent margins, and that global warming over 100 years will cause ocean warming by 2° C, the warming will drop the gas hydrate stability field by ~70 m. Near the seafloor methane hydrate concentration is lower than close to the BSR if we assume that only 10% of the convergent margins hydrate reservoir of ~0.65 x 10^{19} g C will decompose, at least 0.65 x 10^{18} g of methane-related C will be released.

- If released over 100 years this will equal 0.65 x 10^{16} g C yr^{-1}. This is about 13 times the present atmospheric methane flux.
- If 99% of this methane flux will oxidize in the ocean, the 1% residual equals ~13% of the modern atmospheric flux.

A realistic scenario may be a release somewhere between the two above end members.

Most recent attempts to model the early geologic history of atmospheric methane do not involve methane hydrates.

MAIN CONCLUSIONS

1. Most gas hydrates recovered at convergent margins are from the 2-phase stability region, from ≥100 to <400 mbsf. Gas hydrates at and near the seafloor have been recovered from offshore northern California (Eel River Basin) and Cascadia (offshore central Oregon and Vancouver Island) convergent margin.
2. The occurrence of methane hydrate near or at the seafloor indicates upward advection of methane-rich fluids or methane gas, therefore this occurrence should be more widespread in convergent than in rifted and sheared margins;
3. At convergent margins BSRs are mostly discontinuous and patchy (in conventional seismic reflection records), and where observed the amplitude is rather strong. At rifted and sheared margins BSRs are continuous, extend over large regions, are less distinct, and have lower amplitudes.
4. The BSRs may occur above or below the base of the equilibrium stability field. Occurrence above the predicted BSR depth suggests limited supply of methane, and occurrence below it indicates ample supply of methane leading to transient intervals of overpressure below the base of the equilibrium stability field.
5. At both convergent and rifted and sheared margins *in situ* biogenic methane generation and upward transport of methane into the stability field, induced by compaction as well as methane recycling at the BSR, occur. At convergent margins, however, pore fluid expulsion and methane-rich fluid migration is intense, widespread, and important, therefore the pore volume occupancy by methane hydrate is generally higher and the spatial distribution is more irregular than at rifted and sheared margins.
6. Except for massive gas hydrate occurrences associated with faults and fractures, common in convergent margins, an overall increase in methane hydrate pore volume occupancy toward the BSR must evolve at both types of margins; the pore volume occupancy and gradient are, however, steeper at convergent margins where migration of methane is important. Whereas, in rifted and sheared margins *in situ* methane production is the main source of methane therefore the pore volume occupancies are low, from minor amounts (1-2%) to <10%.
7. Estimates of sediment pore volume occupancies by methane hydrate in convergent margin drilled and

cored sediments, that range from a few percent to >25%, include inferences of the amount dissociated during recovery operations; the inferences are based on remote sensing geophysical methods and comprehensive pore fluid chemical data. Chloride concentrations in conjunction with at least Cl, Sr (and Li) isotopic ratios, provide a rather accurate estimate of the amount of dissociated gas hydrate. The estimates are similar to those based on organic matter content and type and methane-rich fluid migration.

8. Convergent margins have several distinct tectonic and hydrologic characteristics that together with the physical-chemical constraints of methane hydrate stability, enhance the formation of gas hydrate relative to rifted and sheared margins, and hence provide a more favorable environment for gas hydrate formation. Based on global areal distribution of margin types, tectonics, hydrology, sedimentology, and organic matter content and type, thus potential of methane yield, it is estimated that, exclusive of shelf and permafrost areas, approximately two-thirds (60-65%) of the vast marine gas hydrate reservoir is situated in convergent margins.
9. Regional continuous or episodic extensive venting of methane through the seafloor is more distinctive at convergent margins. It influences seawater chemistry locally, primarily by oxygen consumption and CO_2 production via microbial oxidation of methane. Catastrophic events of methane release from hydrate, by massive slumping, for example, may cause oceanic anoxia and somewhat reduce ocean's capacity of CO_2 incorporation. The non-oxidized methane will be released to the atmosphere and enhance global warming.
10. Because of the relatively large volume of gas hydrate situated in convergent margins, this environment is of particular importance for considerations of the role of methane hydrate in the oceanic C cycling, and of the potential responses of gas hydrate to environmental stresses.

Acknowledgements. I thank Roland von Huene and Steve Eittreim for their helpful suggestions and comments, and Peter Lonsdale, Wolfgang Berger, and Bill Hay for useful discussions. This research and synthesis was partially supported by NSF Grants OCE 97-12135 and OCE 98-17612.

REFERENCES

Baker, V.R., Strom, R.G., Gulick, V.C., Kargel, J.S., Komatsu, G., and Kale, V.S., Ancient oceans, ice sheets and the hydrological cycle on Mars, *Nature*, 352, 589-594, 1991.

Behl, R.J., and Kennett, J.P., Brief interstadial events in the Santa Barbara basin, NE Pacific, during the past 60 kyr, *Nature*, 379, 243-246, 1996.

Blunier, T., Chappellaz, J., Schwander, J., Dällenbach, A., Stauffer, B., Stocker, T.F., Raynaud, D., Jouzel, J., Clausen, H.B., Hammer, C.U., and Johnsen, S.J., Asynchrony of Antarctica and Greenland climate change during the last glacial period, *Nature*, 394, 739-743, 1998.

Bonham, L.C., Migration of hydrocarbons in compacting basins, *AAPG Bulletin*, 64, 549-567, 1980.

Booth, J.S., Rowe, M.M., and Fischer, K.M., Offshore gas hydrate sample database with an overview and preliminary analysis, *USGS Open-File Report*, 96-272, 1996.

Brewer, P.G., Gas hydrates and global climatic change, in: *Gas Hydrates, Challenges for the Future*, Holder, G.P. and Bishoni, P.R., eds., Annals, N.Y. Acad. Sci., 912, 195-199, 1999.

Broecker, W.S., and Peng, T-H., Evaluation of the ^{13}C constraint of the uptake of fossil fuel CO_2 by the ocean, *Global Biogeoch. Cycl.*, 7, 619-626, 1993.

Brooks, J.M., Field, M.E., and Kennicut, M.C., Observations of gas hydrates in marine sediments, offshore northern California, *Mar. Geol.*, 96, 103-109, 1991.

Buffet, B.A., and Zatsepina, O.Y., Formation of gas hydrate from dissolved gas in natural porous media, *Mar. Geol.*, 164, 69-77, 2000.

Carson, B., Seke, E., Paskevich, V., and Holmes, M.L., Fluid expulsion sites on the Cascadia accretionary prism: Mapping diagenetic deposits with processed GLORIA imagery, *J. Geophys. Res.*, 99, 11,959-11,969, 1994.

Claypool, G.E., and Kaplan, I.R., The origin and distribution of methane in marine sediments, in: *Natural Gases in Marine Sediments*, Kaplan, I.R., ed., Plenum Press, 99-139, 1974.

Claypool, G., and Kvenvolden, K.A., Methane and other hydrocarbon gases in marine sediments, *Ann. Rev. Earth Planet. Sci.*, 11, 299-327, 1983.

Claypool, G.E., Threlkeld, C.N., Mankiewicz, P.N., Arthur, M.A., and Anderson, T.F., Isotopic composition of interstitial fluids and origin of methane in slope sediments of the Middle America Trench, Deep Seas Drilling Project Leg 84, in: *Initial Reports of the Deep Sea Drilling Project*, 84, von Huene, R., Aubouin, J., et al., Washington, DC, U.S. Government Printing Office, 683-691, 1985.

Clennell, M.B., Hovland, M., Booth, J.S., Henry, P., and Winters, W.J., Formation of natural gas hydrates in marine sediments, Part 1: Conceptual model of gas hydrate growth conditioned by host sediment properties, *J. Geophys. Res.*, 104, 22,985-23,003, 1999.

Craig, H., and Hom, B., Relationship between deuterium oxygen-18 and chlorinity in the formation of sea ice, *Trans. Am. Geophys. Union*, 49, 216-217, 1968.

Davidson, D.W., El-Defrawy, M.K., Fuglem, M.O., and Judge, A.S., Natural gas hydrates in northern Canada, *Proc. 3rd Int. Conf. Permafrost*, 3, 937-943, 1978.

de Angelis, M.A., Lilley, M.D., Olson, E.J., and Baross, J.A., Methane oxidation in deep-sea hydrothermal plumes of the Endeavour segment of the Juan de Fuca Ridge, *Deep Sea Res.*, 40, 1169-1186, 1993.

Dickens, G.R., Paull, C.K., Wallace, P., and ODP Leg 164

Scientific Party, Direct measurement of *in situ* methane quantities in a large gas-hydrate reservoir, *Nature*, 385, 426-428, 1997.

Dickens, G.R., and Quinby-Hunt, M.S., Methane hydrate stability in pore water: a simple theoretical approach for geophysical applications, *J. Geophys. Res.*, 102, 773-783, 1997.

Edwards, R.N., On the resource evaluation of marine gas hydrate deposits using seafloor transient electric dipole-dipole methods, *Geophysics*, 62, 63-74, 1997.

Englezos, P., and Bishnoi, P.R., Predictions of gas hydrate formation conditions in aqueous solutions, *Am. Inst. Chem. Eng.*, 34, 1718-1721, 1988.

Field, M.E., and Kvenvolden, K.A., Gas hydrates on the northern California margin, *Geology*, 13, 517-520, 1985.

Gieskes, J.M., Vrolijk, P., and Blanc, G., Hydrogeochemistry, ODP Leg 110: An overview, *Proc. Ocean Drill. Program Sci. Results*, 110, 395-408, 1990.

Ginsburg, G.D., and Soloviev, V., Methane migration within the submarine gas hydrate stability zone under deep-water conditions, *Marine Geol.*, 37, 49-57, 1997

Gornitz, V., and Fung, I., Potential distribution of methane hydrates in the World's oceans, *Global Biogeoche. Cycles*, 8, 335-347, 1994.

Graedel, T.E., and Crutzen, P.J., *Atmospheric Change, An Earth System Perspective*, Freeman, 1993.

Handa, Y.P., Effect of hydrostatic pressure and salinity on the stability of gas hydrates, *J. Phys. Chem.*, 94, 2652-2657, 1990.

Harrison, W.E., Hess, R., and Gieskes, J.M., Relationship between sedimentary facies and interstitial water chemistry of slope, trench, and Cocos Plate sites from the Middle America Trench transect, active margin off Guatemala, Deep Sea Drilling Project Leg 67, in: *Initial Reports of the Deep Sea Drilling Project*, 67, Aubouin, J., von Hume, R. et al., Washington, DC, U.S. Government Printing Office, 603-613, 1982.

Hesse, R., and Harrison, W.E., Gas hydrates (clathrates) causing pore-water freshening and oxygen isotope fractionation in deep-water sedimentary sections of terrigenous continental margins, *Earth Planet. Sci. Lett.*, 55, 453-462, 1981.

Holder, G.D., Zetts, S.P., and Pradhan, N., Phase behavior in systems containing clathrate hydrates, *Rev. Chem. Eng.*, 5, 1-70, 1988.

Hyndman, R.D., Quantitative estimates of deep sea gas hydrate and underlying free gas, *EOS*, 78, 340, 1997.

Hyndman, R.D., and Davis, E.E., A mechanism for the formation of methane hydrate and seafloor bottom simulating reflectors by vertical fluid expulsion, *J. Geophys. Res.*, 97, 7025-7041, 1992.

Kastner, M., Elderfield, H., and Martin, J.B., Fluids in convergent margins: What do we know about their composition, origin, role in diagenesis, and importancefor oceanic chemical fluxes?, *Philosophical Trans. Royal Soc. London*, A335, 243-259, 1991.

Kastner, M., Kvenvolden, K.A., and Lorenson, T., Chemistry, isotopic composition, and origin of methane hydrogen sulfide hydrate at the Cascadia subduction zone, *Earth Planet. Sci. Lett.*, 156, 173-183, 1998.

Kastner, M., Kvenvolden, K.A., Whiticar, M.J., Camerlenghi, A., and Lorenson, T.D., Relation between pore fluid chemistry and gas hydrate associated with bottom-simulating reflectors at the Cascadia margin, sites 889 and 892, in: *Proc. Ocean Drilling Program Sci. Res.*, Carson, B., Westbrook, R.J., Musgrave, R.J., and Suess, E., eds., 146, 175-187, 1995.

Kastner, M., Paytan, A., Volmer, M., Ransom, B., Bidle, K., Bartlett, D., Robertson, G., Jannasch, H., Plant, J., Carson, B. and Jaeger, J., Methane advection, methane hydrate formation and possibly dissociation at the Cascadia margin off central Oregon, *Trans. Amer. Geophys. Union*, 79, 419, 1998.

Kennett, J.P., Cannariato, K.G., Hendy, I.L., and Behl, R.J., Carbon isotope evidence for methane hydrate instability during Quaternary Interstadials, *Science*, 288, 128-133, 2000.

Khalil, M.A.K., and Shearer, M.J., Sources of methane: An overview, in: *NATO Series*, 113, 180-198, 1993.

Kimura, G., Silver, E., Blum, P. et al., in: *Proc. ODP Initial Reports* , 170, 458 p., Texas A&M University, Ocean Drilling Program, College Station, TX, USA, 1997.

Kulm, L.D., Suess, E., Moore, J.C., Carson, B., Lewis, B.T.R., Ritger, S.D., Kadko, D.C., Thornberg, T.M., Embley, R.M., Rugh, W.D., Massoth, G.J., Langseth, M.R., Cochrane, G.R., and Scamman, R., Oregon subduction zone: venting, fauna, and carbonates, *Science*, 231, 561-566, 1986.

Kvenvolden, K.A., Methane hydrate - a major reservoir of carbon in the shallow geosphere?, *Chem. Geol.*, 71, 41-51, 1988.

Kvenvolden, K.A., Gas hydrate as a potential energy source - a review of the methane content, in: *The Future of Energy Gases, U.S. Geol. Survey Prof. Paper* 1570, 555-561, 1993.

Kvenvolden, K.A., Gas hydrate - Geological perspective and global change, *Review of Geophysics*, 31, 173-187, 1993.

Kvenvolden, K.A., Potential effects of gas hydrate on human welfare, *Proc. Natl. Acad. Sci.*, 96, 3420-3426, 1999.

Kvenvolden, K.A., and Barnard, L.A., Gas hydrates of the Blake Outer Ridge, Deep Sea Drilling Project Site 553, Leg 76, in: *Initial Reports of the Deep Sea Drilling Project*, Sheriden, R. and Gradstein, F., eds., 75, 353-366, Washington, DC, U.S. Government Printing Office, 1983.

Kvenvolden, K.A., and Barnard, L.A., Hydrates of natural gas in continental margins, in: Watkins, J.S. and Drake, C.L., eds., Studies in Continental Marine Geology: *Amer. Assoc. of Petrol. Geol. Memoir* No. 34, 631-640, 1983.

Kvenvolden, K.A., Claypool, G.E., Threlkeld, C,N., and Sloan, E.D., Geochemistry of a naturally occurring massive marine gas hydrate, *Organic Geochemistry*, 6, 703-713, 1984.

Kvenvolden, K.A., Ginsburg, G.D., and Soloviev, V.A., World-wide distribution of subaquatic hydrates, *Geo-Mar. Lett.*, 13, 32-40, 1993.

Kvenvolden, K.A., and Kastner, M., Gas hydrate of the Peruvian outer continental margin, in: *Proc. Ocean Drill. Prog. Sci. Results*, Suess, E. and von Huene, R., eds., 112, 517-526. College Station, Ocean Drilling Program, 1990.

Kvenvolden, K.A., and MacDonald, T.J., Gas hydrates of the Middle America Trench-Deep Sea Drilling Project Leg 84, in: *Initial Reports Ocean Drilling Project*, 84, von Huene, R., Aubuoin, J., et al., Washington, DC, U.S. Government Printing Office, 667-682, 1985.

Kvenvolden, K.A., and McMenamin, M.A., Hydrates of natural gas: a review of their geologic occurrence, *U.S. Geol. Surv. Circ.*, 825, 11, 1980.

Larson, R.L., Latest pulse of Earth: evidence for mid-Cretaceous superplume, *Geology*, 19, 547-550, 1991.

Lashof, D.A., and Ahuja, D.R., Relative contributions of greenhouse gas emissions to global warming, *Nature*, 344, 529-539, 1990.

Le Pichon, X., Foucher, J.-P., Boulégue, J., Henry, P., Lallement, S., Benedetti, M., Avedik, F., and Mariotti, A., Mud volcano field seaward of the Barbados accretionary complex: A submersible survey, *J. Geophys. Res.*, 95, 8931-8943, 1990.

Levitus, S., Antonov, J.I., Boyer, T.P., and Stephens, C., Warming of the world ocean, *Science,* 287, 2225-2229, 2000.

Lücker, A., Kastner, M., Littke, R., and Cramer, B., Gas in marine sediments from the Costa Rica subduction zone, ODP leg 170: primary composition and secondary alteration, "in preparation."

Lunine, J.I., and Stevenson, D.J., Thermodynamics of clathrate hydrate at low and high pressure with application to the outer solar system, *Astrophys. J. Suppl. Ser.*, 58, 493-531, 1985.

Lunine, J.I., and Stevenson, D.J., Clathrate and ammonia hydrates at high pressure: application to the origin of methane on Titan, *Icarus*, 70, 61-77, 1987.

MacDonald, G.J., Role of methane clathrates in past and future climates, *Clim. Change*, 16, 247-281, 1990.

MacDonald, G.J., The future of methane as an energy resource, *Annual Review of Energy*, 15, 53-83, 1990.

MacDonald, I.R., Guinasso Jr., N.L., Sassen, R., Brooks, J.M., Lee, L., and Scott, K.T., Gas hydrate that breaches the sea floor on the continental slope of the Gulf of Mexico, *Geology*, 22, 699-702, 1994.

MacKay, M.E., Jarrard, R.D., Westbrook, G.K., Hyndman, R.D., and the Shipboard Scientific Party of ODP Leg 146, Origin of bottom simulating reflectors: Geophysical evidence from the Cascadia accretionary prism, *Geology*, 22, 459-462, 1994.

Martin, J.B., and Kastner, M., Chemical and isotopic evidence for sources of fluids in a mud volcano field seaward of the Barbados accretionary wedge, *J. Geophys. Res.*, 101, 20,325-20,345, 1996.

Metha, A.D., and Stoan, E.D., Thermodynamic model for structure H hydrates, *AICHE J.*, 40, 312-316, 1994.

McIver, R.D., Role of naturally occurring gas hydrates in sediment transport, *Amer. Assoc. Petrol. Geol. Bull.*, 66, 789-792, 1982.

Menard, H.W., and Smith, S.M., Hypsometry of ocean basin provinces, *J. Geophys. Res.*, 71, 4305-4325, 1966.

Miller, S.L., The nature and occurrence of clathrate hydrates, in: *Natural Gases in the Marine Sediments*, Kaplan, I.R. ed., Plenum Press, 151-177, 1974.

Miller, S.L., Clathrate hydrates in the solar system, in: *Ices in the Solar System*, Klinger, J., et al., Reidel Publ. Co., 59-79, 1985.

Miller, S.L., and Smythe, W.D., Carbon dioxide clathrate in the Martian ice cap, *Science*,170, 531-533, 1970..

Milton, D.J., Carbon dioxide hydrates and floods on Mars, *Science*, 183, 654-656, 1974.

Nisbet, E.G., The end of the ice age, *Can. J. Earth Sci.,* 27, 148-157, 1990.

Nisbet, E.G., and Piper, D.J.W., Giant submarine landslides, *Nature*, 392, 329-330, 1998.

O'Neil, J.R., Hydrogen and oxygen isotope fractionation between ice and water, *J. Phys. Chem.*,72, 3683-3684, 1968.

Paull, C.K., Matsumoto, R., Wallace, P.J., et al., in: *Initial Reports of the Ocean Drilling Program 146A*, College Station, Texas, 623 p, 1996.

Paull, C.K., and Ussler, W., III, Are low salinity anomalies below BSRs a consequence of interstitial gas bubble barrier?, *EOS,* 78, 339, 1997.

Paull, C.K., and Ussler, W., III and Borowski, W.S., Source of methane to form marine gas hydrates, in: *Natural Gas Hydrates*, Sloan, E.D., Happel, Jr., J., and Hnatow, M.A., eds., 161-176. New York: NY Acad. Sci., 1994.

Pflaum, R.C., Brooks, J.M., Cox, H.B., Kenicutt, M.C., and Sheu, D.D., Molecular and isotopic analysis of core gases and gas hydrates, in: *Initial Reports, Deep Sea Drilling Project Leg 96*, Bouma, A.H., Coleman, J.M., Meyer, A.W., et al., eds., U.S. Government Printing Office, Washington, DC, 781-784, 1996.

Premuzic, E.T., Benkovitz, C.M., Gaffney, J.S., and Walsh, J.J., The nature and distribution of organic matter in the surface sediments of world oceans and seas, *Organic Geochemistry*, 4, 63-77, 1982.

Ransom, B., Spivack, A.J., and Kastner, M., Stable Cl isotopes in subduction-zone pore waters: implications for fluid-rock reactions and the cycling of chlorine, *Geology*, 23, 715-718, 1995.

Raynaud, D. and Chappellaz, J., The record of atmospheric methane, in*: NATO Series*, 113, Khalil, M.A.K., ed., 38-61, Springer-Verlag, 1993.

Rea, D.K. and Ruff, L.S., Composition and mass flux of sediment entering the World's subduction zones: Implications for global sediment budgets, great earthquakes, and volcanism. *Earth Planet. Sci. Lett.*, 140, 1-12, 1996.

Reeburgh, W.S., Whalen, S.C., and Alperin, M.J., The role of methylotrophy in the global methane budget, in: *Microbiology of C compounds*, Murrell, J.C. and Kelly, D.P. eds., Intercept, Andover UK, 1-14, 1993.

Rempel, A., and Buffet, B., Formation and accumulation of gas hydrate in porous media, *J. Geophys. Res.*, 102, 10,151 - 10,164, 1997.

Rempel, A.W., and Buffett, B.A., Mathematical models of gas hydrate accumulation, in: Gas Hydrates*: Relevance to World Margin Stability and Climate Change*, Henriet, J.P. and Meinert, J., eds., 137, 63-74. London: Geological Society, 1998.

Revelle, R.R., Methane hydrates in continental slope sediments and increasing atmospheric carbon dioxide, in: *Changing Climate: Report of the Carbon Dioxide Assessment Committee*, 252-261, National Academy Press, 1983.

Ritger, S., Carson, B., and Suess, E., Methane-derived authigenic carbonates formed by subduction-induced pore-water expulsion along the Oregon/Washington margin, *Geol. Soc. Amer. Bull.*, 98, 147-156, 1987.

Sassen, R., and MacDonald, I.R., Evidence for the structure H hydrate, Gulf of Mexico continental slope, *Organic Geochem.,* 22, 1029-1032, 1994.

Schoell, M., The hydrogen and C isotopic composition of methane from natural gases of various origins, *Geochim. Cosmochim. Acta*, 44, 649-666, 1980.

Severinghaus, J.P., and Brook, E.J., Abrupt climate change at the end of the last glacial period inferred from trapped air in polar ice, *Science*, 286, 930-934, 1999.

Shipboard Scientific Party, Site 808, in*: Proceedings, Ocean Drilling Program, Initial Reports*, 131, Taira, A., Hill, I., Firth, J.C. et al., eds., 71-269, Ocean Drilling Program, College Station, TX, 1991.

Shipley, T.H., and Didyk, B.M., Occurrence of methane hydrate offshore southern Mexico, in: *Initial Reports of the Deep Sea Drilling Project*, 66, 547-555, Watkins, J.S., Moore, J.C., et al., eds., Washington DC, U.S. Government Printing Office, 1982.

Sloan , E. D. Jr., *Clathrate Hydrates of Natural Gases*, Marcel Dekker, New York, 641 p, 1990.

Sloan, E.D. Jr., *Clathrate Hydrates of Natural Gases* (second edition), New York, Marcel Dekker, 730 p, 1998.

Southam, J.R., and Hay, W.H., Global sedimentary mass balance and sea level change, in: *The Sea*, vol. 7, Emiliani, C. ed., 1617-1684, Wiley.

Spence, G., Riedel, M., Novosel, I., Solem, C., Hyndman, R., Chapman, R., and Kastner, M., Seafloor gas hydrate on an active vent offshore Vancouver Island: seismic, coring, and heat flow investigations, *Trans. Amer. Geophys. Union*, 81, 603, 2000.

Stauffer, B., Lohbronner, E., Oeschger, H., and Schwander, J., Methane concentration in glacial atmosphere was only half that of the preindustrial Holocene, *Nature*, 332, 812-814, 1988.

Suess, E., Torres, M.E., Bohrmann, G., Collier, R.W., Greinert, J., Linke, P., Rehder, G., Trehu, A., Wallmann, K., Winckler, G., and Zuleger, E., Gas hydrate destabilization: enhanced dewatering, benthic material turnover and large methane plumes at the Cascadia convergent margin, *Earth Planet Sci. Lett.*, 170, 1-15, 1999.

Tucholke, B.F., Bryan, G.M., and Ewing, J.I., Gas-hydrate horizons detected in seismic-profiler data from the western North Atlantic, *Amer. Assoc. of Petrol. Geol. Bulletin*, 61, 698-707, 1977.

Valentine, D.L., Blanton, D.C., Reeburgh, W.S., and Kastner, M., Water column methane oxidation adjacent to an area of active hydrate dissociation, Eel River Basin, *Geochim. Cosmochim. Acta*, "in press."

von Huene, R., and Pecher, I.A., Vertical tectonics and the origins of BSRs along the Peru margin, *Earth Planet. Sci. Lett.*, 166, 47-55, 1999.

von Huene, R., and Scholl, D.W., Observations at convergent margins concerning sediment subduction, subduction erosion, and the growth of continental crust, *Rev. of Geophys.*, 29, 279-316, 1991.

Ward, B.B., Kilpatrick, K.A., Novelli, P.C., and Scranton, M.I., Methane oxidation and methane fluxes in the ocean surface layer and deep anoxic waters, *Nature*,327, 226-229, 1987.

Westbrook, G.K., Carson, B., Musgrave, R.J., et al., in: *Proc. of the ODP, Initial Reports*, 146 (Part 1), College Station, TX (Ocean Drilling Program), 1994

Westbrook, G.K., and Smith, M.J., Long décollements and mud volcanoes: Evidence from the Barbados ridge complex for the role of high pore fluid pressure in the development of an accretionary complex, *Geology*, 11, 279-283, 1983.

Whitcar, M.J., Faber, E., and Schoell, M., Biogenic methane formation in marine and freshwater environments: CO_2 reduction vs. acetate fermentation - isotope evidence, *Geochim. Cosmochim. Acta*, 50, 693-709, 1998.

Whiticar, M.J., Hovland, M., Kastner, M., and Sample, J.C., Organic geochemistry of gases, fluids, and hydrates at the Cascadia accretionary margin, *Proceedings of the Ocean Drilling Program, Scientific Results, Leg* 146 , 385-397, 1995.

Xu, W., and Rupple, C., Predicting the occurrence, distribution, and evolution of methane gas hydrate in porous sediments, *J. Geophys. Res.*, 104, 5081-5085, 1999.

Yamano, S., Alcauskas, J.B., and Crozier, T.E., Solubility of methane in distilled water and seawater, *J. Chem. Engineer. Data*, 21, 78-80, 1976.

Yamano, M., Uyeda, S., Aoki, Y., and Shipley, T.H., Estimates of heat flow derived from gas hydrates, *Geology*, 10, 339-343, 1982.

Yuan, T., Hyndman, R.D., Spence, G.D., and Desmons, B., Seismic velocity increases and deep-sea gas hydrate concentration above a bottom-simulating reflector on the northern Cascadia continental slope, *J. Geophys. Res.*, 101, 13,655-13,671, 1996.

Zatsepina, O.Y., and Buffet, B.A., Thermodynamic conditions for the study of gas hydrate in the seafloor, *J. Geophys. Res.*, 103, 24,127 - 24,139, 1998.

M. Kastner, Scripps Institution of Oceanography, 9500 Gilman Drive, MC#0212, La Jolla, CA 92093 (mkastner@ucsd.edu)

Sea Floor Methane Hydrates at Hydrate Ridge, Cascadia Margin

E. Suess[1], M.E. Torres[2], G. Bohrmann[1], R.W. Collier[2], D. Rickert[1], C. Goldfinger[2], P. Linke[1], A. Heuser[1], H. Sahling[1], K. Heeschen[1,2], C. Jung[1], K. Nakamura[3], J. Greinert[1], O. Pfannkuche[1], A. Trehu[2], G. Klinkhammer[2], M.J. Whiticar[4], A. Eisenhauer[1], B. Teichert[1], and M. Elvert[1,5]

The crest of Hydrate Ridge harbors a variety of gas hydrates in near-surface sediments. Hydrate formation and destruction continuously shape the ridge topography. Interstitial Cl-anomaly patterns in conjunction with video-guided sampling have established that the uppermost sediment column contains several distinct layers of gas hydrate which are exposed at the sea floor. A methane-oxidizing bacterial consortium populates the exposures of hydrate; colonies of vent macro-fauna are abundant as well. Discharge of methane from destabilized hydrate at the seafloor stimulates high rates of benthic oxygen consumption. These rates, however, vary by many orders of magnitude spatially and temporally, highlighting the need for implementing seafloor observatories at gas hydrate sites. Two types of hydrate fabrics were observed: A highly porous fabric with an estimated pore space of approx. 60 vol.-% and a massive type, with no visible pore space. Both types contain varying amounts of chloride, which need to be taken into account when estimating hydrate volumes from Cl-depletion of pore waters. The porous hydrate has low bulk density, which may cause periodic release of large chunks of hydrate from the sea floor. They float to the surface and leave behind a chaotic topography of mounds and depressions. These pieces of floating hydrates constitute an important transport mechanism for methane from the seafloor directly to the atmosphere.

[1]GEOMAR Research Center for Marine Geosciences, Kiel, Germany

[2]College of Ocean and Atmospheric Sciences (COAS), Oregon State University, Corvallis, Oregon

[3]Geological Survey of Japan, Tsukuba, Japan

[4]Earth and Ocean Sciences, University of Victoria, Victoria, Canada

[5]Max-Planck Institute for Marine Microbiology, Bremen, Germany

Natural Gas Hydrates: Occurrence, Distribution, and Detection
Geophysical Monograph 124

1. INTRODUCTION

The biogeochemical processes associated with gas hydrate dynamics and fluid venting in continental margins constitutes a first order scientific objective (Suess and Thiede; 1999; Pisias and Delaney, 2000). More than a decade of research on the Cascadia convergent margin has

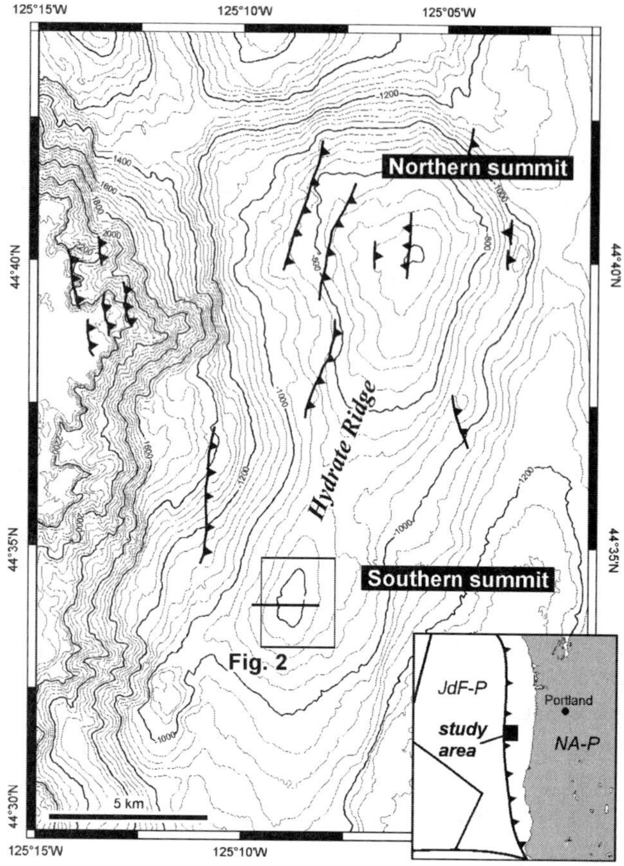

Figure 1. Bathymetry of the second accretionary ridge, Hydrate Ridge, and area of TECFLUX-work; box = southern summit; E-W line = 18 kHz section shown in Plate 1a; inset = plate tectonic setting of the Cascadia convergent margin.

documented active venting of fluids and gases and exposures of methane hydrates at the seafloor (Fig. 1). Therefore, an understanding of gas hydrate dynamics and the effects on benthic fluxes and diagenetic pathways of carbon and trace gases in the specific tectonic setting of the Cascadia margin could ultimately shed light on several broader issues. Foremost among these are climate-forcing by gas hydrate eruptions on global and regional scales (MacDonald, 1990; Dickens et al., 1997; Kennett et al., 2000); the role of methane hydrate in the carbon cycle (Dickens, 1999); the partitioning of the methane carbon pool into biological tissue, respired carbon dioxide, carbonate minerals, and free methane (Suess et al., 1999a); and the need for better estimates of the magnitude of the fossil fuel-based energy resource, which is believed to reside in marine and terrestrial gas hydrates (Kvenvolden, 1993). Some of the sea-floor hydrates are porous and less dense than seawater. As they grow downward towards the rising stream of methane, their buoyancy eventually exceeds the lithostatic loading by sediments and chunks of hydrate detach and float to the sea surface. This scenario is reconstructed here based on high-resolution small-scale mapping of vent fields and hydrate exposures and on sampling of the complete hydrate–pore water–sediment system at the interface. In piecing together this scenario, which represents an efficient mechanism for this greenhouse gas to escape into the atmosphere, extensive direct observations from DSR ALVIN were available as well and sea floor video-surveys and sampling from aboard RV SONNE (Torres et al., 1999; Bohrmann et al., 2000; Linke et al., 2001).

Methane hydrates were recognized in seismic reflection data at the Cascadia convergent margin (MacKay et al.1994; Trehu et al., 1999). They were encountered in cores by deep sea drilling (Westbrook et al., 1994; Kastner et al., 1998) and were first sampled from the sea floor during the RV SONNE Cruise 110 (Suess et al., 1999a). Water depth and temperature place these seafloor gas hydrates close to their stability limit, hence minor changes in the seafloor conditions may have a major effect on the destabilization of hydrates and in turn have profound environmental consequences. Part of the methane from destabilization of hydrate is injected into the water column to form extensive plumes. The behavior of methane plumes is critical in order to evaluate how much, if any, of this potent greenhouse gas escapes into the atmosphere or is utilized within the interior of the ocean. Another part of the methane provides an enormously rich carbon and energy reservoir for benthic life. Efficient methane turnover from gas hydrates is accomplished by the recently discovered highly specialized archaeal-bacterial consortium where cells of archaea are surrounded by sulfate reducing bacteria forming structured aggregates (Boetius et al., 2000). These aggregates are abundant in the sediments of Hydrate Ridge and mediate anaerobic methane oxidation. This process drives the formation of authigenic carbonates that surround the near-surface hydrates and are known as chemoherms (Bohrmann et al., 1998).

The fault pattern on Hydrate Ridge which provides the pathways for methane and fluid venting is well known (Kulm et al., 1986; MacKay et al., 1992; Trehu et al., 1999) and is related to the evolution and growth of the accretionary complex of the Cascadia convergent margin. The faults extend through the accreted sediments to below the gas hydrate phase transition. At depth, they tap a fluid reservoir, which contains free methane. The faults serve as conduits and channel methane up to the seafloor where it either escapes into the water column or forms secondary gas hydrates, as is the case on the summits of Hydrate Ridge.

Here we summarize several of the general findings from past studies on sea floor hydrates and present new observations and data obtained during the TECFLUX'99 and 2000 campaigns. Our discussion covers manifestations of sea floor hydrates as well as biota, authigenic carbonate lithologies, flow rates and benthic material turnover stimulated by emissions of methane. In this volume the authigenic carbonate lithologies resulting from venting and oxidation of methane at Hydrate Ridge are dealt with in a separate contribution (Greinert et al., this volume) as are aspects of carbon isotopes in biomarkers and the microbial consumption of gas hydrate methane from the same location (Elvert et al., this volume). Variability and magnitudes of flow rates and benthic material turnover have been reported elsewhere (Linke et al., 1994; Suess et al., 1999a; Tryon and Brown, 1999). One of the major implications from the injection of gas hydrate methane to the sediment-water interface concerns the benthic oxygen consumption. The oxidation of chemically reduced species, such as methane, requires oxygen that far exceeds the consumption normally recorded at the deep seafloor. The total rate of supply of methane, hydrogen sulfide, and ammonia in oxidation equivalents from a gas hydrate vent on Hydrate Ridge was 1450±350 mmol m^{-2} d^{-1} and the measured oxygen demand 1600±300 mmol m^{-2} d^{-1} illustrating the close quantitative relationship in benthic material turnover (Suess et al., 1999a). For comparison, the sediment oxygen demand for deep-sea and organic-rich coastal environments, which are not influenced by venting, is generally between 1-10 mmol m^{-2} d^{-1}. Clearly though, such comparisons are in need of further study since no geochemical time-series have yet been available. Our data collected during TECFLUX'99 showed that the largest uncertainty is related to the non-steady flow from vents (Linke et al., 1999; Tryon and Brown, 1999). Flux measurements indicate that rates vary by many orders of magnitude spatially, due to the heterogeneity in fluid expulsion and underlying plumbing network. Temporal fluctuations due to tidal forcing and other superimposed frequencies have been recorded during deployments of instruments lasting from hours to weeks. Even episodes of flow reversals were recorded (Tryon et al., 1999). These changes in flux behavior give a unique insight into the complex dynamics behind the hydrate formation/ dissociation and gas venting.

Other new findings at Hydrate Ridge demonstrate gas hydrate occurrences in discrete layers at the sediment water interface only millimeters below the seafloor as well as in more massive layers at depths. Recovery of these deposits allowed us to analyze the physical properties and chemical composition of the hydrate water, which characterizes these near-surface hydrates. The gas hydrate fabric is highly variable with porous to massive gradations. These hydrates contain varying amounts of sea salt apparently trapped during formation. Estimates of amounts of gas hydrate in sediments, which are usually based on the Cl-depletion pattern of pore waters, may be more complex when considering these trapped chlorides and the low bulk density as will be shown in the following sections.

2. METHANE HYDRATE MANIFESTATIONS

A pavement consisting of carbonate crusts, bacterial mats, and hydrates were imaged on Hydrate Ridge by a deep-towed video survey in 1996. During this cruise a TV-guided grab retrieved hydrate samples. Hydrate was found at the southern summit just beneath the thinly sediment-covered surface and at the northern summit in sediments accumulating in tensional fractures between blocks of the chemoherm carbonates (Suess et al. 1999a,b; Bohrmann et al., 1998). In June of 1999 a new side-scan sonar survey of the area, supplementing earlier work (Carson et al., 1994; Zhou et al., 1999), showed in detail that the ridge flank, the saddle, and crestal region are covered by acoustically hard, highly reflective materials related to fluid venting. New targets were identified by this survey, either as "white spots", "scars", "pock marks", "pinnacles" or "mud diapirs", which indicated anomalous environments or manifestations of hydrates and/or fluid venting. Several of these were investigated by submersible during RV ATLANTIS Cruise AT3-35b (Torres et al., 1999). Observations from the submersible showed methane bubbles escaping from the top of the ridge. An acoustic image recorded aboard RV SONNE using a narrow beam 18-kHz echosounder shows a >200 m high anomaly in the water column emanating from the southern summit (Plate 1a). The simultaneous observation of vigorous bubbling at the summit confirms that it is the source of the acoustic anomaly.

Supplementing the submersible dives, the top of the summit was surveyed during RV SONNE Cruises 143 and 148 by a deep-towed camera system (OFOS) and sampled using video-guided multicorer and grab sampling devices. Submersible and deep-towed images showed the sea floor at the southern summit to resemble a chaotic landscape dotted with pockmarks, sediment drifts, carbonate pinnacles, hydrate exposures, bacterial mats, and clam colonies; all varying on a meter-scale (Plate 2, insets a & b). The chaotic appearance is amplified by dune-like sediment bodies, barren of any biota, alternating with depressions containing multi-colored patches of prolific vent biota. The OFOS-observations revealed a zonation of the vent activity with clam colonies and carbonate pinnacles at the outer perimeter and bacterial mats and patches of mixed vent-biota towards the center (Fig. 2). The landscape is most characteristically

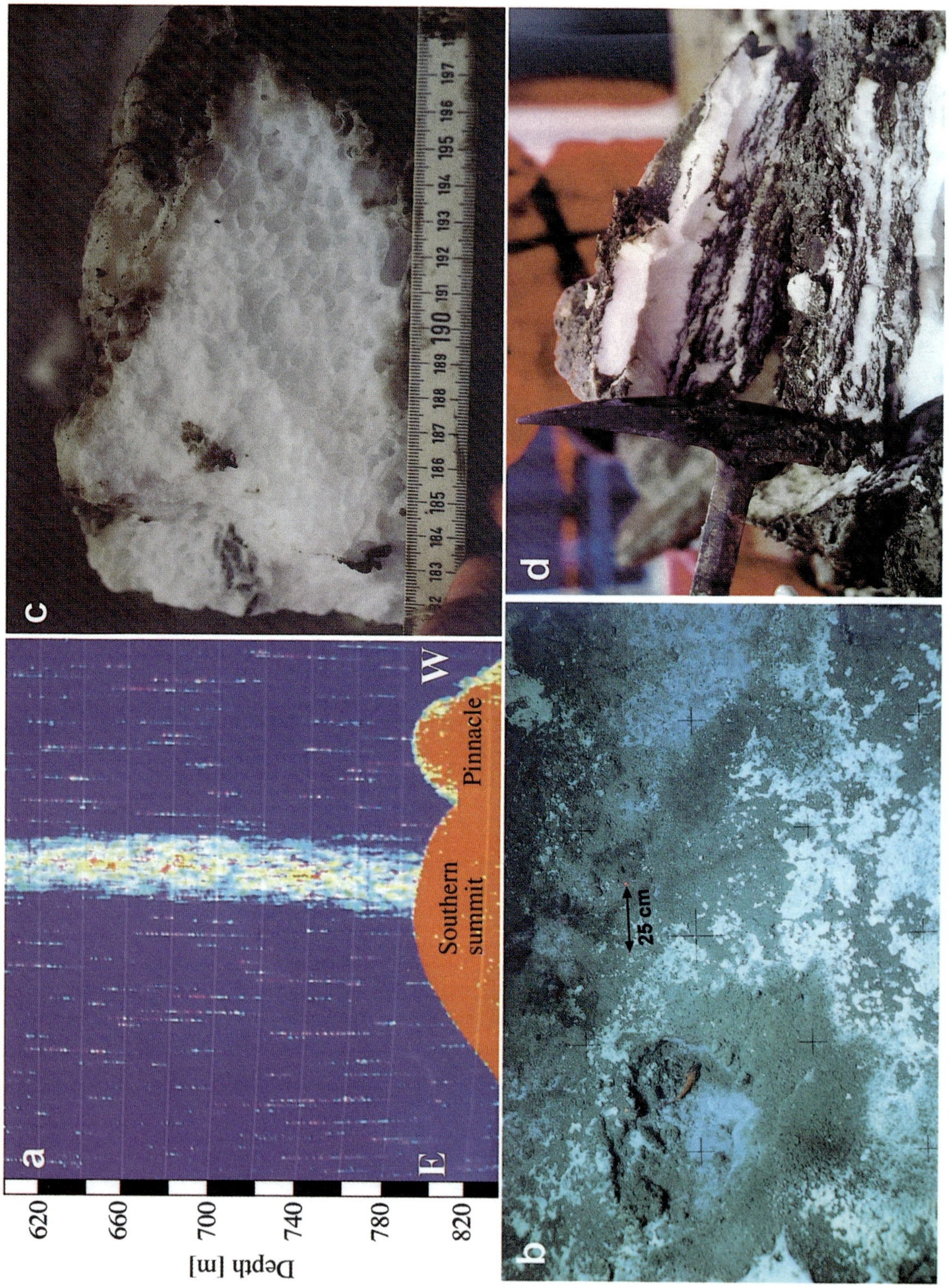

Plate 1. (a) Plume image above southern summit and pinnacle; (b) Sea floor image of microbial mat and orifice of vent (circular depression); (c) Hydrate bubble-fabric; (d) Massive hydrate fabric; scale bar = 25 cm; all photos GEOMAR

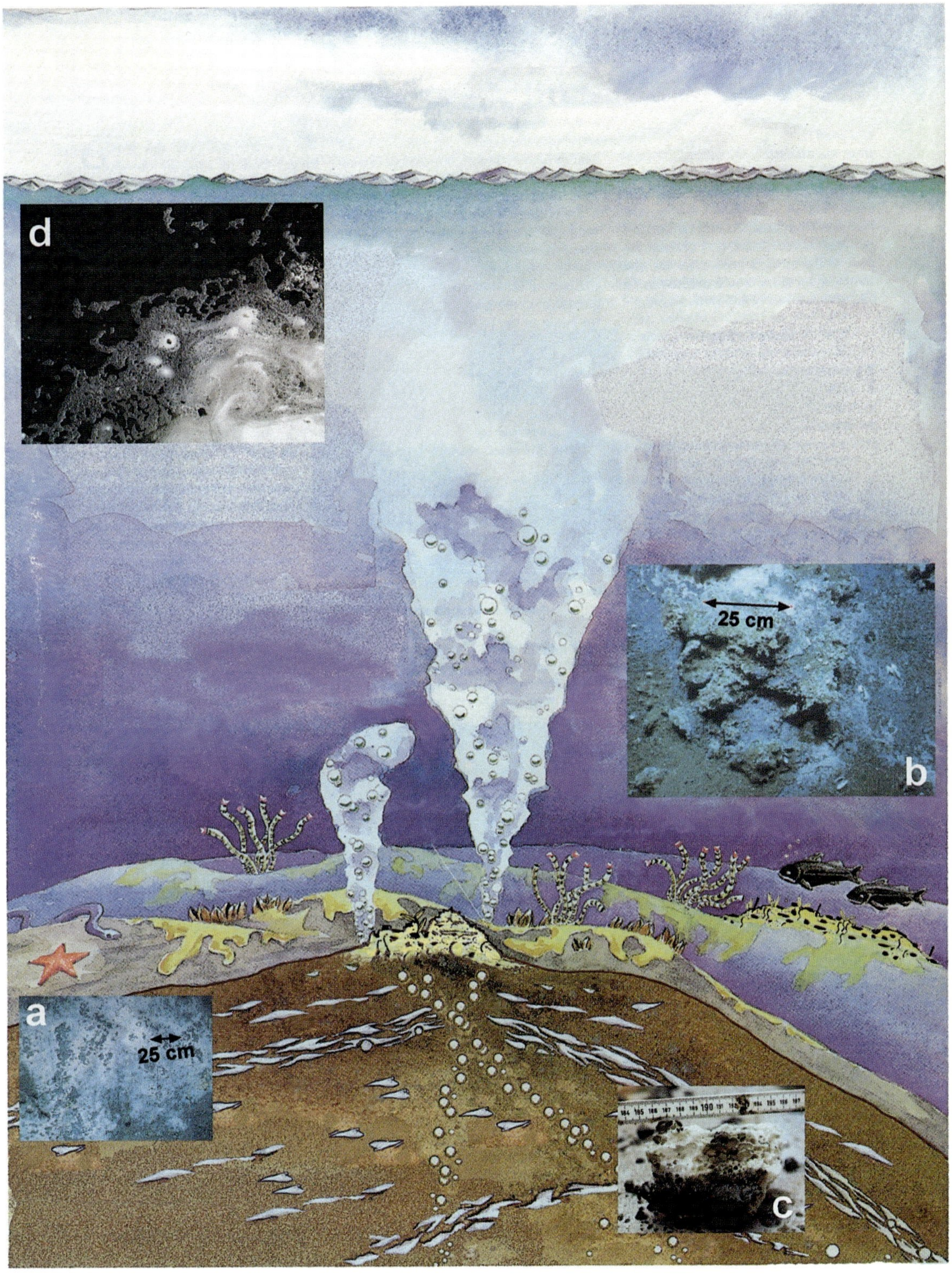

Plate 2. Scene at sea floor of southern summit; modified from Suess et al., (1999b); (a) Microbial mat, (b) Carbonate pinnacle, (c) Hydrate with bubble-fabric, (d) Floating hydrate at the sea surface; scale bar = 25 cm; all photos GEOMAR

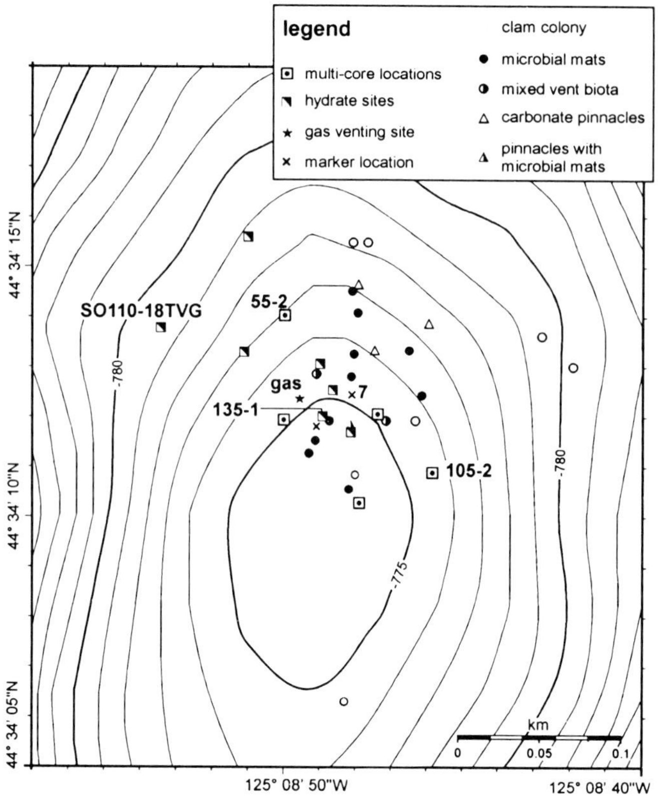

Figure 2. Southern summit vent activities, gas hydrate and core locations.

developed towards the northern end of the southern summit where the submersible ALVIN observed the vigorously flowing methane vent. The location of this gas vent was marked with a reflector labeled #7 (Fig. 2). This marker served as a navigation point during surveys and remains deployed for future work. Also shown in Fig. 2 are all locations from which hydrate was sampled. The initial gas hydrate find of 1996 is labeled as SO110-18TVG.

The impression of a chaotic landscape is best captured by the artistic view of the southern summit region reproduced in Plate 2. The tube worm colonies shown here are not present on the southern summit but in deeper waters off the northern summit of Hydrate Ridge. Highly characteristic features are the orange and white microbial mats with and without dense colonies of clams, the methane-plumes, carbonate pinnacles, and hydrate layers at the surface and in the sub seafloor. Plate 2 summarizes the results from Hydrate Ridge (Suess at al., 1999a, b) which are supported here by several new images showing the ropy appearance of strands of microbial mats (Plate 2, image a and Plate 1, image b). This larger image of white and orange-colored microbial mats also includes an actual vent, about 30 cm in diameter with a red-colored fish inside. The far wall of the almost circular vent depression appears horizontally bedded with crust-like layers. Most probably these are hydrate layers but could also be a rigid framework of carbonate crusts, hydrate layers and sediment held together by microbial mats. Also shown in detail is a carbonate pinnacle (30-50 cm high) partially covered by microbial mats and individual clamshells at its base (Plate 2, image b).

Below the sediment surface in the artist's view of Plate 2 the porous hydrate fabric is shown (image 2c) and an enlarged image of that fabric in Plate 1c. Macro-pores, several mm in diameter and slightly elongated, are shown. These pores are identical to those observed from the submersible ALVIN at the seafloor. A more massive hydrate fabric is displayed by the hand specimen in Plate 1d with alternating sediment and hydrate layers. The gas hydrate develops preferentially along bedding planes several millimeters to decimeters thick, but sometimes also cuts across bedding planes. As previously reported (Suess et al., 1999a) there was no instance observed where the hydrate occupied the original sediment pore space and cemented the sediment framework on a micro-scale. Rather, it filled large secondary pore space either as fractures or seemed to create its own pore space during growth by fracturing or pushing apart the original sediment framework.

During investigations in 1999 and 2000 the top of the southern summit was sampled and hydrate-containing sediment and different types of hydrate fabrics were recovered. A video-guided multi-corer device specifically designed to recover the uppermost hydrate layers accomplished this. The locations of the samples are shown in Fig. 2. In all more than 50 cores were retrieved covering the entire range of microenvironments encountered on the southern summit.

3. CHLORIDE-ANOMALY PATTERN

The direct retrieval of gas hydrates to the sea surface, without maintaining *in situ* conditions, is fortuitous and restricts the precise characterization of their composition and physical properties. Nevertheless advances have been made relying on indirect evidence for the presence of gas hydrates through the well-known artifact of negative Cl-anomalies in pore fluids (Hesse, 1990; Suess, von Huene et al. 1998; Paull et al., 1996). Choride anomalies result from the release of hydrate water during core retrieval with the dilution of the normal pore water chlorinities. Today we still rely largely on this artifact as the best evidence for the presence of hydrates, although our new data suggest that the calculation of the amount of hydrate present is complicated by residual chloride present within the hydrate pore space.

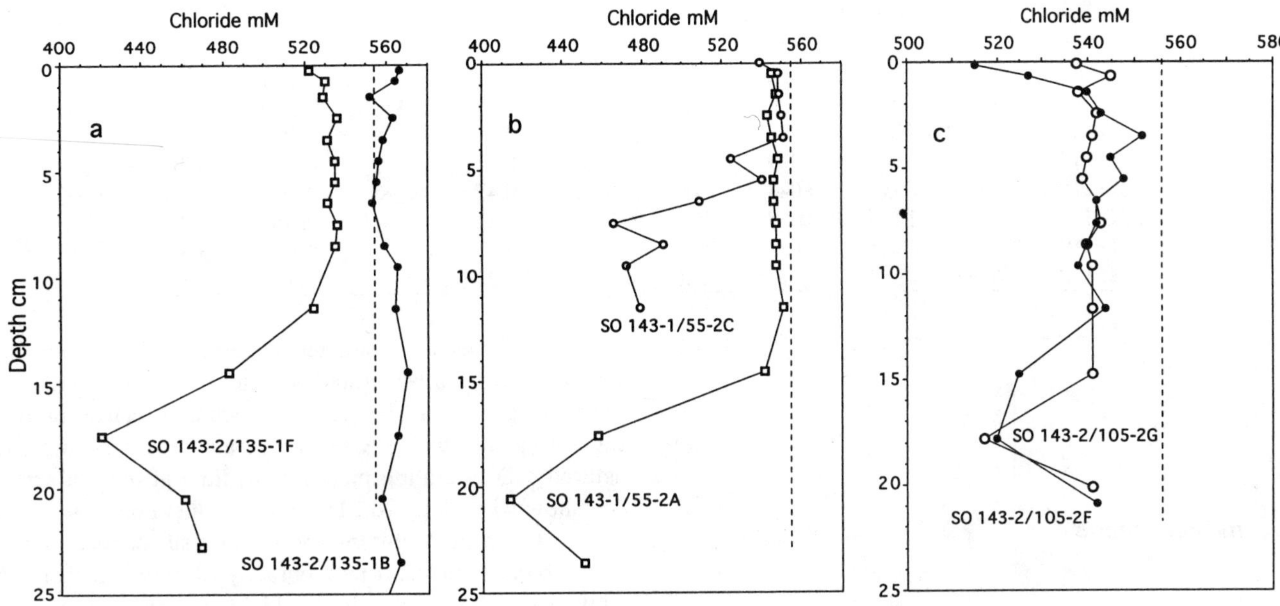

Figure 3. Composite Cl-anomaly patterns in near-surface sediment cores (a-c); discrete Cl-depletions indicate depth of hydrate layers.

We have mapped negative Cl-anomalies in pore fluids of near-surface sediments on the southern summit and estimated their hydrate content taking into account the porosity of the hydrates as well as the fact that chloride from interstitial water becomes trapped in the hydrate fabric. Combined with the visual observations of methane ebullition and video-documentation of hydrate exposures intimately associated with biota and carbonates, these data provide convincing evidence for the highly dynamic, small-scale, and patchy distribution of seafloor hydrates.

Variable Cl-depletions of up to 20% lower than seawater chlorinity occur at distinct depth horizons as well as directly at the sediment-water interface. They indicate the presence of discrete hydrate layers. Examples illustrated in Figs. 3a-c are from cores SO143-1/55-2A and -2C, from cores 143-2/105-2G and -2F, and from core 143-2/135-1F. In most of these cases actual pieces of hydrates were found at the depth intervals where the depletion occurred, although in various stages of decomposition. The depletions are quite significant considering that the analytical precision of Cl-determinations is better than ±1%.

The Cl-depletions in the sediment pore waters indicate the presence of several gas hydrate layers between the surface and 25 cm below sea floor. In a straightforward approach the degree of Cl-dilution would indicate that between 10-30 vol.-% of the total sediment sampled consists of gas hydrates. However, our analyses of more than 25 hydrate waters, which is the solution that remains after pure hydrate is experimentally dissociated in a clean environment, show that they contains between 30-70 mM of chloride as well as other seawater ions (Table 1). A frequency plot of the Cl-analyses (Fig. 4) indicates that waters from the porous hydrates appear to have higher chloride contents than those from more massive hydrates. This would be expected if pore waters became trapped during hydrate formation. The trapped chloride combined with the high porosity of the near-surface hydrates strongly affects the hydrate volume calculated from Cl-dilution.

This is illustrated in Fig. 5 which shows selected theoretical Cl-dilution curves of interstitial water through addition of hydrate water of zero chloride (massive fabric) and 70 mM chloride (porous fabric), respectively. For simplicity four cases are calculated whereby the two types of hydrate fabrics are each assumed to release hydrate water within sediments of initial *in situ* porosities of 80 vol.-% (a and b) and 40 vol.-% (c and d). As is expected the porous hydrates release less water than the massive hydrates and, taking into account the Cl-content of the decomposing porous hydrate, the resulting estimate of hydrate volume can vary by as much as a factor of 2. This finding should deserve attention when estimating gas hydrate reservoir sizes based on interstitial Cl-anomalies.

4. FLOATING HYDRATES: EVIDENCE FROM HYDRATE FABRIC AND LOW BULK DENSITIES

Massive and catastrophic release of methane hydrates from the seafloor has been suggested as the trigger for

Table 1. Hydrate water composition

Sample ID	Cl mM	BmM	Na mM	Mg mM	Ca mM	K mM	SO_4 mM	Sr µM
SO 110-18TVG#2	44,34	0,033	34,9	3,89	0,44	0,78	0.100	4,07
SO 110-18TVG#3	35,17	0,029	27,7	3,1	0,93	0,7	0.350	6,35
SO 110-18TVG#4	55,06	0,041	44	4,91	0,47	0,96	0.700	4,84
SO 110-18TVG#5	32,61	0,025	25,1	2,88	0,32	0,54	0.700	2,8
SO 110-18TVG#6	38,28	0,029	30,7	3,47	0,3	0,7	0.360	3,42
SO 110-18TVG#7	28,04	0,021	21,8	2,55	0,28	0,5	n.d.	2,61

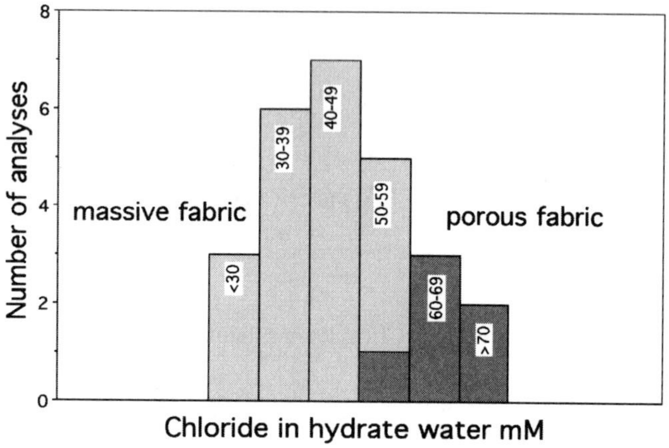

Figure 4. Chloride content of residual hydrate water; frequency of analyses of Cl show porous hydrate fabric to have higher Cl-contents than massive fabric; mean Cl-content = 50 mM.

transient past warm climates or accelerated evolution of biota (e.g. Kennett and Stott, 1991; Zachos et al., 1993; Katz et al., 1999). A locality from which such release of methane hydrates, preceding the Late Paleocene Thermal Maximum (LPTM), has been identified on the Blake Spur, based on chaotic and brecciated sediment fabric from disrupted strata left behind (Norris and Röhl, 1999). Supposedly the myth surrounding the hazards of the Bermuda triangle invokes catastrophic release of methane hydrate to the atmosphere, yet no direct evidence for such a process transferring methane from the deep seafloor to the atmosphere has ever been presented. Here we document for the first time that floating hydrates of up to a cubic meter in size exist and constitute a mechanism for continuous or episodic transport of methane from the sea floor directly to the atmosphere. Previously, 'hydrate rafting' has been postulated as a means for generating methane plumes in the water column (Paull et al., 1995) and for sediment transport (MacDonald et al., 1994), leaving pockmarks on the ocean floor.

The floating hydrates off the Cascadia margin were spotted around the RV SONNE on several occasions while working over the southern summit. Such chunks, measuring more than 1 meter in diameter, were reported by the vessels watch about 30-60 minutes after bottom contact by sampling gear, but also after considerable elapse of time, suggesting that the release of hydrates could be triggered naturally. Dissociating pieces of hydrate at the sea surface are shown in Plate 2d. The size of the image is 3-5 m across. The whitish clumps are hydrates surrounded by froth of methane bubbles. In moderately stormy weather with white caps developing it would be nearly impossible to identify floating hydrate at the sea surface.

Hydrates at the southern summit show a characteristic porous fabric, which develops from the accumulation of methane bubbles as they become encased in a hydrate skin. At a depth of 770m bubbles were observed rising from below the seafloor. They either escaped into the water column or, when arrested turned into hydrate. As a result they form positively buoyant interlayered strata which include hydrate, sediment, carbonate, microbial mats, free

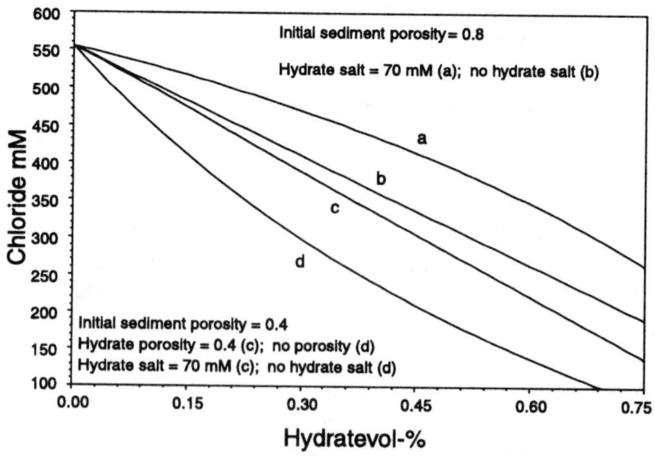

Figure 5. Relationship between degree of Cl-depletion and amount of hydrate in sediments; estimates of hydrate vary significantly whether or not residual Cl-content and low bulk density are considered; selected cases are (a) and (b) are for initial sediment porosity of 80 vol.-% and cases (c) and (d) for 40 vol.-%, respectively; different properties of hydrate are assumed as indicated in the figure legend.

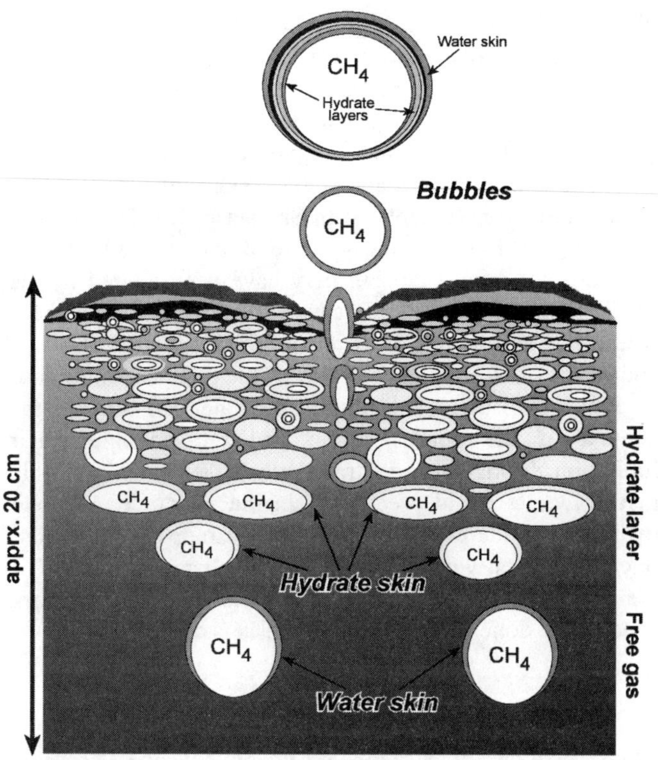

Figure 6. Development of bubble fabric and evolution of seafloor hydrate layers; rising methane bubbles are trapped below sediment and/or microbial mats and converted to hydrate; diffusion of water into the methane bubble is slowed by increasing thickness of hydrate skin; hence availability of water becomes the limiting factor and permits free gas to exist even though P-T-conditions favor solid gas hydrate.

gas, and water as schematically shown in Fig. 6. The porous hydrate fabric consists of macroscopic pores, which are connected by a thin veneer of hydrate. The instantaneous development of a hydrate skin around bubbles rising from a vent was observed by the submersible. Growth of a hydrate skin impedes further reaction between methane and water, even though the P-T-conditions favor the solid gas hydrate phase. The growth of hydrate and the persistence of gas bubbles occur because the system becomes water limited. With the exhaustion of water during hydrate formation the pore water ions are concentrated and may precipitate analogous to evaporites.

As the porous hydrate grows downward within the sediment towards rising methane bubbles, the buoyancy of the interlayered strata increases until eventually, and presumably at random, large chunks detach and float to the sea surface. Supporting evidence is provided from submersible observations. The ALVIN Dive 3430 recorded approx. 50 m off the bottom above the southern summit of Hydrate Ridge a very strong radar reflection of an elongated 20m-large object rising to the surface. This object is believed to be a floating hydrate chunk.

Central to the development of floating hydrate chunks is the role of the porous hydrate fabric. Coring from aboard RV SONNE recovered on numerous occasions 1-50 cm thick layers of porous gas hydrate. They consist of macroscopic pores connected by a skin of hydrate. Such fabric has been experimentally produced by releasing methane from a tank at water depths and temperatures well inside the stability field of hydrates (Brewer et al., 1997).

Another issue related to the floating of hydrate pieces is the question of whether the pores *in situ* are filled with gaseous methane or with interstitial water. If they were filled with gas, the bulk density of hydrate layer would be very much lower than that of the massive hydrate substance. This lowered density would give rise to a strong positive buoyancy needed to effectively float the hydrate pieces. On the other hand if the pores were filled with water, the density contrast and hence buoyancy would be considerably less.

We offer three lines of evidence that the bubble fabric contains free gas and only traces of pore water or possibly sea salt formed when pore water is consumed during hydrate formation: (1) The sea surface around the giant TV-grab sampler literally appears to boil from escaping gas when the instrument is brought up from depth. The amount and rate of gas escaping in this case appears orders of magnitude stronger than when observing a piece of solid hydrate degas as the result of destabilization (Plate 2d). The implication is that the apparent boiling is caused by the escape of trapped gas from the porous fabric. (2) In all our experience with sampling, processing, and cleaning of numerous pieces of natural porous hydrate, not once did we observe water flowing out of the pores or even detect moisture other than the hydrate water liberated during destabilization. (3) Pieces of a porous and massive hydrate layers were used to obtain bulk densities onboard ship. Of 76 measurements taken during RV SONNE Cruise 148 (Linke et al., 2001) all but 7 samples had densities of approximately 0.89 g cm^{-3} whereas most values clustered around maxima of 0.4 g/cm^3 and 0.7 g cm^{-3} (Fig. 7). The theoretical material density of pure hydrate is 0.91 g/cm^{-3}; Davidson et al., 1998). We also made estimates of pore volume by image analyses of the macro-pores over selected areas of the sample shown in Plate 1c. We found consistently that 55±5 % of the area is occupied by void space. If this 2-dimensional ratio can be assumed to be representative for three dimensions of the hydrate fabric, and if this volume is occupied by gaseous methane and the remainder by hydrate, then the bulk density contrast between the unit volume of hydrate layers and seawater would be >0.60 g cm^{-3}. This would result in an enormous

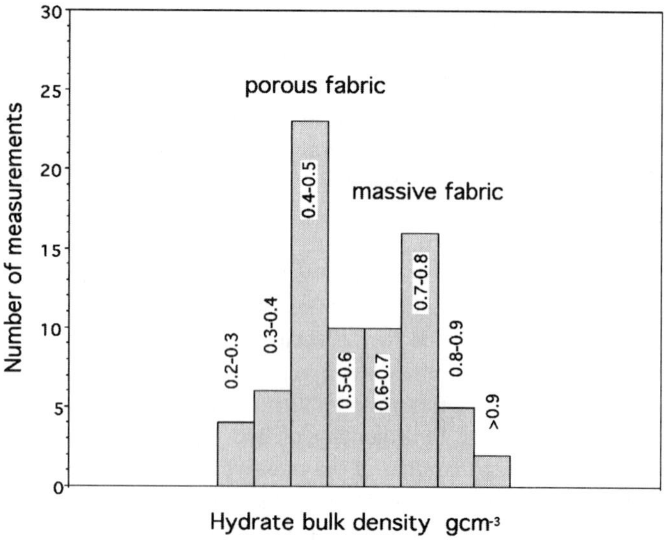

Figure 7. Bulk density of naturally occurring gas hydrates; frequency of measurements cluster around 0.4 and 0.7 g/cm^3 with porous fabric having lower densities than massive fabric; theoretical bulk density of hydrates = 0.9 g/cm^3.

positive buoyancy force, which could lift off and float pieces of hydrate even if covered by thick layers of sediment. Quantification and more detailed observations on the development of the floating hydrates wait to be attempted during future research.

5. HYDRATE WATER COMPOSITION

A consistent feature of the hydrates recovered is the residual chloride and seawater ions found in the hydrate water. This has implications for underestimating the hydrate volumes of sediments as just presented above and may also provide information as to the origin of the trapped ions. Initially, based on the Cl-, Na-, K-, and Mg-concentrations found in the residual hydrate water, we assumed that they came from contamination by about 10% seawater (Suess et al., 1999a). But consistent and reproducible salt compositions found in all hydrate waters since indicate to us that these may be an integral part of the natural gas hydrate fabric. The very low SO_4-content of the samples (Table 1) as well as in other samples analyzed since then (Linke et al. 2001) and the characteristic depletion of Ca and Sr indicate that the source of the salt is not pure seawater but rather pore water altered by aragonite precipitation. Pore waters of Hydrate Ridge sediments are strongly depleted in SO_4 (Boetius et al. 2000) and show varying Ca- and Sr-depletions caused by aragonite precipitation. The ratio of Ca-to-Sr in pore waters consistently indicates preferential removal of Ca over Sr through carbonate mineral formation, of mostly low and high Mg-calcite and some aragonite and approaches a molar ratio of Ca/Sr = 450 (Greinert, 1998). This value differs significantly from that found in the hydrate waters. Figure 8 shows hydrate waters of samples from site 110-18TVG and samples obtained during 1999 and 2000 cruises (Bohrmann et al., 2000; Linke et al., 2001) to be more strongly depleted in Sr than in Ca. This implies that the ions found in the hydrate water are neither pure seawater nor pore water but may have been altered by the precipitation of aragonite. Because aragonite is a solid phase where the partitioning coefficient $[Sr/Ca]_{arag}/[Sr/Ca]_{sw} > 1$, and hence causes the Ca/Sr-ratio of the residual solution to be lower than that of the initial seawater solution. In the context of the porous fabric and the possible water limitation during rapid hydrate growth, it seems plausible that aragonite could precipitate in the interstices of solid hydrates and thereby alter the initially trapped pore water. Such a reaction would explain the peculiar composition of the residual hydrate water; clearly however, more work needs to be done before this hypothesis is confirmed.

6. CONCLUSIONS

Gas hydrate in surface sediments of the Cascadia accretionary margin contains new detailed information on the dynamics of methane supply, of hydrate formation and dissociation, and of the variability of physical and chemical hydrate properties. Of these, the highly porous fabric combined with altered pore water, which apparently is trapped in the hydrate interstices, has the most far-reaching

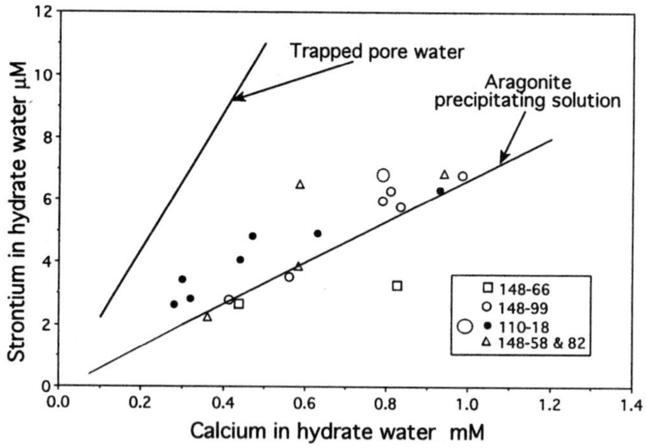

Figure 8. Strontium and calcium contents of hydrate waters; change in Sr/Ca-ratio from that of trapped pore water towards preferentially depleted Sr-solution might indicate removal of Sr via aragonite precipitation; hydrate water compositions cluster around removal of approx. 1% Sr in aragonite; data for samples 110-18 are listed in Tab. 1.

implications. As one of the consequences of the porous fabric, the bulk density of natural hydrates is considerably lower than that of pure, theoretical and experimentally produced hydrates. The lower bulk density is responsible for the efficient transfer of methane from the seafloor to the surface ocean and hence into the atmosphere via floating chunks of hydrate. The implications of low bulk density, if it persists also downward throughout the sediment column, for slope stability and seismic velocity structure are profound.

There is also a largely unknown series of chemical reactions possible when hydrates form through rapid supply of gaseous methane, as with the vent systems encountered at Hydrate Ridge. By consuming most, if not all, of the water available in the sediment pore space, hydrate formation may cause interstitial brines to precipitate transient solid phases, which remain enclosed in the porous fabric. Such reactions are at present entirely hypothetical, but future sampling of hydrates, designed to preserve *in situ* conditions for the study of and experiments with gas hydrates are essential in order to advance our knowledge.

The newly recognized methane-oxidizing bacterial consortium which populates the exposures of hydrate and hydrate vents at the Hydrate Ridge provides the driving force for the exceedingly high benthic biological activity which is evident in the sediment oxygen consumption. The microbial assemblages and their rôle in this activity are only beginning to be understood. Such high microbial activity is fueled by the discharge of methane from destabilized hydrate at the seafloor or from below, if faults open up pathways for free methane to escape upwards from below the hydrate stability zone. Hence, accretionary tectonics provide unique settings through continuous supply of methane, for AMO-communities (= **a**naerobic **m**ethane **o**xidation) to develop. Thus processes at hydrate vents may provide a window to the deep biosphere. The methane supply rates, however, vary by many orders of magnitude spatially and temporally, which highlights the need for implementing of seafloor observatories at gas hydrate sites.

Acknowledgements. The TECFLUX program (= TECtonically induced FLUXes) addresses processes associated with fluid venting from the Cascadia continental margin. An international group of scientists from institutions in Germany, the USA, Canada, and Japan established the program based on the initial discovery of methane hydrate exposures at Hydrate Ridge. Expeditions aboard RV ATLANTIS and RV SONNE were jointly planned, coordinated and carried out by the GEOMAR Research Center for Marine Geosciences at Kiel and the College of Ocean and Atmospheric Sciences of Oregon State University, Corvallis, with investigator groups from the following institutions: Scripps Institution of Oceanography, La Jolla; Large Lake Observatory, Duluth; Humboldt State University, Eureka; University of Southern California, Los Angeles; Monterey Bay Aquarium Research Institute, Moss Landing; University of Victoria, Victoria; Geological Survey of Japan, Tsukuba; Max-Planck Institute for Marine Microbiology, Bremen; Bundesanstalt für Geowissenschaften und Rohstoffe, Hannover; Technische Universität, Berlin, and Geoforschungszentrum, Potsdam. The projects were funded in Germany by the Federal Ministry of Science and Technology (Fkz 03G0143A and 03GO148A) and in the USA by the National Science Foundation (OCE-9731-157; OCE-9811-471, and OCE-9731-23). This manuscript benefited from the comments of 2 anonymous reviewers and the editors. We wish to thank all individuals for their input and support.

REFERENCES

Boetius, A., K. Ravenschlag, C. J. Schubert, D. Rickert, F. Widdel, A. Giesecke, R. Amann, B. B. Jorgensen, U. Witte, and O. Pfannkuche, A marine microbial consortium apparently mediating anaerobic oxidation of methane, *Nature, 407*, 623-626, 2000.

Bohrmann G., J. Greinert, E. Suess, and M. E. Torres, Authigenic carbonates from Cascadia subduction zone and their relation to gas hydrate stability, *Geology, 26*, 1998.

Bohrmann, G., P. Linke, E. Suess, and O. Pfannkuche, RV SONNE Cruise Report SO143, *GEOMAR Report, 93*, 2000.

Brewer P.G., M.J. Franklin, G. Friederich, K.A. Kvenvolden, D.I. Orange, J. McFarlane and W. Kirkwood, Deep-ocean field test of methane hydrate formation from a remotely operated vehicle, *Geology 25*, 407-410, 1997.

Carson, B., E. Seke, V. Paskevich, and M.L. Holmes, Fluid expulsion sites on the Cascadia accretionary prism: Mapping diagenetic deposits with processed GLORIA imagery, *J. Geophys. Res. 99* (B6) 11, 959-969, 1994.

Davidson, D.W., D.G. Last, and R. Hesse, Oxygen-18 enrichment in the water of a clathrate hydrate, *Geochem. Cosmoch. Acta, 47*, 2293-2295, 1998.

Dickens, G.R., C. Paull, P. Wallace and ODP Leg 164 Scientific Party, Direct measurement of in situ methane quantities in a large gas-hydrate reservoir, *Nature, 385*, 426-428, 1997.

Dickens, G.R., Carbon cycle: The blast in the past, *Nature, 401*, 752-753, 1999.

Greinert, J., Rezente submarine Mineralbildungen: Abbild geochemischer Prozesse an aktiven Fluidaustrittstellen im Aleuten- und Cascadia-Akkretionskomplex, Dissertation, 1998.

Hesse, R., Pore water anomalies in gas gydrate-bearing sediments of the deeper continental margins: Facts and Problems. *J. Inclusion Phenomena and Molecular Recognition in Chemistry, 8*, 117-138, 1990.

Kastner, M., K.A. Kvenvolden, and T. Lorenson, Chemistry, isotopic composition, and origin of methane hydrogen sulfide hydrate at the Cascadia subduction zone, *Earth Planet Sci. Lett., 156*, 173-183, 1998.

Katz, M., D. Pak , D. Dickens, and K. Miller, The Source and fate of massive carbon input during the latest Paleocene Thermal Maximum, *Science, 286*, 1531-1533, 1999.

Kennett, J.P. and L.D. Stott, Abrupt deep-sea warming, palaeo-oceanographic changes and benthic extinctions at the end of the Palaeocene, *Nature, 353*, 225-229, 1991.

Kennett, J.P., K.G. Cannariato, I.L. Hendy, and R.J. Behl, Carbon isotopic evidence for methane hydrate instability during Quaternary interstadials, *Science, 288*, 128-133, 2000.

Kulm, L.D., E. Suess, J.C. Moore, B. Carson, B.T Lewis, S.D. Ritger, D.C. Kadko, T.M. Thornburg, B.W. Embley, W.D. Rugh, M.G. Langseth, G.R. Cochran, and R.L. Samman, Oregon subduction zone: Venting, fauna, and carbonates, *Science, 231*, 561-566, 1986.

Kvenvolden, K.A., Gas hydrates - Geological perspective and global change, *Rev. Geophys., 31*, 173-183, 1993.

Linke, P., E. Suess, M. Torres, V. Martens, W.D. Rugh, W. Ziebis, and L.D. Kulm, In situ measurement of fluid flow from cold seeps at active continental margins, *Deep Sea Res., 41*, 721-739, 1994.

Linke, P., O. Pfannkuche, M.E. Torres, R.W. Collier, U. Witte, J. McManus, D.E. Hammond, K.M. Brown, M.D. Tryon, and K. Nakamura, Variability of benthic flux and discharge rates at vent site determined by in situ instruments, *EOS transactions, 80*(46), Fall Meet. Suppl., F509, 1999.

Linke, P., E. Suess and Scientific Party, RV SONNE Cruise Report SO148 TECFLUX-II-2000, *GEOMAR Report, 98*, 2001.

MacDonald, I.R., Role of methane clathrates in past and future climates, *Climate Change, 16*, 247-281, 1990.

MacKay, M.E., G.F. Moore, G.R. Cochrane, J.C. Moore, and L.D. Kulm, Landward vergence and oblique structural trends in the Oregon margin accretionary prism: Implications and effect on fluid flow, *Earth Planet. Sci. Lett., 109*, 477-491, 1992.

MacKay, M.E., R.D. Jarrard, G.K. Westbrook, and R.D. Hyndman Origin of bottom simulating reflectors: Geophysical evidence from the Cascadia accretionary prism, *Geology, 22*, 459-462, 1994.

MacDonald, I.R., N.L. Guinasso Jr., R. Sassen, J.M. Brooks, L. Lee, K.T. Scott, Gas hydrate that breaches the sea floor on the continental floor of the Gulf of Mexico, *Geology, 22*, 699-702, 1994.

Norris, R., and U. Röhl, Carbon cycling and chronology of climate warming during the Paleocene/Eocene transition, *Nature, 401*, 775-778, 1999.

Paull, C.K., W.U. Ussler III, W.S. Borowski, and F.N. Spiess, Methane-rich plumes on the Carolina continental rise: Associated with gas hydrates, *Geology, 23*, 89 - 92, 1995.

Paull, C.K., R. Matsumoto, P. Wallace, et al, *Proc. ODP, Init. Repts, 164*, College Station TX (Ocean Drilling Program) 277-318, 1996.

Pisias, N.G. and M.L. Delaney, COMPLEX: Conference on Multiple Platform Exploration of the Ocean Vancouver Brithish Columbia, 25-29 May 1999, *Joint Oceanographic Institutions*, Washington DC, 210 pp., 2000.

Suess, E., R. von Huene, et al, *Proc. ODP, Init. Repts, 112*, College Station TX (Ocean Drilling Program) 1015 pp., 1998.

Suess, E., M.E. Torres, G. Bohrmann, R.W. Collier, J. Greinert, P. Linke, G. Rehder, A. Trehu, K. Wallmann, G. Winckler, and E. Zuleger, Gas hydrate destabilization: enhanced dewatering, benthic material turnover, and large methane plumes at the Cascadia convergent margin, *Earth Planet. Sci. Lett., 170*, 1-15, 1999a.

Suess, E., G. Bohrmann, J. Greinert, and E. Lausch, Flammable ice, *Scientific American, 11*, 76-83, 1999b.

Suess, E., and J. Thiede, Gashydrate im Geosystem, Forschungszentrum Jülich GmbH, Projektträger BEO, Jülich Rostock-Warnemünde, 64 pp., 1999.

Torres, M.E., G. Bohrmann, K. Brown, M.A. de Angelis, D. Hammond, G. Klinkhammer, J. McManus, E. Suess, and A.M. Trehu, Geochemical observations on Hydrate Ridge, Cascadia margin, July 99. *OSU Data Report 174*, ref 99-3. 87 pp., 1999.

Trehu, A.M., M.E. Torres, G.F. Moore, E. Suess, and G. Bohrmann, Temporal and spatial evolution of a gas-hydrate-bearing accretionary ridge on the Oregon Continental margin, *Geology, 27*, 939-942, 1999.

Tryon, M.D., K.M. Brown, M.E. Torres, A.M. Trehu, J. McManus, and R.W. Collier, Measurements of transience and downward fluid flow near episodic methane gas vents, Hydrate Ridge, Cascadia, *Geology, 27*, 1075-1078, 1999.

Tryon, M.D. and K.M. Brown, Transient fluid flow patterns on Hydrate Ridge, Cascadia: A response to methane gas hydrate discharge? *EOS transactions, 80*(46), Fall Meet. Suppl., F528, 1999.

Westbrook, G.K., B. Carson and R.T. Musgrave et al, *Proc. ODP, Init. Repts, 146*, College Station TX (Ocean Drilling Program), 609 pp., 1994.

Zachos, J.C., K.C. Lohmann, J.C.G. Walker, and Wise, S.W., Abrupt climate change and transient climates during the Paleogene: A marine perspective, *Geology, 101*, 191-213, 1993.

Zhou, Y., C. Goldfinger, J.E. Johnson, M.E. Torres, A.M. Trehu, D. Clague, C. Paull and G. Bohrmann, Distribution and morphology of venting-related carbonates near Hydrate Ridge, Oregon Margin, based on sidescan sonar and multibeam imagery. *EOS transactions, 80*(46), Fall Meet. Suppl., F510, 1999.

Gas Hydrate-Associated Carbonates and Methane-Venting at Hydrate Ridge: Classification, Distribution, and Origin of Authigenic Lithologies

Jens Greinert, Gerhard Bohrmann, and Erwin Suess

GEOMAR Research Center for Marine Geosciences, Kiel, Germany

Hydrate Ridge is part of the accretionary complex at the Cascadia margin and is an area of widespread carbonate precipitation induced by the expulsion of methane-rich fluids. All carbonates on Hydrate Ridge are related to the methane-carbon pool either through anaerobic methanotrophy or through methanogenesis. Several petrographically distinct lithologies occur in boulder fields or in massive autochtonous *chemoherm complexes* which include methane-associated diagenetic mudstones and venting-induced breccias. The mudstones result from methane diagenesis in different sediment horizons and geochemical environments related to very slow methane venting. Cemented bioturbation casts occur as fragments, complex framework or as clasts together with bivalve shells as part of intraformational breccias, which are restricted to chemoherm complexes. Here, fluids ascend from the sub-seafloor and support aragonite-dominated carbonate precipitation near or at the sediment surface. Voids within mudclast breccias are either aragonite-rich indicating a formation near the surface at vent sites or are cemented by dolomite, which indicates formation in deeper parts of the sediment column. Brecciation is caused by tectonic or slump processes. In addition, we recognized a direct relationship between gas hydrates and sediment fracturing as well as the oxygen isotope composition of carbonate lithologies. Such gas hydrate-associated carbonates either show layered megapores and veins as relics of the original gas hydrate fabric or consist of aragonite-cemented intraclast breccias formed by growing and decomposing gas hydrate near the sediment surface. Both rock fabrics and the enrichment of ^{18}O in high Mg-calcite demonstrate *carbonate-forming mechanisms of gas hydrate*.

1. INTRODUCTION

Authigenic carbonates are a common feature at cold vent sites where methane-enriched or bicarbonate-oversaturated cold fluids escape from the seafloor. Cold vents and associated lithologies were discovered at the accretionary complex of the Cascadia margin in 1984 [*Suess et al.*, 1985; *Kulm et al.* 1986; *Ritger et al.*, 1987]. Due to the observation of active methane ebullition [*Linke et al.* 1994], the documentation of a well-developed bottom simulating reflector (BSR) [*Westbrook et al.*, 1994; *Trehu et al.*, 1999], and the recovery of carbonates and gas hydrate at ODP-Site 892 [*Sample and Kopf*, 1995; *Kastner et al.*, 1995], Hydrate Ridge (Fig. 1) became the focus of intense investigations of GEOMAR Research Center during RV SONNE cruises SO109 and SO110 in 1996. One goal of this international and interdisciplinary program was to investigate the carbonates near or at vent sites to elucidate their formation mecha-

Natural Gas Hydrates: Occurrence, Distribution, and Detection
Geophysical Monograph 124

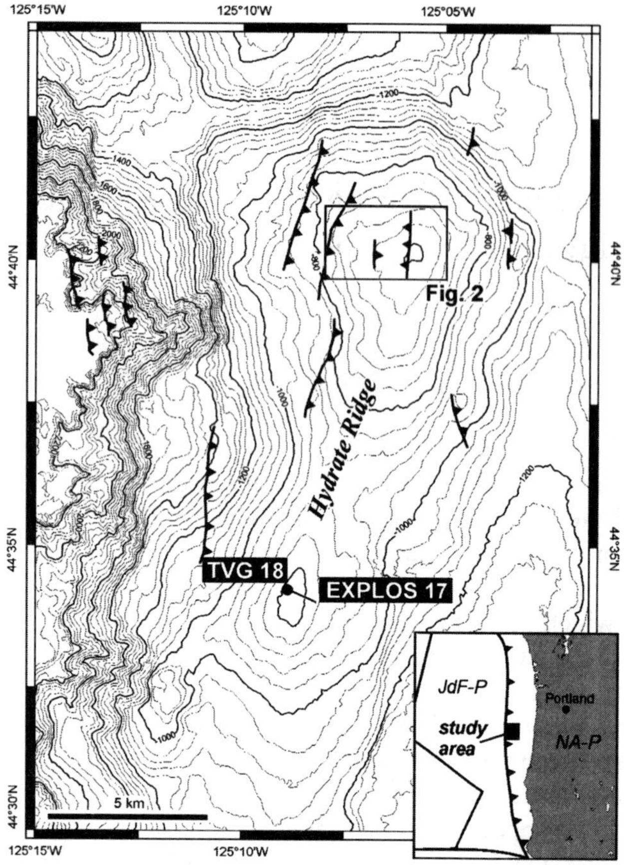

Figure 1 The map shows the location of the second accretionary ridge at the Cascadia margin called Hydrate Ridge. The northern summit is enlarged in Fig. 3 (square). Line EXPLOS 17 and the open circle TVG 18 at the southern summit indicate an observation track and sampling site of video-guided systems. JdF-P = Juan de Fuca Plate; NA-P = North American Plate.

nism. The objectives included studies of the formation environment *(1)* at or near the sediment surface with carbonates showing the impact of slumping, bioturbation and the presumed influence of gas hydrates; *(2)* in deeper sediment horizons at or near fracture zones with carbonates exhibiting fissures and veins from post-consolidation stress conditions; *(3)* in different diagenetic zones and from various fluid sources.

These geochemically distinct zones are *(a)* the oxic zone, *(b)* the sulfate reduction zone, *(c)* the upper part of the methanogenic zone, *(d)* the lower part of the methanogenic zone. These zones can be condensed and displaced towards the sediment surface due to the upward migration of fluids, but they still leave characteristic authigenic lithologies behind.

Previous work on Hydrate Ridge has shown that the prominent features covering the northern ridge crest are extensive morphological complexes made up of a large variety of carbonate lithologies. *Kulm and Suess* [1990] have presented a classification of these authigenic carbonate structures that distinguishes between slabs, irregular edifices and four types of chimneys, which vary in shape from doughnuts to conical chimneys of more than 1 m in height. Earlier, *Ritger et al.* [1987] classified block- and slab-like specimens as mudstones. In addition, *Sample and Reid* [1998] described mudstones and mudstone breccias from the northern summit of Hydrate Ridge. The observation of gas hydrate-associated carbonates [*Bohrmann et al.*, 1998] was an important impetus for the attempt to classify the carbonate lithologies at Hydrate Ridge in greater detail and explain their origin.

2. GEOLOGICAL SETTING AND GEOCHEMICAL ENVIRONMENT OF CARBONATE FORMATION

Hydrate Ridge is part of the second ridge of the north-south striking accretionary ridge system of the continental margin off Oregon (Fig. 1). Tectonic uplift and thrusting driven by the plate convergence causes thrust faults, extensional fractures and breached folds to develop along the ridge crest [*Carson et al.*, 1995]. The faults extend through the accreted Pliocene terrigenous silty-clayey sediments with interlayered sand to down below the BSR. Upward displacements in the BSR, which underlies the entire Hydrate Ridge [*Trehu et al.*, 1999], indicates a focused upward flow of warm fluids along the fault [*Westbrook et al.*, 1994] which serves as conduit [*Kastner et al.*, 1995 b]. The ascending warmer fluids stimulate the decomposition of gas hydrate, which leads to further dewatering [*Sample and Kopf*, 1995; *Suess et al.*, 1999] and generally drives the fluid venting phenomena and carbonate precipitation at cold vents.

Important diagenetic processes for the pore water and fluid composition are the formation and dissolution of minerals (clay minerals, carbonates, zeolithes) and biogenic components (e.g. foraminifers) in the sediment which modify the concentration of major elements (e.g. Ca, Mg, Si, K) and isotopic compositions (e.g. oxygen, strontium) in the pore water. Gas hydrate decomposition decreases the salt content as can be often recognized in lower chloride values [e.g. *Kvenvolden*, 1998] and affects the oxygen isotope composition due to the release of ^{18}O-enriched water ($\varepsilon_{gas\ hydrate/H2O}$ ~ 3‰) [*Davidson et al.*, 1983]. Methane from gas hydrate decomposition or from the methanogenic zone can be oxidized to HCO_3^- via sulfate reduction, which increases total alkalinity and partly controls whether aragonite, Mg-calcite or dolomite precipitates. Further, the formation of authigenic carbonates at cold vents is influenced by other reactions during the diagenetic turnover of organic matter, which also influences the total alkalinity and the isotopic composition of ΣCO_2 [*Claypool and Kaplan*, 1974; *Whiticar*, 1996]. Methane generation via CO_2 reduction lowers the $\delta^{13}C_{CH4}$ to values of less than -80‰ PDB just below the sulfate reduction zone [*Whiticar*, 1996; *Elvert et al.*, 1999], whereas with increasing burial depth the residual CO_2 pool of the methanogenic zone becomes isotopically heavier and methane generated from this CO_2-pool becomes isotopically heavier, too. This ^{13}C enrichment in the dissolved CO_2 species is the reason for the extremely heavy $\delta^{13}C$ values of up to +26‰ PDB observed in dolomite mudstones and reported by *Sample and Reid* [1998] from samples found at Hydrate Ridge. However, typical $\delta^{13}C$ values from carbonates formed at methane vents show light values between -70 and -35 ‰ PDB [*Ritger et al.*, 1987; *Matsumoto*, 1990; *Kulm and Suess*, 1990; *Bohrmann et al.*, 1998]. These values are generated by methane oxidation via sulfate reduction ($CH_4 + SO_4^{2-} \rightarrow H_2O + HS^- + HCO_3^-$), which is thought to be the most important mechanism to increase alkalinity and trigger the authigenic precipitation of carbonates at cold vents [*Ritger et al.*, 1987; *Han and Suess*, 1989; *Kulm and Suess*, 1990; *Paull et al.*, 1992; *Whiticar*, 1996].

Apart from the availability of carbonate ions and the diagenesis of carbon, carbonate precipitation and dissolution is strongly influenced by the concentration of the cations Ca and Mg (Fe, Mn), complex forming anions like SO_4^{2-} or $PO_4^{3-}$and temperature [*Burton*, 1993; *Fernández-Díaz et al.*, 1996; *Morse et al.*, 1997]. These factors determine the development of authigenic carbonates in general; hence, lithologies can be seen as a record of the geological and geochemical conditions under which they formed. With regard to cold vents, *Burton* [1993] and *Savard et al.* [1996] pointed out that aragonite seems to be favored in more oxic environments (high SO_4^{2-}) with higher total alkalinity concentrations. The crystallization of Mg-calcite preferentially occurs under slightly more anoxic conditions with lower SO_4^{2-} and total alkalinity concentrations.

3. CARBONATE COMPLEXES AND SAMPLING SITES

In 1996 we used a towed video sled (EXPLOS) to locate potential sample sites and map the distribution of carbonate lithologies during SO109 and SO110. Based on this observations, we started sampling at the carbonate complex found by DSRV ALVIN in 1987 formerly called 'Bioherm', 300 m west of ODP-Site 892 [*Westbrook et al.*, 1994]. Samples were taken by a video-guided grab (television grab, TVG) at four locations near active cold vent fields (Fig. 2). Because of the large grab size carbonates of up to 70 cm in size and > 40 kg in weight were retrieved. During SO109 a second complex was found approximately 800 m west of the northern summit of Hydrate Ridge. Again four samples were taken by TV-grab (Fig. 2). Despite the abundant clam and gastropod shells in the carbonates of these two complexes we use the term *'Chemoherm'* for these both features [*Bohrmann et al.*, 1998]. In the following, the former 'Bioherm' will be called *'ALVIN-Chemoherm'* and the newly discovered complex west of the summit *'SONNE-Chemoherm'* (Fig. 2). Samples from the summit itself were available from three TV-grabs. Because of the massive occurrence of carbonate boulders which pave the seafloor in several layers and form a 20 m high and 400 m wide structure at the summit, we call this area *'Boulder-Complex'*. In addition, samples from TVG 115 broadened the variety of carbonate lithologies from the northern summit of Hydrate Ridge.

Along *EXPLOS track 5* (Fig. 2), the seafloor around the Boulder-Complex and the SONNE-Chemoherm is characterized by soft sediment, sometimes with ripple marks and single or clusters of carbonate blocks and edifices. These carbonates vary in shape from well-rounded to subangular disk-like slabs and boulders, doughnut-shaped chimneys, and cemented coiled bioturbation trails. In between the equivalent blocks and edifices at the Boulder-Complex (Plate I A and B), active venting is indicated by small fields of living chemoautotrophic clams (*Calyptogena*), whitish bacterial mats (Plate I B) and rising gas bubbles [*Suess et al.*, 1999] as observed in the western part of the Boulder-Complex. Based on the interpretation of seismic line OR 8 [*Westbrook et al.*, 1994], the area of active fluid venting at the Boulder-Complex is probably fuelled by a seaward-dipping thrust fault which crops out at or near the western edge. At the SONNE-Chemoherm, very irregular and porous carbonate lithologies appear as a massive autochthonous carbonate complex incised by gullies several meters in depth and width. Only a thin sediment layer covers the rocks, but sediment accumulates inside the gullies, which allows clams to live. These gullies are probably induced tectonically by faults which cut the massive carbonate complex and also focus the fluid flow. The carbonates themselves contain clam and gastropod shells and typically show yellow-colored edges; the latter could be identified as pure aragonite precipitations.

On the southern summit of Hydrate Ridge randomly distributed single carbonate blocks were found during the track of *EXPLOS 17*. (Fig. 1) Bacterial mats and irregular white gas hydrate patches of up to 10 m in diameter paving the seafloor together with shells and living clams of *Calyptogena* indicating venting activity (Plate I C) could be observed. The sampling at Station TVG 18 (Fig. 1) yielded more than 50 kg of pure white gas hydrate intercalated with strongly reduced sediment and carbonate breccias [*Bohrmann et al.*, 1998]. Additionally, angular edges of light yellow colored layers cropping out at the seafloor (Plate I

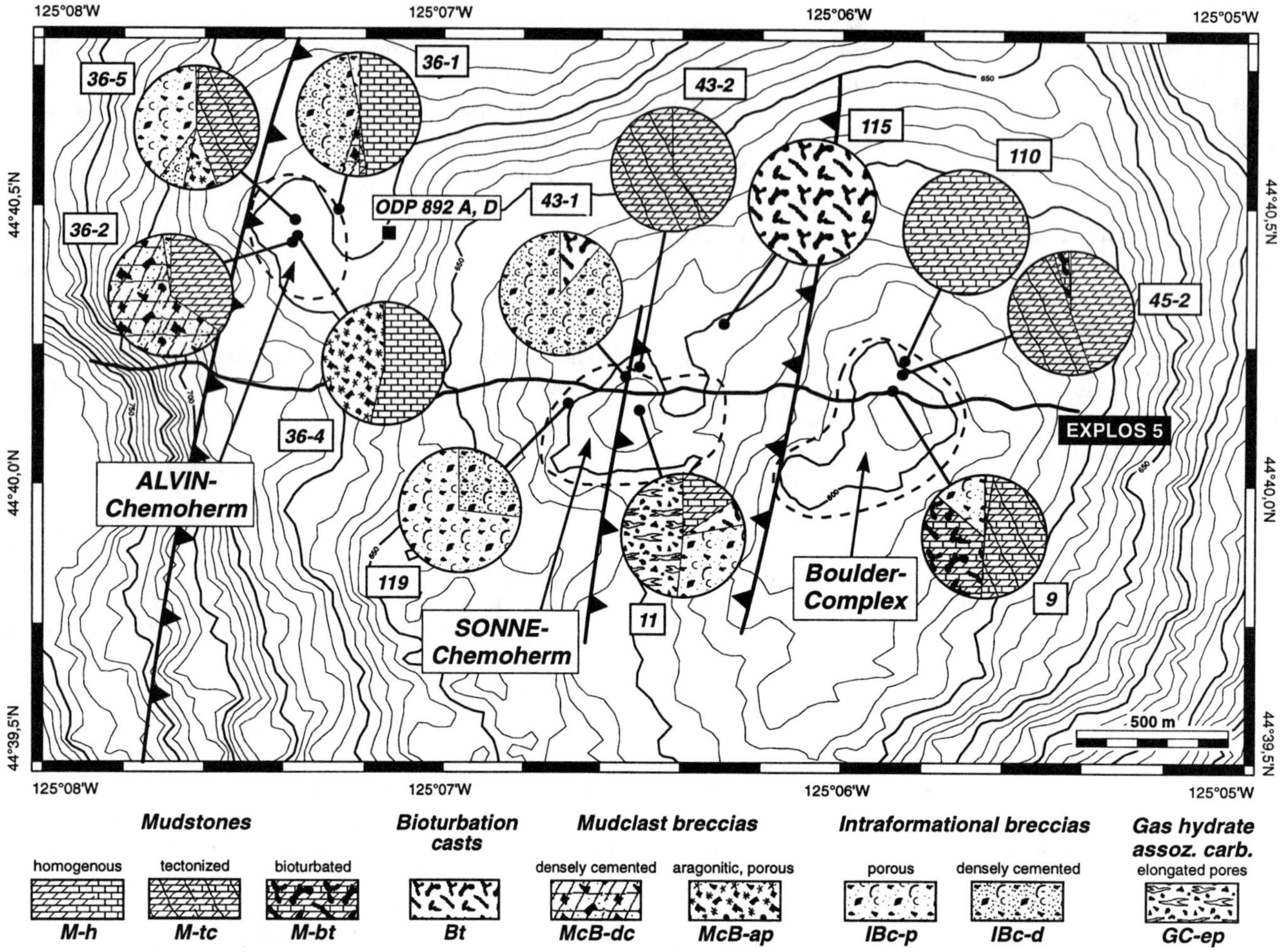

Figure 2 Distribution of the carbonate lithologies based on weight percent of the samples from the northern summit of Hydrate Ridge. Dashed lines mark the estimated outline of the carbonate complexes; the solid line corresponds to the EXPLOS 5 track of Fig. 2. Numbers represent the station number of SONNE Cruise 109 (TVG 36-1, 36-2, 36-4, 36-5, 43-1, 43-2, 45-2, 110, 115, 119) and SONNE Cruise 110 (TVG 9, 11).

C) were identified as aragonite crusts; they occurred in direct contact with and even inside gas hydrates.

4. ANALYTICAL METHODS

The carbonates were examined microscopically in thin sections and analyzed by x-ray diffraction (XRD). Subsamples were mechanically separated with a microdrill for mineral and isotope investigations. Samples for XRD analyses were prepared by a standard procedure adding corundum as internal standard. XRD patterns were obtained from 20° to 60° 2θ (0.01° 2θ steps) to calculate semiquantitative abundances of the carbonate phases aragonite (arag), low Mg-calcite (lMc, <8 Mol% $MgCO_3$), high Mg-calcite (hMc, 8 - 20 Mol% $MgCO_3$), protodolomite (proto, 30 - 40 Mol% $MgCO_3$) and dolomite (dolo, 40 - 55 Mol% $MgCO_3$) using standard calibration curves of different mineral mixtures (accuracy ± 5 wt.%). Using the shift in d-spacing of the (104) reflection, we estimated the $MgCO_3$-content of the trigonal carbonate phases. Neither chemical analyses (ICP-OES; HCl and acetic acid leaching) nor cathodoluminescence microscopy of 10 samples yielded any evidence for the presence of Fe or Mn in the carbonate phases. The carbonate and organic carbon content was calculated from the carbon difference between bulk and HCl-treated subsamples measured by a Carlo-Erba NA-1500 Element Analyzer.

The CO_2 extraction for $\delta^{13}C$ and $\delta^{18}O$ measurements was carried out with pure H_3PO_4 at 75°C with a Carbo-Kiel online device connected to a Finnigan MAT 252. Replicate

Plate I Seafloor images (locations see Fig. 3) taken during EXPLOS 5 (A, B) and EXPLOS 17 (C); scale: horizontal length = 2m, length of bottom-weight = 20 cm. A: block- and slab-like mudstones (M) with brown Fe/Mn coatings and bioturbation casts (Bt); B: irregular chemoherm block (center, IBc), white bacterial overgrowth on a mudstone surrounded by white clams (left). C: white patches of outcropping gas hydrate, bacterial mats, and aragonite crusts in direct contact with gas hydrate, dead and living clams on the sediment. Mudclast breccias McB-dc TVG 36-2, McB-ap TVG 36-4; mudstones M-bt TVG 9, M-tc TVG 36-1, M-h TVG 36-1. The McB-ap sample is fixed in resin, which darkens the rims of the mudclasts. Color plates can be downloaded at http://www.geomar.de/sci_dpmt/publications/

analyses of a laboratory standard show a standard deviation better than 0.03‰ for $\delta^{18}O$ and 0.02‰ for $\delta^{13}C$. Because of the unequal oxygen fractionation during the reaction to CO_2, dolomite has an analytical offset of +1.63 ‰ $\delta^{18}O$ [*Rosenbaum and Sheppard*, 1986]. Our data are not corrected for this offset, which has to be taken into account in temperature calculations for dolomitic samples. For reasons outlined by *Greinert* [1999], we used the equations by *Friedman and O'Neil* [1977] for Mg-calcite, *Hudson and Anderson* [1989] for aragonite, *Irwin* [1968] for protodolomite and *Clayton et al.* [1968] for dolomite to calculate equilibrium formation temperatures or to determine a possible influence of "heavy water" from decomposing gas hydrate on the $\delta^{18}O$ signal of the carbonate.

Strontium isotope analyses were performed on acid leachates of carbonate precipitates with bulk or microdrilled samples. Following *Sample and Kopf* [1995], we used ca. 2 ml of 1.0 N acetic acid for 2 h to leach the samples after gently grinding and rinsing them with micropure water. Sr was separated on cation exchange columns with 2.5 N HCl and dried at 120 - 150°C. Analyses were carried out with a *Finnigan* MAT 262.

5. RESULTS AND DISCUSSION

5.1 Petrographic Description and Classification

Our carbonate samples (totaling 450 kg and > 130 individual hand specimens) were visually pre-classified into mudstones (M), mudclast breccias (McB), intraformational breccias (IB), cemented bioturbation trails (Bt) and gas hydrate-associated carbonates (GC). Shore-based analyses show that homogeneous, tectonized and bioturbated mudstones can be subdivided into different types (Plate I).

Homogeneous mudstones (M-h): Samples of this type are gray to light brown in color but often coated by dark brown Mn/Fe oxi-hydroxides (Plate I A). They are characterized by a dense homogeneous matrix of terrigenous sediment (quartz, feldspar, amphibole, and clay minerals) and foraminifers. They are completely cemented by micrite to micro sparite (>67 wt% carbonate) consisting of pure dolomite or mixtures of hMc, proto and dolomite (Table 1). Dark specks of framboidal pyrite are common; enrichments of pyrite can also be found in dark layers sub-parallel to the rock surface (Plate I). Frequent light-colored rims of up to 2 cm in thickness indicate a secondary oxidation which is supported by oxidized pyrite and decreasing organic carbon contents. Some samples are enriched in dolomite (35-80 wt.%) in the outer rim suggesting dolomite recrystallization. *Ritger et al.* [1987] report a decrease in carbonate content for such rims, suggesting they are undergoing dissolution.

Tectonized mudstones (M-tc): These samples have a matrix similar to type M-h but are systematically enriched in dolomite (Table 1). The surface mirrors the internal texture of in situ brecciation with angular edges. Dark fissures and veins (<1 mm) represent clay mineral enrichment (Plate I). Other veins are cemented by dolomite-rich micrite to sparite formed by crystallization in an unconfined space. Occasionally these veins show different cement generations as described by *Sample and Reid* [1998]. Hand specimens of strongly tectonized mudstones show the transition to shear breccias.

Bioturbated mudstones (M-bt): Typical for this type are oval-shaped gray areas of bioturbation casts of 0.5 to 3 cm in diameter covering more than 80% of the rock cross-sections (Plate I). A wave-like surface is shaped as relief of the bioturbation trails. The terrigenous matrix is similar to that of other mudstones but is cemented by lMc, hMc and less than 35 wt.% by protodolomite (Table 1). No significant trends, neither in mineral composition nor in organic carbon content, were found between the bioturbation casts and the undisturbed beige-colored sediment matrix. Some samples showed sheared casts, indicating a secondary strain of the rigid carbonate block (Plate I).

Cemented bioturbation casts (Bt): These features frequently occur as elongated or bent single specimens on the seafloor (Plate I A) or build frameworks of crossed burrows (in particular at TVG 115). They are rich in glauconite (< 30%) and colored from light gray to dark green. Single casts usually can be classified as matrix-supported calciarenitic breccias of sub-angular clay-clasts (hMc to dolomite cemented) within a silt- to sand-sized matrix cemented by aragonite. Others are composed of a silt to clay matrix with sand-sized quartz or feldspar grains (< 10%) and are cemented by aragonite, hMc or protodolomite.

Mudclast breccias (McB): These arenitic to ruditic breccias consist exclusively of intraclasts or have few extraclasts in addition (Plate I). A densely cemented type (McB-dc) can be differentiated from a porous, light and extremely aragonite-cemented type (McB-ap). Both types are grain- to matrix-supported. Hand specimens of type McB-ap are composed exclusively of internally uncemented angular mudclasts within a calcisiltite matrix which is cemented by sparitic aragonite (Table 1). In some samples aragonite cementation culminates in cm-wide white areas which are composed entirely of radially grown aragonite needles (10 to 40 μm across, up to 1 mm in length). Only a few samples are enriched in lMc and show transitions to the McB-dc type. This type consists of weakly cemented mudclasts in a dense matrix of terrigenous material with monomineralic or polyphased cements (Table 1). Dark veins and fractures provide evidence for post-cementation tectonization.

Table 1. Mineral composition and isotope data of carbonates from Hydrate Ridge.

Lithology	arag[a] wt.%	lMc[b] wt.%	hMg[c] wt.%	proto[d] wt.%	dolo[e] wt.%	$\delta^{13}C$ ‰ PDB	$\delta^{18}O$ ‰ PDB
M-h n=31	-	10[f] (8-70)[g] n=10	33 (28-93) n=13	13 (7-73) n=17	44 (5-100) n=20	-37.0[f] -54.4 to +26.3[g]	6.1 4.6 to 13.1
M- tc n=15	-	7 (36-66) n=2	20 (14-100) n=5	5 (19-35) n=3	68 (34-100) n=12	-37.9 -50.0 to -25.5	6.2 4.5 to 7.3
M-bt n=9	-	51 (38-79) n=7	25 (25-87) n=4	29 (13-34) n=9	-	-47.4 -50.9 to -38.1	5.0 4.7 to 5.3
McB-dc n=28	3 (93) n=1	33 (13-100) n=11	20 (6-87) n=10	2 (1-16) n=4	42 (51-100) n=13	-32.3 -51.8 to +26.2	6.3 3.3 to 14.9
McB-ap n=13	94 (61-99) n=13	1 (4) n=1	4 (1-40) n=6	1 (8) n=1	-	-52.6 -55.2 to -39.1	4.2 3.4 to 4.8
IBc-p n=23	65 (61-99) n=17	1 (1-11) n=10	27 (2-85) n=21	5 (1-21) n=12	1 (1-8) n=9	-49.4 -54.8 to -38.0	4.0 3.0 to 5.1
IBc-d n=31	59 (9-100) n=23	7 (1-100) n=12	20 (1-82) n=23	7 (1-60) n=16	7 (2-100) n=7	-48.9 -55.1 to -40.1	5.2 3.5 to 12.2
Bt n=9	41 (23-94) n=5	4 (1-16) n=5	36 (1-78) n=9	11 (3-30) n=6	8 (3-38) n=5	-45.9 -49.8 to -39.5	4.8 3.3 to 6.3
GC-ep n=6	57 (0-97) n=4	-	38 (3-91) n=6	5 (2-15) n=4	-	-53.8 -54.4 to -53.0	3.8 3.2 to 4.4
GC-Ba n=54	77 (0-100) n=46	-	23 (0-100) n=40	-	-	-48.5 -54.2 to -40.6	3.9 3.2 to 4.9

(-) Not detected by XRD; n = number of samples; [a] aragonite; [b] low Mg-calcite; [c] high Mg-calcite; [d] protodolomite; [e] dolomite; [f] mean; [g] range.

Intraformational breccias, chemoherm-typical (IBc): These polymict lithologies contain chaotically orientated ruditic components of matrix-like intraclasts (> 1 cm), dark extraclasts (< 5%), bioturbation casts, shells and poorly lithified mudclasts (Plate II). They are the dominant carbonates at the chemoherm complexes and identical to the *irregular edifices* reported by *Kulm and Suess* [1990], exhibiting extremely irregularly shaped and randomly interconnected plumbing network of mesopores. The IBc carbonates investigated can be divided into a porous type with a pore volume of up to 20% (IBc-p) and a more densely cemented type (IBc-d; Plate II). Samples of the IBc-p type consist of more than 70% intraclasts, which are composed similarly to the surrounding matrix. This calciarenitic, clast-supported matrix of mudclasts (0.2 - 1 cm) and glauconite grains (ca. 5%; < 1 mm) in clay- to silt-sized terrigenous sediment is cemented by micritic and randomly orientated sparitic aragonite and little hMc (Table 1). The amount of bioclasts observed (*Calyptogena*, *Acharax* and gastropod shells) was less than 15%; mesopores (0.1 to 3 mm) usually make up 5% to 10% of the volume. Within these mesopores, botryoidal, white aragonite cements occur (Plate II). Both clasts and matrix contain dispersely distributed pyrite (~1%) which also occurs inside foraminifers. Samples of type IBc-d show zonal rim cements of botryoidal aragonite (<7mm thickness) and an advanced, cloudy matrix cementation (slightly enriched in lMc, Table 1) which leads to a pseudobrecciated texture (Plate II).

By simultaneously sampling massive gas hydrate layers and carbonates [*Bohrmann et al.*, 1998], we could classify at least two *gas hydrate-related rock structures* (Plate II). Surface-near growing gas hydrate at Hydrate Ridge builds massive, horizontal gas hydrate layers of sponge-like bubble fabric [*Bohrmann et al.*, 1998]. This fabric is formed by the crystallization of gas hydrate at the rim of those gas bubbles which were trapped in the sediment [*Greinert*, 1999; *Suess et al.*, 1999]. Both ascending bubbles and growing gas hydrate layers cause sediment brecciation. The resulting angular sediment clasts build an irregular, clast-supported fabric with high amounts of meso- and megapores. Gas hydrate layers or gas hydrate-cemented sediment parts in the host sediment prevent cementation by carbonate. In those cases the gas hydrate-free host sediment is cemented, the hydrate-bearing parts build pores after the gas hydrate has decomposed. Decomposition may occur in situ or during sample recovery. Lithologies with one of the above-mentioned rock structures were classified as *gas hydrate carbonates* (GC).

Aragonitic collapse breccia, formed by gas hydrate (GC-Ba): This type of breccia was recovered in direct contact to gas hydrate at Station TVG 18 (Fig. 1) [*Bohrmann et al.*,

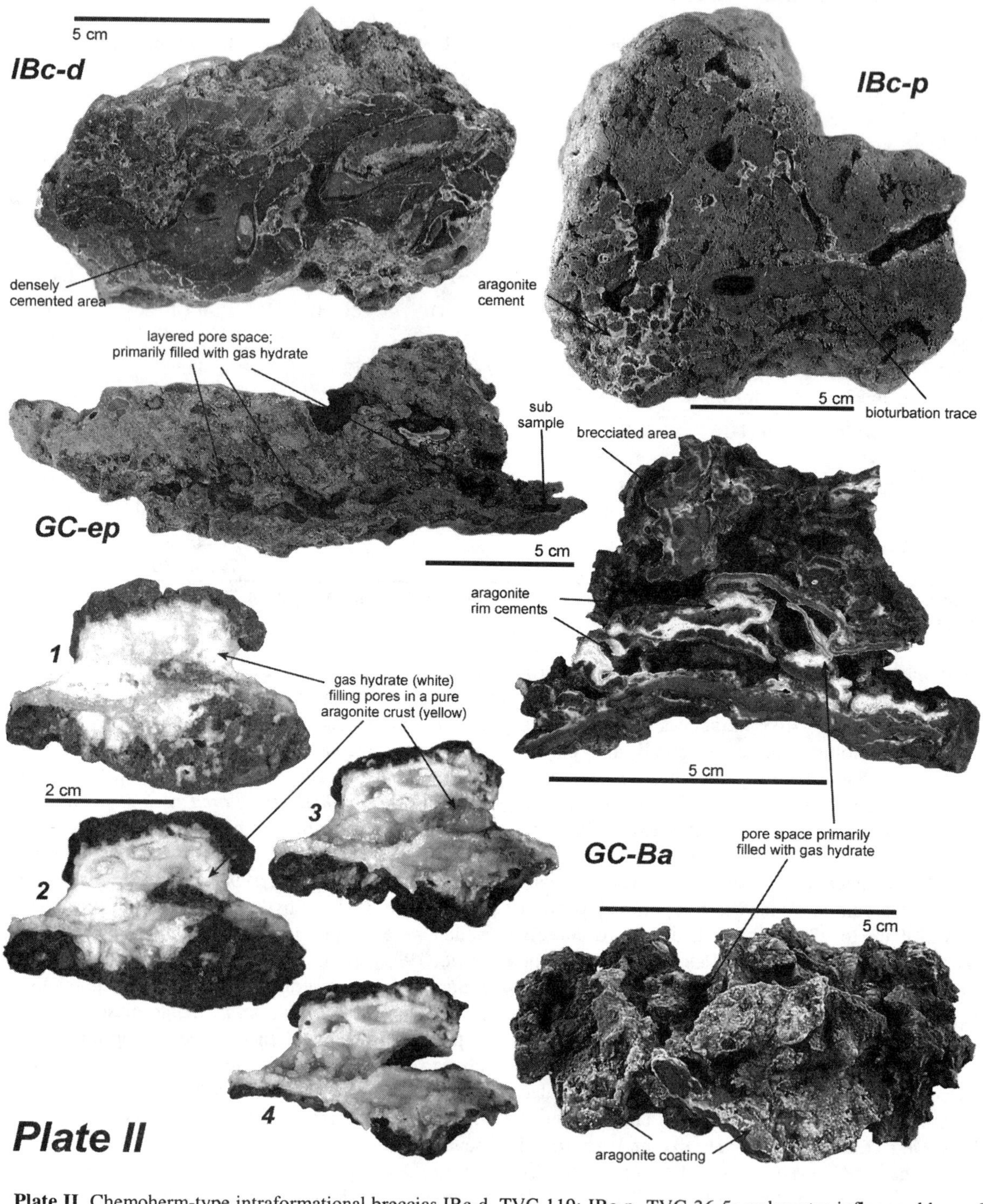

Plate II Chemoherm-type intraformational breccias IBc-d TVG 119; IBc-p TVG 36-5; carbonates influenced by and formed in contact with gas hydrate GC-ep TVG 11, GC-Ba TVG 18. The images 1 to 4 show the process of gas hydrate dissolution in the pores of an aragonite crust.

1998]. Samples show angular rudite- to arenite-sized clasts (Plate II). These monomict clasts are micrite-cemented, predominantly by aragonite, but consist also exclusively of hMc (Table 1). Due to gas hydrate decomposition the clasts build a grain-supported texture equivalent to collapse breccias with up to 35% pore space, which is cemented by botryoidal aragonite of several mm thickness (Plate II). Accessory, isolated layers of yellow aragonite - also observed at the seafloor surface during EXPLOS 17 (Plate I C) - were found in direct contact to gas hydrate and are interlayered with it (Plate II). Some layers show casts generated from the precipitation in the sponge-like bubble structure of hydrate layers [*Suess et al.*, 1999]. Biomarker investigations of those layers show extremely high amounts of components typical for methane-consuming and sulfate-reducing microorganisms (*Elvert et al.*, this volume). Those aragonite layers very likely represent a direct encrustation of biofilms.

Elongated pores, relics of gas hydrate cementation (GC-ep): Significant for this petrographically-unrestricted structural feature are elongated megapores of irregular shapes (Plate II) which look similar to gas hydrate layers observed in the sediment at site TVG 18 [*Bohrmann et al.*, 1998; *Suess et al.*, 1999]. We found such pores (up to 2 x 12 cm) in a mudclast breccia from SONNE-Chemoherm, but they can be expected in other lithologies (e.g. mudstones) as well, if environmental conditions favors surface-near gas hydrate formation.

5.2 Stable Isotope Signatures of C and O

Carbon and oxygen isotopes are widely used to distinguish between 'normal' marine carbonates and vent-related ones which are influenced by the sedimentary diagenetic carbon cycle. Whereas carbon isotopes reflect the carbon source, oxygen isotopes are effected by the formation temperature, the isotopic composition of the source water and the respective fractionation of the carbonate phases [*Friedman and O'Neil*, 1977].

The widest range in carbon and oxygen isotope data was observed in samples from ALVIN-Chemoherm with strongly positive $\delta^{13}C$ and $\delta^{18}O$ values up to +26.2‰ and +14.8‰ PDB, respectively, and extremely negative values also for $\delta^{13}C$ (down to -56‰ PDB, Fig. 3). Based on these different isotope values and the corresponding carbonate lithologies and mineral compositions, we can discriminate six groups which represent different geochemical environments for carbonate formation.

Specimens from ALVIN-Chemoherm exclusively belong to *group A,* but similar isotope data were reported by *Sample and Reid* [1998], *Kopf et al.* [1995], and *Sample and Kopf* [1995] from ODP Site 892. The mudstones and one clast of a mudclast breccia consist of pure dolomite or mixtures of dolo - proto - lMc ($\delta^{13}C$ <10‰). The diagenetic mechanism generating such positive $\delta^{13}C$-values is most likely caused by CO_2 reduction in methanogenic zones [*Piscioto and Mahoney*, 1981; *Mozley and Burns*, 1993; *Kopf et al.*, 1995; *Whiticar*, 1996]. Additionally, the temperature calculation for the formation of dolomite in water with 0‰ SMOW predicts crystallization at temperatures between 15.3 and 30.5 °C (corrected for the analytical offset of dolomite). Taking into account the temperature gradient of 51°C/km [*Carson et al.*, 1995] and the decrease in $\delta^{18}O$ in the pore water of -1‰ per 220m at ODP site 892 [*Kastner et al.*, 1995], samples from group A theoretically should have formed either between 150 to 400 mbsf or between 0 and 220 mbsf depending on the mineral phase dolomite or protodolomite (Fig. 4).

Samples which fall into *group B* are tectonized mudstones; they consist of dolomite (78 to 100 wt.%), hMc and protodolomite. Formation temperatures in equilibrium with water of 0‰ SMOW $\delta^{18}O$ suggest a crystallization at 7°C for protodolomite or 17°C for dolomite. The conditions observed at ODP Site 892 predict a formation depth between 20 and 80 mbsf or 170 to 240 mbsf depending on the primary carbonate phase (Fig. 4). This indicates a formation of group B carbonates at temperature conditions equivalent to greater sediment depth, too

Samples with isotope data falling into *group C* (McB-dc, M-tc / h and Bt) were found at both chemoherms and the Boulder-Complex (Fig. 3). Caused by the dolomite-enriched mineralogy the oxygen isotope values are between 6.3 and 7.5 ‰ PDB. For dolomitic samples of TVG 45-2, formation in equilibrium with pore water of 0‰ SMOW requires 16.9°C or water of -3.5‰ SMOW at 4.55°C, which is the in situ bottom water temperature at the northern summit [*Greinert*, 1999]. Such isotopically negative pore water in surface-near sediments is unlikely, therefore higher temperatures must be assumed equivalent to a deeper origin (190 mbsf, Fig. 4). For the origin of primary protodolomite at 4.55°C the water would have to be -0.44‰ SMOW; such values were measured at TVG 18 several cm below the surface [*Suess et al.*, 1999] but only below 80 mbsf at ODP Site 892 [*Kastner et al.*, 1995]. Based on the $\delta^{13}C$ values from -34.6 to -46.7 ‰ PDB and an estimated amount of methane-derived carbon of more than 30%, using Rayleigh fractionation of a mixture of methane-derived carbon and CO_2 of anaerobically degraded organic matter of -20‰ PDB (*Greinert*, 1999), formation of group C carbonates in the sulfate reduction zone is very likely.

The samples of *group D* can be distinguished from group E by heavier $\delta^{13}C$ values of up to -29.2 ‰ PDB and the absence of aragonite. The $\delta^{13}C$ data likely indicate a mixture of CO_2 generated by decomposition of organic matter

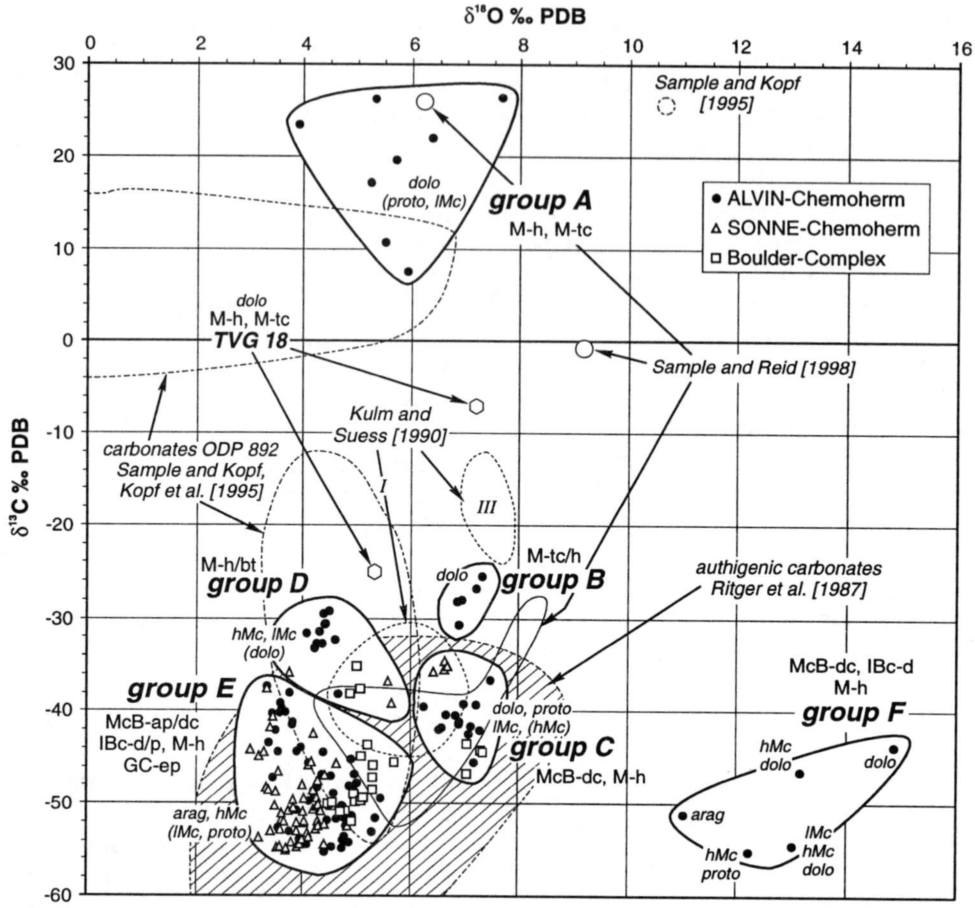

Figure 3 Carbon and oxygen isotope values of carbonates from the northern summit of Hydrate Ridge and two dolomitic mudstones from station TVG 18 (southern summit). Isotope data from GC-Ba type carbonates of TVG 18 fall exclusively into group F. Six groups (A to F) were distinguished which differ in their dominating carbonate cement phases as well as in their carbon sources and conditions of formation. Also shown are fields which mark analytical results taken from literature.

via sulfate reduction ($2CH_2O + SO_4^{2-} \rightarrow 2HCO_3^- + H_2S$) and methane-derived HCO_3^-, latter in smaller amounts than for group C. The lighter $\delta^{18}O$ values, compared to group C, results from the dominance of high Mg-calcite (12 mol% $MgCO_3$). The estimation of the $\delta^{18}O$-value of the source water in equilibrium with pure hMc carbonates (4.3 ‰ $\delta^{18}O$ PDB) at a bottom water temperature of 4.55°C yielded 0.81‰ SMOW. However, waters below 510 m water depth show values between -0.15 and -0.24 ‰ SMOW, which indicates an apparently ^{18}O-enriched source water or lower formation temperatures of around 1.1°C (water 0‰ SMOW) for the hMc phase of group D. We assume that hydrate water might be the reason for this effect.

Group E contains most of our samples as was the case with the samples described by *Ritger et al.* [1987] and *Kulm and Suess* [1990] (Fig. 3). Typical carbonate lithologies are mudclast and intraformational breccias, but bioturbation casts and some mudstones have similar isotopic values. The mineralogy is strongly dominated by aragonite and hMc with minor amounts of lMc and protodolomite. With $\delta^{13}C$ data ranging from -38.2 to -55.2 ‰ PDB carbonates are typical for methane-induced carbonate precipitations. With $\delta^{18}O$ values of up to 6‰ PDB (Fig. 3), most samples of group E are too heavy to have formed from ambient seawater at 4.55°C at an assumed value of 0 ‰ SMOW. Linear correlations exist between the $\delta^{18}O$ values and wt.% aragonite, which indicates an intercept of 3.34 ± 0.1 ‰ PDB $\delta^{18}O$ for pure aragonite (Fig. 5; TVG 36-5, 11, 119). This supports the hypothesis that aragonite forms in equilibrium with ambient sea water temperatures and $\delta^{18}O$ values between 0 and -0.27 ‰ SMOW at the northern summit of Hydrate Ridge. The corresponding high Mg-calcite phase (Fig. 5) shows $\delta^{18}O$ values ranging from 4.86 to 7.3‰ PDB. This indicates formation in equilibrium with isotopically

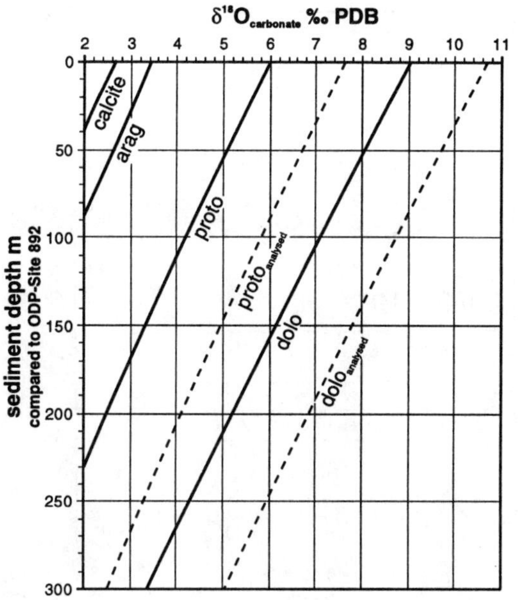

Figure 4 Calculated $\delta^{18}O$ values for different carbonate phases in equilibrium with the temperature gradient and oxygen isotope values found at ODP Site 892 (solid lines). Dashed lines for protodolomite and dolomite represent the analytical offset of 1.63‰ for dolomitic samples. They were used for the depth estimation of isotope investigations reported here.

heavy water of around 3.7‰ SMOW at 4.55°C or of around 0‰ SMOW at -8.7°C at station TVG 36-5. For the hMc end members at TVG 11 (5.6‰ $\delta^{18}O$ PDB, 16 mol% $MgCO_3$) and TVG 119 (6.5‰ $\delta^{18}O$ PDB, 14 mol% $MgCO_3$), formation temperatures of -2.7°C and -6.1°C (0‰ SMOW) or 1.9‰ and 2.9‰ SMOW at 4.55°C represent equilibrium conditions. The good correlations strongly support that the geochemical and isotopic conditions during the hMc precipitation are significantly different from those during aragonite formation. As temperatures below 0°C are truly unrealistic, heavier pore water has to be assumed during the formation of hMc. This is also shown by isotopic analyses of gas hydrate-related carbonates of type GC-Ba from station TVG 18 which fall exclusively into group E. As before, the $\delta^{18}O$ values correlate with the aragonite content and 3.66‰ for pure aragonite and 4.86‰ for pure hMc (17 mol% $MgCO_3$; Fig. 5). Calculations of the formation temperature in water with 0‰ SMOW resulted in 3.82°C for aragonite and 0.68°C for hMc. Using the bottom water temperature of 4.12°C from the southern summit [*Greinert*, 1999], aragonite should form from water with 0‰ SMOW $\delta^{18}O$, but the hMc-phase should form in "heavier" water of 0.86‰ SMOW. This further supports the assumption that gas hydrate influence hMc formation.

Still enigmatic are the extremely high $\delta^{18}O$ values of *group F* samples, whereas their carbon isotopes are typical for a methane-generated carbon source. Taking into account the dolomite fractionation, even pure dolomite as found at station TVG 36-2 (14.9‰ $\delta^{18}O$ PDB) would have to form at -8.3°C in water of 0‰ SMOW or at 4.55°C in water of 4.1‰ SMOW. Hydrate water as we know it from Hydrate Ridge [*Suess et al.*, 1999] would not cause such a significant shift toward heavy oxygen isotopes.

5.4 Sr-Isotopes and Seawater Influence

Sr-analyses of pure aragonite cements (McB-ap; IBc-p/d; GC-Ba), aragonite-rich clasts and matrixes (IBc-d/p; GC-Ba), and hMc, proto- and dolo-containing mudstones (M-h/tc) show values between 0.709191 and 0.709120 $^{87}Sr/^{86}Sr$. Hence, all these carbonates are formed from water with Sr-values similar to present day seawater (0.709175) [*Paytan et al.*, 1993]. Only one hMc- and one dolomite-rich sample from the ALVIN-Chemoherm indicate a different source with $^{87}Sr/^{86}Sr$ values of 0.708320 and 0.708853. Such values are equivalent to pore water analyses at ODP-Site 892 at around 10 mbsf [*Kastner et al.*, 1995 b].

5.5 Carbonates of Diagenetic Origin in Sediments of Very Slow Venting: Mudstones

Mudstones are formed as micrite-cemented concretions in the sediment pore space in carbonate-oversaturated sedimentary environments. They are generated in the sediment column and can precipitate whenever supersaturation is induced by the general diagenetic environment. It is still not clear if they form exclusively by diagenesis or require very slow ascending fluids. However, that the carbon mass balances cannot be maintained in a closed system suggest at least very slow venting. Typically, concretions seem to grow as horizontal slabs or round boulders, which can be broken by tectonic stress conditions by vertically oriented fault planes (which may serve as fluid pathways). Evidence for authigenic carbonate formation in deeper sediment horizons was found in samples of type M-tc/h (group A). This is shown by the dominance of dolomite (Table 1), which requires high Mg/Ca ratios, an extreme ^{13}C-enrichment and a calculated formation temperature of up to 30°C. Appropriate environmental conditions occur at ODP-Site 892 between 10 to 20 mbsf [*Kastner et al.*, 1995 b]. Methanogenesis via CO_2-reduction and the fractionation in a closed system pushes the $\delta^{13}C$ values up to +18‰ between 19 and 24 mbsf [*Whiticar et al.*, 1995]. Thus, group A carbonates are supposed to represent the deepest genesis of any carbonates found at Hydrate Ridge (Fig. 6), which form in environmental conditions close to the primary precipitation of dolomite [*Lippmann*, 1973].

Specimens of group B are suggested to have formed near carbonates of group A because of the same dolomitic car-

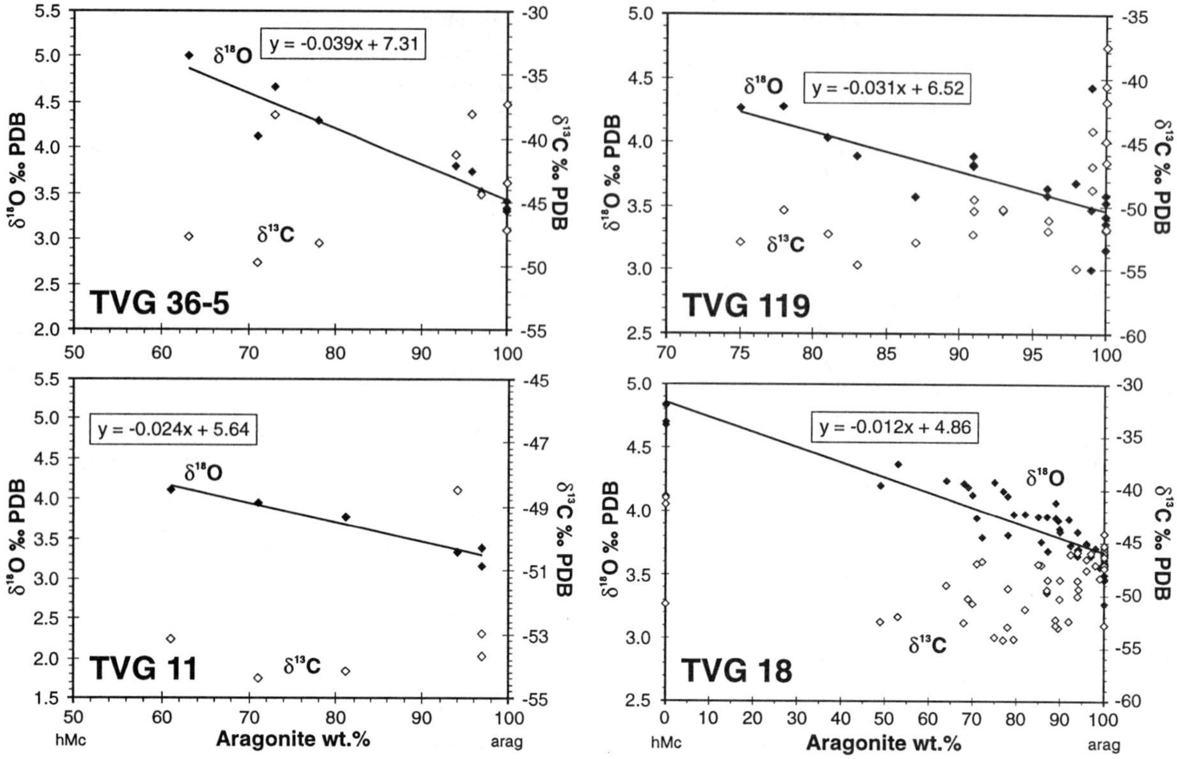

Figure 5 Group E samples (Fig. 4) reveal linear correlation between the arag/hMc ratio and the $\delta^{18}O$ values (black symbols) with an ^{18}O-enriched hMc phase. Pure aragonite yields 3.5 ± 0.1 ‰ $\delta^{18}O$ PDB, whereas ^{18}O is strongly enriched in the high Mg-calcite phase (hMc). There is no clear correlation between the mineralogy and $\delta^{13}C$ values (open symbols). TVG 36-5 from ALVIN-Chemoherm; TVG 11 and 119 from SONNE-Chemoherm; TVG 18 from the southern summit of Hydrate Ridge.

bonate types, equivalent $\delta^{18}O$ values and also higher formation temperatures (up to 17°C). The lighter carbon isotope values compared to group A can be explained by admixtures of small amounts of methanotrophic CO_2 with CO_2 derived from methane oxidation. Thus, the geochemical environment appears to be the deepest part of the sulfate reduction zone (Fig. 6) where geochemical conditions (Mg/Ca ratio) strengthen the crystallization of Mg-enriched carbonate phases [*Kastner et al.*, 1995 b]. Pore water data from ODP-Site 892 show that the base of the sulfate reduction zone occurs at 20 mbsf. This depth of the methanogenesis sulfate reduction-boundary depends on the geochemical conditions which are themselves influenced by ascending fluids (Fig. 6) and can be pushed upward up to some cm below the seafloor surface [*Boetius et al.*, 2000].

Group C carbonates are still dolomite-dominated and crystallized at elevated temperatures, which implies a deeper environment than the lMc- and hMc-dominated mudstones of group D and E. Because of the suggested higher methane-derived carbon content ($\delta^{13}C$ <34‰ PDB) relative to group B, we assume a formation in the Mg-enriched sulfate reduction zone, which is not influenced by the methanogenic zone, for the mudstones of group C. Such geochemical conditions are probably present above the transition zone (Fig. 6) and thus above group B carbonates.

Indicated by non-dolomite-dominated cements and even "too heavy" oxygen isotope values compared to the ambient bottom water temperature, we suggest a formation of group D mudstones in an geochemical environment with decreased Mg/Ca ratios relative to group C carbonates and thus at a shallower depth. A mixture of methane-derived carbon and carbon from degraded organic matter can be assumed from the $\delta^{13}C$ values (Fig. 3) which again indicate a formation in the sulfate reduction zone (Fig. 6).

Significant negative $\delta^{13}C$ values as in group E support the formation of M-h and M-bt lithologies from methane-derived carbon as main carbon source. Because of their dolomite-free, not-sheared occurrence and the high amount of bioturbation features in mudstones, they very likely represent a formation environment near the sediment surface (Fig. 6) but still buried in anaerobic conditions, which are characterized by methane oxidation.

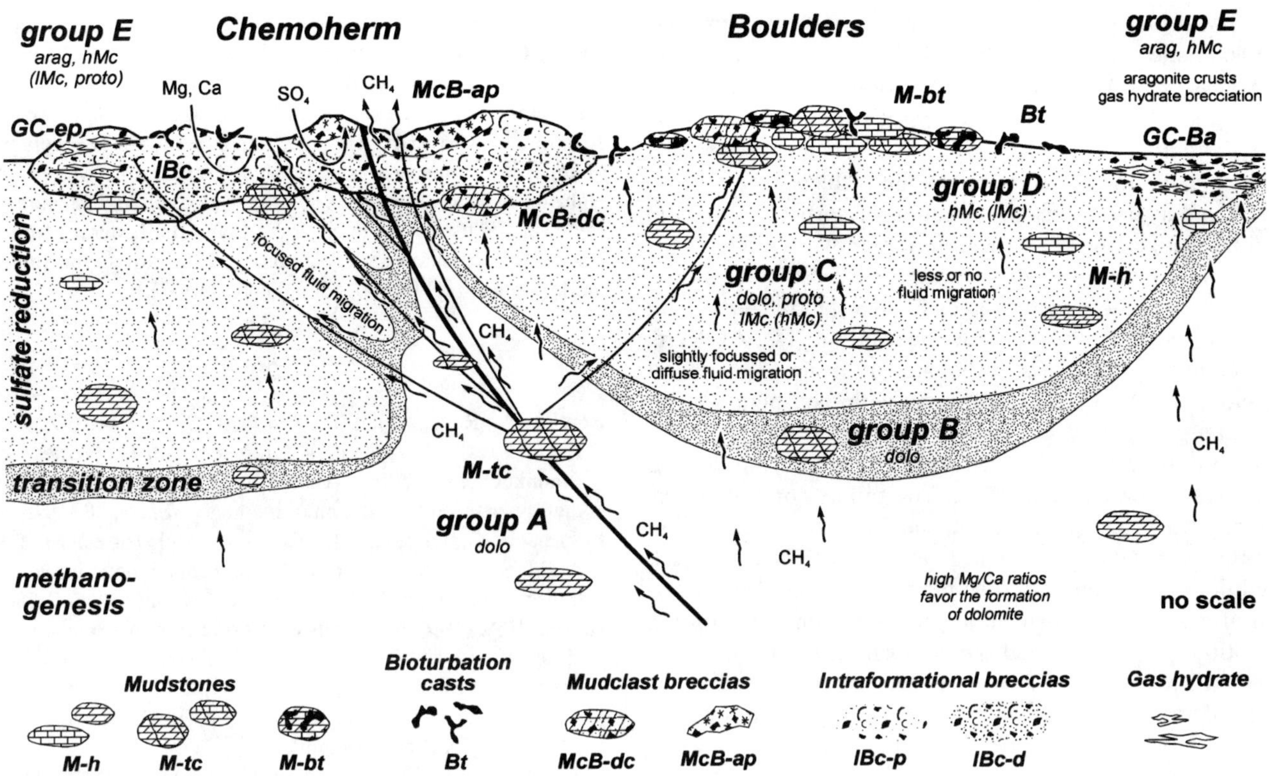

Figure 6 Sketch of the origin of petrographically and isotopically different lithologies relative to the controlling diagenetic environment. Group A is restricted to the deep methanogenic zone resulting in the formation of dolomite in M-h and M-tc carbonates. Group B, also dolomite-dominated, is representative of the transition between sulfate reduction and methanogenesis. Group C lithologies, with typical methane-derived $\delta^{13}C$ values but protodolomite and dolomite as carbonate phases, have formed in the sulfate reduction zone. Group D contains HCO_3^- from a mixture of degraded organic matter and methane-derived carbon; carbonates are generated close to the surface which can be deduced from the absence of dolomite and the calculated formation temperatures. The aragonite-rich carbonates of group E represent a formation near to or at the surface, influenced by methane oxidation via sulfate reduction; typical are irregular brecciated fabrics of chemoherm carbonates as well as the gas hydrate-related carbonates GC-Ba and GC-ep.

5.6 Methane Venting-Induced Carbonates: Mudclast and Intraformational Breccias

The uncompacted and porous texture of the aragonite-rich mudclast (McB-ap) and intraformational breccias (IBc-p/d; group E) exclude formation far below the sediment surface. Angularly shaped, uncemented mudclasts in McB-ap and -dc lithologies suggest in situ brecciation of semi-consolidated sediment. Brecciation may be triggered by sediment slumps or tectonic- and venting-induced disturbances like the "sudden" release of trapped fluids or gas [*Matsumoto*, 1990]. In general, carbonates of the IBc and McB-ap lithologies represent par-autochthonous formations restricted to areas close to cold vent sites and elevated rates of channeled fluid flow. They constitute the typical lithologies at both chemoherms. The significant enrichment of aragonite in IBc, McB-ap and Bt type carbonates of group E supports the formation in a sulfate-bearing environment of slightly reducing conditions, where aragonite precipitation is favored against high Mg-calcite crystallization [*Burton*, 1993; *Savard et al.*, 1996]. Such conditions are realized very close to or directly at the seafloor where methane-derived HCO_3-rich fluids mix with bottom or surface-near pore water (Fig. 6) as also shown by present day Sr-isotope values. A precipitation into the bottom water seems confirmed by pure sinter-like aragonite crusts found at both chemoherm complexes [*Greinert*, 1999].

5.7 Carbonate Lithologies Generated by Gas Hydrate

The influence of gas hydrate on carbonates on the one hand is illustrated by the irregular and flattened 'gas hydrate pores' in a mudclast breccia (GC-ep; Plate II) from the northern summit. On the other hand, gas hydrate growth and decomposition produce the aragonitic collapse breccias characteristic for TVG 18 (GC-Ba, Plate II). As shown in the linear correlation between the aragonite content and $\delta^{18}O$ values (Fig. 5) and in agreement with observations by *Sample and Reid* [1998], we propose that the oxygen isotope data of hMc phases reflect the influence of decomposed gas hydrates on the formation of carbonates at both Hydrate Ridge summits. The dominance of aragonite in the matrix, the occurrence of pure aragonite crusts (GC-Ba) on the seafloor (Plate I C) or in gas hydrate fabric [*Bohrmann et al.*, 1998] indicates a surface-near origin (Fig. 6). Here, the conditions typical for carbonates of group E in general stimulate the aragonite precipitation with their high alkalinity and higher sulfate concentrations [*Burton*, 1993; *Savard et al.*, 1996]. In contrast, Mg-calcite formation is favored at a relatively lower alkalinity and low sulfate concentrations. For hMc of group E, this has to be conform to a sediment environment with ^{18}O-enriched pore water (Fig. 5). Such conditions are suggested to be present after the decomposition of gas hydrate when sulfate has been consumed for the oxidation of hydrate-methane via sulfate reduction. The released hydrate water further changes the isotopic composition of the pore water and decreases the alkalinity by dilution.

6. CONCLUSIONS

Petrographic investigations and detailed isotope analyses allow to characterize the formation environment of varied venting-related carbonate lithologies. Excluding dolomitic mudstones ($\delta^{13}C$ +10 to +26 ‰PDB), which represent the deepest formation environment within the methanogenic zone, all analyzed carbonates from Hydrate Ridge contain methane-derived carbon. Suggested by the mineralogical and isotopic composition, other mudstones are formed within the sulfate reduction zone from mixed carbon sources (residual CO_2 from the methanogenic zone, oxidized methane, decomposed organic matter) at different depths. Surface-near growing carbonates show increased lMc, hMc and dominantly aragonite contents, whereas carbon mainly derived from methane ($\delta^{13}C$ below -38‰ PDB). Oxygen isotope data indicate low formation temperatures and influences of decomposed gas hydrate during the hMc crystallization. Intraformational breccias directly formed at vent sites near the sediment surface are the typical carbonates building massive carbonate complexes, "Chemoherms". Structural signs of carbonate-forming mechanisms by growing and decomposing gas hydrate are given by aragonite-cemented collapse breccias and elongated megapores which can occur in different lithologies. Again, oxygen isotope data of the hMc phase suggest an influence of decomposed gas hydrate during the formation of these "gas hydrate carbonates".

Acknowledgement. This contribution is a part of the Ph.D. thesis of J.G., he wishes to thank all persons who helped from start to finish. We all acknowledge in particular the excellent technical assistance by B. Domeyer and J. Heinze (GEOMAR) as well as M. Joachimski (University Erlangen) and H. Strauß (University Bochum) for isotope analyses. Further thanks to E. Hütten and T. Nähr for helpful discussions as well as C. Paull and an anonymous reviewer for useful comments. We thank the Federal Ministry of Education, Research and Technology, Bonn for supporting these work (Grants 03G0110A/B and 03G0109A/C).

REFERENCES

Boetius, A., Ravenschlag, K., Schubert, C.J., Rickert, D., Widdel, F., Gieskes, A., Amann, R., Jørgensen, B.B., Witte, U. and Pfannkuche, O. (2000): A microbial consortium apparently mediating anaerobic oxidation of methane. *Nature*, 407, 623-626.

Bohrmann, G., Greinert, J., Suess, E., and Torres, M. (1998): Authigenic carbonates from the Cascadia subduction zone and their relation to gas hydrate stability. *Geology*, 26, 647-650.

Burton, E.A. (1993): Controls on marine carbonate cement mineralogy: review and reassessment. *Chem. Geol.*, 105, 163-179.

Carson, B., Westbrook, G.K., Musgrave, R.J., and Suess, E. (eds.) (1995): *Proc. ODP, Sci. Results, 146* (Pt. 1): College Station, TX (Ocean Drilling Program).

Claypool, G.E. and Kaplan, I.R. (1974): The Origin and Distribution of Methane in Marine Sediments. In: Kaplan, I.R. (ed.): *Natural Gases in Marine Sediments*. Plenum Press, New York and London. 99-139.

Clayton, R.N., Jones, B.F., and Berner, R.A. (1968): Isotope studies of dolomite formation under sedimentary conditions. *Geochim. Cosmochim. Acta*, 32, 415-432.

Davidson, D.W., Leaist, D.G., and Hesse, R. (1983): Oxygen-18 enrichment in the water of a clathrate hydrate. *Geochim. Cosmochim. Acta*, 47, 2293-2295.

Elvert, M, Suess, E., and Whiticar, M. J. (1999): Anaerobic methane oxidation associated with marine gas hydrates: superlight C-isotopes from saturated and unsaturated C_{20} and C_{25} irregular isoprenoids. *Naturwissenschaften*, 86(6), 295-300.

Fernández-Díaz, L., Putnis, A., Prieto, M., and Putnis, C.V. (1996): The role of magnesium in the crystallization of calcite and aragonite in a porous medium. *J. Sed. Res.*, 66, 482-491.

Friedman, I. and O'Neil, J.R. (1977): Compilation of stable isotope fractionation factors of geochemical interest. In: M. Fleischer (ed.): *Data of Geochemistry*, 6th ed., U.S.G.S Professional Paper, 440-KK

Greinert, J. (1999): Rezente submarine Mineralbildungen: Abbild geochemischer Prozesse an aktiven Fluidaustrittsstellen im Aleuten- und Cascadia-Akkretionskomplex. Dissertation, Christian-Albrechts-University, Kiel; GEOMAR Report 87, 217 pp.

Han, M.W. and Suess, E. (1989): Subduction-induced pore fluid venting and the formation of authigenic carbonates along the Cascadia continental margin: Implications for the global Ca-cycle. *Palaeogeogr., Palaeoclimatol., Palaeoecol.*, 71, 97-118.

Hudson, J.C. and Anderson, T.F. (1989): Ocean temperatures and isotopic compositions through time. In: Clarkson, E.N.K.,

Curry, G.B., and Rolfe, W.D.I. (eds.): *Environments and physiology of fossil organisms*, Trans. R. Soc. Edinburgh, 80, 183-192.

Irwin, H. (1980): Early diagenetic precipitation and pore fluid migration in the Kimmeridge Clay of Dorset, England. *Sedimentology*, 27, 577-591.

Kastner, M., Kvenvolden, K.A., Whiticar, M.J., Camerlenghi, A., Lorenson, T.D. (1995): Relation between pore fluid chemistry and gas hydrate associated with bottom-simulating reflectors at the Cascadia margin, Sites 889 and 892. In: Carson et al. (eds.) (1995): *Proc. ODP, Sci. Results, 146* (Pt. 1): College Station, TX (Ocean Drilling Program), 175-187.

Kastner, M., Sample, J.C., Whiticar, M.J., Hovland, M., Cragg, B.A., Parkes, J.R. (1995 b): Geochemical evidence for fluid flow and diagenesis at the Cascadia convergent margin. In: Carson, B. et al. (eds.) (1995), *Proc. ODP, Sci. Results, 146* (Pt. 1): College Station, TX (Ocean Drilling Program), 375-384.

Kopf, A., Sample, J.C., Bauer, P., Behrmann, J.H., and Erlenkheuser, H. (1995): Diagenetic carbonates from Cascadia Margin: Textures, chemical composition, and oxygen and carbon stabile isotope signatures. In: Carson, B. et al. (eds.) (1995), *Proc. ODP, Sci. Results, 146* (Pt. 1): College Station, TX (Ocean Drilling Program), 117-136.

Kulm, L.D., Suess, E., Moore, J.C., Carson, B., Lewis, B.T., Ritger, S.D., Kadko, D.C., Thornburg, T.M., Embley, R.W., Rugh, W.D., Massoth, G.J., Langseth, M.G., Cochrane, G.R., and Scamman, R.L. (1986): Oregon subduction zone: Venting, fauna and carbonates. *Science*, 231, 561-566.

Kulm, L.D. and Suess, E. (1990): Relationship between carbonate deposits and fluid venting: Oregon accretionary prism. *J. Geophys. Res.*, 95, B6, 8899-8915.

Kvenvolden, K.A. (1998): A primer on the geological occurrence of gas hydrates. In: Henriet, J.-P. and Mienert, J. (eds.): *Gas hydrates, Relevance to world margin stability and climatic change*. Geological Soc. Spec. Pub., 137, 9-30.

Linke, P., Suess, E., Torres, M., Martens, V., Rugh, W.D., Ziebis, W., and Kulm, L.D. (1994): In-situ measurements of fluid flow from cold seeps at the active continental margins. *Deep Sea Res.*, 41, 721-739.

Lippmann, F. (1973): *Sedimentary carbonate minerals*. Berlin, Heidelberg: Springer, 228 pp.

Matsumoto, R. (1990): Vuggy carbonate crust formed by hydrocarbon seepage on the continental shelf of Baffin Island, northeast Canada. *Geochemical Journal*, 24, 143-158.

Morse, J.W., Wang, Q., and Tsio, M.Y. (1997): Influence of temperature and Mg:Ca ratio on $CaCO_3$ precipitates from seawater. *Geology*, 25;, 85-87.

Mozley, P.S. and Burns, S.J.(1993): Oxygen and Carbon isotopic composition of marine carbonate concretions: An Overview. *J. Sed. Pet.*, 63, 73-83.

Paull, C.K., Chanton, J.P., Neumann, A.C., Coston, J.A., and Martens, C.S. (1992): Indicators of methane-derived carbonates and chemosynthetic organic carbon deposits: Examples from the Florida Escarpment. *Palaios*, 7, 361-375.

Paytan, A., Kastner, M., Martin, E.E., MacDougall, J.D., and Herbert, T. (1993): Marine barite as a monitor of seawater strontium isotope composition. *Nature*, 366, 445-449.

Pisciotto, K.A. and Mahoney, J.J (1981): Isotopic survey of diagenetic carbonates, Deep Sea Drilling Project Leg 63. In: Orlofsky, S. (ed.), *Initial Reports of Deep Sea Drilling Project 63*, Texas A. & M. University, Ocean Drilling Program College Station, Texas, United States, 595-609.

Ritger, S., Carson, B., and Suess, E. (1987): Methane-derived authigenic carbonates formed by subduction-induced porewater expulsion along the Oregon/Washington margin. *Geo. Soc. Am. Bull.*, 98, 147-156.

Rosenbaum, J. and Sheppard, S.M.F. (1986): An isotopic study of siderites, dolomites and ankerites at high temperatures. *Geochim. Cosmochim. Acta*, 50, 1147-1150.

Sample, J.C. and Kopf, A., 1995, Isotope geochemistry of syntectonic carbonate cements and veins from the Oregon margin: Implications from the hydrogeologic evolution of the accretionary wedge. In: Carson, B. et al. (eds.) (1995): *Proc. ODP, Sci. Results, 146* (Pt. 1): College Station, TX (Ocean Drilling Program), p. 137-147.

Sample, J.C. and Reid, M.R. (1998): Contrasting hydrogeologic regimes along strike-slip and thrust faults in the Oregon convergent margin: Evidence from the chemistry of syntectonic carbonate cements and veins. *Geol. Soc. Am. Bull.*, 110, 48-59.

Savard, M.M., Beauchamp, B., and Veizer, J. (1996): Significance of Aragonite Cements around Cretaceous Marine Methane Seeps. *J. Sed. Res.*, 66, 430-438.

Suess, E., Carson, B., Ritger, S.D., Moore, J.C., Jones, M.L., Kulm, L.D., and Cochrane, G.R., (1985): Biological communities at vent sites along the subduction zone off Oregon. In: Jones, M.L. (ed.): The hydrothermal vents of the eastern Pacific: An Overview. *Biol. Soc. Wash. Bull.*, v. 6, p. 475-484.

Suess, E., Torres, M., Bohrmann, G., Collier, R.W., Greinert, J., Linke, P., Rehder, G., Trehu, A., Wallmann, K., Winckler, G., and Zuleger, E. (1999): Gas hydrate destabilization: enhanced dewatering, benthic material turnover and large methane plumes at the Cascadia convergent margin. *Earth Planet. Sc. Lett.*, 170, 1-15.

Trehu, A.N., Torres, M.E., Moor., G.F., Suess, E. and Bohrmann, G. (1999): Temporal and spatial evolution of a gas hydrate-bearing accretionary ridge on the Oregon continental margin. *Geology*, 27, 939-942.

Westbrook, G.K., Carson, B., Musgrave, R.J., et al (1994): *Proc. ODP, Init. Repts., 146* (Pt.1): College Station, TX (Ocean Drilling Program), 55-125.

Whiticar, M.J., Hovland, M., Kastner, M., Sample, J.C. (1995): Organic geochemistry of gases, fluids, and hydrates at the Cascadia accretionary margin. In: Carson, B. et al. (eds.) (1995): *Proc. ODP, Sci. Results, 146* (Pt. 1): College Station, TX (Ocean Drilling Program), 385-397.

Whiticar, M.J. (1996): Isotope tracking of microbial methane formation and oxidation. *Mitt. der intern. Verein. für theoretische und angewandte Limnol.*, 25, 39-54.

G. Bohrmann, J. Greinert, and E. Suess, GEOMAR Research Center for Marine Geosciences, Wischhofstrasse 1-3, 24148 Kiel, Germany. Contact: jgreinert@geomar.de.

Carbon Isotopes of Biomarkers Derived from Methane-Oxidizing Microbes at Hydrate Ridge, Cascadia Convergent Margin

Marcus Elvert[1], Jens Greinert, and Erwin Suess

Department of Marine Environmental Geology, GEOMAR Research Center for Marine Geosciences, Kiel, Germany

Michael J. Whiticar

Biogeochemistry Facility, School of Earth and Ocean Sciences, Victoria, Canada

Abstract - The microbial community inhabiting anaerobic sediments of Hydrate Ridge, an area characterized by extensive methane-venting and gas hydrate occurrence on the Cascadia accretionary prism, is dominated by a complex consortium of archaea and bacteria which performs anaerobic methane oxidation (AMO). The process is indicated by the dominance of archaeal-derived biomarkers such as crocetane, pentamethylicosane, and archaeol in anaerobic sediments and authigenic carbonates. The biomarkers are highly depleted in ^{13}C relative to their methane carbon source, which results in carbon isotopic compositions as low as -128‰ PDB. The essentially limitless reservoir of methane available to the archaea in this gas hydrate environment allows the kinetic carbon isotope effect ($\varepsilon_{CH4\text{-}biomarker}$) associated with AMO to be fully developed (ε_{max} = 72‰). No substrate depletion effects were detected. Varying isotope effects and concentrations of the biomarkers on and off Hydrate Ridge indicate that the abundance of these compounds is essentially dependent on the supply of methane from this special environmental setting.

1. INTRODUCTION

Recent studies on the stable carbon isotope composition of individual organic compounds present in sediments and petroleum have significantly improved the identification of substrate sources [e.g., *Freeman et al.*, 1990; *Kohnen et al.*, 1992]. The carbon isotope composition of these compounds, most often called biomarkers because of their specific structural relationship to a certain biological source, reflects the carbon source utilized and the isotopic fractionation accompanying carbon fixation and biosynthesis [*Hayes*, 1993]. Where biomarkers exhibit extreme isotopic values, their origin can be identified with confidence. For example, biomarkers highly depleted in ^{13}C are indicative of either aerobic or anaerobic methane oxidation [*Freeman et al.*, 1990; *Collister et al.*, 1992; *Elvert et al.*, 1999; *Hinrichs et al.*, 1999; *Thiel et al.*, 1999; *Elvert et al.*, 2000; *Pancost et al.*, 2000]. Depletions in ^{13}C result from isotopic fractionation of the methane substrate during oxidation due to kinetic isotope effects that enrich the residual methane in ^{13}C and deplete the carbon fixed in biomass in ^{13}C [*Barker and Fritz*, 1981; *Coleman et al.*, 1981; *Summons et al.*, 1994; *Jahnke et al.*, 1999]. Considering that the initial production of methane from methanogens is accompanied by a

[*] Present address: Department of Biogeochemistry, Max Planck Institute for Marine Microbiology, Bremen, Germany

Natural Gas Hydrates: Occurrence, Distribution, and Detection
Geophysical Monograph 124

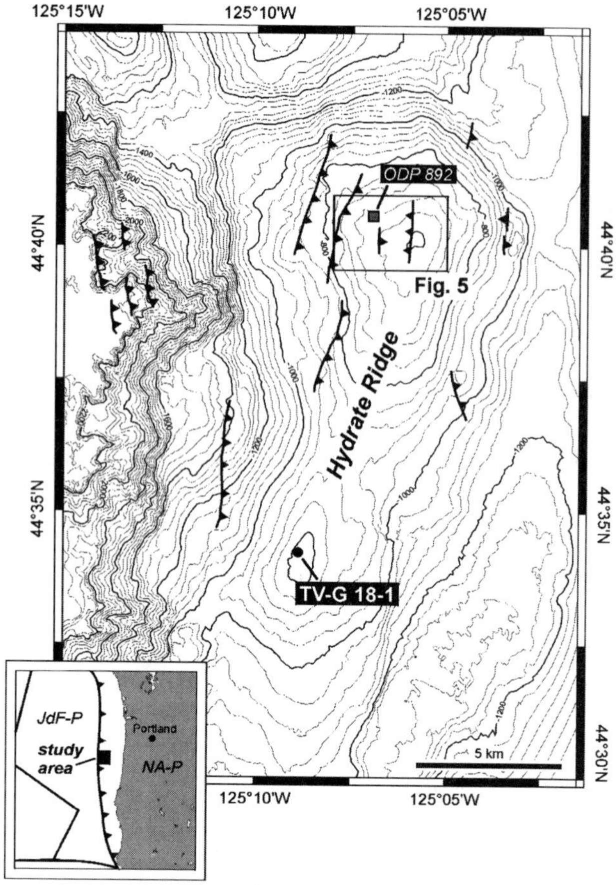

Figure 1. Bathymetry of Hydrate Ridge (in meters; contour interval 40 m) and plate tectonic setting of the Cascadia convergent margin with the study area Hydrate Ridge located on the accretionary complex. An enlarged view of the northern summit is shown in Figure 5.

large fractionation in carbon isotopes, resulting in $\delta^{13}C$ values as negative as -110‰ PDB [*Whiticar*, 1999], oxidation processes may therefore produce $\delta^{13}C$ values even lower than that.

Anaerobic methane oxidation (AMO) is estimated to consume between 20 and 100 Tg of methane per year [Tg = 10^{12} g], thus limiting net transport of methane from marine environments into the water column and into the atmosphere [*Reeburgh et al.*, 1991]. Therefore, detection and characterization of specific biomarkers related to the consumption of methane under anaerobic conditions provide important environmental information. Certain biomarker patterns may be associated with gas hydrate environments [*Elvert et al.*, 1999; *Hinrichs et al.*, 1999], whereas others may be specific for recent or ancient methane vent settings [*Thiel et al.*, 1999; *Elvert et al.*, 2000], or methane-rich mud volcanoes [*Pancost et al.*, 2000]. Although the mechanism of AMO is still poorly understood and the specific organisms involved have not been isolated, biomarker evidence points to archaeal microbes involved in the consumption of methane and probably capable of using methane as a sole carbon source under anaerobic conditions. Microorganisms which could perform this process are either methanogenic archaea operating in reverse in a syntrophic consortium with sulfate-reducing bacteria (SRB) [*Hoehler et al.*, 1994; *Elvert et al.*, 1999; *Thiel et al.*, 1999; *Pancost et al.*, 2000] and/or previously unrecognized, obligate methanotrophic archaea [*Hinrichs et al.*, 1999].

In an earlier study, we observed carbon isotopic compositions of irregular C_{20} and C_{25} isoprenoids (e.g., crocetane, penatmethylicosane) between -102 and -124‰ PDB in association with exposed methane-sulfide gas hydrates on the second accretionary ridge (Hydrate Ridge) off the coast of Oregon [*Elvert et al.*, 1999]. These exposures have since become a prime location to study the characteristics of gas hydrate decomposition induced by active fluid-venting along faults which are produced as a result of the collision of the Juan de Fuca plate with the North American plate (Fig. 1) [*Suess et al.*, 1999]. Extensive methane-venting and gas hydrates were observed at the northern and southern summit at 585 m and 780 m water depth, respectively. The episodic occurrence of rising bubbles generate enormous methane plumes in the lower water column with methane concentrations exceeding 100,000 nL/L. Video surveys of both summits revealed prolific vent communities, thick bacterial mats, and authigenic carbonate lithologies, all of which are based on the oxidation of reduced compounds such as methane and hydrogen sulfide. Pore water analyses showed that all near-surface cores were strongly anaerobic with sulfate essentially exhausted at the sediment surface and sulfide concentrations and alkalinity exceeding 20 mM and 30 mM, respectively. Moreover, these analyses also provided evidence for gas hydrate occurrence throughout the near-surface sediments by lowered chloride concentrations between 515 and 540 mM compared to an overlying seawater value of 555 mM. Such anomalies indicate gas hydrate decomposition during sample recovery due to the pure water content of the gas hydrates [*Hesse and Harrison, 1981*].

The differences of over 20‰ PDB in the stable carbon isotopic composition of irregular isoprenoids specific for AMO at Hydrate Ridge [*Elvert et al.*, 1999] might indicate that substrate availability dictates the wide range of isotopic compositions observed. Furthermore, the detection of a four-times unsaturated C_{25} isoprenoid by Elvert and coworkers [1999] at one station, which contains the pentamethylicosane carbon skeleton, suggests that the process of AMO is reflected in various biomarker patterns. Additionally, by detailing the crocetane structure it should be considered that co-elution with its widespread isomer phytane, which is mostly derived from photoautotrophic sources, results in ^{13}C-enriched carbon isotope values. This emphasizes that carbon isotope values of biomarkers specific for AMO detected in sedimentary records may be highly variable. The present study thus extends our earlier results from a limited number of samples over a wider area of Hydrate Ridge with different geochemical features. Such information is essential for assessing the isotopic compositions and patterns of biomarkers associated with anaerobic methane-oxidizing microbes in marine environments and will lead to a better differentiation of future biomarker analyses tracing the anaerobic cycling of methane back in time.

2. MATERIALS AND METHODS

2.1. Samples

Anaerobic sediments and authigenic carbonates from depth intervals between 6 and 22 cm were collected from Hydrate Ridge during R/V SONNE expeditions in 1996 (SO 109-1, SO 109-2, and SO 110-1a). At both summits of Hydrate Ridge, TV-guided grab samplers recovered gas hydrate from just beneath the sediment-covered surface [*Suess et al.*, 1999]. Additional TV-guided grab samplers on the northern summit were distributed over a larger area surrounding the hydrate location and covering the most active vent sites as well as less active sites. These samples are the basis for the investigations reported here. The authigenic carbonates used for this study were pure aragonites (samples Arag A and D) of light yellow color from a 2-phase mineral assemblage known as gas hydrate carbonates [*Greinert et al.*, this volume]. The aragonites represent botryoidal rim cements resulting from crystallization into an open pore space of clast-supported breccias, which unambiguously arise after complete or partial removal of gas hydrates. $^{87}Sr/^{86}Sr$ ratios between 0.709145 and 0.709180 indicate that precipitation occurs near the sediment-water interface in contact with recent seawater [*Greinert*, 1999; *Greinert et al.*, this volume]. For comparison, strontium isotopic ratios containing significantly less ^{87}Sr are indicative of pore fluids from the deeper sedimentary section, thus implying mixing between seawater and a deep fluid source enriched in ^{86}Sr [*Kastner et al.*, 1995].

2.2. Biomarker Extraction and Fractionation from Sediments and Carbonates

Biomarker fractions were obtained from the sediments and carbonates as described previously [*Elvert et al.*, 1999; *Elvert et al.*, 2000]. Briefly, sediments and aragonite residues were extracted ultrasonically with organic solvents of different polarity. Aragonite residues were obtained from the carbonates by dissolution of the carbonate phase in 1 N HCl, centrifugation, and decantation of the supernatant. Separation of the extracts into different fractions was performed by column chromatography on silica gel using solvents and solvent mixtures of increasing polarity. The compounds of interest eluted in the hydrocarbon and alcohol fractions.

Hydrogenation of selected hydrocarbon fractions was carried out by saturation of 50 μl *n*-hexane with H_2 in fusible glass ampoules, pre-filled with 10 mg of PtO_2 (Adam's catalyst) and subsequent adding of 50 μl of sample (½ aliquot in *n*-hexane). After flushing with pure nitrogen, the ampoules were sealed and stored at room temperature for one hour.

To facilitate gas chromatographic analysis of alcohols, trimethylsilyl (TMS) derivatives were produced. Alcohol fractions were evaporated under a stream of pure nitrogen to near dryness, mixed with 100 μl BSTFA (bis(trimethylsilyl)trifluoroacetamide; Supelco), and heated in sealed glass ampoules for 2 h at 80°C. Following evaporation to near dryness under nitrogen, the residue was taken up in *n*-hexane.

2.3. Bulk Parameters

Organic carbon (C_{org}) weight percentages were determined using a Carlo Erba NA-1500 Element Analyzer. For carbonate removal of the sediments, 3 g of wet sediment were treated overnight with 15 ml of 10% HCl. After freeze-drying, samples were homogenized by using an agate ball mill. C_{org}-contents of the carbonates were calculated from the carbon difference between bulk and HCl-treated sub-samples. Standard deviations of both methods were 0.02%. Interstitial sulfate was measured by ion chromatography and detected by conductivity. Sulfate values are reported in mM and were calibrated with IAPSO-standard seawater of different concentrations. Using duplicate measurements, standard deviations were within 1.5%.

2.4. Gas Chromatography (GC)

Analyses of hydrocarbons were performed using a 30 m apolar DB-5 fused silica capillary column (0.25 mm internal diameter (ID), film thickness 0.25 μm; J & W Scientific) in a Carlo Erba 5160 GC equipped with an on-column injector and a flame ionization detector. The samples were injected at 60°C. After a 1 min hold time, the oven temperature was raised to 140°C at 10°C/min, then to 310°C at 5°C/min and finally kept at 310°C for 25 min. The carrier gas was H_2 at a flow rate of 2.5 ml/min. Concentrations for each compound were determined by adding internal standards (3-methylnonadecane, 2-methylicosane, 5β(H)-cholane) of known concentrations prior to GC analysis and are reported in μg/g C_{org}. Standard deviations for single compounds are below $\sigma = 2$ μg/g C_{org}. Loss of material during analysis was monitored by adding a recovery standard (*n*-

C_{40}) prior to the overall analytical procedure. Typical recoveries were 70 to 80% relative to n-C_{40}.

2.5. Gas Chromatography-Mass Spectrometry (GC-MS)

Hydrocarbons and alcohols (as TMS-derivatives) were identified by GC-MS using a Carlo Erba 8000 gas chromatograph interfaced to a Fisons MD 800 mass spectrometer operated in electron impact (EI-) mode at 70 eV (cycle time 0.9 s; resolution 1000) with a mass range of m/z 40-600 for hydrocarbons and m/z 40-800 for alcohols. The GC was equipped with a DB-1 fused silica capillary column (30 m, 0.25 mm ID) coated with cross-linked methyl silicone (film thickness 0.25 μm; J & W Scientific) using He as carrier gas. The samples were injected in splitless mode (hot needle technique; injector temperature: 285°C) and subjected to the same temperature program given for GC measurements (see 2.4.).

2.6. Stable Carbon Isotope Analysis

Carbon isotope compositions of individual hydrocarbons and alcohols (carbonate residues only) were determined by a coupled gas chromatograph-combustion line-isotope ratio mass spectrometer (GC-C-IRMS). The mass spectrometer (Finnigan MAT 252) was connected to a Varian 3300 GC equipped with a 50 m CP-Sil 5 CB-MS (0.25 mm ID, 0.4 μm stationary phase; Chrompack). The carrier gas was He at a flow rate of 1.5 ml/min. The samples were on-column injected at 60°C and after 1 min the oven temperature was raised to 140°C at a rate of 10°C/min, then to 230°C at rate of 3°C/min, and finally to 310°C at a rate of 2°C/min at which it was held for 35 min. Carbon isotope ratios are reported in delta (δ) notation as per mil (‰) deviation from the Pee Dee Belemnite standard (PDB). Internal standards (5β(H)-cholane and n-C_{36} for hydrocarbons; n-C_{20} and n-C_{36} for alcohols) of known isotopic composition were co-injected with each sample for monitoring reproducibility and precision during the project. Analytical reproducibility was on average within ± 0.2 to 0.3‰ PDB for an n-alkane standard (n-C_{13} to n-C_{38}) with no background, but was much more variable for the complex mixtures analyzed here (up to ± 1.6‰ PDB). This low precision is especially obvious for the complex cluster of C_{25} isoprenoids eluting very close to each other (see Figs. 2a and 3). Isotopic compositions of alcohols were measured in the form of TMS-derivatives and corrected for the isotopic shift associated with the addition of carbon atoms during derivatization according to *Huang et al.* [1995].

The isotopic composition of $\delta^{13}C_{org}$ was measured on carbonate-free sub-samples by elemental analysis-isotope ratio mass spectrometry (EA-IRMS) using a Carlo Erba Elemental Analyzer connected via a ConFlo™ interface to the Finnigan MAT 252. Analytical reproducibility for duplicate runs was below ± 0.1‰ PDB. For carbon isotope analysis of the aragonite precipitates see *Greinert et al.* [this volume].

3. RESULTS AND DISCUSSION

3.1. Biomarkers: Abundance and Structures

Irregular saturated and unsaturated C_{20} and C_{25} isoprenoids such as crocetane (2,6,11,15-tetramethylhexadecane) and pentamethylicosane (2,6,10,15,19-pentamethylicosane; PMI) are the dominant compounds of hydrocarbon fractions obtained from the anaerobic sediments and aragonite precipitates (Fig. 2a). These abundant biomarkers in such an environmental setting are specific for archaea related to known methanogenic archaea, which grow on methane as their carbon source rather than forming it as a metabolic product [*Elvert et al.*, 1999; *Thiel et al.*, 1999; *Elvert et al.*, 2000; *Pancost et al.*, 2000]. In particular, the predominance of a cluster of unsaturated PMIs (pentamethylicosenes) in many of the samples is highly diagnostic for methanogenic archaea such as *Methanosarcina mazei* or *Methanolobus bombayensis*, which were found to contain significant amounts of these compounds [*Schouten et al.*, 1997]. To illustrate the presence of the different C_{25} isomers containing the same number of double bonds (between one and five), an enlarged view of the cluster of pentamethylicosenes is shown in Figure 3. Mass spectra of some of the prominent 2,6,10,15,19-pentamethylicosenes are shown in Figure 4. The carbon skeleton of these compounds was confirmed by hydrogenation of the hydrocarbon fraction, which yielded PMI as the only irregular C_{25} isoprenoid. Identification was by comparison with the mass spectrum and retention index of PMI from the untreated fraction. Biosynthesis of the set of unsaturated irregular C_{25} isoprenoids is probably performed through tail-to-tail coupling of farnesyl pyrophosphate and geranyl pyrophosphate and partial reduction of the double bonds [*Sinninghe Damsté et al.*, 1997]. This pathway would be analogous to the well-known squalene formation in bacteria and eukarya through tail-to-tail coupling of two farnesyl pyrophosphate units. Similar to this, the biosynthesis of unsaturated irregular C_{20} isoprenoids (crocetenes) as found in deep-sea sediments of the eastern Aleutian subduction zone [*Elvert et al.*, 2000] and to some degree in Hydrate Ridge sediments (Cr:1) may be initially performed through tail-to-tail coupling of two geranyl pyrophosphate units [*Elvert*, 1999].

Absolute concentrations of over 50 μg/g C_{org} obtained for crocetane, PMI, and pentamethylicosenes in anaerobic sediments are by far higher than those obtained for compounds derived from allochthonous sources such as n-alkanes (Table 1). The abundance of these compounds in

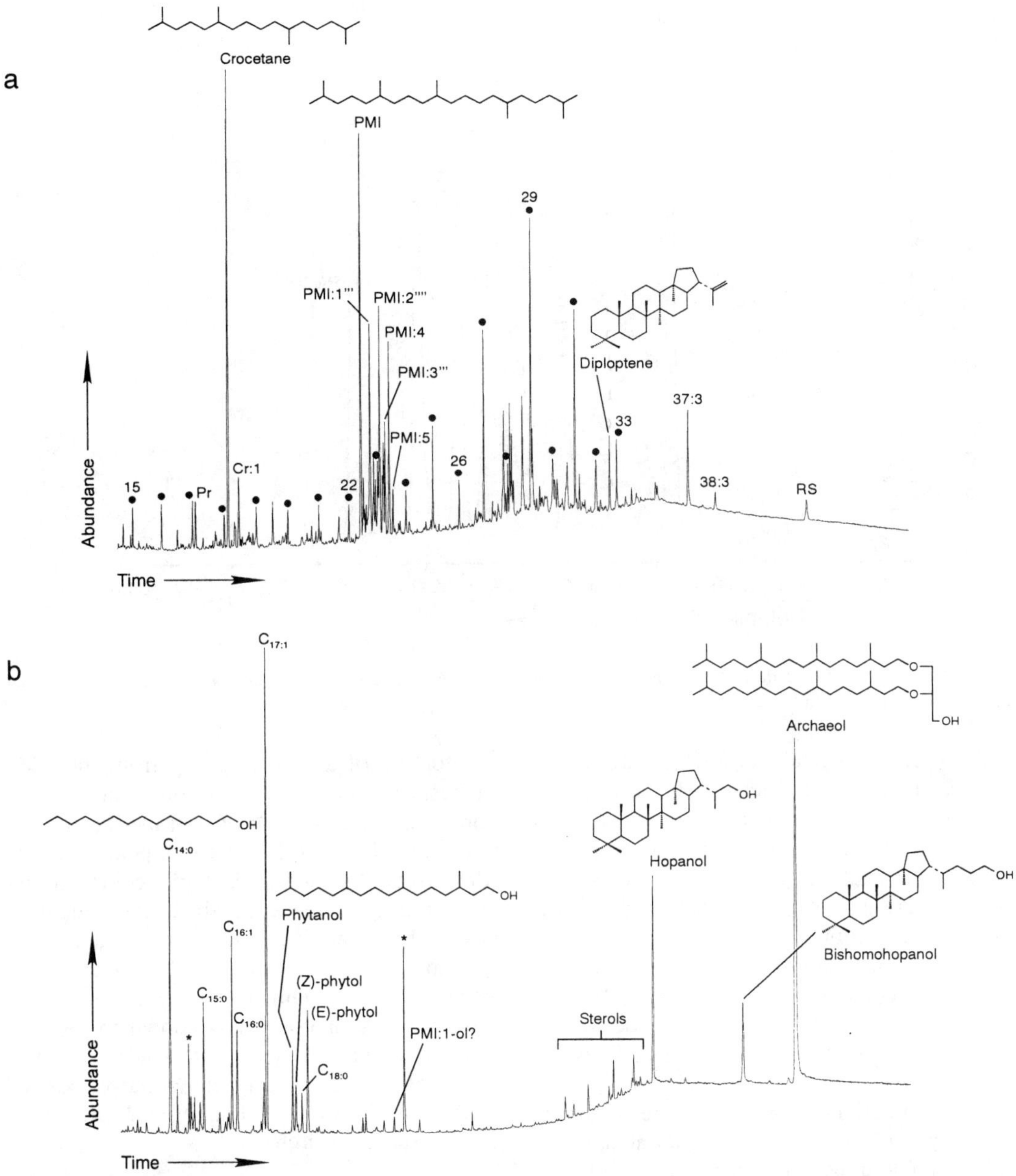

Figure 2. GC-MS chromatograms of a) a typical hydrocarbon fraction obtained from anaerobic sediments of Hydrate Ridge (station TV-G 18-1) and b) a trimethylsilylated alcohol fraction obtained from a methane-derived aragonite (TV-G 18-1 Arag A). *n*-Alkanes are indicated by dots with numbers representing total number of carbon atoms. (Pr [pristane]: 2,6,10,14-tetramethylpentadecane; Cr:1: compound with a crocetane structure containing one double bond; PMI:1''', PMI:2'''', PMI:3''', PMI:4, and PMI:5: compounds with a PMI structure containing one, two, three, four, or five double bond(s), respectively; RS: recovery standard; $C_{x:y}$: *n*-alcohols containing x carbon atoms and y double bond(s); *: contaminant).

the aragonite precipitates even exceeds concentrations higher than 1,000 μg/g C_{org}. Other microbial-derived biomarkers such as the ubiquitous hopanoid hop-22(29)-ene (diploptene), known to be produced by various aerobic bacteria such as cyanobacteria, ammonia-oxidizing bacteria, methylotrophs, and methanotrophs [*Rohmer et al.*, 1984; *Ourisson et al.*, 1987; *Summons et al.*, 1994] but not reported to be derived from anaerobic bacteria, are of minor abundance. Specifically, diploptene concentrations are decreased on Hydrate Ridge, but increase in sediments not lo-

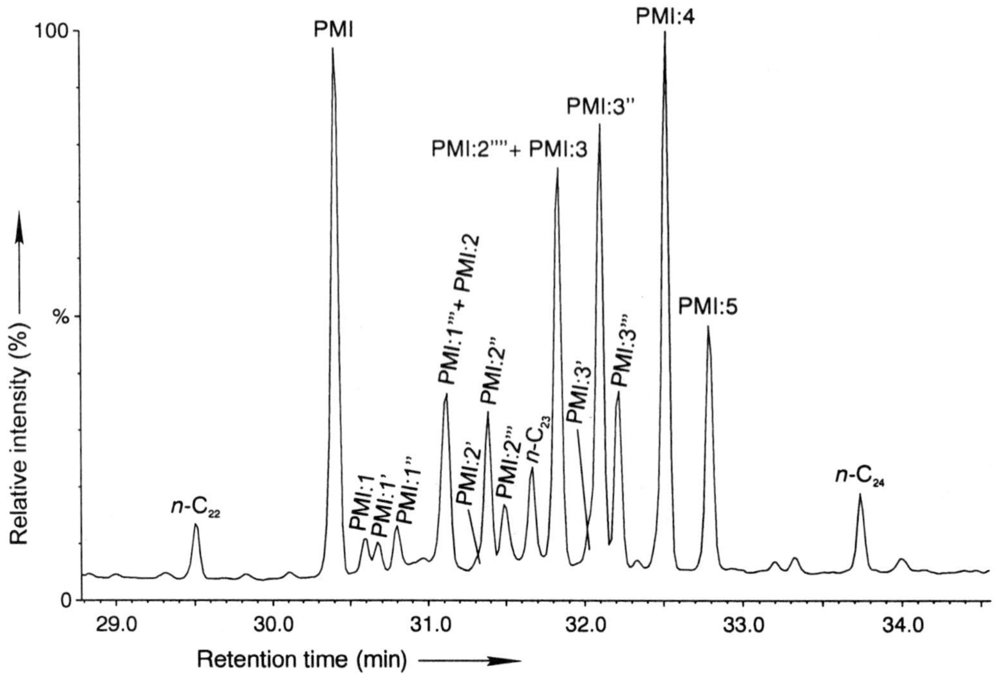

Figure 3. Partial GC-MS chromatogram showing the broad range of irregular saturated and unsaturated C_{25} isoprenoids isolated at station TV-G 45-2.

cated on the ridge; whereas crocetane or PMI are at highest concentration on the ridge but significantly decrease away from the ridge (Table 1). This reciprocal behavior strongly suggests derivation from different microbial sources. Moreover, unsaturated irregular isoprenoids are not detectable at stations which are not exactly located on the ridge and therefore not directly influenced by active methane-venting and gas hydrate decomposition. We propose that the suite of biomarkers composed of crocetane, PMI, and pentamethylicosenes are indicative of AMO occurring within the Hydrate Ridge sediments whereas this process is less evident off the ridge.

Analysis of alcohol fractions isolated from the aragonite residues yielded a series of saturated and unsaturated, short chain *n*-alcohols; isoprenoid alcohols such as (Z)- and (E)-phytol (phytol = 3,7,11,15-tetramethylhexadec-2-en-1-ol), and phytanol (3,7,11,15-tetramethylhexadecan-1-ol); and the phytanyl glycerol ether archaeol (2,3-di-*O*-phytanyl-*sn*-glycerol) (Fig. 2b). Archaeol is a biomarker specific for archaea and known to occur in anaerobic sediments surrounding methane seeps associated with marine gas hydrates, recent methane vent settings, or methane-rich mud volcanoes [*Hinrichs et al.*, 1999; *Elvert et al.*, 2000; *Pancost et al.*, 2000]. In contrast, hydroxyarchaeols as observed by *Hinrichs et al.* [1999] and *Pancost et al.* [2000] were not detected. We also observed a compound containing a similar mass spectrum as that known for phytol, but showing an obvious fragment ion at m/z 423 (phytol m/z 353), probably due to loss of a methyl group from the TMS-ether. This suggests the presence of a compound containing 5 additional carbon atoms. On this basis, the compound is tentatively identified as 3,7,11,15,19-pentamethylicos-2-en-1-ol (PMI:1-ol). Its carbon isotopic composition falls in the same range as those obtained for irregular isoprenoids, phytanol, and archaeol and therefore a common source organism is very likely (see 3.2.). In addition, the alcohol fractions also contained non-specific hopanoid biomarkers (hopanol, bishomohopanol) known to be present in many different strains of bacteria [*Rohmer et al.*, 1984; *Summons et al.*, 1994], e.g., cyanobacteria, purple non-sulfur bacteria, methylotrophs, and methanotrophs; but which have not been observed in archaea.

3.2. Carbon Isotopes and Isotope Effects Associated with AMO at Hydrate Ridge

Carbon isotopic compositions of specific biomarkers found in anaerobic sediments and methane-derived aragonites from Hydrate Ridge and the first accretionary ridge are shown in Tables 2 and 3. On Hydrate Ridge, isotope values of irregular isoprenoids as low as -128.0‰ PDB (PMI:2'') were encountered and clearly demonstrate the presence of archaea growing on methane derived from extensive methane seepage in this special gas hydrate environment [*Elvert et al.*, 1999] (see Table 2). With increasing distance from the ridge, isotope values are more enriched in

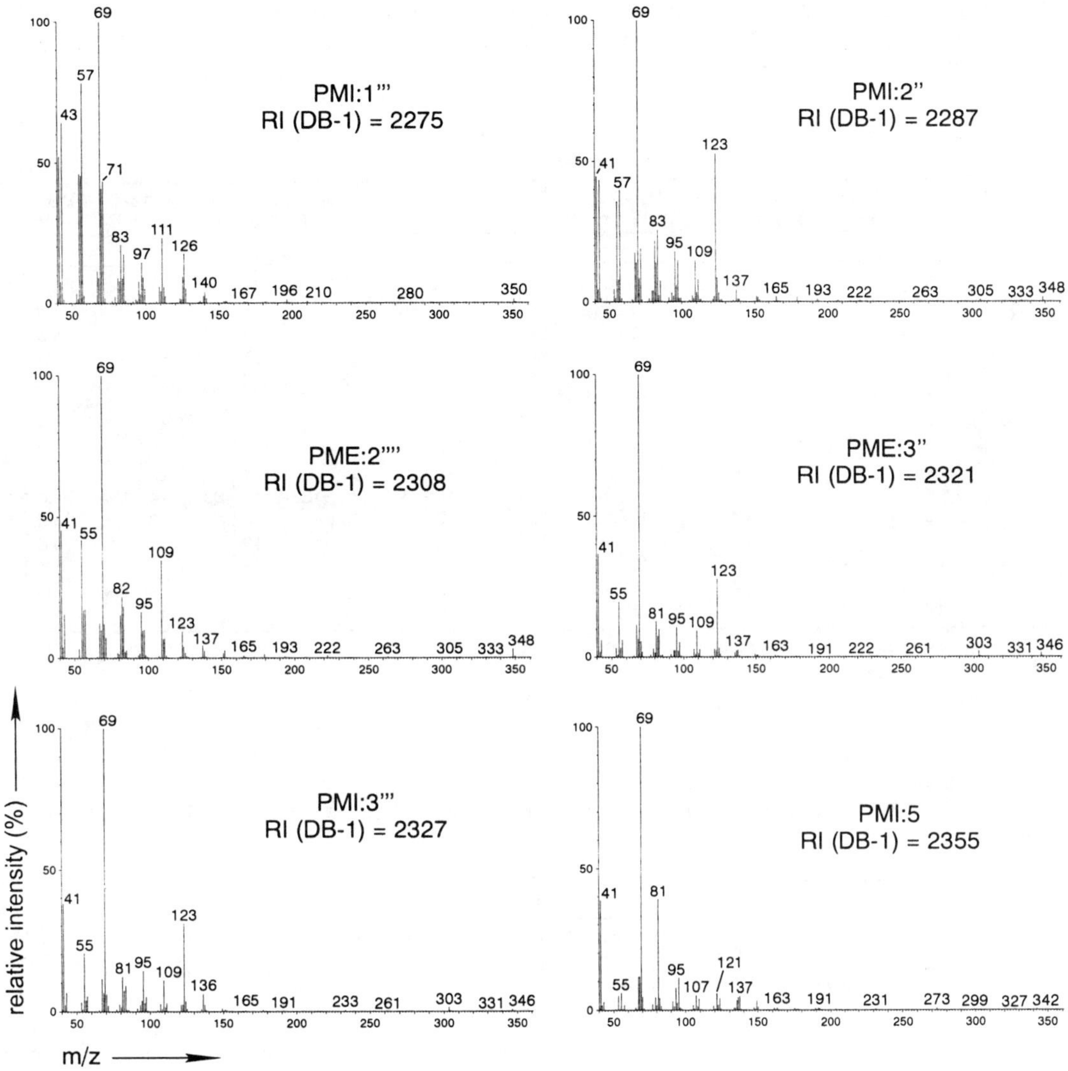

Figure 4. Electron impact mass spectra and corresponding retention indices (RI) on an apolar DB-1 column of pentamethylicosenes (PMIs) obtained from anaerobic sediments at Hydrate Ridge containing one (PMI:1'''), two (PMI:2'' and PMI:2''''), three (PMI:3'' and PMI:3'''), and five (PMI:5) double bond(s), respectively. d.b.: double bond(s).

^{13}C and therefore, with the simultaneous concentration decrease, very strongly indicate the dependence of archaea on a methane carbon source. It is very likely that only a fairly high concentration of methane and with continuous delivery can account for the extreme ^{12}C-enrichment. Especially for crocetane and PMI, isotopic values are highly depleted in ^{13}C on the ridge (-112.6 to -125.0‰ PDB), become more enriched in ^{13}C on the slope and the east flank (-87.7 to -107.6‰ PDB), and are almost indistinguishable from other microbial sources on the first accretionary ridge (-63.3 and -37.2‰ PDB) (Fig. 5). Moreover, at the first accretionary ridge, concentrations of irregular isoprenoids are so low that carbon isotope analysis for these compounds were not feasible (TV-G 30-2).

The observed range in carbon isotopic compositions of irregular isoprenoids is apparently influenced by the activity of anaerobic methane-oxidizing microbes, which depend heavily on available methane substrate. Methane concentrations in the lower water column are highest on the northern and southern summit of the ridge [*Suess et al.*, 1999] and readily decrease with distance from the ridge [*Kulm et al.*, 1986; *Carson et al.*, 1990]. This concentration gradient appears to be directly reflected in the abundance of irregular isoprenoids (Table 1) and their enrichment in ^{12}C (Fig. 5). This finding corresponds to carbon isotopic compositions of dissolved CO_2 obtained from vent fluids at the sediment-water interface which are less depleted in $^{13}C_{CO2}$ without a gas hydrate component and more depleted in

Table 1. Concentrations of hydrocarbons obtained from anaerobic sediments and two methane-derived aragonite precipitates (in μg/g C_{org}) at the first (TV-G 25-1 and TV-G 30-2) and the second accretionary ridge (Hydrate Ridge, all other TV-G) of the Cascadia margin. Note that crocetane and phytane are not resolvable by conventional gas chromatographic conditions [*Thiel et al.*, 1999; *Elvert*, 1999] and thus are given as the sum of both.

	Sediments									Aragonites	
	TV-G 25-1 (10-13 cm)	TV-G 30-2 (8-10 cm)	TV-G 36-5 (13-16 cm)	TV-G 45-2 (10-13 cm)	TV-G 115 (16-19 cm)	TV-G 119 (19-22 cm)	TV-G 9-1 (13-16 cm)	TV-G 11-1 (13-16 cm)	TV-G 18-1 (6-8 cm)	TVG 18-1 Arag A	TVG 18-1 Arag D
Hydrocarbon	(μg/g C_{org})										
n-C_{17}	5	7	4	4	6	5	8	19	7	n.d.	24
Pristane	6	10	6	5	6	5	12	25	9	n.d.	n.d.
n-C_{18}	4	6	3	3	4	3	5	5	4	n.d.	n.d.
Crocetane (+ Phytane)	3	3	5	27	51	27	34	20	51	214	1368
Cr:1[a]	n.d.[d]	n.d.	n.d.	4	4	2	5	3	10	8	98
PMI[b]	5	3	5	25	28	41	52	28	42	166	1083
PMI:1‴[c]	n.d.	n.d.	n.d.	10	20	13	16	10	20	8	229
PMI:2″	n.d.	n.d.	n.d.	8	4	6	11	4	6	n.d.	83
PMI:2″″	n.d.	n.d.	n.d.	19	34	15	23	14	29	8	221
PMI:3″	n.d.	n.d.	n.d.	21 (Σ 108)	9 (Σ 131)	19 (Σ 99)	26 (Σ 133)	11 (Σ 63)	14 (Σ 132)	n.d. (Σ 16)	n.d. (Σ 533)
PMI:3‴	n.d.	n.d.	n.d.	10	35	8	13	6	20	n.d.	n.d.
PMI:4	n.d.	n.d.	n.d.	27	25	25	33	12	33	n.d.	n.d.
PMI:5	n.d.	n.d.	n.d.	13	4	13	11	6	10	n.d.	n.d.
n-C_{24}	4	9	4	4	5	5	8	7	7	17	26
Diploptene	18	18	11	5	5	6	7	6	9	n.d.	146

[a]Cr:1: compound with a crocetane structure containing one double bond

[b]PMI: 2,6,10,15,19-pentamethylicosane

[c]PMI:x: compounds with a PMI structure containing x double bond(s)

[d]n.d.: not detected

Table 2. C_{org} (weight %) and carbon isotopic compositions (‰ PDB) of C_{org}, methane-derived aragonite precipitates, and hydrocarbons obtained from anaerobic sediments and aragonite crusts of the Cascadian margin.

	Sediments									Aragonites	
	TV-G 25-1 (10-13 cm)	TV-G 30-2 (8-10 cm)	TV-G 36-5 (13-16 cm)	TV-G 45-2 (10-13 cm)	TV-G 115 (16-19 cm)	TV-G 119 (19-22 cm)	TV-G 9-1 (13-16 cm)	TV-G 11-1 (13-16 cm)	TV-G 18-1 (6-8 cm)	TVG 18-1 Arag A	TVG 18-1 Arag D
C_{org} (%)	0.52	0.70	2.11	1.72	1.05	1.85	1.62	1.57	1.73	0.12	0.05
$\delta^{13}C$ (‰)	-24.5	-22.6	-23.9	-36.7	-36.9	-39.8	-25.1	-33.1	-26.5	-46.5	-45.2
Hydrocarbon	$\delta^{13}C$(‰)										
n-C_{17}	-31.0	-30.1	-31.1	-46.0	-38.5	-59.7	-39.9	n.m.	-51.7	n.d.	-81.2
Pristane	-26.8	-27.3	-25.7	-27.3	n.d.	-29.2	-26.0	-27.2	-28.5	n.d.	n.d.
n-C_{18}	-26.7	-28.7	-26.9	-39.7	-76.8	-49.0	n.m.	n.m.	-47.8	n.d.	n.d.
Crocetane (+ Phytane)	-63.3	n.m.[e]	-87.7	-126.3	-119.3	-124.7	-120.3	-118.4	-124.8	-124.9	-125.0
Cr:1[a]	n.d.[d]	n.d.	n.d.	-58.3	n.m.	n.m.	n.m.	n.m.	-73.5	n.m.	-122.0
PMI[b]	-37.2	n.m.	-88.9	-125.5	-115.0	-125.3	-121.4	-115.5	-116.4	-121.0	-121.7
PMI:1‴[c]	n.d.	n.d.	n.d.	-127.2	-116.4	-124.0	-127.9	n.m.	-121.7	n.m.	-123.4
PMI:2″	n.d.	n.d.	n.d.	-124.6	-120.4	-123.3	-128.0	n.m.	-120.0	n.d.	n.m.
PMI:2″″	n.d.	n.d.	n.d.	-126.0	-122.0	-124.9	-125.1	-123.6	-124.6	n.m.	-124.4
PMI:3″	n.d.	n.d.	n.d.	-125.1	-119.3	-120.6	-120.6	-110.7	-119.1	n.d.	n.d.
PMI:3‴	n.d.	n.d.	n.d.	-127.6	-122.8	-125.5	-127.0	n.m.	-125.3	n.d.	n.d.
PMI:4	n.d.	n.d.	n.d.	-122.9	-115.7	-118.7	-116.8	-101.7	-116.7	n.d.	n.d.
PMI:5	n.d.	n.d.	n.d.	-122.8	-101.8	-119.6	-116.9	n.m.	-102.5	n.d.	n.d.
n-C_{24}	-27.0	-27.7	-26.3	-26.0	-25.0	-32.2	-31.2	-27.6	-31.1	n.m.	n.m.
Diploptene	-36.3	-33.6	-34.9	-85.5	-76.6	-87.7	-77.3	-74.7	-61.0	n.d.	-80.2

[a]Cr:1: compound with a crocetane structure containing one double bond

[b]PMI: 2,6,10,15,19-pentamethylicosane

[c]PMI:x: compounds with a PMI structure containing x double bond(s)

[d]n.d.: not detected

[e]n.m.: not measured, insufficient amount for carbon isotope analysis

Table 3. Carbon isotopic compositions (‰ PDB) and isotope effects $\varepsilon_{CH4\text{-}biomarker}$ (‰ PDB; shown in parentheses) of prominent alcohols isolated from methane-derived aragonites at station TV-G 18-1 (Fig. 1). Calculations of isotope effects is explained in the text, assuming a $\delta^{13}C$ value of substrate methane of -65.0‰ PDB [*Greinert*, 1999]. Other studies on methane hydrate from Hydrate Ridge showed $\delta^{13}C_{CH4}$ between -62.4 and -71.5‰ PDB [*Hovland et al.*, 1995; *Kastner et al.*, 1998; *Suess et al.*, 1999].

	Alcohol									
	n-$C_{14:0}$[a]	n-$C_{15:0}$	n-$C_{16:1}$	n-$C_{16:0}$	n-$C_{17:1}$	Phytanol	PMI:1-ol?	Hopanol	Bishomo-hopanol	Archaeol
Aragonite sample	$\delta^{13}C$ (‰) [$\varepsilon_{CH4\text{-}biomarker}$(‰)]									
TVG 18-1 Arag A	-98.6 [37.3]	-104.5 [44.1]	-105.8 [45.7]	-97.4 [35.9]	-108.0 [48.2]	-122.0 [64.9]	-118.8 [61.1]	-95.4 [33.6]	-96.7 [35.1]	-118.8 [61.1]
TVG 18-1 Arag D	-105.3 [45.1]	-108.9 [49.3]	-107.7 [47.9]	-100.0 [38.8]	-108.9 [49.3]	-121.2 [64.0]	-122.9 [66.1]	-93.3 [31.2]	-96.4 [34.8]	-119.8 [62.3]

[a] n-$C_{x:y}$: n-alcohols containing x carbon atoms and y double bond(s)

$^{13}C_{CO2}$ containing a gas hydrate component [*Suess et al.*, 1999]. Therefore, a larger portion of the CO_2 from vent fluids encountered on the first accretionary ridge derives from the oxidation of sedimentary organic matter [*Suess and Whiticar*, 1989], whereas fluids from vents on Hydrate Ridge mainly represent CO_2 derived from the oxidation of methane.

In contrast to irregular isoprenoids, diploptene on the ridge reveals ^{13}C-enriched carbon isotope values between -61.0 and -87.7‰ PDB (Table 2); off the ridge diploptene shows carbon isotope values between -33.6 and -36.3‰ PDB. Diploptene on the ridge has thus to be considered, at least to some part, as a biomarker derived from a microbial source involved in the oxidation of methane, perhaps aerobic methanotrophs that are known to be present in cold seep environments, to contain diploptene in suitable amounts [*Rohmer et al.*, 1984], and to have carbon isotope values which match the ones found here [*Jahnke et al.*, 1995]. Because the amount of diploptene found in anaerobic sediment samples on the ridge is small compared to the content of irregular isoprenoids (see Table 1), diploptene in these samples may not to be formed *in situ* but rather be encountered here due to sedimentation or transport processes. Nevertheless, *Pancost et al.* [2000] observed that hopanoid compounds increase with sediment depth in methane-rich mud volcanoes and therefore are probably generated under anaerobic conditions. The inferred source organisms were acetogenic bacteria, which could possibly grow heterotrophically on ^{13}C-depleted archaeal biomass [*Pancost et al.* 2000] or use ^{13}C-depleted CO_2 as a carbon source.

Analyses of hydrocarbons and alcohols released from the aragonite residues yielded chromatograms that contained specific biomarkers related to methane oxidation largely undiluted by other organic compounds (Figs. 2b and 6). Archaeal-derived alcohols such as phytanol, PMI:1-ol, and archaeol are highly depleted in ^{13}C, having carbon isotope values between -118.8 and -122.0‰ PDB (Table 3), whereas biomarkers such as short-chain n-alcohols and hopanols show distinct $\delta^{13}C$ values ranging from -97.4 to -108.9‰ PDB and from -93.3 to -96.7‰ PDB, respectively. The low $\delta^{13}C$ values obtained for the n-alcohols and hopanols strongly suggest that these biomarkers are also derived from microbial sources involved in the oxidation of methane. Whereas hopanols may be very likely produced by methanotrophs, the source of n-alcohols highly depleted in ^{13}C is difficult to elucidate. They have never been reported as biomarker components in archaea. A possible source of such ^{13}C-depleted n-alcohols may therefore be SRB living in syntrophy with methane-oxidizing archaea [*Elvert et al.*, 2000], which was also suggested from the coexistence of highly ^{13}C-depleted archaeal biomarkers with less-depleted alcohols such as *anteiso*-$C_{15:0}$, mid-chain branched $C_{16:0}$, and n-$C_{16:0}$ in a Miocene limestone [*Thiel et al.*, 1999]. The observed n-alcohols may not be produced directly by the SRB but could rather be breakdown products of precursor molecules such as non-isoprenoid alkyl glycerol ethers, known to be present in thermophilic and mesophilic clostridia and in the thermophilic sulfate reducer *Thermodesulfotobacterium commune* [*Langworthy and Pond*, 1986]. Until recently, reports of non-isoprenoid alkyl glycerol ethers in natural environments have been restricted to hydrothermal environments [*Zeng et al.*, 1992a; 1992b]. Sedimentary non-isoprenoid alkyl glycerol ethers have now been observed in different active cold seep environments offshore California [*Hinrichs et al.*, 2000]. The biomarkers are especially prominent in Santa Barbara Basin sediments and show strong depletions in ^{13}C. They are accompanied by a set of archaeal-derived biomarkers that are slightly more depleted in ^{13}C and therefore may provide further evidence for SRB growing syntrophically with methane-oxidizing archaea.

The low $\delta^{13}C$ values of the aragonite carbon indicate that the carbonates initially precipitated in the pore space of anaerobic sediments [*Greinert*, 1999; *Greinert et al.*, this

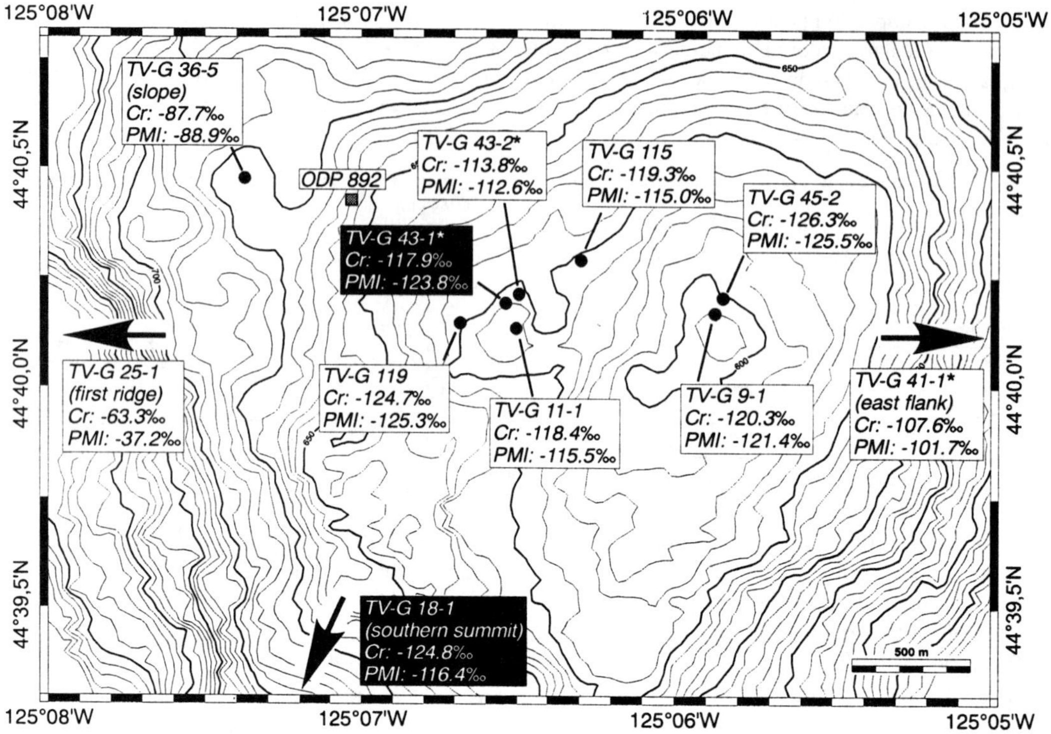

Figure 5. Northern summit of Hydrate Ridge (bathymetry in meters; contour interval 10 m) with carbon isotopic compositions obtained for crocetane and PMI isolated from anaerobic sediments. Black rectangles indicate that these TV-guided grab samplers recovered gas hydrates. *: carbon isotopic signatures of crocetane and PMI from stations TV-G 41-1, 43-1, and 43-2 are taken from [*Elvert et al.*, 1999]; also shown is ODP site 892.

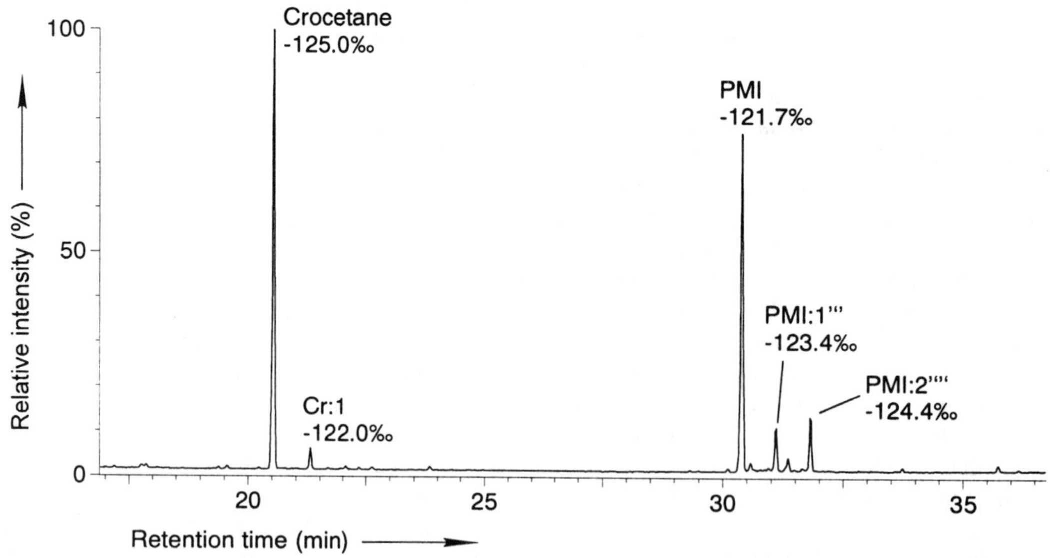

Figure 6. Partial GC-MS chromatogram of hydrocarbon fraction obtained from a methane-derived aragonite (Arag D) at station TV-G 18-1 with $\delta^{13}C$ values for each compound.

Table 4. Isotope effects $\varepsilon_{CH4\text{-biomarker}}$ (‰ PDB) of irregular C_{20} and C_{25} isoprenoids and diploptene associated with methane oxidation at Hydrate Ridge. Sample locations are shown in Figures 1and 5.

	Sediments							Aragonites	
	TV-G 36-5 (13-16 cm)	TV-G 45-2 (10-13 cm)	TV-G 115 (16-19 cm)	TV-G 119 (19-22 cm)	TV-G 9-1 (13-16 cm)	TV-G 11-1 (13-16 cm)	TV-G 18-1 (6-8 cm)	TVG 18-1 Arag A	TVG 18-1 Arag D
Hydrocarbon	$\varepsilon_{CH4\text{-biomarker}}$(‰)								
Crocetane (+ Phytane)	24.9	70.1	61.7	68.2	62.9	60.6	68.3	68.4	68.5
Cr:1		-7.1					9.2		64.9
PMI[a]	26.2	69.5	56.5	69.0	64.2	57.1	58.2	63.7	64.6
PMI:1'''[b]		71.3	58.2	67.4	72.1		64.6		66.6
PMI:2''		68.1	63.0	66.5	72.2		62.5		
PMI:2''''		69.8	64.9	68.4	68.7	66.9	68.1		67.8
PMI:3''		68.7	61.7	63.2	63.2	51.4	61.4		
PMI:3'''		71.8	65.9	69.2	71.0		68.9		
PMI:4		66.0	57.3	60.9	58.7	40.9	58.5		
PMI:5		65.9	41.0	62.0	58.8		41.8		
Diploptene	-31.2	22.4	12.6	24.8	13.3	10.5	-4.3		16.6

[a]PMI: 2,6,10,15,19-pentamethylicosane; Cr:1: compound with a crocetane structure containing one double bond
[b]PMI:x: compounds with a PMI structure containing x double bond(s)

volume]. Assuming non-limiting substrate availability, the carbon isotope values of archaeal biomarkers found within the aragonites directly reflect the carbon isotopic fractionation associated with AMO (i.e. open system fractionation). Carbon isotopic fractionations are expressed using the epsilon (ε) notation [*Hayes*, 1993]:

$$\varepsilon_{CH_4-\text{biomarker}}(‰) = (\alpha_{CH_4-\text{biomarker}} - 1) \times 10^3$$

$$\text{with: } \alpha_{CH_4-\text{biomarker}} = \frac{(\delta^{13}C_{CH_4} + 10^3)}{(\delta^{13}C_{\text{biomarker}} + 10^3)}$$

In these equations, ε is the kinetic carbon isotope effect and α is the fractionation factor between the carbon of the substrate methane ($\delta^{13}C_{CH4}$ = -65.0‰ PDB; *Greinert*, 1999) and the carbon of the resultant biomarker. The kinetic carbon isotope effects obtained for specific biomarkers at each location are presented in Tables 3 and 4. Isotope effects calculated for archaeal biomarkers in sediments and aragonites on Hydrate Ridge are generally higher than 57‰ and are clearly distinct from isotope effects encountered for irregular isoprenoids detected on the slope of the ridge (TV-G 36-5). In contrast to isotope effects of archaeal-derived biomarkers, isotope effects of *n*-alcohols are shifted by approximately 15‰ and of hopanoids are shifted by at least 26‰ (Table 3). Smaller isotope effects than 57‰ found for irregular isoprenoids are suggestive of a co-elution of the respective biomarker with compound derived from sources other than methane-oxidizing microbes. For example, considering that conventional gas chromatographic conditions on apolar capillary columns can only partially separate crocetane from its widespread photoautotrophic-derived isomer phytane (2,6,10,14-tetramethylhexadecane) [*Thiel et al.*, 1999; *Elvert*, 1999], co-elution is especially virtual for ^{13}C-enriched crocetane obtained on the slope of Hydrate Ridge (TV-G 36-5). Carbon isotopic compositions of the biomarkers can also be affected by the availability and the initial carbon isotopic composition of the methane substrate. For example, *Pancost et al.* [2000] observed carbon isotopic compositions of archaeol as high as -63.1‰ PDB in methane-rich mud volcano sediments (Mediterranean Ridge) suggesting either methane substrate limitation or local ^{13}C-enrichment of the available methane pool.

The large isotope effects observed for archaeal biomarkers as high as 72‰ fall in the same range as those associated with biomarkers isolated from cultures of methylotrophic methanogens grown under non-limiting substrate conditions [*Summons et al.*, 1998]. However, culture conditions, providing extensive substrate carbon such as methanol, dimethylsulfide, or trimethylamine for methylotrophic methanogenesis, are not likely to occur in a dynamic natural setting such as Hydrate Ridge. We believe that at Hydrate Ridge the essentially limitless reservoir of methane available to the archaea due to extensive methane-venting and gas hydrate dissolution allows the kinetic isotope effect associated with the process of AMO to be fully expressed. Moreover, the significant difference between the δ^{13}C value of archaeal biomarkers and their methane substrate found in all sediment samples from Hydrate Ridge demonstrates that carbon partitioning in anaerobic methane-oxidizing archaea leads to ^{12}C-enrichment greater than observed in any other kind of methane-consuming organism.

3.3. Microbial Community Structure and the Process of AMO in Hydrate Ridge Sediments

The abundance of specific biomarkers and carbon isotopic evidence identifies AMO as an important process in

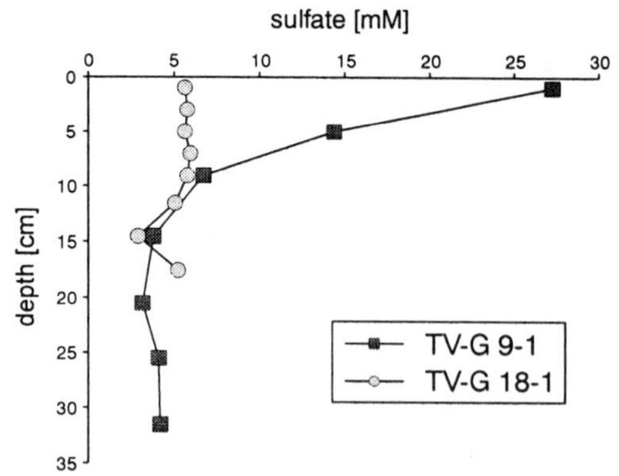

Figure 7. Comparison of pore water sulfate profiles in anaerobic sediments at the northern (TV-G 9-1) and the southern summit (TV-G 18-1) of Hydrate Ridge.

the anaerobic sediments of Hydrate Ridge. Microbial communities inhabiting anoxic environments usually include SRB and methanogenic archaea [*Munson et al.*, 1997]. These microbes control the terminal steps in the transformation of organic carbon to carbon dioxide and methane. Geochemically, Hydrate Ridge sediments are characterized by bacterial sulfate reduction extending from the surface to 20 meters below seafloor (mbsf) at ODP site 892 [*Kastner et al.*, 1995]. In our samples, intense sulfate reduction is well documented at both the northern and the southern summit (Fig. 7). Sediments at the northern summit (TV-G 9-1) show decreasing sulfate concentration to as low as ~5 mM within the first 15 centimeters, sediments at the southern summit (TV-G 18-1) are sulfate-depleted immediately below the seafloor. Sediments at the southern summit are often covered with bacterial mats and accompanied by gas hydrate layers occurring as close as 2 mm to the sediment-water interface. Methane concentrations as high as 130,000 ppmv as detected in shallow sediments at ODP site 892 [*Hovland et al.*, 1995] strongly suggest a methane source which is not related to *in situ* methanogenesis. Indeed, at Hydrate Ridge ascending fluids enhance methane hydrate decomposition, contributing free methane to the water column [*Suess et al.*, 1999]. Carbon isotope analyses of isoprenoid biomarkers such as crocetane and PMI indicate that a part of the methane is consumed by AMO probably performed by a syntrophic consortium of methane-oxidizing archaea and SRB using water as the formal oxidant [*Elvert et al.*, 1999]. The activity of the overall process is also evident from $\delta^{13}C$ values of C_{org} as low as -39.8‰ PDB in our samples (Table 2) and of dissolved CO_2 as low as -35‰ PDB in the gas hydrate-bearing zone at ODP site 892 [*Whiticar et al.*, 1995]. Moreover, a recent study performed by Boetius and co-workers (2000) identified highly abundant clusters of archaeal cells surrounded by layers of SRB in sulfate-depleted surface sediments of the southern summit, adding another piece of evidence that AMO is active and apparently mediated by a specific consortium of archaea and SRB.

Further evidence that AMO is the major process contributing to the low $\delta^{13}C$ values of CO_2 in the gas hydrate zone is provided by aragonite precipitates found at the southern summit of Hydrate Ridge where gas hydrate is exposed on the seafloor. These aragonite precipitates appear often in yellow layers of remarkable purity and grow directly within gas hydrate layers parallel to stratification, exhibiting a surface morphology that is suggestive of the bubble fabric of gas hydrate [*Bohrmann et al.*, 1998]. The aragonites show carbon isotope values of about -46‰ PDB (Table 2) and harbor biomarkers highly enriched in ^{12}C reminiscent of archaea and bacteria utilizing methane as a carbon source (see Figs. 2b and 6). Considering the observed biomarker pattern in combination with their respective carbon isotope values, it is likely that irregular isoprenoids and alcohols such as phytanol, PMI:1-ol, and archaeol are indicative of anaerobic methane-oxidizing archaea, whereas less ^{13}C-depleted short chain *n*-alcohols are derived from SRB. We did not observe more specific biomarkers such as mid-chain, *iso*-, and *anteiso*- branched alcohols [*Thiel et al.*, 1999] or intact non-isoprenoid alkyl glycerol ethers [*Hinrichs et al.*, 2000] that have been attributed to SRB. But considering that the $\delta^{13}C$-values of the *n*-alcohols reveal a distinct offset to that of archaeal biomarkers, we propose that short-chain *n*-alcohols found in this environment are derived from SRB using an isotopically depleted intermediate substrate for biosynthesis. Such a substrate could be for instance ^{13}C-depleted CO_2 directly produced by the archaea or acetate delivered by acetogenic bacteria which use ^{13}C-depleted CO_2 as their carbon source or grow heterotrophically on archaeal biomass [*Pancost et al.* 2000]. The partial involvement of acetogenic bacteria might explain why all samples analyzed on Hydrate Ridge also contain hopanoid compounds with depleted $\delta^{13}C$ values. However, hopanoid compounds have not been observed yet in any kind of anaerobic bacteria. It is also quite possible that hopanoids, at least in some part, are derived from aerobic methanotrophs. At the southern summit of Hydrate Ridge the concentration of diploptene in the sediment samples is higher than anywhere else on the ridge (Table 1), suggesting that degradation of hopanol may contribute much more effectively to the diploptene pool where gas hydrates and extreme anaerobic conditions are observed right below the sediment surface. Ultimately, it is very likely that the overall methane oxidation at such an "extreme" environmental setting includes both aerobic and anaerobic microbes growing on methane. This association suggests that aerobic methanotrophs use methane as a carbon source at or very close to the sediment-water interface, whereas archaea, performing methane oxidation in a syn-

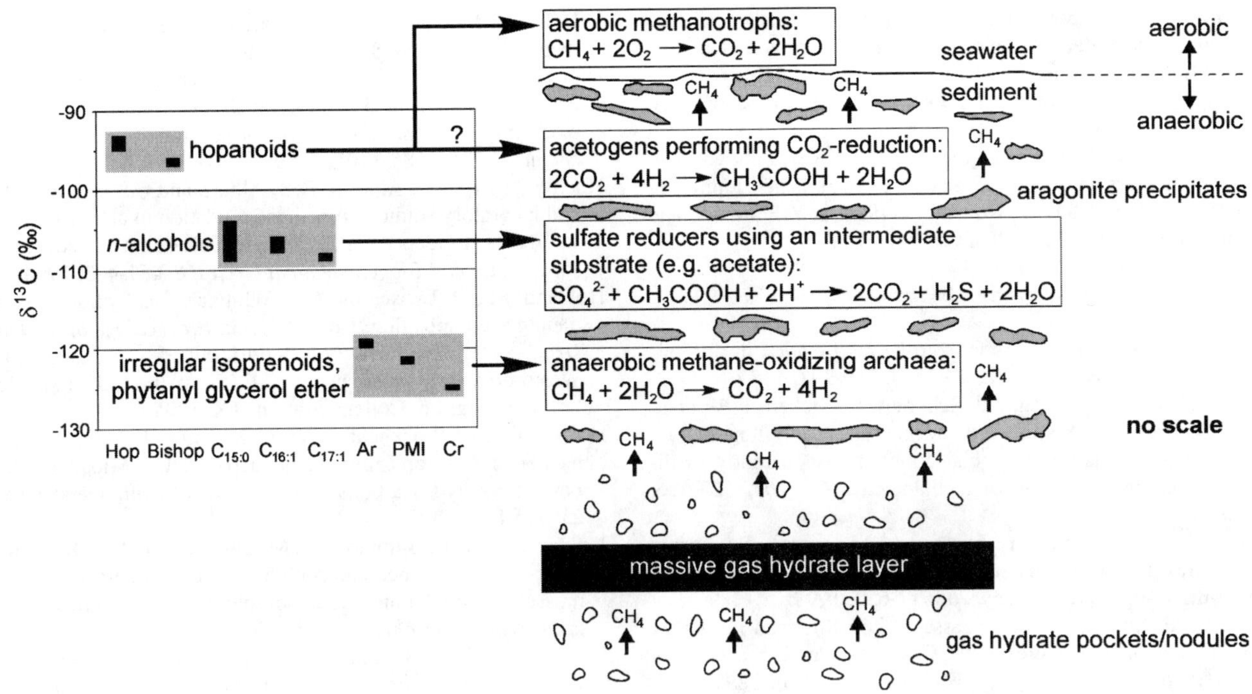

Figure 8. Schematic view of the different microbes thriving on methane, inferred from the analysis of sediments and aragonite residue at TV-G 18-1. The variation in the carbon isotopic composition of selected biomarkers is indicated by bars (Hop: hopanol; Bishop: bishomohopanol; $C_{x:y}$: *n*-alcohols containing x carbon atoms and y double bond(s); Ar: archaeol; PMI: pentamethylicosane; Cr: crocetane).

trophic consortium with SRB (and acetogens?), are restricted to deeper subsurface anaerobic sediment zones (Fig. 8).

4. CONCLUSIONS

Detailed studies on microbial biomarkers extracted from sediments and authigenic carbonates contain important information about biogeochemical processes associated with methane-venting and marine gas hydrates, especially when accompanied by pore water and carbonate analyses. Molecular and stable carbon isotope analyses of highly ^{13}C-depleted irregular isoprenoids demonstrate the activity of anaerobic methane-oxidizing archaea. The abundance of these biomarkers is essentially dependent on the supply of methane in such an environmental setting, which is also reflected in various isoprenoid biomarker patterns. Furthermore, the analysis of biomarkers isolated from aragonite precipitates reveals that microbial communities performing methane oxidation are very complex, probably involving a very close interaction between archaea and bacteria. Microbes implicated in methane carbon cycling are methanotrophic bacteria, SRB, acetogenic bacteria, and methane-oxidizing archaea. Evidence for the activity of these microbial players is most obvious at locations where methane gas hydrate breaches the seafloor. Our preferred scenario is: (1) bacteria such as aerobic methanotrophs use methane as a carbon source at or near the sediment-water interface, whereas (2) archaea, performing methane oxidation in a syntrophic consortium with SRB (and acetogens?), are restricted to subsurface anaerobic sediment zones beneath the aerobic seafloor regime. Without the processes of highly efficient methane oxidation, the escape of methane into the water column would be even greater, possibly allowing methane to enter the atmosphere. Thus, microbial methane oxidation, occurring either anaerobically or aerobically, is an important sink in the global methane budget.

Acknowledgments. We gratefully acknowledge M. Eek and M. McQuoid (University of Victoria) for their help with the $\delta^{13}C_{org}$ measurements. B. Domeyer and A. Bleyer (GEOMAR) are acknowledged for C_{org} measurements and for sulfate measurements, respectively. We thank D. Schulz-Bull and G. Petrick (Institute of Marine Science, University of Kiel) for providing laboratory space at the Sonderforschungsbereich 313 and for the help with the hydrogenation and mass spectrometric analysis of selected hydrocarbon fractions. Critical comments by two anonymous reviewers were highly appreciated and helped to improve the quality of the final manuscript. We also thank R. Heath (Graduate School at GEOMAR) for comments on an earlier version of this manuscript. Finally, all people of the Graduate School at GEOMAR are acknowledged for their personal help and support during this project. Financial support was provided by the Federal Ministry of Education, Research, and Technology (Bonn, Germany; grant 03G0109B and 03G0110A/B) and by a research stipendium from the German Research Foundation to M. Elvert. This paper is part of the Ph. D. thesis of M. Elvert entitled "Microbial biomarkers in anaerobic marine environments dominated by methane" completed in November 1999 at GEOMAR and submitted to the faculty at the University of Kiel.

REFERENCES

Barker, J.F., and P. Fritz, Carbon isotope fractionation during microbial methane oxidation, *Nature, 293*, 289-291, 1981.

Boetius, A., K. Ravenschlag, C.J. Schubert, D. Rickert, F. Widdel, A. Giesecke, R. Amann, B.B. Jørgensen, U. Witte, and O. Pfannkuche, A marine microbial consortium apparently mediating anaerobic oxidation of methane, *Nature, 407*, 623-626, 2000.

Bohrmann, G., J. Greinert, E. Suess, and M. Torres, Authigenic carbonates from Cascadia subduction zone and their relation to gas hydrate stability, *Geology, 26*, 647-650, 1998.

Carson, B., E. Suess, and J.C. Strasser, Fluid flow and mass flux determinations at vent sites on the Cascadia margin accretionary prism, *Journal of Geophysical Research, 95* (B6), 8891-8897, 1990.

Coleman, D.D., J.B. Risatti, and M. Schoell, Fractionation of carbon and hydrogen isotopes by methane-oxidizing bacteria, *Geochimica et Cosmochimica Acta, 45*, 1033-1037, 1981.

Collister, J.W., R.E. Summons, E. Lichtfouse, and J.M. Hayes, An isotopic biogeochemical study of the Green River oil shale, *Organic Geochemistry, 19*, 265-276, 1992.

Elvert, M., E. Suess, and M.J. Whiticar, Anaerobic methane oxidation associated with marine gas hydrates: superlight C-isotopes from saturated and unsaturated C_{20} and C_{25} irregular isoprenoids, *Naturwissenschaften, 86*, 295-300, 1999.

Elvert, M., Microbial biomarkers in anaerobic marine environments dominated by methane, Ph.D. thesis, 110 pp., Christian-Albrechts-Universität Kiel, Germany, 1999.

Elvert, M., E. Suess, J. Greinert, and M.J. Whiticar, Archaea mediating anaerobic methane oxidation in deep-sea sediments at cold seeps of the eastern Aleutian subduction zone, *Organic Geochemistry, 31*, 1175-1187, 2000.

Freeman, K.H., J.M. Hayes, J.-M. Trendel, and P. Albrecht, Evidence from carbon isotope measurements for diverse origins of sedimentary hydrocarbons, *Nature, 343*, 254-256, 1990.

Greinert, J., Rezente submarine Mineralbildungen: Abbild geochemischer Prozesse an aktiven Fluidaustrittsstellen im Aleuten-Cascadia Akkretionskomplex, *GEOMAR Rep. 87*, GEOMAR Forschungszentrum für marine Geowissenschaften, Christian-Albrechts-Universität, Kiel, Germany, 1999.

Hesse, R., and Harrison, W.E., Gas hydrates (clathrates) causing pore-water freshening and oxygen isotope fractionation in deep-water sedimentary sections of terrigenuos continental margins, *Earth and Planetary Science Letters, 55*, 453-462, 1981.

Hayes, J.M., Factors controlling ^{13}C contents of sedimentary organic compounds: Principles and evidence, *Marine Geology, 113*, 111-125, 1993.

Hinrichs, K.-U., J.M. Hayes, S.P. Sylva, P.G. Brewer, and E.F. DeLong, Methane-consuming archaebacteria in marine sediments, *Nature, 398*, 802-805, 1999.

Hinrichs, K.-U., R.E. Summons, V. Orphan, S.P. Sylva, and J.M. Hayes, Molecular and isotopic analyses of anaerobic methane-oxidizing communities in marine sediments, *Organic Geochemistry, 31*, 1685-1701, 2000.

Hoehler, T.M., M.J. Alperin, D.B. Albert, and C.S. Martens, Field and laboratory studies of methane oxidation in an anoxic marine sediments: Evidence for a methanogenic-sulphate reducer consortium, *Global Biogeochemical Cycles, 8*, 451-463, 1994.

Hovland, M., D. Lynse, and M.J. Whiticar, Gas hydrate and sediment gas composition, Hole 892A, in *Proceedings of the Ocean Drilling Program, Scientific Results*, edited by B. Carson, G.K. Westbrook, R.J. Musgrave, and E. Suess, pp. 151-161, Ocean Drilling Program, College Station, TX, 1995.

Huang, Y., M.J. Lockheart, J.W. Collister, and G. Eglinton, Molecular and isotopic biogeochemistry of the Miocene Clarkia Formation: hydrocarbons and alcohols, *Organic Geochemistry, 23*, 785-801, 1995.

Jahnke, L.L., R.E. Summons, L.M. Dowling, and K.D. Zahiralis, Identification of methanotrophic lipid biomarkers in cold-seep mussell gills: Chemical and isotopic analysis, *Applied and Environmental Microbiology, 61*, 576-582, 1995.

Jahnke, L.J., R.E. Summons, J.M. Hope, and D.J. Des Marais, Carbon isotopic fractionation in lipids from methanotrophic bacteria II: The effects of physiology and environmental parameters on the biosynthesis and isotopic signatures of biomarkers, *Geochimica et Cosmochimica Acta, 63*, 79-93, 1999.

Kastner, M., J.C. Sample, M.J. Whiticar, M. Hovland, B.A. Cragg, and J.R. Parkes, Geochemical evidence for fluid flow and diagenesis at the Cascadia convergent margin, in *Proceedings of the Ocean Drilling Program, Scientific Results*, edited by B. Carson, G.K. Westbrook, R.J. Musgrave, and E. Suess, pp. 375-384, Ocean Drilling Program, College Station, TX, 1995.

Kastner, M., K.A. Kvenvolden, and T.D. Lorenson, Chemistry, isotopic composition, and origin of a methane-hydrogen sulfide hydrate at the Cascadia subduction zone, *Earth and Planetary Science Letters, 156*, 173-183, 1998.

Kohnen, M.E.L., S. Schouten, J.S. Sinninghé Damste, J.W. de Leeuw, D.A. Merritt, and J.M. Hayes, Recognition of paleobiochemicals by a combined molecular sulfur and isotope geochemical approach, *Science, 256*, 358-362, 1992.

Kulm, L.D., E. Suess, J.C. Moore, B. Carson, B.T. Lewis, S.D. Ritger, D.C. Kadko, T.M. Thornburg, R.W. Embley, W.D. Rugh, G.J. Massoth, M.G. Langseth, G.R. Cochrane, and R.L. Scanman, Oregon subduction zone: Venting, fauna, and carbonates, *Science, 231*, 561-566, 1986.

Langworthy, T.A., and J.L. Pond, Membranes of lipids and thermophiles, in *Thermophiles: general, molecular and applied microbiology*, edited by T.D. Brock, pp. 107-135, Wiley, New York, 1986.

Munson, M.A., D.B. Nedwell, and T.M. Embley, Phylogenetic diversity of *Archaea* in sediment samples from a coastal salt marsh, *Applied and Environmental Microbiology, 63*, 4729-4733, 1997.

Ourisson, G., M. Rohmer, and K. Poralla, Prokaryotic hopanoids and other polyterpenoid sterol surrogates, *Annual Reviews of Microbiology, 41*, 301-333, 1987.

Pancost, R.D., J.S. Sinninghe Damsté, S. de Lint, M.J.E.C. van der

Maarel, J.C. Gottschal, and The Medinaut Shipboard Scientific Party, Biomarker evidence for widespread anaerobic methane oxidation in Mediterranean sediments by a consortium of methanogenic archaea and bacteria, *Applied and Environmental Microbiology, 66*, 1126-1132, 2000.

Reeburgh, W.S., B.B. Ward, S.C. Whalen, K.A. Sandbeck, K.A. Kilpatrick, and L.J. Kerkhof, Black Sea methane geochemistry, *Deep-Sea Research, 38*, S1189-S1210, 1991.

Rohmer, M., P. Bouvier-Nave, and G. Ourisson, Distribution of hopanoid triterpenes in prokaryotes, *Journal of General Microbiology, 130*, 1137-1150, 1984.

Schouten, S., M.J.E.C. van der Maarel, R. Huber, and J.S. Sinninghe Damsté, 2,6,10,15,19-Pentamethylicosenes in *Methanolobus bombayensis*, a marine methanogenic archaeon, and in *Methanosarcina mazei*, *Organic Geochemistry, 26*, 409-414, 1997.

Sinninghe Damsté, J.S., S. Schouten, N.H. van Vliet, R. Huber, and T.A.J. Geenevasen, A polyunsaturated irregular acyclic C_{25} isoprenoid in a methanogenic archaeon, *Tetrahedron Letters, 38*, 6881-6884, 1997.

Suess, E., and M.J. Whiticar, Methane-derived CO_2 in pore fluids expelled from the Oregon subduction zone, *Palaeogeography, Palaeoclimatology, Palaeoecology, 71*, 119-136, 1989.

Suess, E., M.E. Torres, G. Bohrmann, R.W. Collier, J. Greinert, P. Linke, G. Rehder, A. Trehu, K. Wallmann, G. Winckler, and E. Zuleger, Gas hydrate destabilization: enhanced dewatering, benthic material turnover and large methane plumes at the Cascadia convergent margin, *Earth and Planetary Science Letters, 170*, 1-15, 1999.

Summons, R.E., L.L. Jahnke, and Z. Roksandic, Carbon isotopic fractionation in lipids from methanotrophic bacteria: Relevance for interpretation of the geochemical record of biomarkers, *Geochimica et Cosmochimica Acta, 58*, 2853-2863, 1994.

Summons, R.E., P.D. Franzmann, and P.D. Nichols, Carbon isotopic fractionation associated with methylotrophic methanogenesis, *Organic Geochemistry, 28* (7-8), 465-475, 1998.

Thiel, V., J. Peckmann, R. Seifert, P. Wehrung, J. Reitner, and W. Michaelis, Highly isotopically depleted isoprenoids: molecular markers for ancient methane venting, *Geochimica et Cosmochimica Acta, 63*, 3959-3966, 1999.

Whiticar, M.J., M. Hovland, M. Kastner, and J.C. Sample, Organic geochemistry of gases, fluids, and hydrates at the Cascadia accretionary margin, in *Proceedings of the Ocean Drilling Program, Scientific Results*, edited by B. Carson, G.K. Westbrook, R.J. Musgrave, and E. Suess, pp. 385-397, Ocean Drilling Program, College Station, TX, 1995.

Whiticar, M.J., Carbon and hydrogen isotope systematics of bacterial formation and oxidation of methane, *Chemical Geology, 161*, 291-314, 1999.

Zeng, Y.B., D.M. Ward, S.C. Brassell, and G. Eglinton, Biogeochemistry of hot spring environments 2. Lipid compositions oy Yellowstone (Wyoming, U.S.A.) cyanobacterial and *Chloroflexus* mats, *Chemical Geology, 95*, 327-345, 1992a.

Zeng, Y.B., D.M. Ward, S.C. Brassell, and G. Eglinton, Biogeochemistry of hot spring environments 3. Apolar and polar lipids in the biologically active layers of a cyanobacterial mat, *Chemical Geology, 95*, 347-360, 1992b.

Marcus Elvert*, Jens Greinert, Erwin Suess, Department of Marine Environmental Geology, GEOMAR Research Center for Marine Geosciences, Wischhofstr. 1-3, D-24148 Kiel, Germany.

Michael J. Whiticar, Biogeochemistry Facility, School of Earth and Ocean Sciences, University of Victoria, British Columbia, V8W 2Y2, Victoria, Canada.

*Present address: Department of Biogeochemistry, Max Planck Institute for Marine Microbiology, Celsiusstr. 1, D-28359 Bremen, Germany. (Contact: melvert@mpi-bremen.de)

Stability of Thermogenic Gas Hydrate in the Gulf of Mexico: Constraints on Models of Climate Change

Roger Sassen, Stephen T. Sweet, Alexei V. Milkov, Debra A. DeFreitas, and Mahlon C. Kennicutt II

Geochemical and Environmental Research Group, Texas A&M University, College Station, Texas

Harry H. Roberts

Department of Oceanography and Coastal Sciences, Louisiana State University, Baton Rouge, Louisiana

Thermogenic (structure II) gas hydrate is abundant on the Gulf of Mexico continental slope because hydrocarbon gases from a deep, hot subsurface petroleum system vent prolifically to the sea floor in regions of low temperatures and high pressures that are within the gas hydrate stability zone (GHSZ). The distribution of gas hydrate is controlled by geologic structure. Molecular distributions of gases venting from hydrate-bearing sediments show that gas hydrate decomposition is not significant in comparison to flux from the subsurface petroleum system. Structure II gas hydrate is commonly encountered, and contains C_1-C_5 hydrocarbons that crystallize from relatively unaltered vent gases. Most surficial structure II gas hydrate sites (0-6 m) studied from the ~540-1930 m water depth range are stable, and gas hydrate is thought to be accumulating to considerable depth in sediment because of the prolific, ongoing venting of gas. Bacterial methane hydrate is also abundant in the Gulf of Mexico. A leaky petroleum system is the larger contributor of thermogenic greenhouse gases to the ocean and atmosphere from the Gulf of Mexico, not gas hydrate decomposition. Gas hydrate crystallization in the Gulf of Mexico sequesters large volumes of greenhouse gases in sediments, deforming soft sediments, and is proposed as an agent that stabilizes or dampens climatic change under the present climatic regime.

1. INTRODUCTION

The northern Gulf of Mexico continental slope may be of global significance as a large reservoir of hydrate-bound greenhouse gases. Gas hydrate is abundant on the Gulf slope because hydrocarbon gases vent to a sea-floor temperature and pressure regime in which gas hydrate is

Natural Gas Hydrates: Occurrence, Distribution, and Detection
Geophysical Monograph 124

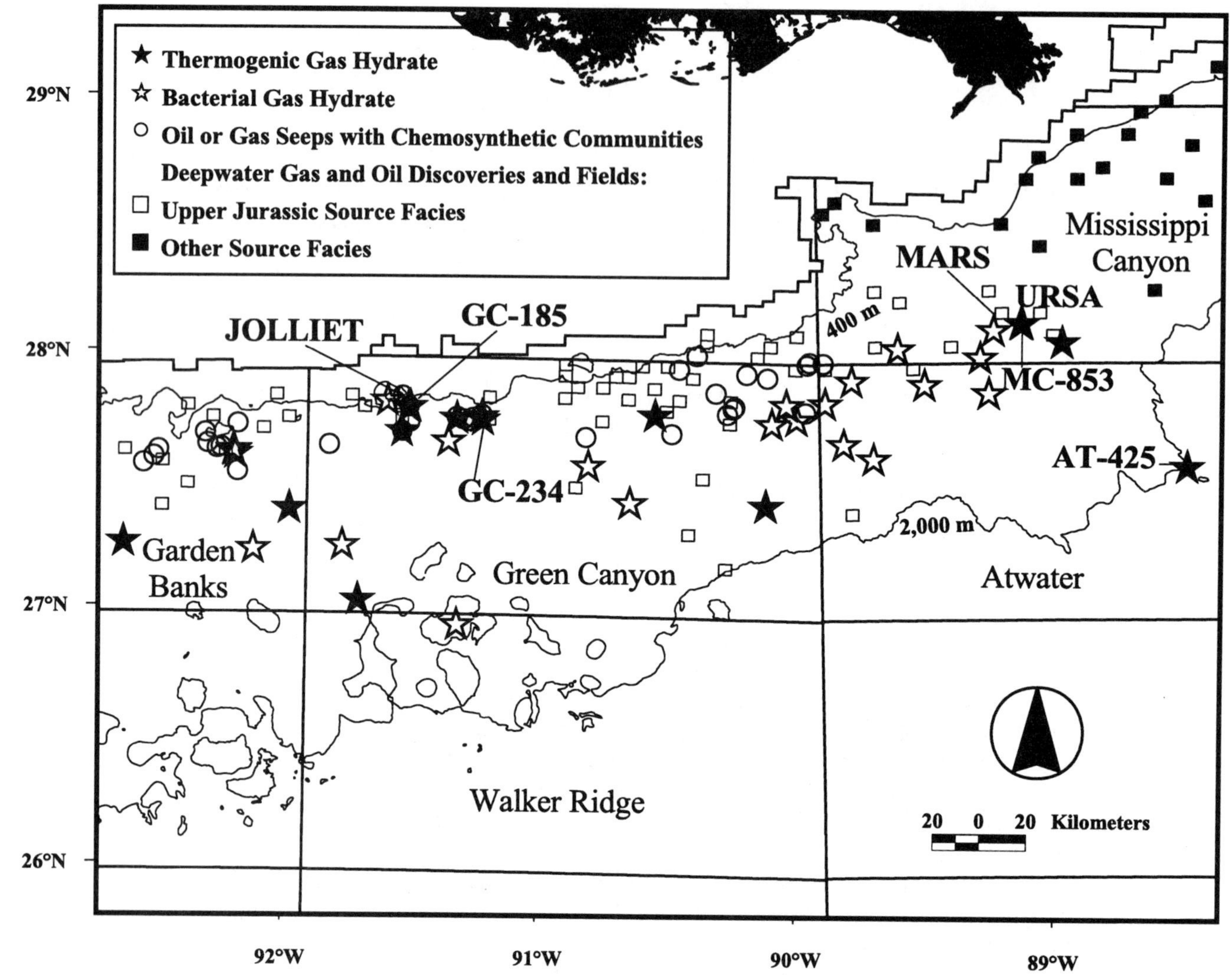

Figure 1. Sketch map showing distribution of known gas hydrate sites, seeps with chemosynthetic communities, and oil and gas fields in the central Gulf of Mexico. Note study areas (GC 185, GC 234, MC 853, AT 425) and closely associated oil fields (Jolliet, Mars, Ursa).

stable. Gas hydrate is known to outcrop on the sea floor and occur in shallow sea floor sediments in the Gulf of Mexico Salt Basin in greater volumes than documented in any other basin [*Sassen et al., 1999a*]. Intact gas hydrate samples have been recovered from sediments by piston coring (<6 m sediment depth) and by research submersibles from >50 localities in the Gulf slope (Figure 1). The minimum observed water depth of gas hydrate in the Gulf of Mexico is ~440 m and the maximum water depth is >2,400 m (Figure 2). The thickness of the gas hydrate stability zone (GHSZ) increases with water depth, and models suggest that gas-hydrate bearing sediment could be ~1 km thick in ultra-deep (~2 km) water [*Milkov and Sassen, 2000*].

Gas hydrates are ice-like crystalline minerals in which hydrocarbon gases and non-hydrocarbon gases are held within rigid cages of water molecules. Structure I gas hydrate has a body-centered cubic lattice, structure II gas hydrate has a diamond lattice, and structure H gas hydrate has a hexagonal lattice [*Sloan, 1998*]. Structure I gas hydrate, which occurs in the Gulf of Mexico and many other basins, is usually predominately bacterial methane [*Kvenvolden, 1995*]. Theoretically, Structure I gas hydrate could contain either thermogenic or bacterial methane, but no examples of thermogenic structure I hydrate have been discovered in the Gulf of Mexico. The $\delta^{13}C$ and δD of bacterial methane from seafloor gas vents sampled by research submersible and from deep-water structure I gas hydrate is thus far consistent with an origin from methanogenesis via CO_2 reduction [*Sassen et al., 1999a,b*]. The origin of bacterial methane (structure I) gas hydrate appears relatively straightforward in the Gulf of Mexico and will not be discussed in this paper.

Both structure II and structure H gas hydrate are believed to co-exist in the Gulf at water depths as shallow as ~540 m [*Sassen and MacDonald, 1997*]. Structure II gas hydrate generally includes C_1-C_4 hydrocarbons (methane through isobutane) whereas structure H gas hydrate generally includes C_1-C_5 hydrocarbons (methane through isopentane) as significant components [*Sloan, 1998*]. In contrast to simple bacterial methane hydrate, thermogenic gas hydrate preserves complex information on origin and stability because numerous hydrocarbon molecules of varying properties are held within the crystal lattices.

Temperature, pressure, and the availability of gas hydrate-forming molecules are fundamental factors, among others, that control the stability of gas hydrate [*Sloan, 1998*]. Transient changes in seawater temperature are thought to impact the stability of outcropping gas hydrate in the Gulf of Mexico [*MacDonald et al., 1994, Roberts et al., 1999*]. Variable or decreasing rates of free gas migration affect concentrations and thus impact gas hydrate stability, especially gas hydrate that crops out on the sea floor [*Egorov et al., 1999*]. Bacterial oxidation of hydrate-bound methane and free hydrocarbon gases in adjacent sediments also could decrease the stability of structure II gas hydrate by removing key hydrate-forming hydrocarbons (e.g. propane) that are necessary to maintain equilibrium [*Sassen et al., 1998, 1999b*]. Less is known about the stability and composition of thermogenic gas hydrate on the middle and lower Gulf slope, at greater water depths where temperatures are lower and temperature fluctuations are less important. Few intact thermogenic gas hydrate samples have been recovered from the ultra-deep waters of the Gulf of Mexico (Figure 1).

Massive decomposition of gas hydrate and concomitant release of greenhouse gases including methane are suggested as one possible agent of rapid climate change [e.g., *MacDonald, 1990; Dickens, 1995, 1997; Bains et al., 1999; Max et al., 1999*]. Some authors believe this climate change hypothesis is overstated [*Kvenvolden, 1999*]. Free gas from hydrate decomposition is thought, on the basis of indirect evidence, to contribute to the formation of sea floor seep features [e.g., *Vogt et al., 1994; Paull et al., 1995*] and alter sediment fabrics [*Kennet and Fackler-Adams, 2000*]. Direct geochemical evidence of significant gas hydrate decomposition is thus far lacking and other explanations are possible.

Recent piston coring has presented us the opportunity to obtain intact sediment-free samples of thermogenic gas hydrate from new and deeper sites in the Gulf of Mexico. Detailed C_1-C_5 molecular and isotopic analyses of the intact samples have been conducted. The primary objective of the present paper is to characterize thermogenic gas hydrate and associated sediment from shallow to ultra-deep water within the GHSZ of the Gulf of Mexico. A secondary objective is to address the basic geologic question of whether there is net accumulation or decomposition of gas hydrate in the Gulf of Mexico at present. Thirdly, linkages between global climate change and sea floor seeps are considered.

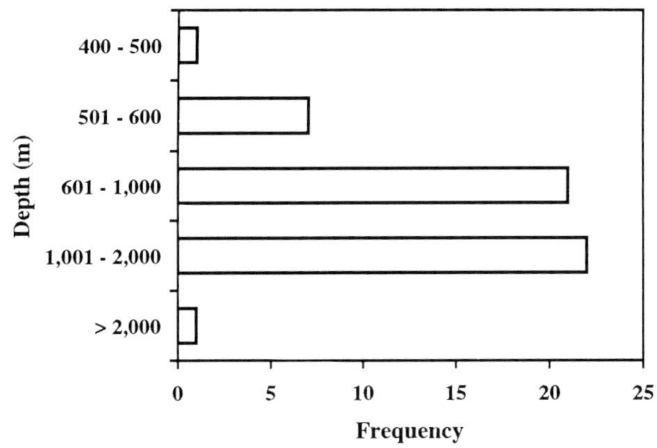

Figure 2. Histogram of water depth distribution of gas hydrate sites in the Gulf of Mexico.

1.1 Geologic Overview

The main structural features of the northern Gulf of Mexico are salt basins including the large Gulf of Mexico Salt Basin and a series of smaller interior salt basins that extend from south Texas to Alabama. The basins formed during Late Triassic rifting and were floored by salt (Louann/Werner formations) during Middle Jurassic marine incursions [*Salvador, 1987*]. The structural style of these basins is profoundly influenced by the effects of salt movement driven by rapid deposition of siliciclastic sediment. The Gulf of Mexico Salt Basin was most affected by salt deformation during the Tertiary. An important effect of rapid sediment influx was massive deformation of the underlying salt. The Gulf of Mexico continental shelf is now characterized by numerous salt domes, whereas the continental slope is more affected by large sheet-like salt thrusts extending from the shelf edge across the continental slope to the Sigsbee Escarpment, near the edge of the abyssal plain [*Worrall and Snelson, 1989*]. The Mississippi Fan Foldbelt extends northeast-southwest for ~350 km along the lower continental slope of the Gulf of Mexico [*Weimer and Buffler, 1992; Wu, 1993; Rowan, 1997*]. Some salt pierced through the overlying section to form allochthonous salt bodies at shallow depths, providing localized conduits for upward migration of fluids to the sea floor. The Mississippi Fan Foldbelt represents the downdip limit of the Gulf of Mexico Salt Basin, and undeformed salt-free sediments of the abyssal plain are thought to occur immediately to the south [*Weimer and Buffler, 1992; Wu, 1993; Rowan, 1997*].

The geology of the Gulf of Mexico slope is conducive to seepage from deeply buried petroleum systems to the sea floor because hydrocarbon generation is geologically recent within deep sediment sections of large salt withdrawal basins [*Wenger et al., 1994*]. Fracture zones associated with moving salt sheets, as well as active faults, provide efficient migration conduits for fluid flow to the sea floor including hydrocarbons and brines. The sea floor shows irregular bathymetry from structural deformation, faulting, fracturing, and slumping [*Roberts and Carney, 1997*]. Massive hydrocarbon seepage manifests itself at the Gulf sea floor as gas hydrate, oil-stained sediments, authigenic carbonate rock with carbon depleted in ^{13}C, and unique seep-associated chemosynthetic communities [e.g., *Aharon et al.,1997; Roberts and Aharon, 1994; MacDonald et al., 1989; Roberts and Carney, 1997*]. Because the fluid flow is structurally focused, gas hydrate accumulations are concentrated along the rims of salt-withdrawal basins, over salt ridges, and near the faulted and folded margin of the Sigsbee Escarpment near the downdip limit of the Gulf of Mexico Salt Basin [*Sassen et al., 1999a*]. Seep-related features such as gas hydrate are only infrequently observed within salt withdrawal basins themselves because such sediments are less deformed and usually lack major migration conduits to shallow sediments or the sea floor. It should be emphasized that because of structural control on gas hydrate, no unambiguous geophysical evidence exists from the northern Gulf of Mexico of the Bottom Simulating Reflectors (BSRs) often associated with gas hydrate in different geologic settings.

2. SAMPLES AND METHODS

2.1 Study Sites

The GC 185 and 234 sites are on the upper slope of the Gulf of Mexico in ~540 m water depth (Figure 1), near the upper limit of the GHSZ. The most recent episodes of hydrocarbon venting in this area appear to have commenced during the late Pleistocene [*Aharon et al., 1997*]. Structure II gas hydrate has been identified in oily sediments of the Gulf of Mexico [*Brooks et al., 1986; Sassen and MacDonald, 1997*] and from the Caspian Sea [*Ginsburg and Soloviev, 1998*]. Gas hydrate samples from the shallow GC 185 and GC 234 sites, at ~540 m water depth (Table 1), have C_1-C_5 hydrocarbon molecular distributions that are consistent with structure II gas hydrate. Methane, ethane, propane, and butanes are significant components diagnostic of structure II gas hydrate [*Sloan, 1998*]. The structural assignment of structure II gas hydrate based on molecular distribution was confirmed by solid state nuclear magnetic resonance (NMR) of intact gas hydrate samples from GC 184 and GC 185 [*Davidson et al., 1986, Sassen and MacDonald, 1997*].

The GC 185 site (Bush Hill) is a fault-related seep feature (27° 45.7' N and 91° 30.5' W). An antithetic fault at GC 185 is structurally related to nearby Jolliet Field (GC 184) and the oil and gas venting at the GC 185 fault correlate with reservoirs at ~2-3 km depth in Jolliet Field [*Kennicutt et al., 1988*]. This is consistent with an origin from the same hydrocarbon source system, or from Jolliet Field itself. Research submersible observations show that sea-floor gas hydrate mounds are generally associated with active gas vents (Plate 1). Samples of vent gas and associated gas hydrate were collected during several submersible dives (Table 1), and analyses show little variation in composition over time.

Multiple gas hydrate mounds at GC 185 have persisted for >6 years, but mounds show deformation on scales of months to years [*MacDonald et al., 1994*]. The characteristic features of vent mounds, such as bubble trains venting into the water column and sediment-covered gas hydrate, are shown in Figure 3. Gas hydrate is capped by a thin layer of deformed hemipelagic mud that contains bacterially oxidized crude oil, free gas, dispersed gas hydrate nodules, authigenic carbonate depleted in ^{13}C, and H_2S [*Sassen et al., 1993,1994, 1999a,b*]. The gas hydrate mounds are covered by bacterial mats (*Beggiatoa*), which oxidize sulfide, and are surrounded by complex chemosynthetic communities including tube worms [*Sassen et al., 1999a,b*].

Outcropping structure II gas hydrate at GC 185 (~540 m) is thought to be transiently unstable because of natural fluctuations in seawater temperature, mainly from warm-cored eddy currents [*MacDonald et al., 1994; Roberts et al., 1999*]. Gas hydrate instability from transient seawater temperature variations is estimated to extend only a few meters into sediments at this water depth [*Milkov et al., 2000*]. Structure II hydrate at the GC 185 site is estimated to be stable to ~450 meters below the sea floor (mbsf) [*Milkov and Sassen, 2000*].

The GC 234 site (27° 44.8' N and 91° 13.3' W) at ~543 m water depth is a fault related seep area over shallow salt (Figure 1). When the site was first identified by research submersible in 1997, a small gas hydrate vein-filling surface (~1 m across) cropped out on the surface of an upthrown fault scarp associated with active gas vents. The exposure was being actively undercut by gas expulsion. Samples of vent gas and associated gas hydrate were

Plate 1. Photograph of an active gas hydrate mound ~1.5 m across in 543 m water depth at GC 185 site. Note individual gas bubble trains from copious venting to water column (~60 L/minute STP). (photo by Ian MacDonald)

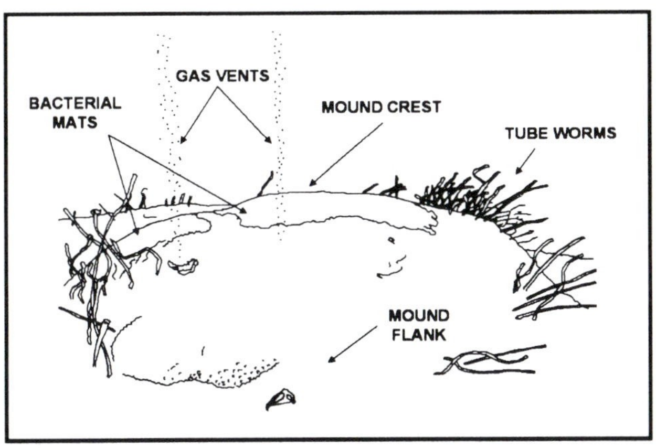

Figure 3. Sketch showing main elements of a typical gas hydrate mound (1-3 m across) formed around thermogenic gas vents in a chemosynthetic community. From Sassen et al. (1999b).

Table 1
Molecular and isotopic compositions of vent gases and gas hydrates from the Gulf of Mexico

Sample Location	Sample	$\delta^{13}C$ C_1	δD C_1	% C_1	$\delta^{13}C$ C_2	% C_2	$\delta^{13}C$ C_3	% C_3	$\delta^{13}C$ i-C_4	% i-C_4	$\delta^{13}C$ n-C_4	% n-C_4	$\delta^{13}C$ i-C_5	% i-C_5	$\delta^{13}C$ n-C_5	% n-C_5	% neo-C_5	$\delta^{13}C$ CO_2
Green Canyon 185 and 234 Sites (~540 m)																		
GC 185[b]	Vent Gas	-46.0	-198	95.9	-29.7	2.4	-25.0	1.2		<0.1	-22.6	0.3	-26.1	0.2		<0.1		-4.9
GC 185[b]	Vent Gas	-44.1	-200	90.4	-30.2	4.5	-26.3	3.7	-27.9	0.6	-23.3	0.6	-26.1	0.2		<0.1		-5.4
GC 185[a]	Vent Gas	-44.9	-171	93.4	-29.4	4.1	-26.2	1.5	-29.8	0.3	-24.4	0.5		0.3		<0.1		+13.2
GC 234[c]	Vent Gas	-48.7	-203	93.7	-29.2	2.4	-26.4	2.4	-27.9	0.5	-25.2	0.6	-26.5	0.2	-25.0	0.1	<0.1	+2.9
GC 234	Vent Gas	-49.0	-162	92.3	-28.6	4.0	-26.3	1.9	-28.3	0.4	-25.6	0.7	-27.3	0.3	-25.4	<0.1	<0.1	-7.3
GC 234	Vent Gas	-47.9		93.1	-27.8	4.1	-25.9	1.7	-27.6	0.3	-25.0	0.5	-26.5	0.2	-27.3	0.2	<0.1	
GC 185[b]	Gas Hydrate	-42.9	-163	83.1	-28.6	7.6	-24.9	8.1	-27.2	0.9	-22.1	0.2						-27.8
GC 185[b]	Gas Hydrate	-42.2	-190	71.7	-29.0	10.6	-25.5	12.6	-27.6	2.6	-22.8	1.7	-26.2	0.8				-20.0
GC 185[b]	Gas Hydrate	-43.5	-177	80.2	-29.7	9.4	-25.5	7.3	-27.9	1.6	-23.0	1.2	-24.8	0.3				-23.7
GC 185[b]	Gas Hydrate	-42.5	-193	72.1	-29.2	12.4	-25.7	11.4	-27.8	2.3	-22.7	1.6	-24.7	0.3				-21.6
GC 185[c]	Gas Hydrate	-42.9	-115	85.7	-28.6	6.3	-25.6	6.1	-26.8	1.1		0.8						
GC 185[a]	Gas Hydrate	-43.6	-167	72.1	-29.8	10.5	-26.1	12.4	-28.1	2.5	-24.0	1.7		0.7		<0.1		+17.5
GC 234[c]	Gas Hydrate	-48.4	-203	73.8	-29.4	12.6	-26.4	9.9	-28.4	1.6	-24.9	1.8		0.2		0.1		
Mississippi Canyon 853 Site (~1060 - 1070 m)																		
MC 853-1[a]	Gas Hydrate	-44.7	-161	76.2	-29.3	9.0	-23.7	9.6	-27.8	4.7	-24.1	0.4	-26.0	0.1		<0.1		
MC 853-2	Gas Hydrate	-46.5	-160	71.8	-28.4	6.5	-25.8	15.5	-27.0	5.6	-24.6	0.6	-26.1	<0.1	-21.9	<0.1		-16.8
MC 853-3	Gas Hydrate	-49.9	-156	77.4	-28.4	6.8	-24.5	11.3	-27.0	3.4	-23.4	0.7		0.3		<0.1		-4.3
MC 853-4	Gas Hydrate	-45.1	-166	75.3	-29.8	7.8	-20.5	8.4	-27.1	8.1	-22.9	0.3		0.1		<0.1	0.1	+11.6
Atwater Valley 425 Site (~1930 m)																		
AT 425-1	Gas Hydrate	-51.6	-173	99.0	-36.6	1.0												-7.1
AT 425-2	Gas Hydrate	-49.4	-183	88.4	-38.1	3.4	-32.1	5.4	-32.4	1.3	-28.6	1.3	-29.6	0.2	-30.5	<0.1	<0.1	+12.9
AT 425-3	Gas Hydrate	-50.1	-171	94.9	-37.6	1.8	-31.8	2.1	-32.5	0.5	-28.6	0.6		<0.1		<0.1		+10.6
AT 425-4	Gas Hydrate	-49.3	-148	77.3	-38.3	5.0	-32.0	12.5	-31.6	3.0	-28.3	1.2	-28.8	0.9	-28.4	0.2	<0.1	+12.4

[a] Data from Sassen et al. (1999a); [b] Data from Sassen et al. (1998); [c] Data from Sassen et al. (1999b)

collected and analyzed (Table 1). The molecular distributions of gas hydrate from GC 234 are consistent with structure II gas hydrate (Sassen et al., 1999b). Bacterial mats (*Beggiatoa*) draped the sediment surface over the exposed gas hydrate. Complex chemosynthetic communities that include tube worms are present at this site [*Sassen et al., 1999b*]. Maximum stability depth in sediments of structure II hydrate at the GC 234 site is ~450 mbsf, similar to the GC 185 site [*Milkov and Sassen, 2000*].

The Mississippi Canyon (MC) 853 site (28° 7.4' N, 89° 8.2' W) at ~1060-1070 m water depth is a large sea-floor mound (~1.5 km across, with ~30-40 m of relief) on the middle Gulf slope (Figure 1). The mound directly overlies shallow salt on the edge of a large salt withdrawal basin near Ursa and Mars fields (Figure 1). The mound is believed to contain large volumes of gas hydrate, free gas, authigenic carbonate, and highly deformed sediments [*Sassen et al., 1999a*]. The molecular properties of the gas hydrate from the MC 853 (Table 1) are consistent with structure II gas hydrate (Sassen et al., 1999). The GC 185, GC 234, and MC 853 sites show rapid hydrocarbon flux related to vertical migration from the same mature Upper Jurassic source facies [*Wenger et al., 1994*] at ~6-10 km burial depth. Structure II hydrate of the MC 853 site is stable to ~860 mbsf [*Milkov and Sassen, 2000*].

The AT 425 study site is an isolated sea-floor vent feature on the lower Gulf of Mexico slope (27° 34.1' N and 88° 29.7' W) in ~1920-1930 m water depth (Figure 1). The site is part of a distinct geologic province near the downdip limit of the Gulf of Mexico Salt Basin known as the Mississippi Fan Foldbelt [*Weimer and Buffler, 1992; Wu, 1993; Rowan, 1997*]. The molecular properties of gas hydrate from the AT 425 site are consistent with structure II gas hydrate (Sassen et al., 2000). Biomarker geochemistry of oil seeps suggest that the source of oil from AT 425 is a different source system than encountered at the other three sites [*Wenger et al., 1994*], explaining the unusual isotopic properties of gas hydrate from the site [*Sassen et al., 2000*]. The AT 425 site is characterized by sea-floor seep features related to faults overlying an allochthonous salt ridge shallow in the sediment [*Sager and Kennicutt, 2000*]. Fluids from depth beneath the AT 425 site pierce a thick apron of gently dipping sediment deposited from the Mississippi Fan Channel. The AT 425 vent is assumed to be a major fluid flow release site from depth in the Mississippi Fan Foldbelt because no other vent or gas hydrate sites are known within tens of kilometers (Figure 1). Structure II hydrate at this site is stable to ~1150 mbsf [*Milkov and Sassen, 2000*].

2.2 Methods

Vent gases and gas hydrate were sampled from the GC 185 and GC 234 sites over a period of several years using research submersibles as described by *Sassen et al.* [*1999a*]. Deeper-water gas hydrate samples were collected at the MC 853 and AT 425 sites in 1998-1999 (Figure 1) from a research vessel using a 6 m (7 cm interior diameter) piston corer equipped with a 900 kg weight. A high-speed winch facilitated rapid core recovery and preservation of solid gas hydrate from deep-water study sites. Solid gas hydrate was removed from cores within minutes of recovery and preserved onboard the vessel by immersion in liquid nitrogen.

Upon removal from liquid nitrogen in the laboratory, aliquots of solid gas hydrate were hand picked to remove sediment and were allowed to decompose under a water-filled bell jar to obtain large volume samples of free gas (0.5 to 1 liter). Aliquots of gas samples were immediately transferred to pre-evacuated metal vacutainers using a 60 ml gas-tight syringe, and were held at –20° C until analysis. Sediment samples were canned and were held at –20° C until analyses for extractable hydrocarbons and headspace gases. Analytical procedures for C_1-C_5 gas chromatography (precision = ± 5%), and measurement of stable isotopic properties of hydrocarbon gases and sediments, are described in detail by Sassen et al. (1999b). The $\delta^{13}C$ values are reported as parts per thousand (‰) relative to the PeeDee Belemnite standard (precision = ± 0.2‰), and the δD values are reported as ‰ relative to Standard Mean Ocean Water (SMOW) (precision = ±5‰).

3. RESULTS

3.1 Vent Gas

The molecular and isotopic properties of vent gases collected by research submersible from the GC 185 and GC 234 localities are shown in Table 1. Methane is the major hydrocarbon component (90.4-95.9%). The $\delta^{13}C$ values of the methane are in the –44.1‰ to –49.0‰ range and δD is in the –162‰ to -203‰ range. Ethane (2.4-4.5%) with $\delta^{13}C$ in the –27.8‰ to –30.2‰ range and propane (1.2-3.7%) with $\delta^{13}C$ in the –25.0‰ to –26.4‰ range, are significant components. Isobutane (<0.1-0.6%) with $\delta^{13}C$ in the –27.6‰ to –29.8‰ range and normal butane (0.3-0.7%) with $\delta^{13}C$ in the –22.6‰ to –25.6‰ range are also present. Isopentane (0.2-0.3%) with $\delta^{13}C$ in the –26.1‰ to –27.3‰ range, normal pentane (<0.1-0.2%), and neopentane (<0.1%) are minor components. The $\delta^{13}C$ values of CO_2 are in the +13.2‰ to -7.3‰ range.

3.2 Gas Hydrate

Analytical results of gas hydrate samples collected by research submersible from the Green Canyon (GC) 185 and GC 234 sites (Figure 1) are summarized in Table 1. Methane is the main component (71.7-85.7%). The $\delta^{13}C$ values of the methane are in the –42.2‰ to –48.4‰ range and δD is in the –115‰ to -203‰ range. Ethane (6.3-12.6%) with $\delta^{13}C$ in the –28.6‰ to –29.8‰ range and propane (6.1-12.6%) with $\delta^{13}C$ in the –24.9‰ to –26.4‰ range are significant components. Isobutane (0.9-2.6%) with $\delta^{13}C$ in the –26.8‰ to –28.4‰ range and normal butane (0.2-1.8%) with $\delta^{13}C$ in the –22.1‰ to –24.9‰ range are also present. Isopentane (0.2-0.8%) with $\delta^{13}C$ in the –24.7‰ to –26.2‰ range and normal pentane (<0.1-0.1%) are minor components. The $\delta^{13}C$ values of CO_2 are in the +17.5‰ to –27.8‰ range.

The C_1-C_5 molecular distributions of gas hydrate collected from the MC 853 site by piston cores are summarized in Table 1. Methane is the main hydrocarbon component of the gas hydrate (71.8-77.4%). The $\delta^{13}C$ values of the methane are in the –44.7‰ to –49.9‰ range and δD is in the narrow –156‰ to -166‰ range. Ethane (6.5-9.0%) with $\delta^{13}C$ in the –28.4‰ to –29.8‰ range and propane (8.4-15.5%) with $\delta^{13}C$ in the –20.5‰ to –25.8‰ range are significant components. Isobutane (3.4-8.1%) with $\delta^{13}C$ in the –27.0‰ to –27.8‰ range, and normal butane (0.3-0.7%) with $\delta^{13}C$ in the –22.9‰ to –24.6‰ range are also present. Isopentane (<0.1-0.3%) with $\delta^{13}C$ from –26.0‰ to –26.1‰, normal pentane (<0.1%), and neopentane (0.1%) are minor components. The $\delta^{13}C$ values of hydrate-bound CO_2 are in the +11.6‰ to –16.8‰ range.

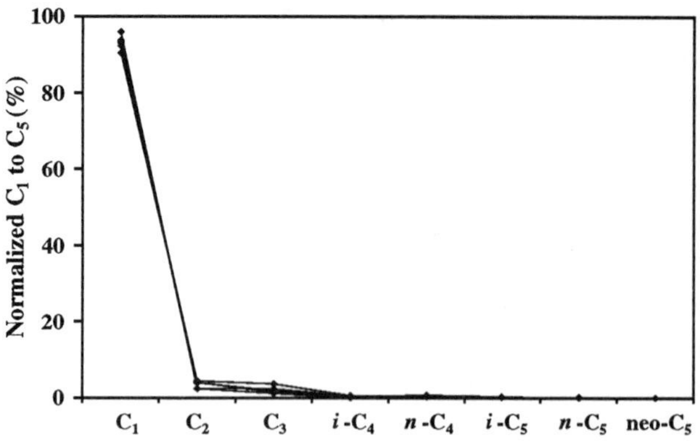

Figure 4. Molecular distributions of hydrocarbons sampled from gas vents over gas hydrates at the GC 185 and GC 234 sites. They are typical of relatively unaltered hydrocarbons from a thermally mature subsurface petroleum system, or oil-related reservoir gases.

The molecular and isotopic properties of 4 gas hydrate samples from the AT 425 site are shown in Table 1. The AT 425 site shows large variations in gas hydrate compositions. The AT 425-1 gas hydrate sample consists mainly of methane (99.0%) with $\delta^{13}C$ of –51.6‰ and δD of –173‰. Ethane is a lesser component (1.0%) with a $\delta^{13}C$ of –36.6‰. Other hydrocarbons are absent or below detection limits. The $\delta^{13}C$ of hydrate-bound CO_2 is –7.1‰.

Gas hydrate samples collected at the AT 425 site by piston cores have complex C_1-C_5 molecular distributions. Methane is the main hydrocarbon component (77.3-99.0%). The $\delta^{13}C$ values of the methane are within the –49.3‰ to –51.6‰ range and δD is in the –148‰ to -183‰ range. Ethane (1.0-5.0%) with $\delta^{13}C$ in the –36.6‰ to –38.3‰ range and propane (2.1-12.5%) with $\delta^{13}C$ in the –31.8‰ to –32.1‰ range are significant components. Isobutane (0.5-3.0%), with $\delta^{13}C$ in the –31.6‰ to –32.5‰ range and normal butane (0.6-1.3%) with $\delta^{13}C$ in the –28.3‰ to –28.6‰ range are also present. Isopentane (<0.1-0.9%) with $\delta^{13}C$ in the –28.8‰ to –29.6‰ range, normal pentane (<0.1-0.2%) with $\delta^{13}C$ values of –28.4‰ and –30.5‰, and neopentane (<0.1%) occur as minor components. The $\delta^{13}C$ values of hydrate-bound CO_2 are in the +12.9‰ to –7.1‰ range.

4. DISCUSSION

4.1 Vent Gases

Thermogenic gas vents that have been sampled thus far overlie hydrate-bearing sediments at the GC 185 and GC 234 sites on the Gulf of Mexico slope at ~540 m water depth (Table 1and Figure 4). The water depth of the sites is near the observed upper limit of the GHSZ in the Gulf of Mexico [*Milkov et al., 2000*]. Such gas hydrate outcrops appear to be episodically unstable, among other factors, because of changing seawater temperature [*MacDonald et al., 1994; Roberts et al., 1999*]. However, the thickness of sediment affected by transient changes in water temperature is only a few meters [*Milkov et al., 2000*], whereas a much larger volume hydrate is postulated to be stable buried at greater depths (~450 m) in sediment at these sites [*Milkov and Sassen., 2000*]. It is important to objectively assess the significance of gas hydrate decomposition at these sites using hydrocarbon geochemistry.

The molecular and isotopic properties of gases that vent directly to the water column from gas hydrate mounds on the sea floor can be interpreted to determine if the gases are related to gas hydrate decomposition. Globally, thermogenic vent gas from subsurface hydrocarbon systems is characterized by an abundance of methane with lesser percentages of C_2-C_5 hydrocarbons that decrease in relative abundance with increasing carbon number [*James and Burns, 1984*].

The molecular properties of structure II gas hydrate are very different. The change in phase from vent gas to solid structure II gas hydrate involves molecular fractionation that favors gases of appropriate molecular diameter [*Sloan, 1998*]. Methane decreases in abundance relative to vent gas and the C_{2+} hydrate-forming gases (especially ethane and propane) increase in abundance in structure II gas hydrate [*Sloan, 1998*]. On this basis, vent gas from decomposition of structure II gas hydrate is predicted to display a molecular distribution with lesser percentage of methane and higher percentages of higher carbon number hydrocarbons (such as ethane, propane, and butanes). If meaningful gas hydrate decomposition was occurring, the gas venting from hydrate bearing sediment would display the distinct molecular properties of structure II gas hydrate (Figure 5), and these features (anomalous abundance of ethane, propane, and butanes) are not observed in vent gas (Figure 4).

The molecular distributions of gas venting from gas hydrate mounds show generally decreasing relative abundance of C_1-C_5 hydrocarbons with increasing carbon number (Table 1 and Figure 4), as commonly observed in relatively unaltered reservoir gases [e.g., *James and Burns, 1984*]. There may be small effects on vent gas composition from stripping of hydrate-forming components during gas hydrate crystallization, resulting in higher than expected percentages of incompatible molecules such as pentanes in vent gas (Table 1). However, we can only speculate concerning hydrate stripping of the vent gas because we lack pristine samples of deep subsurface gas for purposes of comparison. The molecular distributions of vent gas appear incompatible with significant flux from gas hydrate decomposition because there is no anomalous abundance of ethane, propane, and butanes. Based on available molecular data, flux of hydrocarbons from the deep subsurface is quantitatively significant, and there is no evidence of gas hydrate decomposition. In any dynamic situation, some decomposition of gas hydrate is likely to occur, but its products are not meaningful in terms of volume within this geologic setting.

The molecular and the isotopic properties of vent gas are consistent with direct migration of relatively unaltered hydrocarbon gases from deeply buried source rocks, or with leakage from nearby subsurface oil reservoirs [*Kennicutt et al., 1988; Sassen and MacDonald, 1997; Sassen et al., 1998; Sassen et al., 1999a,b*]. This assessment of vent gas properties over shallow hydrate-bearing sediments in the Gulf of Mexico has important implications. At gas hydrate study sites in the Gulf of Mexico, rapid flux of thermogenic greenhouse gas to the water column appears to be dominated by a leaky subsurface hydrocarbon system that is charging subsurface reservoirs with oil and gas [*Wenger et al., 1994*]. Similarly, most other seep features at the Gulf of Mexico sea floor are related to rapid flux from the subsurface hydrocarbon system, and are not related to gas hydrate decomposition.

4.2 Gas Hydrate and Sediment Fabrics

Interpreting observed gas hydrate fabrics and textures helps to address whether net gas hydrate accumulation or decomposition is occurring in the Gulf of Mexico. Deformed vein-filling gas hydrates and gas hydrate nodules are observed exposed on the flanks of gas hydrate mounds at the GC 185 and GC 234 sites (~540 m) in the Gulf of Mexico [*Sassen et al., 1998; 1999a*]. The basic morphology of gas hydrate recovered from piston cores at the deep-water MC 853 and AT 425 sites consists of vein fillings and nodules at 1.8 to 4.4 m sediment depth. The vein-filling fabrics we observe both in shallow and ultra-deep water are typical of gas hydrate settings with high flux rates of hydrocarbon gases [*Ginsburg, 1998; Ginsburg and Soloviev, 1998*]. The rapid flux of hydrate-forming hydrocarbons favors high gas concentrations and gas hydrate accumulation within the GHSZ of the Gulf of Mexico.

4.3 Structure II Gas Hydrate

Molecular data indicate that structure II gas hydrate is common at the GC 184, GC 234, MC 853, and AT 425 sites (Table 1 and Figure 5). It is assumed that the C_1-C_5 structure II gas hydrate from all sites crystallized from relatively unaltered C_1-C_5 hydrocarbon gases at or near

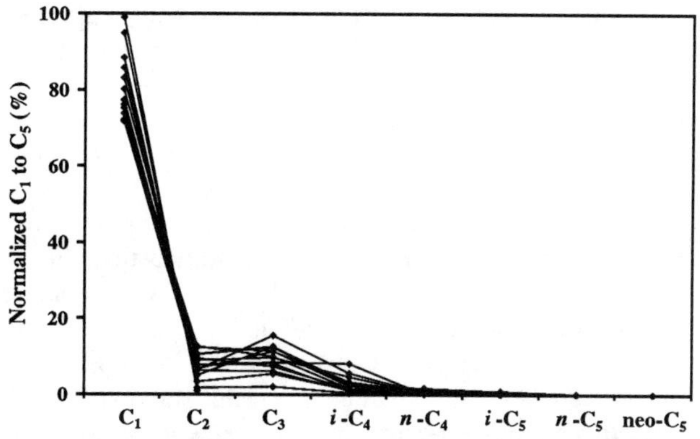

Figure 5. Molecular distributions of hydrocarbons of gas hydrates from the GC 185, GC 234, MC 853, and AT 425 sites are typical of gas hydrates crystallized from relatively unaltered vent gases.

natural gas vents exhibiting high gas flux rates. The general similarity of $\delta^{13}C$ and δD of hydrate-forming hydrocarbons in vent gases and associated gas hydrate from the GC 185 and GC 234 sites indicates that most hydrocarbon molecules are protected within the crystal lattice from bacterial oxidation [*Sassen et al., 1999a, 1999b*].

Rapid crystallization of gas hydrate affects the molecular distribution of gases in associated sediments by exclusion of incompatible molecules [*Sassen et al., 2000*]. Crystallization of structure II gas hydrate preferentially removes hydrate-forming gases (i.e. methane, ethane, propane, butanes), and enriches residual gases in incompatible hydrocarbons (i.e. pentanes) that are too large to fit within structure II hydrate cages. Isopentane is present in anomalous abundance (as much as 13.8%) in sediments near active gas hydrate mounds at GC 185 and GC 234 [*Sassen et al., 1999b*]. Anomalous isopentane is also observed adjacent to structure II gas hydrate at the AT 425 site [*Sassen et al., 2000*]. The isopentane in hydrate-bearing sediments from several Gulf of Mexico sites is interpreted as evidence that gas hydrate crystallization is recent or ongoing at the time of sampling and that gas hydrate is stable. The free isopentane in hydrate-bearing sediments furthermore contributes to explaining the observed association of structure H gas hydrate, rich in isopentane, with structure II gas hydrate in the Gulf of Mexico [*Sassen and MacDonald, 1994*].

The presence of the C_5 hydrocarbons in structure II gas hydrate is enigmatic because their molecular radii are considered too large to fit within the crystal lattice [*Sloan, 1998*]. The work of *Udachin and Ripmeester* [*1999*] implies that the samples analyzed are not simple end-member structure II gas hydrate, but could contain some crystal intergrowths with structure H symmetry capable of encaging larger molecules. At least some fraction of these C_5 hydrocarbons could also occur as free gas within crystal inclusions. In either case, the natural structure II gas hydrate displays more complex molecular distributions than anticipated.

One gas hydrate sample from AT 425 consists only of methane and ethane (Table 1). It is assumed that this gas hydrate (AT 425-1) crystallized from bacterially-altered vent gas containing only methane and ethane. The $\delta^{13}C$ of methane is depleted in ^{13}C by as much as 2.3‰ when compared to other gas hydrate from the same site (Table 1). Methanogenesis occurs in hydrate-bearing sediments [*Sassen et al., 1999b*], and could explain depletion of ^{13}C in the gas hydrate bound methane. The $\delta^{13}C$ of ethane is enriched in ^{13}C by as much as 1.7‰ relative to other gas hydrate from the AT 425 site. Enrichment of ^{13}C is a kinetic isotope effect of bacterial hydrocarbon oxidation caused by preferential utilization of ^{12}C from the reactant pool [*Barker and Fritz, 1981; Coleman et al., 1981, 1996*]. This methane-ethane gas hydrate could be structure II since even minor ethane appears to preclude crystallization of structure I gas hydrate and to favor the crystallization of structure II gas hydrate [*Subramanian et al., 2000*].

5. CONCLUSIONS

At this point in geologic time, the major source of thermogenic greenhouse gases under current climatic conditions in the Gulf of Mexico is a leaky subsurface hydrocarbon system that is also charging subsurface reservoirs of oil and gas. Gulf of Mexico gas hydrate accumulations are a significant reservoir of greenhouse gases that could be released from sediments if the gas hydrates decomposed. One basic question is assessment of the relative rates of gas hydrate accumulation and decomposition in the Gulf of Mexico. Several lines of reasoning suggest that structure II gas hydrate is forming in the Gulf of Mexico at a significantly greater rate than it is decomposing. Vent gases over hydrate-bearing sediments and gas hydrate mounds show no evidence that gas hydrate decomposition is quantitatively significant because the venting gas lacks the abundant ethane, propane, and butanes characteristics of structure II gas hydrate. All study sites (~540-1930 m water depth) are high-flux settings where hydrocarbon migration conduits intersect the sea floor deep within the GHSZ. Structure II gas hydrate is widespread in sediment and associated sediment fabrics, which is consistent with the rapid crystal growth caused by high flux rates of hydrocarbon gases. The C_{2+} hydrocarbons of gas hydrate are protected from bacterial oxidation when locked within the stable structure II gas hydrate crystal lattice [*Sassen et al., 1998,1999a,b*]. The abundance of incompatible isopentane in sediments adjacent to structure II gas hydrate provides additional evidence of recent or ongoing gas hydrate crystallization and also contributes to explaining the observed association of structure II and isopentane-rich (structure H) gas hydrate in the Gulf of Mexico [*Sassen and MacDonald, 1994*].

The molecular and isotopic properties of free hydrocarbon gases from hydrate-associated sediments show effects consistent with bacterial oxidation. This results in the conversion of hydrocarbons to CO_2 and accumulation of authigenic carbonate rock depleted in ^{13}C [*Sassen et al., 1999b*]. It is suggested that bacterial oxidation rapidly alters and depletes available free hydrocarbons in hydrate-associated sediments. Some fraction of the bacterially altered gas could crystallize as methane-ethane hydrate, but most free hydrocarbons from decomposition of buried gas hydrate are likely to be oxidized before they enter the water column [*Kvenvolden, 1999*].

The process of generation and migration of oil and gas from source rocks can last for as long as millions of years. It is concluded that gas hydrate is sometimes unstable in the Gulf of Mexico, particularly when exposed to the water column at shallow water depths near the upper limit of the GHSZ, where water temperatures are most variable [*Roberts et al., 1999*]. Nevertheless, overall geochemical results are consistent with net accumulation of gas hydrate as a consequence of recent and ongoing crystallization. Thermogenic gas hydrate crystallization is driven by prolific flow of hydrocarbon gas from an active subsurface hydrocarbon system.

The three-dimensional aspect is important. Maximum thickness of the GHSZ increases with increasing water depth in the Gulf of Mexico [*Milkov and Sassen, 2000*]. Models of stability suggest that thermogenic gas hydrate could accumulate within sediments at depths >1 km beneath the sea floor in the ultra-deep Gulf of Mexico [*Milkov and Sassen, 2000*]. Deeply buried gas hydrate is probably well insulated from the transient effects (such as temperature changes) that impact gas hydrate at or near the sea floor. Large volumes of gas hydrate in the Gulf of Mexico are probably buried as stable accumulations in sediments in ultra-deep waters where the maximum thickness of the GHSZ is greatest (Figure 6).

The molecular distributions of thermogenic gas vents suggest that the flux of hydrocarbon greenhouse gases from a leaky subsurface hydrocarbon system is far more significant than from gas hydrate decomposition in the Gulf of Mexico at the present time. Insofar as the Gulf of Mexico is a meaningful analog, it appears prudent to constrain or modify speculation on the relation between gas hydrate decomposition and global climate change.

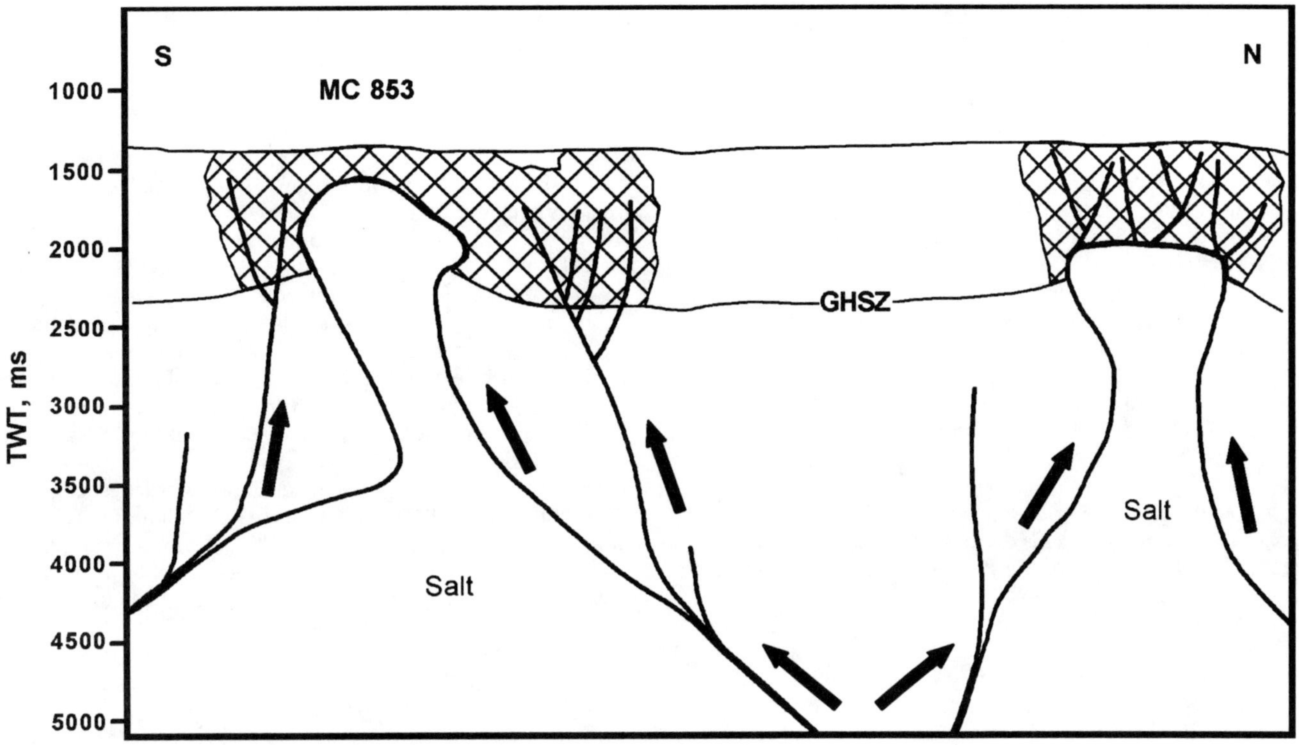

Figure 6. Sketch showing modeled subsurface distribution of gas hydrate at the MC 853 site (hatched areas). Hydrocarbons migrate vertically from deep within salt withdrawal basins along salt and fault conduits (arrows). Much gas hydrate is thought to be deeply buried and stable in sediments over salt where thermogenic hydrocarbon flux from within salt withdrawal basins is continuous for meaningful spans of time. (Modified from Milkov and Sassen, 2000).

Other factors leading to the massive discharge of gas into the water column and atmosphere, such as the coincident maturation of prolific hydrocarbon source rocks in leaky basins on global continental margins could be more important. Because gas hydrate crystallization results in the net trapping of hydrocarbons before subsurface gases can impact the sea and atmosphere, stable or accumulating gas hydrate appears likely to buffer the atmosphere by sequestering some greenhouse gases under current climatic conditions. Thus, massive accumulation of gas hydrate in leaky basins could contribute to global climate stability over meaningful spans of geologic time.

Acknowledgements. We acknowledge the support provided by ten energy companies and the Minerals Management Service that participated in Texas A&M University's Applied Gas Hydrate Research Program, 1999-2000, and support from the National Undersea Research Center at the University of North Carolina at Wilmington for submersible operations. The assistance in obtaining piston core samples of gas hydrates in ultra-deep water by the scientists, graduate students, and crew of the Research Vessel Powell is greatly appreciated.

REFERENCES

Aharon, P. Schwarcz, H.P., Roberts, H.H., 1997. Radiometric dating of hydrocarbon seeps in the Gulf of Mexico. GSA Bulletin 109, 568-579.

Bains, S., Corfield, R.M., Norris, R.D., 1999. Mechanisms of climate warming at the end of the Paleocene. Science 285, 724-727.

Barker, J.F., Fritz, P., 1981, Carbon isotopic fractionation during microbial methane oxidation. Nature 293, 289-291.

Brooks, J.M., Cox, H.B., Bryant, W.B., Kennicutt, M.C. II, Mann, R.G., McDonald, T.J., 1986. Association of gas hydrates and oil seepage in the Gulf of Mexico. Organic Geochemistry 10,221-234.

Coleman, D.D., Risatti, J.B., Schoel, M., 1981. Fractionation of carbon and hydrogen isotopes by methane oxidizing bacteria. Geochimica et Cosmochimica Acta 45, 1033-1037.

Coleman, D.D., Chao-Li, Hackley, K.C., Pelphrey, S.R., 1996. Isotopic identification of landfill methane. Environmental Geosciences 2, 95-103.

Davidson, D.W., Garg, S.K., Gough, S.R., Handa, Y.P., Ratcliffe, J.A., Ripmeester, J.A., Tse, J.S., Lawson, W.F., 1986. Laboratory analysis of a naturally occurring gas hydrate from sediment of the Gulf of Mexico. Geochimica et Cosmochimica Acta 50, 619-623.

Dickens, G.R., O'Neil, J.R., Rea, D.K., Owen, R.M., 1995. Dissociation of oceanic methane hydrate as a cause of the carbon isotope excursion at the end of the Paleocene. Paleoceanography 10, 965-971.

Dickens, G.R., Castillo, M.M., Walker, J.C.G., 1997. A blast of gas in the latest Paleocene: Simulating first-order effects of massive dissociation of oceanic methane hydrate. Geology 25, 259-262.

Egorov, A.V., Crane, K., Vogt, P.R., Rozhkov, 1999. Gas hydrate that outcrops on the sea floor: stability models. Geo-Marine Letters 19, 69-75.

Ginsburg, G.D., Soloviev, V.A., 1998. Submarine gas hydrates. VNIIOkeangeologia, St. Petersburg, p. 216.

Ginsburg, G.D., 1998. Gas hydrates in deep-water sediments. In: Henreit, J.-P., Meinert, J. (Eds.), Gas hydrates: Relevance to world margins and climate change. Geological Society, London, pp. 51-62, Special Publication 137.

James, A.T., Burns, B.J., 1984. Microbial alteration of subsurface natural gas accumulations. AAPG Bulletin 68, 957-960.

Kennet, J.P., Fackler-Adams, B.N., 2000. Relationship of clathrate instability to sediment deformation in the upper Neogene of California. Geology 28, 215-218.

Kennicutt II, M.C., Brooks, J.M., Denoux, G.J., 1988. Leakage of deep, reservoired petroleum to the near surface of the Gulf of Mexico continental slope. Marine Chemistry 24, 39-59.

Kvenvolden, K.A., 1995. A review of the geochemistry of methane in natural gas hydrate. Organic Geochemistry 23, 997-1008.

Kvenvolden, K.A., 1999. Potential effects of gas hydrate on human welfare. Proceedings National Academy of Sciences, USA 96, 3420-3426.

MacDonald, G.J., 1990. Role of methane clathrates in past and future climates. Climatic Change 16, 247-281.

MacDonald, I.R., Boland, G.S., Baker, J.S., Brooks, J.M., Kennicutt II, M.C., Bidigare, R.R., 1989. Gulf of Mexico hydrocarbon seep communities II. Spatial distribution of seep organisms and hydrocarbons at Bush Hill. Marine Biology 101, 235-247.

MacDonald, I.R., Guinasso, Jr., N.L., Sassen, R., Brooks, J.M., Lee, L., Scott, K.T., 1994. Gas hydrate that breaches the sea-floor on the continental slope of the Gulf of Mexico. Geology 22, 699-702.

Max, M.D., Dillon, W.P., Nishimura, C., Hurdle, W.P., 1999, Sea-floor methane blow-out and global firestorm at the K-T boundary. Geo-Marine Letters 18, 285-291.

Milkov, A.V., Sassen, R., 2000. Thickness of the gas hydrate stability zone, Gulf of Mexico continental slope. Marine and Petroleum Geology 17, 981-991.

Milkov, A.V., Sassen, R., Novikova, I., Mikhailov, E., 2000. Gas hydrates at minimum stability water depths in the Gulf of Mexico: Significance to geohazard assessment: Trandsactions Gulf Coast Association of Geological Societies 50, 217-224.

Paull, C.K., Ussler, W., III, Borowski, W.S., Spiess, F.N., 1995. Methane-rich plumes on the Carolina continental rise: Association with gas hydrates. Geology 23, 89-92.

Roberts, H.H., Aharon, P., 1994. Hydrocarbon-derived buildups of the northern Gulf of Mexico: A review of submersible investigations. Geo-Marine Letters 14, 135-148.

Roberts, H.H., Carney, R., 1997. Evidence of episodic fluid, gas, and sediment venting on the northern Gulf of Mexico slope. Economic Geology 92, 863-879.

Roberts, H.H., Wiseman Jr., W.J., Hooper, J., Humphrey, G.D., 1999. Surficial gas hydrates of the Louisiana continental slope – Initial results of direct observations and in situ data collection. Offshore Technology Conference, Houston, Texas. OTC 10770, 259-272.

Rowan, M.G., 1997. Three-dimensional geometry and evolution of a segmented detachment fold, Mississippi Fan Foldbelt. Journal of Structural Geology 19, 463-480.

Sager, W.W., Kennicutt II, M.C., 2000, Proposal for Ocean Drilling Program on gas hydrate in the Gulf of Mexico. Offshore Technology Conference, Houston, Texas. OTC 12111, p. 17.

Salvador, A., 1987. Late Triassic-Jurassic paleogeography and origin of Gulf of Mexico basin. AAPG Bulletin 71, 419-451.

Sassen, R., Roberts, H.H., Aharon, A., Larkin, J., Chinn, E.W., Carney, R., 1993. Chemosynthetic bacterial mats at cold hydrocarbon seeps, Gulf of Mexico continental slope. Organic Geochemistry 20, 77-89.

Sassen, R., MacDonald, I.R., 1994. Evidence of structure H hydrate, Gulf of Mexico continental slope. Organic Geochemistry 23, 1029-1032.

Sassen, R., MacDonald, I.R., 1997. Hydrocarbons of experimental and natural gas hydrates, Gulf of Mexico continental slope. Organic Geochemistry 26, 289-293.

Sassen, R., MacDonald, I.R., Guinasso, N.L., Joye, S., Requejo, A.G., Sweet, S.T., Alcala-Herrera, J., DeFreitas, D.A., Schink, D.R., 1998. Bacterial methane oxidation in sea-floor gas hydrate: Significance to life in extreme environments. Geology 26, 289-293.

Sassen, R., Sweet, S.T., Milkov, A.V., DeFreitas, D.A., Salata, G.G., McDade, E.C., 1999a. Geology and geochemistry of gas hydrates, central Gulf of Mexico continental slope. Transactions Gulf Coast Association of Geological Societies 49, 462-468.

Sassen, R., Joye, S., Sweet, S.T., DeFreitas, D.A., Milkov, A.V., MacDonald, I.R., 1999b. Thermogenic gas hydrates and hydrocarbon gases in complex chemosynthetic communities, Gulf of Mexico continental slope. Organic Geochemistry 30, 485-497.

Sassen, R., Sweet, S.T., DeFreitas, D.A., Milkov, A.V. (2000) Exclusion of isopentane during crystallization of structure II gas hydrate in sea-floor sediment, Gulf of Mexico: Organic Geochemistry 31, 1257-1262.

Sloan, E.D., 1998. Clathrate hydrates of Natural Gases, 2nd Ed., Dekker, New York., p. 705.

Subramanian, S., Kini, R.A., Dec, S.F., Sloan, E.D., 2000. Evidence of structure II hydrate formation from methane + ethane mixtures. Chemical Engineering Science 55, 11, 1981-1999.

Udachin, K.A., Ripmeester, J.R., 1999. A complex clathrate showing bimodal guest hydration. Nature 397, 420-423.

Vogt, P.R., Crane, K., Dundvor, E., Max, M.D., Pfirman, S.L., 1994. Methane generated (?) pockmarks on young, thickly sedimented oceanic crust in the Arctic: Vestnesa Ridge, Fram Strait. Geology 22, 255-258.

Weimer, P., Buffler, R.T., 1992, Structural geology and evolution of the Mississippi Fan Foldbelt, Deep Gulf of Mexico. AAPG Bulletin 76, 255-251.

Wenger, L.M., Goodoff, L.R., Gross, O.P., Harrison, S.C., Hood, K.C., 1994. Northern Gulf of Mexico: An integrated approach to source, maturation, and migration. In: Geological Aspects of Petroleum Systems. Proceedings, First Joint American Association of Petroleum Geologists/Asociasion Mexicana de Geologos Petroleos Research Conference, Mexico City, p. 6.

Worrall, D.M., Snelson, S., 1989. Evolution of the northern Gulf of Mexico with emphasis on Cennozoic growth faulting and the role of salt: In: Bally, A.W., Palmer. E.R. (Eds.) The Geology of North America – An Overview, Geological Society of America, pp. A97-A138.

Wu, S., 1993. Salt and Slope Tectonics Offshore Louisiana. Ph.D. Thesis, Rice University, Houston, Texas, p. 251.

Roger Sassen, Debra A. DeFreitas, Mahlon C. Kennicutt, Alexei V. Milkov, and Stephen T. Sweet, Geochemical and Environmental Research Group, Texas A&M University, 833 Graham Road, College Station, TX 77845 USA

Harry Roberts, Coastal Studies Institute, Department of Oceanography and Coastal Sciences, Louisiana State University, Baton Rouge, LA 70803

Fluid and Gas Expulsion on the Northern Gulf of Mexico Continental Slope: Mud-Prone to Mineral-Prone Responses

Harry H. Roberts

Louisiana State University

In the northern Gulf of Mexico slope province, complex structural relationships resulting from dynamic adjustments between large volumes of sediments and salt provide numerous faulted pathways for deep subsurface fluids and gases to be transported to the modern seafloor. Geological response at the seafloor to these hydrocarbon-rich fluids and gases is highly variable and dependent largely on rate and duration of delivery as well as fluid and gas composition. In a qualitative framework, rapid expulsions of fluids (including fluidized sediment) and gases generally result in buildups of sediment in the form of cones (mud volcanoes) that vary from a few meters to several kilometers in diameter and/or sheet-like flows that may extend tens of kilometers downslope. Conversely, slow seepage promotes lithification of the seafloor through precipitation of a variety of mineral species. Most important is the microbial utilization of hydrocarbons and precipitation of ^{13}C-depleted Ca-Mg carbonates as by-products. These carbonates have $\delta\,^{13}C$ values that range between –18% to –55% (PDB), suggesting mixed carbon sources from crude oil to biogenic methane. The ^{13}C-depleted carbonates form mounds and hardgrounds that occur over the full depth range of the slope. Mounded carbonates can have relief of up to 30 m, but mounds of 5-10 m relief are most common at sites thus far investigated. Mound-building carbonates are mixed mineral phases of aragonite, Mg-calcite, and dolomite with Mg-calcite being the most common. Barite is another product that is precipitated from mineral-rich fluids that arrive at the seafloor in low-to-moderate seep rate settings. However, barite precipitation is not as pervasive as that of ^{13}C-depleted carbonates. The Gulf's intermediate flux settings seem best exemplified by areas where gas hydrates occur at the seafloor or in the very shallow subsurface. Intermediate flux environments display considerable variability with regard to surficial geology and on a local scale have elements of both rapid and slow flux settings. However, the intermediate flux environments appear to have the unique set of conditions necessary to support and sustain densely populated communities of chemosynthetic organisms. Since most of these areas are associated with faulting at the edges of intraslope basins,

Natural Gas Hydrates: Occurrence, Distribution, and Detection
Geophysical Monograph 124

surficial or shallow subsurface gas hydrates (accessible by piston coring) are oriented along these faults and not in broad areas characterized by distinct bottom simulating reflectors (BSRs) as is the case in many simpler geologic settings. These shallow gas hydrates are composed of a complex mixture of biogenic-thermogenic methane and other thermogenic gases. Slight variations in near-bottom water temperature resulting from a variety of natural oceanographic processes cause gas hydrate dissociation and out-gassing resulting in the degradation to disappearnace of surficial gas hydrate mounds.

1. INTRODUCTION

Present state of knowledge of both surface and subsurface geology of the Gulf of Mexico continental slope has been directly linked to the search for and the production of hydrocarbons. Oil and gas activity in the northern and northwestern Gulf of Mexico has steadily been moving into deeper and deeper water since the mid-1980s. The northern Gulf continental slope is the most mature deep-water oil and gas province in the world. An enormous database on the continental slope has been and continues to be collected in support of exploration and production activities. Much of these data are scaled for evaluation of the modern seafloor and the shallow subsurface beneath it. These high-resolution data sets are acquired for engineering and regulatory purposes related to choosing drilling sites, pipeline routes, platform locations, and for identifying potential geohazards. Even though they are exceptionally revealing regarding the seafloor and probable processes impacting it, they typically cover small areas. These high resolution data sets are site specific and usually separated by considerable distances. Therefore, the wide variety of bottom features and their links to formative processes have been difficult to determine. However, two relatively recent data types now cover nearly the entire slope: multibeam bathymetry (Plate 1) and 3D-seismic. These two data types have emphasized the complex surface geology of the Gulf's northern slope and put this complexity into a regionally consistent framework. Mapping surface amplitude anomalies from 3D-seismic data [*Roberts et al.*, 1992a; *Roberts,* 1996; *Hill,* 1996] has helped identify fluid and gas expulsion features. Coupled with polarity analysis of the surface reflector, these methodologies can identify hard bottom from soft bottom areas as well as areas charged with gas. Calibrated to "ground truth" data provided by piston cores and direct observation and sampling using manned submersibles, it is now clear that fluid and gas expulsion is an important process affecting both seafloor geology and biology across the entire depth range of the northern and northwestern Gulf of Mexico continental slope. Although most calibration data have been acquired from the upper slope (< 1000 m water depth), 3D-seismic amplitude data, piston cores, and a few ROV and manned submersible observations confirm the importance of fluid and gas expulsion processes to the base of the slope at water depths of about 2500 m.

The computer enhanced multibeam bathymetry image of Plate 1 illustrates the numerous ridges, domes, and intraslope basins that configure the surface of northern Gulf of Mexico continental slope. On the flanks of intraslope basins, complex faulting and associated salt masses focus the flux of subsurface fluids and gases to the modern seafloor resulting in a variety of response features generally superimposed on dome-like regional features supported by underlying salt. Such complex areas are regionally distinct on the multibeam bathymetry image of Plate 1 from smooth bottom areas representative of the intraslope basins. The small-scale geologic features that mantel salt-supported domes and other salt-associated areas of the seafloor are partly the products of fluid and gas expulsion. Most of the examples discussed in this paper are derived from the upper and middle slope area of the northern Gulf immediately down dip of the Plio-Pleistocene deltaic depocenters [*Woodbury et al.,* 1973]. This region is within the tabular salt and minibasin province of *Diegel et al.* [1995] where the geology is dominated by the formation and filling of intraslope basins. These intraslope basins formed within large salt masses that were emplaced during the Oligocene and Miocene [*Diegel et al.,* 1995]. Later, Cenozic deposition of large volumes of siliciclastic sediments mobilized allochthonous salt masses out of the shelf minibasin and detachment provinces into the tabular salt and minibasin province [*Prather et al.,* 1998]. Subsequent loading of the salt has formed the complex array of basins and basin-flanking features that shape today's slope surface (Plate 1). Within this complex geological setting, venting-to-seepage of fluidized sediment, hydrocarbons, formation fluids, and gas create a variety of features that play an important role in the surface geology of the slope. The terms venting and seepage are used in this paper as qualitative expressions of rates at which crude oil, formation fluids, fluidized sediment, and gases are delivered to the modern slope surface. Venting implies rapid delivery while seepage suggests slow delivery. This paper organizes data on surficial geology, and to a lesser extent biology, related to the flux of fluids and gases to the

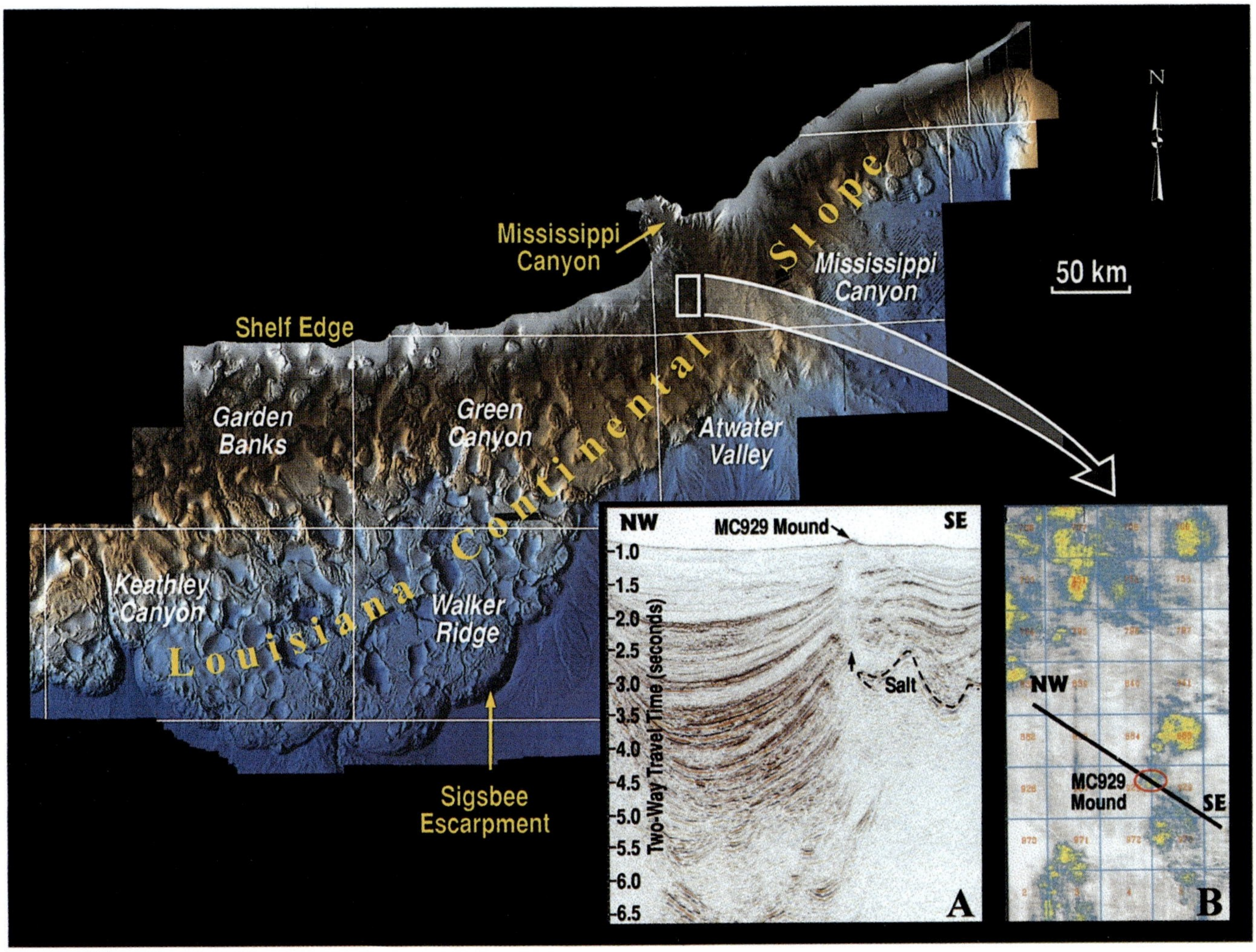

Plate 1. A computer enhanced multibeam bathymetry image of the continental slope opposite Louisiana (northern Gulf of Mexico) showing the complex of intraslope basins and salt-supported domes and ridges (compliments of F. Diegel, Shell Offshore). Insets show a 3-D seismic profile across an intraslope basin illustrating an expulsion feature (Mississippi Canyon 929; 28°01.49'N; 89°43.65'W) with an acoustic zone beneath it and a surface 3D-seismic surface amplitude map which identifies an abundance of high amplitude sites typical of the slope setting. These sites equate to locations characterized by fluid and gas expulsion (each lease block is 4.8 km x 4.8 km). The larger lease areas (e.g. Green Canyon) are shown on the computer enhanced image of the continental slope.

modern surface of the slope. Even though quantification of flux rate has been an illusive target, the present database qualitatively suggests a strong relationship between flux rate and seafloor response. The objective of this paper is to explore a qualitative framework for understanding the links between flux-related processes and geologic-biologic responses at the slope surface. Special emphasis is placed on gas hydrates and authigenic carbonate mounds. The remainder of this paper focuses on a discussion of features that form under rapid, intermediate, and slow flux of fluids and gases to the modern seafloor and the characteristics of each major feature category.

2. FLUID AND GAS EXPULSION FEATURE SPECTRUM

In the Gulf of Mexico, direct observations of fluid and gas expulsion features on the seafloor date back to manned submersible research of the mid-to-late 1980s [*Brooks et al.,* 1985; *Kennicutt et al.,* 1985]. These data sets, coupled with high resolution seismic and side-scan sonar surveys, have provided a new vision of the surface geology and biology of the slope. A direct linkage between fluid and gas expulsion and the occurrence of benthic communities that live on the expulsion products is now clear. Although a quantitative relationship between fluid–gas expulsion and geologic-biologic response has yet to be derived, it is clear that qualitatively geologic-biologic response is not only dependent on fluid and gas type but largely on flux rate and frequency of delivery events [*Roberts and Carney,* 1997]. Data on the frequency, composition, and impacts of expulsion events are nonexistent for the slope. Onshore, however, *Ulrich et al.* [1984] describe sulfide layers in the anhydrite caprock of the Winnfield salt dome resulting from the episodic release of reducing waters from a deep overpressure zone. At the salt-anhydrite interface sulfides are precipitated as faults deliver the reducing waters from the deep subsurface. Paleomagnetic dating suggests that these expulsion events occurred at a frequency of about every 300 years [*Kyle et al.,* 1987]. Modeling results suggest that expulsion events should be expected on much shorter time scales, more consistent with actual observational data [*Roberts et al.,* 1996]. Repeated observations of expulsion sites on the upper continental slope of the northern Gulf support the short-term episodic nature of fluid-gas venting and seepage at a scale still not predicted by numerical models.

Seafloor features that are the consequence of the fluid-gas expulsion process convey information about the types of fluid and gases being delivered, the episodic nature of the delivery process, the duration of expulsion events, and the delivery rate. On both high resolution and lower frequency exploration-scale seismic profiles most expulsion features are accompanied by an acoustically opaque zone (acoustic wipe-out zone) that extends from the seafloor into the subsurface (Plate 1). The general interpretation for this response is the presence of a free gas phase (bubbles) in surface and near-surface sediments. On seismic records, gas-charged sediments are identified by high amplitude reflecting horizons and/or acoustic wipe-out zones below [*Bryant,* 1981]. Although an acoustically opaque response on a seismic record can result from a number of surface and subsurface conditions [*Roberts et al.,* 1999]; expulsion features are associated with free gas as well as fluids. In most cases, these features are clearly linked to faulting as determined from seismic records.

On the northern Gulf of Mexico continental slope faults range from major growth faults that cut thousands of meters of sedimentary section to much smaller compensating faults related primarily to salt movement in the shallow subsurface [Rowan, 1995; Rowan et al., 1999]. In addition to off-setting the seafloor and creating local topography with oversteepened slopes, faults are responsible for numerous constructional seafloor features related to the vertical flux of fluids (including fluidized sediment) and gases. *Roberts* and *Carney* [1997] qualitatively identified a spectrum of seafloor responses to relative changes in the rate of vertical flux and expulsion of fluids (hydrocarbons, formation water, and sediment) and gases on the seafloor (Plate 2). The rapid flux end of the feature spectrum is sediment-prone. Large mud volcanoes and extruded fluidized sediment sheets [*Neurauter* and *Bryant,* 1990; *Neurauter* and *Roberts,* 1992; *Neurauter* and *Roberts,* 1994] are formed by mostly fine-grained sediment forced up faults. At the other end of the spectrum, vertical flux of gases and fluids may be very slow. This end of the spectrum is mineral-prone. Microbial degradation of both hydrocarbon gases and crude oil associated with this process can produce by-products such as calcium-magnesium carbonates that form a variety of seafloor features including hardgrounds and mound-like structures of various dimensions [*Roberts et al.,* 1992 a, b]. This process has been described from other settings, such as accretionary prisms, where salt tectonics is not a factor and biogenic methane is the primary hydrocarbon source [*Ritger et al.,* 1987; *von Rad* et al., 1996; and *Sakai, et al.,* 1992; *Bohrmann et al.*, 1998]. *Paull et al.* [1992] describe this process from the base of the western Florida Escarpment. In addition, mineral-rich fluids may precipitate products such as barite, forming crusts and low-relief cones and chimneys [*Roberts and Aharon,* 1994; Fu et al., 1994; *Torres et al.,* 1996]. Intermediate flux rates are frequently associated with the occurrence of exposed or shallow subsurface gas hydrates. It appears that chemotrophic communities are most common to these intermediate flux environments.

3.1. *Rapid Flux (Mud Prone) Case*

Hedberg [1974] recognizes that the vertical migration of gas and fluids through unconsolidated-to-semiconsolidated sediments may create a slurry-like mixture of sediment, gas, and water that may be extruded on the seafloor to form a variety constructional features. Mud volcanoes and extruded mud sheets have been observed from many different marine geologic settings [*Hovland* and *Judd*, 1988], ranging from degassing and dewatering of accretionary prisms during the process of subduction [*Barber et al.*, 1986; *Zhao et al.*, 1986 *Brown,* 1990; and *Orange and Breen,* 1992] to areas of active salt deformation like the Gulf [*Neurauter and Bryant,* 1990; *Neurauter and Roberts,* 1992; and *Kohl* and *Roberts,* 1994] to breached anticlines in the Caspian Sea [*Bagirov et al.*, 1996; *Corthay,* 1998; *Corthay and Aliyev,* 2000]. These examples share sedimentological and geomorphic evidence supporting episodic, rapid, and sometimes violent expulsion events resulting in the local buildup of sediments. Since data on the frequency of expulsion events associated with submarine mud volcanoes and other mud extrusion features is essentially nonexistent, the rapid flux interpretation of these features is made on geomorphic evidence and over a decade of research involving direct observations from manned submersibles.

On the northern Gulf of Mexico continental slope there are numerous examples of expulsion features that fit the rapid flux part of the feature spectrum illustrated in Plate 2. Most of these features develop into conical mounds of various dimensions that are collectively referred to as mud volcanoes. As the 3D-seismic profile of Figure 1 clearly shows, mud volcanoes are accompanied by a vertical pathway of acoustic attenuation (acoustic wipe-out zone) that is commonly interpreted as an indicator of free gas as well as sediment remolding in a subsurface feeder system. This interpretation is easy to accept since abundant gas can frequently be observed escaping from the central vent of active mud-extrusion features. However, the acoustically opaque nature of the central transport pathway is maintained in systems that are now dormant. Engineering borings into a dormant extrusion feature in the Garden Banks 161 (27°49.83'N; 92°30.24'W) suggest that completely remolded sediment with no distinct acoustic impedance horizons are typical of these systems and therefore they would appear acoustically opaque on seismic profiles even when they become inactive and do not contain free gas [*Roberts et al.*, 1999a].

In the rapid flux setting, exemplified by mud volcano formation, considerable heat as well as fluidized sediment and hydrocarbons is transported to the modern seafloor. Temperature measurements made by the authors in the central vents of mud volcanoes on the northern Gulf slope

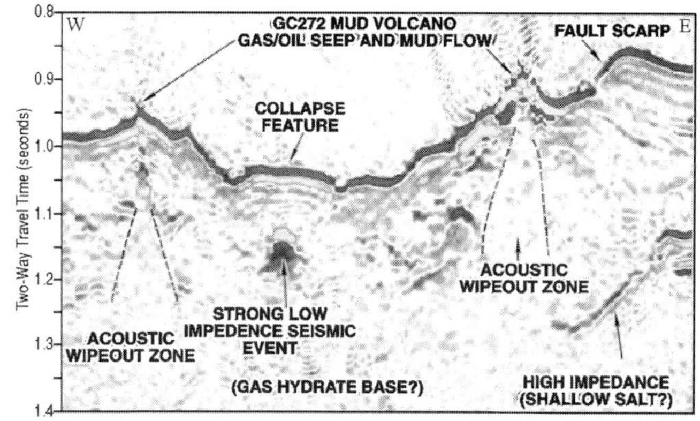

Figure 1. An east-west 3D-seismic profile that cuts through two large mud volcanoes in Green Canyon block 272 (27°41.17'N; 91°32.20'W) on the Louisiana continental slope. Note the vertically oriented acoustic wipe-out zones beneath the mud volcanoes that are the fluid and gas transport routes to the modern seafloor. Also, note the high amplitude event between the two mud volcanoes which may represent free gas below the base of localized gas hydrate in the shallow sediment column.

indicate that fluidized sediment temperature can be considerably above ambient water temperatures (over 20°C, unpublished data acquired by the author). *McDonald et al.* [2000] have just published measurements from a mud volcano from the Garden Banks area (Plate 1) that reached an absolute temperature of 48°C on a part of the slope with ambient water temperatures that range between 4-7°C. The process of rapid vertical flux of fluids and gases suggests that there is little time for the hydrocarbon fraction of the carrier fluids to be microbially oxidized. *Sassen et al.* [1994] note that oil-stained sediments from an active mud volcano in Green Canyon 143 (27°50.14'N, 91°21.93'W) have characteristics (an envelope of n-alkanes and isoprenoids) suggesting that the oil has not been microbially degraded. They conclude that bacterial oxidation is unable to keep pace with high rates of hydrocarbon venting typical of the rapid flux and mud prone setting.

The strong vertical flux of fluidized sediments associated with mud-venting at the seafloor initiates the mixing of sediments of various ages. Figure 2 schematically illustrates processes associated with lateral transport of sediment ejected into the water column and downslope through mud flows. Ejected sediment carries microfossils and other particles such as shale clasts and exotic grains [*Roberts et al.*, 1999a]. *Kohl and Roberts* [1994] found that microfossils as old as Miocene were being extruded onto the seafloor at a former rapid venting site in the Garden Bank 338 (27°37.80'N; 92°28.12'W). At this location Miocene sediments are buried thousands of

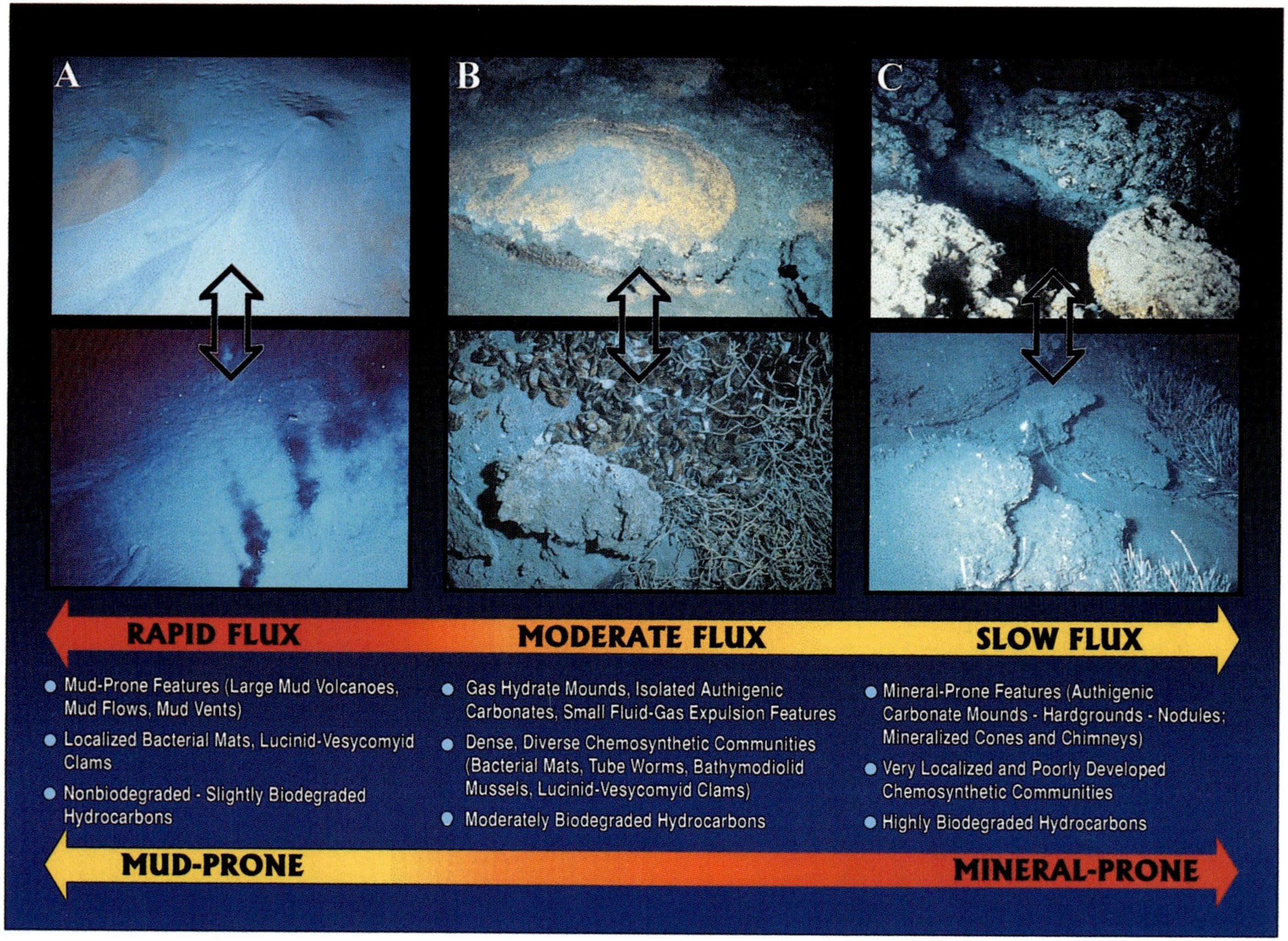

Plate 2. This figure summarizes the qualitative relationships between flux of fluids and gases to the modern seafloor and general geologic-biologic response [modified from *Roberts* and *Carney,* 1997]. Pictures illustrating the (a) rapid flux (mud-prone) end of spectrum are mud volcanoes and a gas-emitting mud vent that has not built a cone, but is extruding fluidized sediment sheets onto the surrounding seafloor. The moderate flux part of the spectrum (b) is characterized by surface exposures of gas hydrate and well-developed chemosynthetic communities as well as local areas of lithified seafloor. At the slow flux (mineral prone) end of the spectrum (c) chemosynthetic communities are sparse or nonexistant and the dominant forms are mounds and other forms of lithified seafloor. Each picture has a field of view 3-4 m across.

meters below the ocean floor, indicating a long transport pathway.

Fined-grained expulsions, both as suspended sediment and as flows, create environments that are hostile for benthic fauna. For filter feeders, high suspended sediment loads overwhelm respiratory systems and flows can bury benthic organisms or make life positioning difficult or impossible. Extruded sediment may derive from sources containing little sulfide. High sediment delivery rates preclude sulfide-producing anaerobic oxidation of hydrocarbons. Mat-forming bacteria *Beggiatoa* may colonize the surface of hydrocarbon-laced fluidized sediments surrounding the venting area or on recently deposited flows. Once the fluidized mud has been deposited, lucinid-vesycomyid clams can colonize as biodegradation of incorporated hydrocarbons produces a sufficient sulfide source. Since expulsion events are episodic, any one flow deposit has a limited hydrocarbon charge and sulfide-producing potential. For this reason infaunal lucinid clam communities establish themselves and then quickly deplete the available sulfide source. Large surficial accumulations of lucinid-vesycomyid clam shells are frequently scattered over the surfaces of extruded muds (Figure 3). Multiple lucinid-vesycomyid shell layers in piston cores through sediments near active expulsion features confirm the episodic nature of mud extrusion events.

As rapid flux systems such as mud volcanoes become progressively less active, these systems start to acquire surface features more characteristic of moderate-to-slow flux cases. For example, dormant or marginally active mud volcanoes may contain gas hydrates, authigenic carbonates and localized chemosynthetic communities [*Ginsburg et al.*, 1999; *Vogt et al.*, 1999].

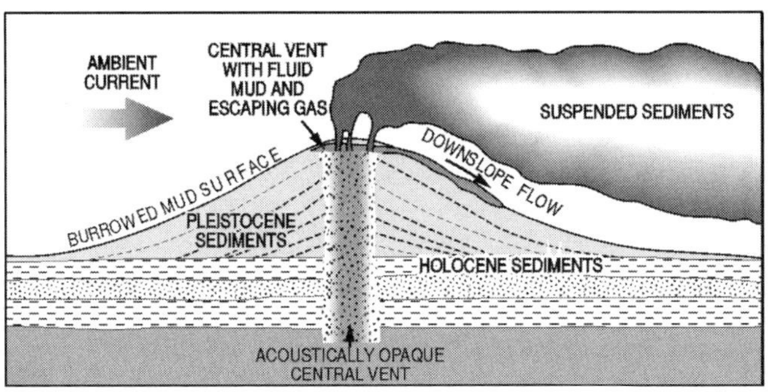

Figure 2. A schematic representation of a submarine mud volcano illustrating the extrusion of sediment and down-slope flows. These processes mix old microfossils from deep within the sedimentary column with modern surface deposits of the continental slope [modified from *Kohl* and *Roberts*, 1994].

Figure 3. Lucinid-vesycomyid clam shells on a sulfide-depleted mudflow surface. Infaunal clams established themselves in mudflow deposits that were quickly depleted of the sulfide reserves needed to sustain a community. This picture was taken on a small mudflow at the apex of a diapiric mound in the Garden Bank 427 (27°33.15'N; 92°25.25'W; WD=620 m).

3.2. Moderate Flux Case

A high degree of geologic and biologic variability is associated with moderate flux sites (Plate 2). In water depths below ~500 m, sites on the upper continental slope of the northern Gulf that fall within this category commonly display gas hydrate exposures at the seafloor. These sites are characterized by dense and comparatively diverse chemosynthetic communities consisting of bacterial mats, lucinid-vesycomyid clams, pogonophoran tube worms, vestimentiferan tube worms, and bathymodiolid mussels. Only infrequent occurrences of chemosynthetic organisms have been reported shallower than the gas hydrate "water depth window" which is dictated by termperature, pressure, and mixture of hydrocarbon gases [*Sloan,* 1990]. This association between gas hydrates and chemosynthetic communities is probably not fortuitous since gas hydrates store the trophic resources necessary to sustain long-term and diverse chemosynthetic communities. Such communities require sustainable sources of hydrocarbon gases and sulfide. As *Carney* [1994] suggests, "solid-phase gas hydrates of methane and water, which are stable at the pressure and temperature conditions below 400-500 m of water depth in the Gulf of Mexico, may be that unique aspect of deep ocean methane and sulfide chemistry that allows for chemosynthetic community development."

Because fluids and gases are vertically focused by salt masses and faults, gas hydrates (particularly those that occur at or near the seafloor) are found around the perimeters of intraslope basins that contain thick

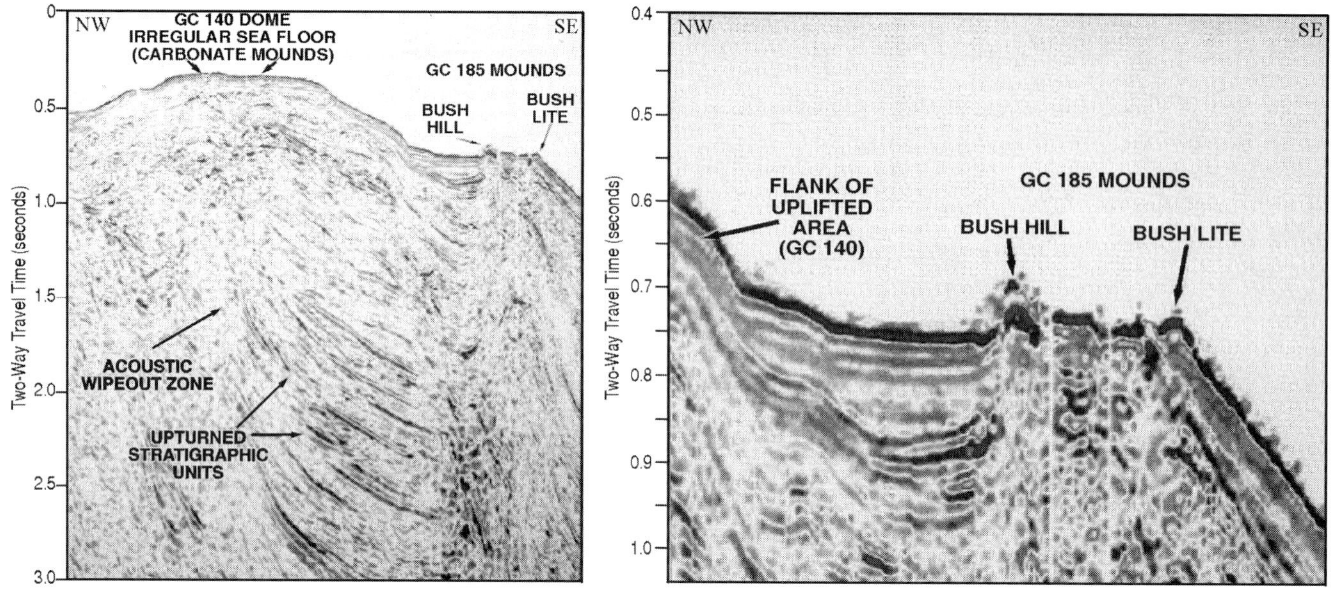

Figure 4. These NW-SE-oriented 3D-seismic profiles from the Green Canyon lease area (Figure 1) illustrate two very different fluid-gas expulsion settings. (a) The Green Canyon 140 dome capped with mounded authigenic carbonates, which represents a slow seepage area. (b) The adjacent mounds, Bush Hill and Bush Lite are sites of moderate expulsion where gas hydrates and diverse-dense chemosynthetic communities are found. Note the well-defined and vertically oriented acoustic wipe-out zones that constitute the fluid-gas migration pathways to the modern seafloor.

sedimentary sections. The many faults on the Gulf slope introduce boundary instabilities causing methane and other hydrocarbon gases to be vertically transported to the shallow subsurface where temperature and pressure conditions favor gas hydrate formation. The Gulf is apparently the first marine setting where gas hydrates were directly observed on the ocean floor. The top picture under moderate flux (Plate 2) represents this first observed outcropping of gas hydrate. This discovery was made in 1991 in the Green Canyon 232 (27°44.49'N; 91°19.07'N). Subsequent dives have determined that surface exposures of gas hydrate are reasonably common to the upper slope and have now been observed by many researchers working on the Gulf of Mexico slope [*MacDonald et al.*, 1994].

Once at or near the seafloor, methane contacts sea water sulfate and microbial communities, and sulfide is generated by methane oxidation-sulfide reduction. *Sassen et al.* [1993] suggest that there is a strong coupling between bacterial mats (*Beggiatoa*) common to the seafloor over gas hydrates and hydrocarbon concentrations in the sediments. The mats act as semipermeable membranes that trap vertically migrating hydrocarbons beneath. Anaerobic microbial oxidation of hydrocarbons depletes oxygen, which encourages sulfate reduction to produce hydrogen sulfide in sediment horizons below needed for *Beggiatoa* growth. So, hydrocarbons associated with intermediate flux environments demonstrate the distinct characteristics associated with biodegradation. *Sassen et al.* [1993] also reemphasizes the point made by other researchers that bicarbonate from anaerobic microbial utilization of hydrocarbons is probably responsible for initiating the precipitation of authigenic carbonates as nodular masses, slabs, hardgrounds, and more massive mound-like buildups.

One of the most thoroughly studied areas in the Gulf where gas hydrates may be directly observed and sampled is a mound-like feature ("Bush Hill") in Green Canyon 185 (27°46.97'N; 91°30.47'W), Figure 4. At this locality, seismic profiles indicate that vertically oriented acoustically opaque corridors connect with the deep subsurface. *Cook and D'Onfro* [1991] indicate that these corridors follow an antithetic fault that connects to a major fault system penetrating deep Pliocene sediments. This conduit to the deeply buried petroleum system provides thermogenic gas and crude oil to the Green Canyon 185 mound. Gas analyses of gas hydrates prove the incorporation of thermogenic hydrocarbons by substantial amounts of hydrocarbons up to n-Butane [*Sassen, et al.*, 1999a; *Sassen, et al.*, 1998]. On 3D-seismic surface amplitude data the mounds shown in Figure 4 stand out as high amplitude features of negative polarity, suggesting free gas in near-surface sediments [*Roberts et al.*, 1992a].

At the Green Canyon 185 site, gas hydrate was first discovered as nodular masses in piston cores (Figure 5). Later, localized exposures were found at the top of the mound (Figure 6). Within a few tens of meters, onc can move from areas of outcropping gas hydrate covered with

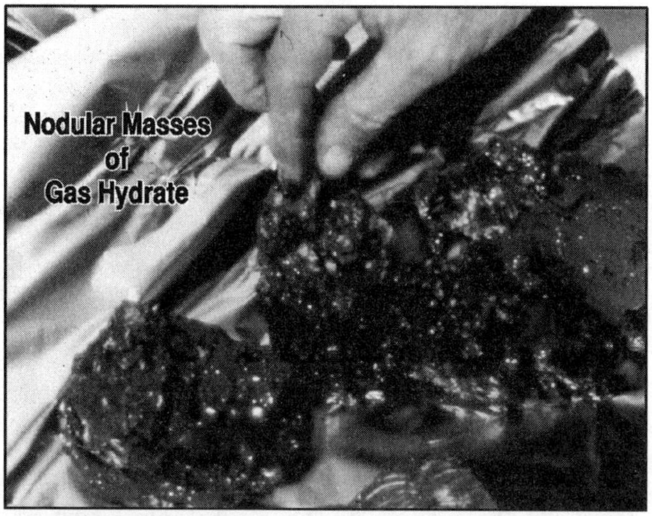

Figure 5. Piston cores taken at the Green Canyon 185 mound complex site frequently contain nodular masses of gas hydrate as illustrated in this figure.

orange *Beggiatoa* mats to complex and densely arranged chemosynthetic communities of mussels and tube worms mixed with small outcrops of authigenic carbonate (Plate 2b). Known gas hydrate features such as the Green Canyon 185 mound complex frequently also have small, localized areas for fluidized mud extrusion accompanied by areas of lucinid-vesycomyid clam concentrations.

Gas hydrate features may occur as small mounds as in the Green Canyon 185 case, large mounds as in Mississippi Canyon 709 (28°13.58'N; 89°42.12'W), elongate exposures associated with fault scarps as in Green Canyon 234 (27° 44.8'N; 91° 13.3'W), oblate in plain-view, but relatively flat as in Green Canyon 240 (27°42.25'N; 91°57.54'W), and perhaps other geomorphic types yet to be discovered. It is apparent in all of these cases that gas is being vented up faults through the gas hydrate complex and into the overlying water column (Figure 7). Minor faults and fractures are the probable conduits for free gas through the hydrate stability zone. It is also apparent through site specific experiments (Green Canyon 185), that surficial gas hydrates decompose and out-gas as a product of water column temperature changes [*Roberts et al.,* 1999b) (Figure 8). Repeated observations of the same sites over nearly a decade of submersible research suggests that some sites go through cycles of decomposition and formation at an interannual frequency. Over a 1-year period of observation, the near-bottom water column temperature data indicate significant variations at the diurnal and 4-20 day frequencies [*Roberts et al.,* 1999b]. The dominant subtidal temperature signals have time scales of 1-2 weeks and are thought to represent mesoscale eddies that have been observed over the upper slope [*Berger et al.,* in press]. Out-gassing from an exposure of gas hydrate is strongly coherent with the water temperature signal [*Roberts et al.,* 1999b] (Figure 8). Even though temperature fluctuations are relatively small during the period when out-gassing and water temperature were simultaneously measured (maximum variation ~ 1.5°C), much more energetic out-gassing is expected during the passage of Loop Current eddies that can elevate water temperatures near the bottom on the upper slope by 4-6°C in water depth to ~1000 m [*Hamilton,* 1990].

3.3. Slow Flux (Mineral Prone) Case

When the rates of delivery of hydrocarbons and their carrier fluids to the seafloor are intermediate-to-low, either through a small primary supply or a systematic decrease

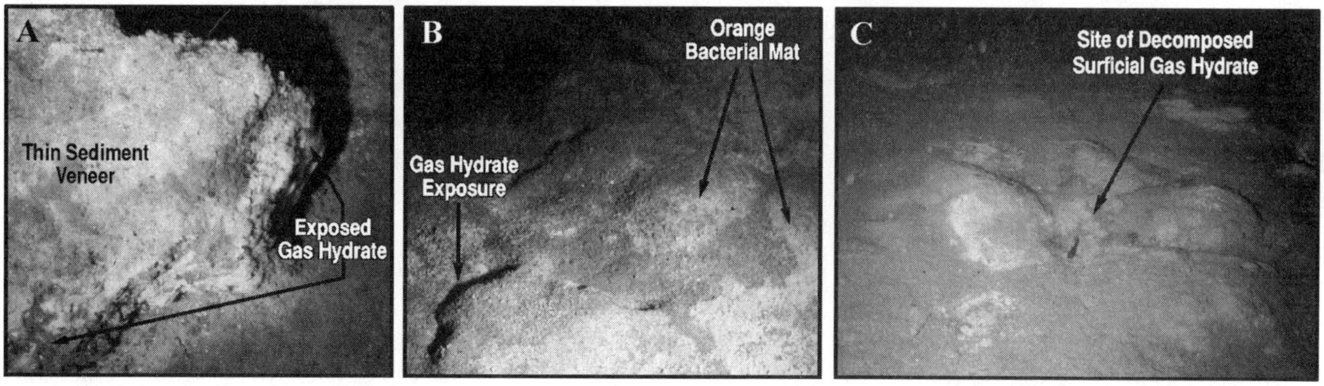

Figure 6. This figure illustrates three submersible-acquired pictures of two gas hydrate exposures at the seafloor and one former exposure site where the gas hydrate mound has degassed and disappeared. The locations and water depths of these exposures are: (a) Green Canyon 185 (27°46.95'N; 91°30.47'W; WD=541 m), (b) Green Canyon 232 (27°44.51'N; 91°19.08'W; WD=569 m), and (c) Green Canyon 232 (27°44.56'N; 91°19.24'W; WD=573 m).

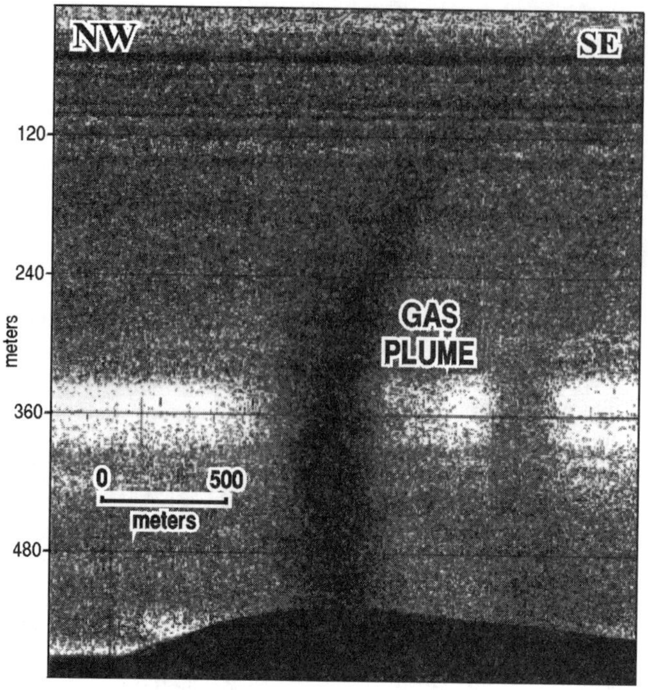

Figure 7. This figure illustrates a gas plume above the Green Canyon 185 (Bush Hill) site of exposed gas hydrate, well-developed chemosynthetic communities, and localized authigenic carbonates. Even though only a few small bubble streams were observed (by manned submersible) at this site directly after this profile was acquired, it is obvious that gas is entering the water column over the entire apex of this feature. The profile was acquired with a 38kHz echo sounder.

from higher rates, significant mineralization within surface and near-surface sediments takes place. Gulf of Mexico slope sites that demonstrate slow vertical fluid and gas flux display abundant forms of seafloor lithification, but they have few macro-scale chemosynthetic organisms. Instead, they are inhabited by microbial communities of sulfate-reducing bacteria, methanogenic bacteria, sulfide- and methane-oxidizing bacteria, and bacteria that oxidize thermogenic hydrocarbon gas and crude oil [*Jannasch,* 1984; *Sassen et al.,* 1993]. The microbial communities are critical to carbon cycling in intermediate-to-slow seepage environments and to producing chemical gradients that result in the precipitation of calcium-magnesium carbonates [*Ritger et al.,* 1987; *Roberts et al.,* 1992a; *Paull et al.,* 1992], and mineral phases such as barite [*Roberts* and *Aharon,* 1994; *Fu et al.,* 1994; *Torres et al.,* 1996].

Parts of the slope surface are characterized by oxygen-poor sediments where a zone of sulfate reduction overlies a shallow subsurface where bacterially-induced fermentation of organic matter produces biogenic methane [*Fang et al.,* 1991]. Biogenic methane is extremely ^{13}C-depleted [*Schoell,* 1982] while thermogenic methane is comparatively enriched in ^{13}C. *Rice* [1980] demonstrates that this enrichment increases with depth and age of producing reservoirs in the Gulf. Values for $\delta\ ^{13}C$ near –80 ‰ are possible for biogenic methane near the seafloor while $\delta\ ^{13}C$ values of around –35 ‰ are expected for deep thermogenic methane [*Sassen et al.,* 1993; *Sassen et al.,* 1999b].

Sulfate-reducing microbial communities in near-surface reduced environments or environments with free oxygen in aerated settings oxidize seep-delivered hydrocarbons and enrich pore waters with CO_2 [*Jorgensen,* 1976]. Under seepage conditions, carbonate precipitation will proceed when the saturation of carbonate minerals is exceeded [*Berner,* 1971]. Authigenic carbonates produced by these bacterially-mediated processes inherit their $\delta\ ^{13}C$ signatures from a carbon pool comprised of carbon that is converted to dissolved inorganic carbon (DIC) from crude oil to biogenic methane as well as DIC from sea water. Therefore, it is unlikely that the carbon in Ca-Mg carbonates is derived from an individual source. However, authigenic carbonates from areas seeping primarily crude oil have $\delta\ ^{13}C$ values that are close to crude oil value (-24 ‰ to –34 ‰ PDB), while similar carbonates from methane seeps are more ^{13}C-depleted. Therefore, the $\delta\ ^{13}C$ values associated with hydrocarbon seep-vent carbonates appear to be a window into the parent carbon sources and the formative microbial processes that initiated precipitation sources [*Aharon et al.,* 1990]. Carbonates derived from sites of fermentation are much closer to $\delta\ ^{13}C$

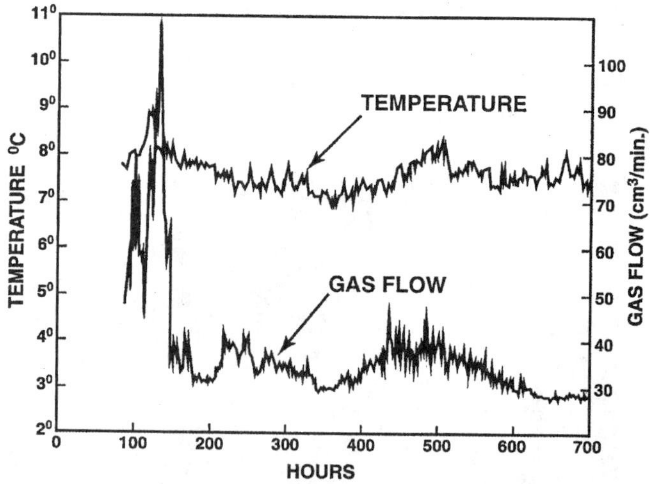

Figure 8. A time-series of near-bottom temperature and gas emitted from decomposing gas hydrate exposed at the seafloor. Gas emission measurements were made with a collector cone containing a Savonius-type rotor in the core throat. Rotor rotations were calibrated to flow before and after deployment of instrumentation. Note the strong coherence between the two data sets.

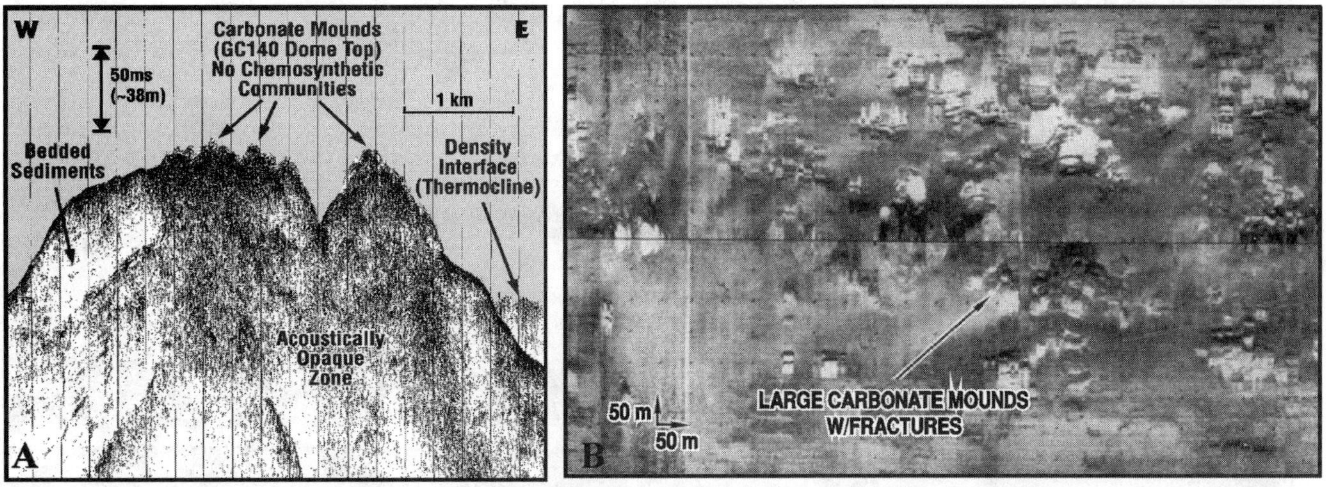

Figure 9. (a) This high resolution seismic profile (15 in^3 water gun) across the Green Canyon 140 dome illustrates the highly reflective and irregular seafloor. (b) The even higher resolution side-scan sonar swath identifies this irregular seafloor as a complex of individual mounds with impressive relief above the bottom (average relief 10-20 m).

values for normal marine biogenic carbonates. The general range of δ ^{13}C values for seep-vent related authigenic carbonates from the nrothern Gulf of Mexico clope is from −18 °/oo to −58 °/oo PDB, while the δ ^{13}C range is about 1.8 °/oo to 5.6 °/oo PDB [*Roberts and Aharon,* 1994].

The mineralogical complexity of these authigenic carbonates from the northern Gulf is greater than counterparts from the base of the Florida Escarpment [*Paull et al.,* 1992], and all three phases-aragonite, Mg-calcite, and dolomite occur [*Roberts et al.,* 1992b]. Magnesian calcites are the most commonly encountered authigenic carbonate mineral phase. Most authigenic carbonates range between 10-15 mol % $MgCO_3$, but *Ferrell* and *Aharon* [1994] indicate that some Mg-calcites contain nearly 20 mol % $MgCO_3$, and the degree of structural disorder as determined by XRD analysis increases with Mg substitution. With the present data base, there seems to be no predictible spatial trend for the occurrence of the three mineral phases. Local environmental conditions such as aragonite/calcite/dolomite saturation state of the precipitating fluids (expulsion products of seepage-venting sites) dictate the character of the mineral species precipitated.

Authigenic carbonates are found as nodular masses within surficial sediment, hardgrounds, slabs, and mound-like buildups of various dimensions. The nodular masses, hardgrounds, and crusts are most common to the intermediate-to-slow flux rate settings. Mounded carbonates and chimneys are also found primarily in slow flux rate environments. Cement morphologies, as well as mineralogies associated with the various authigenic carbonates found in association with seep-vent sites are highly variable. These morphologies range from microcrystalline dolomites and Mg-calcites to void-filling and circum-granular acicular and botryoidal aragonites [*Roberts et al.,* 1992b; *Roberts* and *Aharon,* 1994].

One of the best examples of mounded authigenic carbonates is in Green Canyon 140 (27°48.06′N; 91°W; WD=270 m) (Figure 9). *Cook* and *D'Onfron* [1991] indicate that salt movement and faulting that provide fluid and gas transport routes to the ocean floor occurred in the lower Pleistocene. Over Pleistocene times, sediments filled a basin created by salt withdrawal causing the salt to penetrate to a shallow subsurface depth in the Green Canyon 140 area. These events resulted in the doming of the seafloor in Green Canyon 140 and faulting that eventually produced the Green Canyon 185 mound complex containing gas hydrates.

On the 3D-seismic surface amplitude data the Green Canyon 140 dome crest provides very high amplitude response because of the high reflective areas of lithified seafloor (Figure 4). The positive polarity of the 3D-seismic surface reflector suggests a hard bottom area with limited gas-charge near-surface sediment [*Roberts et al.,* 1992a]. High resolution seismic profiles and side-scan sonar swaths across the dome crest (water depth ~240-280 m) reveal a mounded seafloor. Individual mounds express themselves as hard targets with sharp boundaries, irregular surfaces, and average relief above the surrounding seafloor ranging between 5-10 m (Figure 9).

Direct observations and sampling of these mounds using manned submersibles clearly identifies them as hard, rocky structures with very few attached organisms (e.g., gorgonians). Each mound appears as an accumulation of chaotically tilted and fractured blocks separated by well-

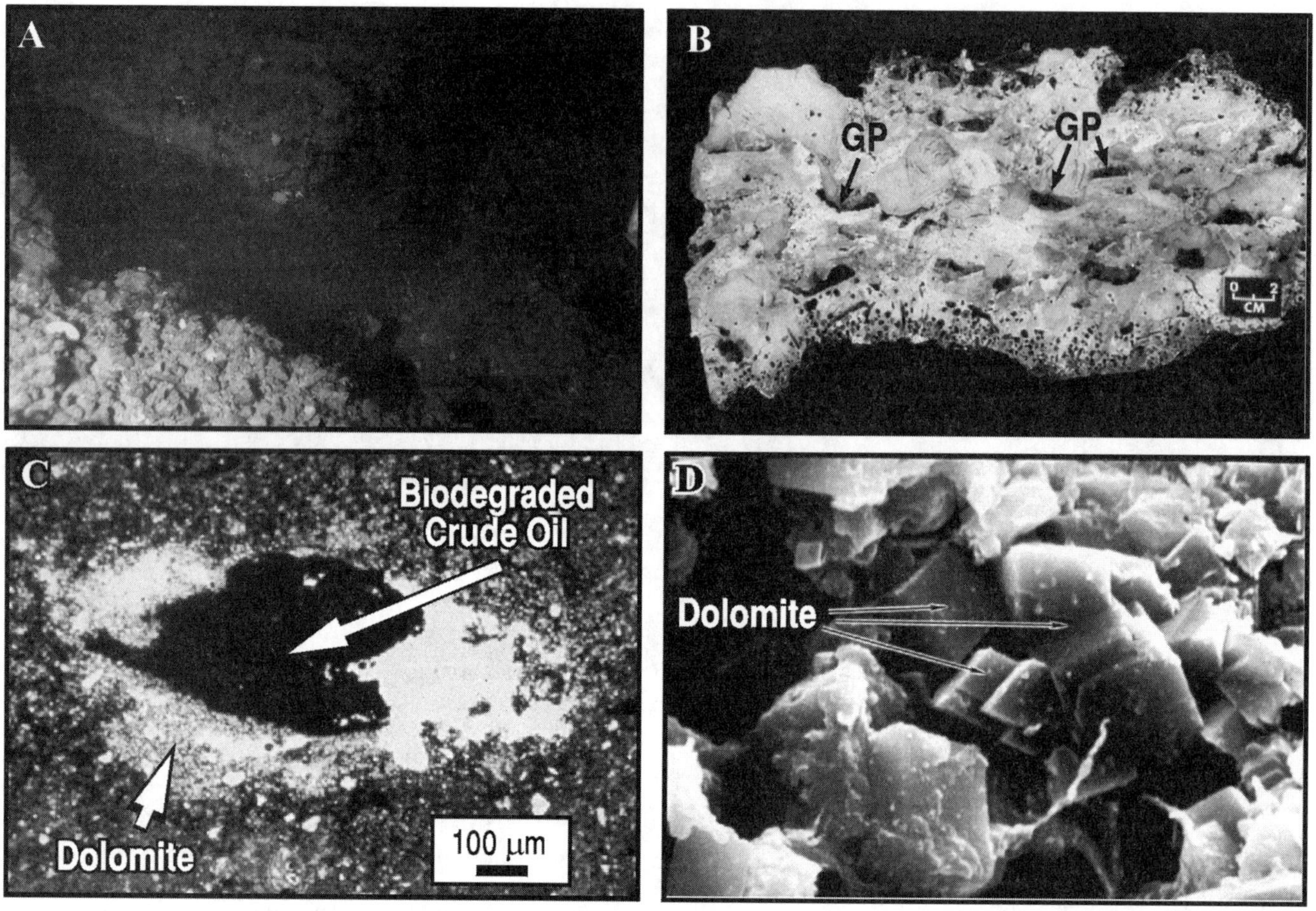

Figure 10. (a) Submersible picture of part of an authigenic carbonate mound in Green Canyon 140 showing carbonate blocks and joints between them. (b) A slab of a authigenic carbonate that was part of the Green Canyon 140 mound illustrating that it is formed of smaller cemented clasts, internal sediment, and cement Note geopetal structures (GP) in various orientations, suggesting multiple attitudes during the development. (c) Thin-section photomicrograph illustrating a dolomitic rock from this mound site containing a pore filled with highly biodegraded crude oil. (d) SEM picture of dolomite rhombs that line one of the pores in the rock slab.

defined crevasses (joints) that apparently result from flexure of underlying shallow salt (Figure 10a). The larger blocks that comprise the mounds are, in themselves, composed of nodular masses and cemented subangular clasts with void spaces partially filled with internal sediment (Figure 10b). Laboratory analyses on these carbonates [*Roberts et al.,* 1992b] indicate that they are extremely ^{13}C-depleted ($\delta^{13}C$ values to –55‰ PDB) and have a mineralogy that is primarily Mg-calcite 1>mole‰ Mg, with dolomite frequently lining pores and fractures (Figure 10c, d). The range of $\delta^{13}C$ values for samples from these mounds suggests a variety of hydrocarbon sources. The $\delta^{13}C$ values range from those associated with crude oil that is found trapped in the pores of many samples from these mounds to the extremely ^{13}C-depleted carbonate cements suggesting that methane is the contributor of isotopically light carbon [*Roberts* and *Aharon,* 1994].

Repeated careful observations suggest that gas is presently being fluxed through these mounds at a very slow rate. Perhaps more episodic and energetic releases of fluids and gases pulse crude oil to the seafloor, but these events have never been observed, and crude oil droplets have never been seen escaping from the pores in these mounded carbonates. However, small *Beggiatoa* mats and a few tube worms and lucinid-vesycomyid clams occur in crevices of some mounds. These marginal chemosynthetic communities are the shallowest (WD=250m) thus far found in the Gulf of Mexico. The slow flux of fluids and gases to the modern seafloor in this setting is thought to be related to the presence of shallow salt and the limited migration pathways provided by faults that originate from the salt and extend above to the domed ocean floor. These complex fault families do not presently have direct access to petroleum system deep within the adjacent sedimentary basin.

3.4. *Transitional Cases*

Not all fluid and gas expulsion features conveniently fit into the rapid flux-to-slow spectrum discussed in this paper and in *Roberts* and *Carney* [1997]. Variations from the simple qualitative relationships illustrated in Plate 2 occur because of (a) differences in the chemistry of fluids and gases being transported to the seafloor; (b) the frequency of expulsion events; (c) duration of expulsion events; (d) the strength of these events; and, (e) the number of expulsion sites within a given area. For example, thin but rapidly deposited mudflows from the crest of a salt-cored diapiric mound in Garden Banks 427 (27°33.59'N; 92°25.53'W) display surface lithification, a slow flux characteristic, because of the infrequency of expulsion events. This surface lithification (primarily Mg-calcite) produces a highly reflective seafloor response on 3D-seismic surface amplitude maps. However, the positive-to-negative polarity shift in the surface seismic reflector suggests the presence of gas-charged sediments [*Roberts* and *Doyle,* 1998]. In settings where expulsion events are more frequent than in the Garden Banks 427 case, such as large and active mud volcanoes, seafloor lithification does not generally occur over broad areas, but still may be found at localized sites [*Roberts* and *Doyle,* 1998].

Through time, sites may shift from one part of the feature spectrum to another, either by a systematic increase or decrease in the rate that fluids and gases are delivered to the seafloor. An example of this systematic evolution occurs upslope of the Garden Banks 427 mound, in Garden Banks 338 (27°37.80'N; 92°28.12'W) and Garden Banks 382 (27°36.49'N; 92°29.05'W). Figure 11a illustrates a map of this part of the continental slope showing that the Garden Banks 382 and 338 sites represent mound-like expulsion centers from which large volumes of fluidized sediment have been transported down-slope. Both of these expulsion centers are in water depths of approximately 500 m and are located on a NE-SW trending salt-supported ridge that forms the northern limit of an intraslope basin containing a thick sedimentary sequence. This salt ridge is directly downslope of a late Pleistocene shelf-edge deltaic depocenter [*Berryhill et al.,* 1987] and is thought to have been activated by sedimentary loading during the sea level fall to the latest Pleistocene glacial maximum. Salt adjustment, associated fault activation, and perhaps decomposition of gas hydrates are probable mechanisms that initiated slope failures and fluidized sediment expulsion from the two centers (Figure 11b, c). During the rise in sea level to its present level, sedimentary loading has been eliminated by the retreat of fluvial systems to their present positions on the inner shelf. In addition, an upper slope hydrostatic pressure and temperature regime has been reestablished. These conditions presently support stability of gas hydrates in the shallow subsurface. Therefore, today the tops of these expulsion centers have evolved from sites that were the sources of large volumes of fluidized sediment to areas where the seafloor is currently being cemented by carbonates and barite through the comparatively slow seepage of mineralizing fluids to the modern seafloor (Figure 11d, e).

3. SUMMARY

When viewed qualitatively, fluid and gas expulsion features of the northern Gulf of Mexico continental slope fall into three general categories that have rather distinct assemblages of geological features and biologic communities (Figure 12). These categories, originally discussed by *Roberts* and *Carney* [1997], have been identified as rapid, moderate, and slow flux settings. Combined data sets from both remotely sensed and in situ data collection methods suggest that the transport of hydrocarbon-rich fluids and gases to the modern continental slope surface occurs at distinctly different rates, each major rate category creating a unique assemblage of seafloor response features.

Rapid flux of fluids (including fluidized sediment) and gases create mud-prone topographic features such as mud volcanoes and/or mud flows in the form of broad sheets or stacked sheets. Mud volcanoes range in scale from local features < 1 m across to large forms with diameters of > 1 km. Mud flows display the same wide range of scales. Existing evidence suggests that the delivery of fluidized sediment to create these features is episodic. During rapid flux events sedimentation rates are so high that diagenetic processes of mineralization and biogenic processes have little opportunity to impact the geologic record. These rapid flux, mud-prone deposits contain hydrocarbons that show little evidence of biodegradation, a conspicuous reduction in hard mineralized surfaces (particularly authigenic carbonates) and few-to-no remains of chemosynthetic organisms. Rapid sedimentation buries sessil organisms and smothers those that are filter feeders. In addition, the weak substrate and finite trophic resources available in each episodic pulse of sediment onto the seafloor are other-community limiting factors. In most cases, the only chemosynthetic organisms present in rapid flux settings are bacterial mats (*Beggiatoa*) and lucinid-vesycomyid clams. Because of the sedimentation rate and heat flux, active mud volcanoes are not places where gas

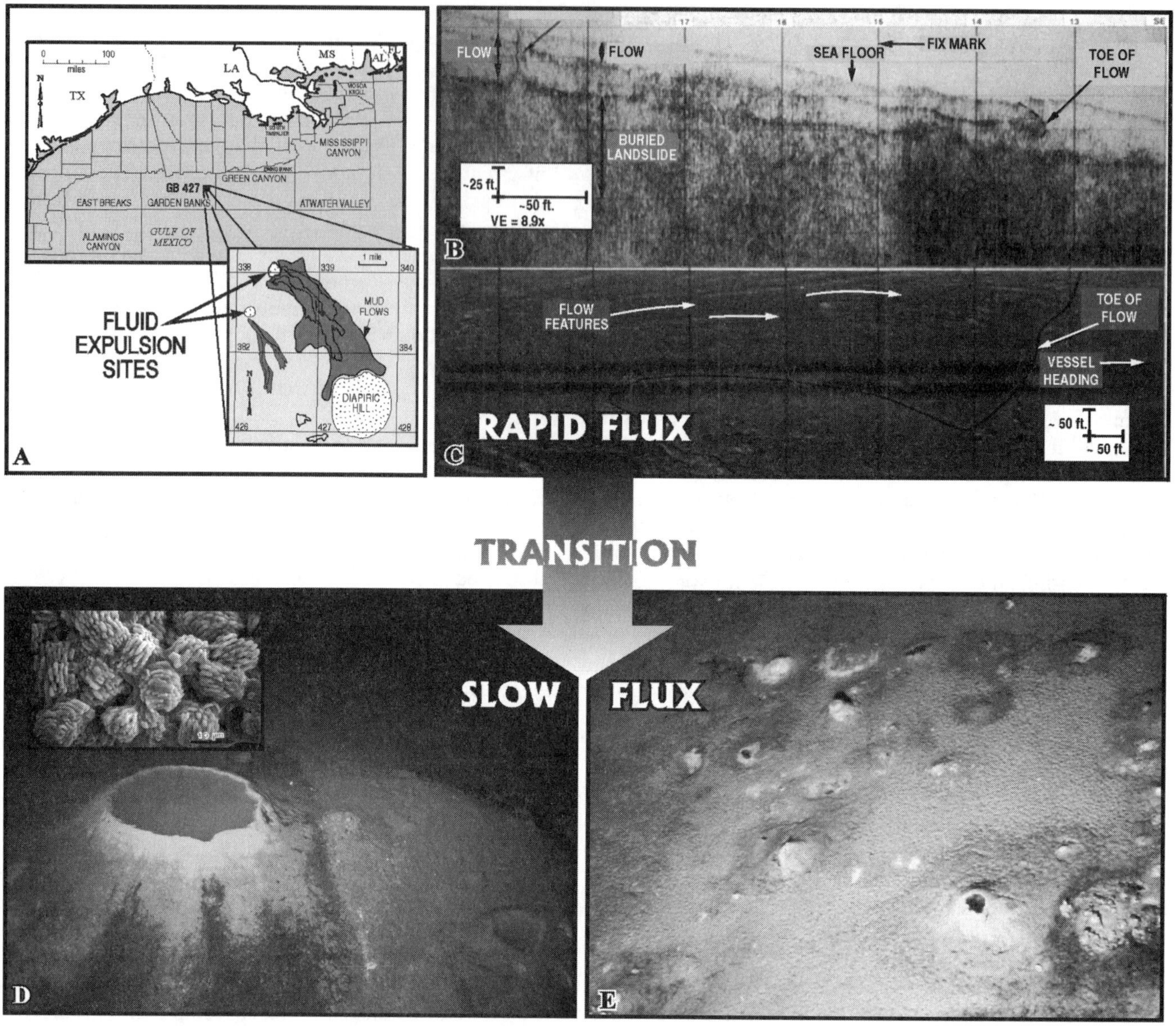

Figure 11. This figure illustrates (a) the locations of the Garden Banks 338 and 382 expulsion centers in relation to the Garden Banks 427 salt-cored mound; (b) a 3.5 kHz profile through the mudflow deposits downslope of the Garden Banks 338 expulsion center and (c) the surface expression of the fluidized flow on a side-scan sonar record; and (d, e) barite cones and chimneys that are now forming at the Garden Banks 338 expulsion center as a product of evolution of this site from a rapid flux (mud-prone) setting to a slow-flux (mineral-prone) environment. Inset picture shows the barite structure of the cone as viewed by SEM (scale bar is 10 microns).

hydrates have been observed at the modern seafloor. However, when the flux of fluids and gases slows, gas hydrate may form in former feeder pipes and flanking sediments. Cores from dormant mud volcanoes confirm the existence of has hydrates in these features.

Moderate flux cases display the most local variability in small-scale geologic features, substrate types, and biologic communities. The rather constant but moderate flux of hydrocarbon gases and to a lesser extent fluids (including moderately biodegraded crude oil) to the seafloor within the hydrate stability zone promotes the formation of gas hydrates at or near the ocean floor (in water depth below ~500 m). Gas hydrates may not be excluded from more rapid and slower flux settings, but in the moderate flux case the hydrocarbon delivery system causes gas hydrate to form in the subsurface and to the

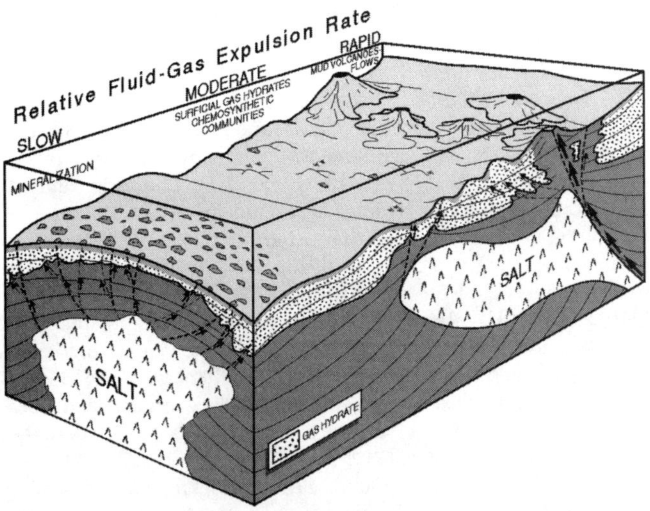

Figure 12. A schematic representation of the relationships between salt, sediments, and faults that produce slow-to-rapid expulsion of fluids and gases at the modern seafloor. Rapid (mud-prone) features derive from deep-penetrating faults that breach overpressure zones of the deep subsurface. Fluids (including fluidized sediment) and gases combine to produce mud volcanoes and mud flows. When active,gas hydrates may be precluded in parts of these sites because of rapid heat-transport from the deep subsurface. Slow flux (mineral-prone) settings are associated with relatively inactive fault networks or those that have limited penetration into the subsurface. Authigenic carbonates and more exotic minerals, such as barite, create nodules, crusts, chimneys, and mounds in these areas. Moderate flux settings promote high spatial variability in expulsion-related features and benthic communities. Surficial gas hydrates and chemosynthetic communities as well as localized authigenic carbonates are associated with moderate flux environments.

sediment-water interface. In this setting, abundant gas is being fluxed through the sediments and by-passing the zones of gas hydrate formation to the water column. In the upper continental slope setting, plumes of gas are commonly documented on echo-sounders above areas where gas hydrates are exposed on the seafloor. Accompanying surface and shallow subsurface concentrations of gas hydrates are densely populated and the most diverse and widely distributed chemosynthetic communities thus far encountered on the northern Gulf of Mexico continental slope. These communities consist of bacterial mats (*Beggiatoa*), lucinid-vesycomyid clams, pogonophoran and vestimentiferan tube worms, bathymodiolid mussels, and a variety of accessory organisms. In addition to gas hydrates and associated chemosynthetic communities, moderate flux settings can display localized outcroppings of authigenic carbonate, nodular masses of carbonate in near-surface sediments, and a variety of small-scale gas-fluid expulsion features (small mud volcanoes, mud vents, and shallow pockmark-like depressions).

Slow flux environments are mineral-prone. As hydrocarbon-rich fluids and gases are slowly transported to the modern seafloor, microbial communities oxidize the hydrocarbons. Products of this process help induce the precipitation of authigenic carbonates (mainly aragonite, Mg-calcite, and dolomite) as by-products. In some cases, barite is precipitated at the seafloor in a variety of forms (crusts, small mounds, and chimneys). The carbonates occur as nodular masses within the sediment, hardgrounds, and lithified slabs, and mound-like buildups that have a wide range of relief ($<$ 1 m to $>$ 20 m above the surrounding seafloor). The carbonate buildups themselves do not generally support abundant chemosynthetic organisms. However, small communities consisting of bacterial mats, pogonophoran tube worms, and lucinid-vesycomyid clams are sometimes found on the mud bottoms surrounding the buildups, in crevasses within the buildups, and adjacent to hardgrounds and slabs.

Finally, the responses to the rate of fluid and gas flux to the modern seafloor discussed in this paper are end-member types. However, because of the episodic nature of expulsion events and the systematic changes in flux rate at a given sites, the seafloor may shift from one part of the feature spectrum to another given sufficient time.

Acknowledgments. I acknowledge the Louisiana Education Quality Support Funds (Contract No. LEQSF (1994-1997) RD-B-12) and NOAA National Underseas Research Center, University of North Carolina at Wilmington (Grants NA80AA-8-00081, NA 88AA-D-UR004, and NA 36RU-0060) for dive time that supported this study. I thank the Minerals Management Service through the Coastal Marine Institute and the Louisiana State University Cooperative Agreement (Contract No. 14-35-0001-30660 and 35-01-99-CA-30951) for supporting studies of the northern Gulf of Mexico continental slope, which involved analysis of 3D-seismic and high resolution acoustic data as well as manned submersible work in evaluating areas where the full spectrum of fluid and gas expulsion features have been investigated. Fugro-McClelland Marine Geosciences Inc. is gratefully acknowledged for providing selected geotechnical data for this study and for participating as partners on two slope studies sponsored by LEQSF. I thank Ms. Mary Lee Eggart for her excellent help with illustrations and Ms. April Hawkins and Ms. Jami Donley for clerical and editorial support. Finally, the CSI Field Support Group is acknowledged for their excellent technical support in all aspects of our slope research, particularly in designing, building, and helping to sample the bottom and to measure critical processes affecting the seafloor.

REFERENCES

Aharon, J.B., H.H. Roberts, R. Sassen, and J. Larkin, Stable isotopes and petrology of chimneys and calcarious fauna from hydrocarbon vents on the Gulf of Mexico slope, Geological Society of American Abstracts with Programs 22:A208, 1990.

Bagirov, E., R. Nadirov, and I. Lerche, Flaming eruptions and ejections from mud volcanoes in Azerbaijan: Statistical risk assessment from the historical record, Energy Exploration and Exploitation, 14:535-583, 1996.

Barber, A.J., S.R. Tjokrosapoetro, and T.R. Charlton, Mud volcanoes, shale diapirs, wrench faults, and melanges in accretionary complexes, eastern Indonesia, American Association of Petroleum Geologists Bulletin 70(11):1729-1741, 1986.

Berger, T.J., P. Hamilton, J.J. Singer, R.R. Leben, Louisiana/Texas shelf physical oceanography program, eddy circulation study, OCS Study MMS 96-00XX, U.S. Department of the Interior, Minerals Management Service, Gulf of Mexico OCS Region, New Orleans, LA, (In Press).

Berner, R.A., Principles of chemical sedimentology, McGraw-Hill, New York, 240 p., 1971.

Berryhill, H.L., J.R. Suter, and N.S. Hardin, Late Quaternary facies and structure, northern Gulf of Mexico, American Association of Petroleum Geologists Studies in Geology #23, 289 pp., 1987.

Bohrmann, G., J. Greinert, E. Suess, and M. Torres, Authigenic carbonates from the Cascadia subduction zone and their relation to gas hydrate stability: Geology 26:647-650, 1998.

Brooks, J.M., B. Cox, W.R. Bryant, M.C. Kennicutt II, R.G. Mann, and T.J. McDonald, Association of gas hydrates and oil seepage in the Gulf of Mexico, Organic Geochemistry, 10: 221-234, 1985.

Brown, K., The nature and hydrogeologic significance of mud diapirs and diatremes for accretionary systems, Journal of Geophysical Research 95:8969-8982, 1990.

Bryant, W. R., Gas charged sediments *in* (A.H. Bouma, ed.) High resolution seismic profiling and side-scan sonar: united States Geological Survey, Office of Marine Geology, p. 139-162, 1981.

Carney, R.S., Considerationof the oasis analogy for chemosynthetic communities at Gulf of Mexico hydrocarbon vents: Geo-Marine Letters, 14:149-159, 1994.

Cook, D. and P. D. D'Onfro, Joliet field thrust fault structure and stratigraphy-Green Canyon block 184, offshore Louisiana: Gulf Coast Association of Geological Societies Transactions, 41:100-121, 1991.

Corthay, J.E., II, Delmeation of a massive seafloor hydrocarbon seep, overpressured acquifer sands, and shallow gas reservoirs, Louisiana continental slope: Proceedings Offshore Technology Conference, OTC Paper 8594, p. 1-20, 1998.

Corthay, J.E. and A.A. Aliyev, Delineation of a mud volcano complex, surficial mudflows, slump blocks, and shallow gas reservoirs, offshore Azerbaijan: Proceedings Offshore Technology Conference, OTC Paper 12066, p. 1-14, 2000.

Diegel, F.A., J.F. Karlo, D.C. Schuster, R.C. Shoup, and P.R. Tauvers., Cenozoic structural evolution and tectono-stratigraphic framework of the northern Gulf Coast continental margin, pp 109-151. In M.P.A. Jackson, D.G. Roberts, S. Snelson, eds. Salt Tectonics: A Global Perspective. AAPG Memoir 65:109-151,1995.

Fang, J., Isotopic evidence for petroleum-derived carbonates in the Gulf of Mexico: Gulf Coastal Association of Geoological Societies Transactions, 41:276-282, 1991.

Ferrell, R.E., Jr. And P. Aharon, Mineral assemblages occurring around hydrocarbon vents in the northern Gulf of Mexico: Geo-Marine, 14:74-80, 1994.

Fu, B., P. Aharon, G.R. Byerly, and H.H. Roberts, Barite chimneys on the Gulf of Mexico slope: Initial report of their petrography and geochemistry: Geo-Marine Letters, 14:81-87, 1994.

Ginsburg, G.D., A.V. Milkov, V.A. Soloviev, A.V. Egorov, G.A. cherkashev, R.P. Vogt, K. Crane, T.D. Lorenson, and M.D. Khutorskoy, Gas hydrate accumulation at the Hakon Mosby mud volcano: Geo-Marine Letters 19:57-67, 1999.

Hamilton, P., Deep currents in the Gulf of Mexico: Journal of Geophysical Research 20:1087-1104, 1990.

Hedberg, H.H., Relation of methane generation to under compacted shales, shale diapirs, and mud volcanoes. American Association of Petroleum Geologists Bulletin 58:661-673, 1974.

Hill, A.W., The use of exploration 3D data in geohazard assessment: Where does the future lie? Proceedings Offshore Technology Conference, OTC Paper 7966, p.113-117, 1996.

Hovland, M., and A.G. Judd, Seabed pockmarks and seepages: Impact on geology, biology and the marine environment. Graham and Trotman. London. 293 pp., 1988.

Jorgensen, N.O., Recent high magnesian calccite/aragonite cementation of beach and submarine sediments from Denmark: Journal of Sedimentary petrology, 46:940-951, 1976.

Jannasch, H.W., Chemosynthesis: The nutritional basis for life at deep-sea vents: Oceanus, 27: 73-78, 1984.

Kennicutt, M.C., II, J.M. Brooks, R.R. Bidigare, R.R. Fay, T.L. Wade, and T.J. McDonald, Vent-type taxa in a hydrocarbon seep region on the Louisiana slope: Nature, 317:351-353, 1985.

Kohl, B. and H.H. Roberts, Fossil foraminifera from four active mud volcanoes in the Gulf of Mexico: Geo-Marine Letters, 14:126-134, 1994.

Kyle, J.R., M.R. Ulrich, and W.A. Gose, Textural and paleomagnetic evidence for the mechanism and timing of anhydrite caprock formation, Winnfield salt dome, Louisiana, in Dynamical Geology of Salt Related Structures, edited by I. Lerche and J.J. O'Brien, Orlando, Florida, Academic Press, pp. 497-542, 1987.

McDonald, I.R., D.B. Buthman, W.W. Sager, M.B. Peccini, and N.L. Guinasso, Jr. Pulsed oil discharge from a mud volcano: Geology, 28:907-210, 2000.

McDonald, I.R., N.L. Guinasso, Jr., R. Sassen, J. M. Brooks, L. Lee and K.T. Scott. Gas hydrate that breaches the sea floor on the continental slope of the Gulf of Mexico. Geolgoy 22:699-792, 1944.

Neurauter, T.W. and W.R. Bryant, Seismic expression of sedimentary volcanism on the continental slope, northern Gulf of Mexico. Geo-marine Letters 10:225-231, 1990.

Neurauter, T.W. and H.H. Roberts, Seismic and visual observation of seepage-related structures on the continental slope, northern Gulf of Mexico. Proceedings 24th Offshore Technology Conference, Houston, Texas, OTC 6850, p. 355-362, 1992.

Neurauter, T., and H.H. Roberts, Three generations of mud volcanoes on the Louisiana continental slope. Geo-marine Letters 14:120-125, 1994.

Orange, D.L., and N.A. Breen, The effects of fluid escape on accretionary wedges, II: Seepage force, slope failure, headless submarine canyons, and vents. Journal of Geophysical Research 97:9277-9295, 1992.

Paull, C.K., J.P. Chanton, A.C. Neumann, J.A. Coston, C.S. Martens, and W. Showers, Indicators of methane-derived carhonates and chemosynthetic organic carbon deposits: Examples from the Florida Escarpment. Palaios 7:361-375, 1992.

Prather, B.E., J.R. Booth, G.S. Steffens, and P.A. Craig, Classification, lithologic calibration, and stratigraphic succession of seismic facies of intraslope basins, deep-water Gulf of Mexico: American Association of Petroleum Geologists Bulletin, 85:701-728, 1998.

Ratcliff, D.W., New technologies improve seismic images of salt bodies. Oil and Gas Journal 91:41-49, 1993.

Rice, D.D., Chemical and isotopic evidence of the origins of natural gases in offshore Gulf of Mexico, Gulf Coast Association of Geological Societies Transactions 30:202-213, 1980.

Ritger, S., B. Carson, and E. Suess, Methane-derived authigenic carbonates formed by subduction-induced pore-water expulsion along the OregonlWashington margin. Geological Society of America Bulletin, 98:147-156, 1987.

Roberts, H.H., D.J. Cook, and M.K. Sheedlo, Hydrocarbon seeps of the Louisiana continental slope: Seismic amplitude signature and seafloor response. Gulf Coast Association of Geological Societies Transactions 42:349-362, 1992a.

Roberts, H.H., P. Aharon, and M.M. Walsh, Cold-seep carbonates of the Louisiana continental slope-to basin floor, pp.95-104. In Rezak, R. and D. Lavoie, eds. Carbonate Microfabrics. Springer-Verlag, Berlin, 1992b.

Roberts, H.H. and P. Aharon, Hydrocarbon-derived carbonate buildups of the northern Gulf of Mexico continental slope: A review of submersible investigations: Geo-Marine Letters, 14:135-148, 1994.

Roberts, S.J., J. A. Nunn, L. Cathles, and F.D. Cipriani, Expulsion of abnormally pressured fluids along faults: Journal of Geophysical Research, 101:28, 231-28,252, 1996.

Roberts, H.H., Surface amplitude data: 3D-seismic for interpretation of seafloor geology (Louisiana slope). Gulf Coast Association of Geological Societies Transactions, 46:353-366, 1996.

Roberts, H.H., and R. Carney, Evidence of episodic fluid, gas and sediment venting on the northern Gulf of Mexico continental slope. Economic Geology, v.92, p.863-879, 1997.

Roberts, H.H. and E. H. Doyle, Seafloor "calibration" of high resolution acoustic data and amplitude rendering of the "diapiric hill," Garden Banks 427: Proceedings Offshore Technology Conference, OTC Paper 8592, 1998.

Roberts, H.H., W.J. Wiseman, Jr., J. Hooper, and G.D. Humphrey, Surficial gas hydrates of the Louisiana continental slope - Initial results of direct observations and in situ data collection. Offshore Technology Conference, Houston, Texas. OTC 10770, 259-272, 1999b.

Roberts, H.H., B. Kohl, D. Menzies, and G.D. Humphrey, Acoustic wipe-out zones – A paradox for interpreting seafloor geologic/geotechnical characteristics (An example for Garden Banks 161): Proceedings Offshore Technology Conference, OTC Paper 10921, p. 1-12, 1999a.

Rowan, M.G., M.P.A. Jackson, and B.D. Trudgill, Salt-related fault families and fault welds in the northern Gulf of Mexico: American Association of Petroleum Geologists Bulletin, 83:1454-1484, 1999.

Rowan, M.G., Structural styles and evolution of allochthonous salt, central Louisiana outer shelf and upper slope, pp.199-228. In M.P.A. Jackson, D.G. Roberts, S. Snelson, eds. Salt Tectonics: A Global Perspective: AAPG Memoir 65, 1995.

Sakai, H., T. Gamo, Y. Ogawa, and J. Boulegue, Stable isotopic ratios and origins of the carbonates associated with cold seepage at the eastern Nankai Trough: Earth and Planetary Science Letters 109:391-404, 1992.

Sassen, R., H.H. Roberts, P. Aharon, J. Larkin, E.W. Chinn, and R. Carney, Chemosynthetic bacterial mats at cold hydrocarbon seeps, Gulf of Mexico continental slope: Organic Geochemistry, 20:77-89, 1993.

Sassen, R., I.R.MacDonald, A.G. Raquejo, N.L. Guinasso, Jr., M.C. Kennicutt II, S.T. Sweet, and J.M. Brooks, Organic geochemistry of sediments from chemosynthetic communities, Gulf of Mexico Slope: Geo-Marine Letters, 14:110-119, 1994.

Sassen, R., I.R. MacDonald, N.L. Guinasso, S. Joye, A.G. Requejo, S.T. Sweet, J. Alcala-Herrera, D.A. DeFreitas, D.R. Schink, Bacteria methane oxidation in sea-floor gas hydrate: Significance to life in extreme environments. Geology 26:289-293, 1998.

Sassen, R., S.T. Sweet, A.V. Milkov, D.A. DeFreitas, G.G. Salata, E.C. McDade, Geology and geochemistry of gas hydrates, central Gulf of Mexico continental slope. Transactions Gulf Coast Association of Geological Societies 49:462-468, 1999a.

Sassen, R.S. Joye, S.T. Sweet, D.A. DeFreitas, A.V. Milkov, I.R. MacDonald, Thermogenic gas hydrates and hydrocarbon gases in complex chemosynthetic communities, Gulf of Mexico continental slope. Organic Geochemistry 30:485-497, 1999b.

Schoell, H.P., Applications of isotope analysis to petroleum and natural gas research: Spectra, 8:32-41, 1982.

Sloan, E.D., Jr., Clathrate hydrates of natural gases: Marcel Dekker, Inc., New York and Basel, 641 p., 1990.

Torres, M., Bohrmann, G., and E. Suess, Authigenic barites and fluxes of barium associated with fluid seeps in the Peru subduction zone: Earth and Planetary Science Letters 144:469-481, 1996.

Ulrich, M.R., J. R. Kyle and P.E. Price. Metalic sulfide deposits in the Winnfield salt dome, Louisiana; evidence for episodic instrodution of metalliferous brines during caprock formation.: Transactions Gulf Coast Association of Geological Societies 34:435-442, 1984,

VonRad, U., H. Rosch, U. Berner, M. Geyh, V. Marchig, and H. Schulz. Authigenic carbonates derived from oxidized methane vented from the Makran accretionary prism off Pakistan: Marine Geology 136:55-77, 1996.

Vogt, P.R., J. Gardner, and K. Crane, The Norwegian-Barents-Svalbard (NBS) continental margin: Introducing a natural laboratory of mass wasting, hydrates, and ascent of sediment, pore water and metane: Geo-Marine Letters 19:2-21, 1999.

Woodbury, H.O., I.B. Murray, Jr., P.J. Pickford, and W.H. Akers, Pliocene and Pleistocene depocenters, outer continental shelf, Louisiana and Texas, American Association of Petroleum Geologists Bulletin 57:2428-2439, 1973.

Zhao, W.L., D.M. Davis, F.A.Dahlen, and J. Suppe, Origin of convex accretionary wedges: Evidence for Barbados. Journal of Geophysical Research 91:10.246-10.258, 1986.

Harry Roberts, Coastal Studies Institute, Department of Oceanography and Coastal Sciences, Louisiana State University, Baton Rouge, LA 70803

Section 3

GEOPHYSICS

Deep-tow Seismic Investigations of Methane Hydrates

Warren T. Wood and Joseph F. Gettrust

Naval Research Laboratory, Stennis Space Center, Mississippi

High-resolution deep-tow multichannel seismic technology, developed by the Naval Research Laboratory, has been used to study natural gas hydrates in the marine environment. The 48-channel deep-tow system uses a 250-650 Hz source; the source and receiver array are towed approximately 300 m above the seafloor. The combination of the geometry achieved and source frequency allows us to study the geologic structure and compressional velocities within the Hydrate Stability Zone (HSZ) with significantly greater resolution than any other available technique. The data taken with this system have revealed features that require modifications of simple models for the HSZ. For example, extensive faulting has been resolved that extends from the seafloor to the base of the HSZ. Further, compressional velocity estimates obtained from these data suggest vertical zones of elevated velocity that appear to be associated with these faults. This is consistent with chemical/hydrologic models that indicate hydrates are created preferentially where fluid flux rates are high enough to exceed saturation. Data acquired at two locations on the Blake Ridge off the east coast of the U. S. and on the Cascadia margin off the Canadian west coast exhibit significantly different seismic character, implying different manifestations of flux through the HSZ. The high-resolution seismic data indicate the need for a 3-D, dynamic model of the HSZ to more fully explain the generation and dissociation of natural gas hydrates.

I. INTRODUCTION

1.1 The Role of the Deep-tow Seismic Technique

Marine geophysical investigations of methane hydrates were effectively begun with the realization that an enigmatic bottom simulating reflector (BSR),[*Markl* et al., 1970] was associated with clathrate hydrates of methane gas [*Stoll* et al., 1971]; which are referred to here as gas hydrates, or simply hydrates [*Sloan*, 1990]. Subsequently, more extensive, and sophisticated analyses of single and multichannel data [*Paull* and *Dillon*, 1981; *Hyndman* and *Spence*, 1992; *Rowe* and *Gettrust*, 1993a; *Singh* et al., 1993; *Wood* et al., 1994; *Katzman* et al., 1994;*Yuan* et al., 1996; *Korenaga* et al., 1997], and even drilling efforts with specific, hydrate related objectives [*Westbrook* et al., 1994; *Paull* et al., 1996], have been used to constrain estimates of the quantity and distribution of hydrates and associated methane gas. Because of the high resolution achievable with deep-tow seismic data, their contribution to the study of hydrates is the understanding of seismic velocities and fine scale structures within the hydrate stability zone (HSZ), on the scale of meters to tens of meters [*Rowe* and

Natural Gas Hydrates: Occurrence, Distribution, and Detection
Geophysical Monograph 124

Published in 2001 by the American Geophysical Union

Gettrust, 1993a,b; *Gettrust* et al., 1999]. As the study of hydrates becomes more interdisciplinary, with biological [*Cragg* et al., 1996; *Wellsbury* et al., 2000], chemical [*Paull* et al., 1991; *Dickens* et al., 1997], and hydrologic [*Xu* and *Ruppel*, 1999; *Clennel* et al., 1999] contributions, the importance of correlating the fine-scale structure (including faults that can act as fluid conduits) with actual physical samples becomes clear. The direction of deep-tow seismic sampling techniques is now headed toward increasingly careful and detailed quantitative seismic analyses. This places chemical and biological analyses in the spatial context of the gas and hydrate accumulations, as well as the local fluid flux regime.

We focus here on high-resolution deep-tow seismic data acquired in two areas in which deep marine hydrates have been studied extensively using seismic reflection as well as drilling techniques (Figure 1), the Blake Ridge [*Paull* et al. 1996], and Cascadia Margin [*Westbrook* et al., 1994]. These areas are distinctly different geologically, the Cascadia site consisting of accreted and slope-fill sediments lying atop an active subduction zone, and the Blake Ridge, a fine grained sediment drift lying atop a passive continental margin. However, the two areas share what may be the most important characteristics for methane hydrate production; a thick sediment column, rich in organic material (within which methane is produced); and upward fluid flux to move the methane into the shallow sediments at concentrations high enough to exceed saturation [*Xu* and *Ruppel*, 1999]. Two smaller deep-tow seismic surveys on the Blake Ridge, (one on the crest and one on the southwestern flank), and one larger deep-tow survey on the Cascadia margin illustrate a wide variety of features likely related to hydrates and fluid flux, as well as interesting anomalies in the BSR itself.

Accurate interpretation of high-resolution deep-tow seismic data requires an understanding of the limitations of the data acquisition, processing, and analyses. Because only some aspects of conventional multichannel seismic acquisition and processing are appropriate for deep-tow data, some nonstandard processing and analysis techniques are required. We discuss some of these techniques here to convey the understanding necessary for proper interpretation of these data.

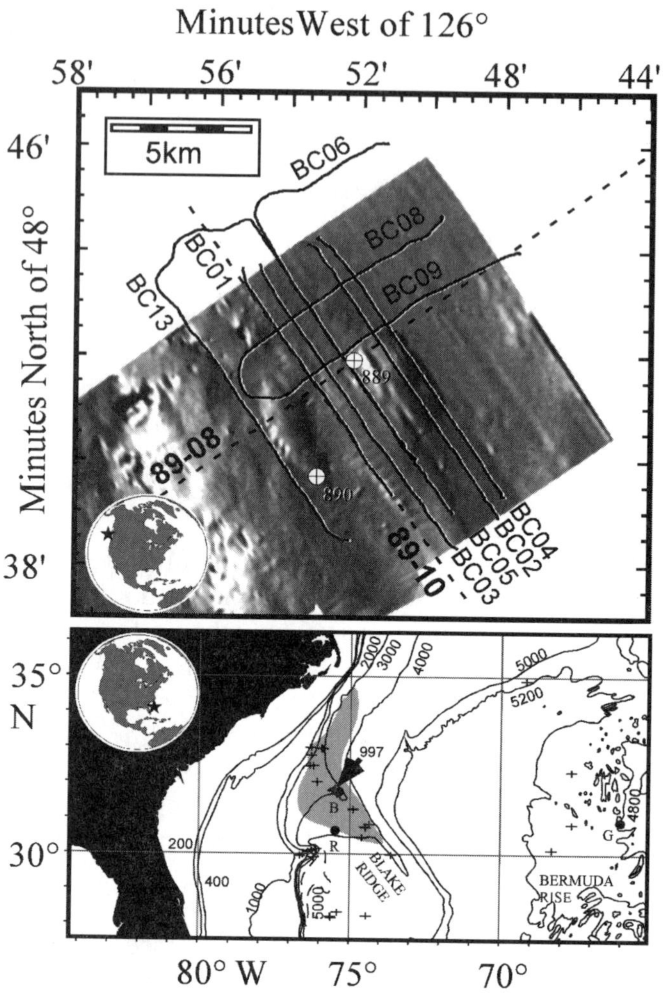

Figure 1. Two areas where methane hydrates have been studied extensively are the Cascadia Margin (upper map, showing DTAGS tracklines), and Blake Ridge (lower) where B (Data from Figure 5), and R (*Rowe* and *Gettrust* [1993 a,b]), mark the sites of short (about 5 km long), DTAGS surveys. G (Gettrust et al. 1988), marks the site of a DTAGS survey on the Bermuda Rise, an area with no hydrates, but similar velocity profiles (see Figure 7). Crosses indicate positions of DSDP and ODP drill sites.

1.2 High-Resolution Deep-Tow Acquisition and Processing

In the 1980's the U. S. Naval Research Laboratory developed a unique, deep-water, high-resolution, multi-channel seismic capability [*Gettrust* et al., 1991]. This deep-tow acoustics/geophysics system (DTAGS, Figure 2) can record 48 seismic channels while operating at full ocean depth. DTAGS is particularly well suited for the investigation of methane hydrates for two reasons. First, the frequency range used, 250-650 Hz is the highest (given current technological constraints) that can effectively penetrate several hundred meters of sediment, and therefore provides the best possible vertical resolution at sediment depths where marine hydrates are found. There are surface-towed systems with this, or even broader frequency range, but most of the seismic investigations of hydrates have employed conventional oil exploration equipment, tuned to

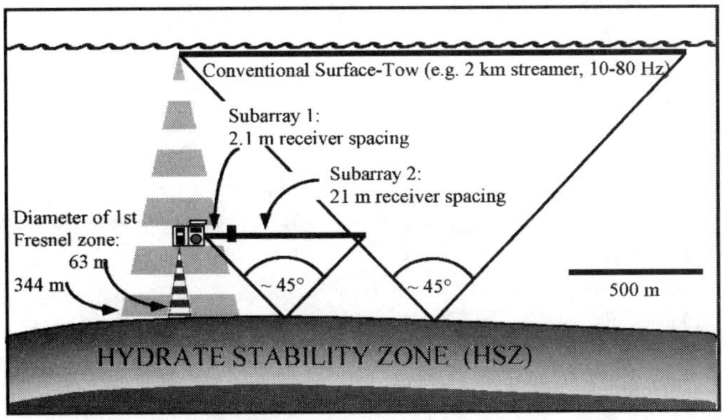

Figure 2. The geometry of DTAGS is similar to a scaled down version of a conventional multichannel system. However, as the water depth increases, the lateral resolution and angular coverage of surface-tow systems decreases, while the lateral resolution and angular coverage of DTAGS remains fixed. DTAGS is outfitted with two arrays, with different receiver spacing to enhance imaging and velocity analysis.

the more deeply penetrating, lower frequencies centered at about 45 Hz. Second, because marine hydrates occur in deep water environments, a deep-tow system can operate much closer to the target (the hydrates and their surroundings) than can a surface-tow system. Not only does this result in a smaller first Fresnel zone, and improved lateral resolution, but it also allows a wider angular coverage for a given offset range resulting in more accurate seismic velocity estimation.

The advantages of a deep-tow system are illustrated in the scaled drawing in Figure 2. In water depths of 1000 m, the first Fresnel zone at the seafloor for mid-range DTAGS frequencies (450 Hz at 300 m altitude) is 63 m, significantly smaller than the corresponding first Fresnel zone for a conventional surface-tow system, (344 m for 45 Hz at 1000 m altitude), or even a high resolution surface-tow system. The difference increases with increasing water depth because the DTAGS altitude above the seafloor remains at approximately 300 m, while that for the surface-tow system increases. Note also that a surface-tow array must be twice as long as the water is deep to achieve the angular coverage of the DTAGS data. Again, the benefits of deep-tow acquisition increase with increasing water depth.

Because DTAGS data are acquired unconventionally, their processing and analysis require unconventional techniques. Conventional common midpoint (CMP) gathering, and subsequent velocity analysis can be performed quite easily when the source and receiver locations for sequential shots are both known to within the quarter wavelength tolerance required for coherent processing. This is usually the case for surface-tow receiving arrays where one-quarter wavelength in water at the highest frequencies is about 5.8 m (65 Hz at 1500 m/s). Applying conventional processing algorithms to DTAGS data would require the data to be acquired with a spatial accuracy of about 0.58 m (650 Hz at 1500 m/s), a nearly impossible task for a 620 m array, towed at hundreds to thousands of meters deep. In addition to geographic positioning issues, the array deviates from the ideal acquisition geometry (horizontal array at the same depth for every shot). These deviations (measured using depth sensors and compasses throughout the array), result in reflection event trajectories which are nonstandard, but do not degrade the quality or information content of the data if properly accommodated in the seismic processing.

1.3. Imaging

One alternative to conventional imaging (CMP gathering, normal moveout correction and stacking) is to use the near 24 channels (subarray 1 in Figure 2, with 2.1 meter spacing) to generate a single fold image, with midpoints spaced at 1.05 m on the seafloor (Figure 3a). When shots are separated by 25 meters, sequential shot foot prints on the seafloor abut, and a continuous image can be formed (Figure 3a, and Figure 4). If the shot gathers are properly corrected for normal move-out, the arrival times on the near channel of one shot match those of the far channel on the next shot. The seismic image section in Figure 4 is formed from 4 sequential shots. Note the significant changes in the reflectors even over the span of a single shot footprint (25 m). Had these data instead been sorted into larger, say 12.5 m bins, this same section of the earth would have been imaged using only 8 traces. Binning, normal moveout (NMO) correcting, and stacking would have enhanced the signal to noise ratio of reflection events which are flat and coherent over each bin, but much of the detail which makes this image remarkable would have been lost. It is important to note that for this imaging technique, neighboring traces have slightly different offset and moveout, and therefore slightly different frequency content from NMO stretch. However, for the small offset to depth ratio of the imaging array portion of the data (25 m offset difference for greater than 300 m depth), this effect is negligible.

Another alternative to conventional imaging was required for the data from the Blake Ridge crest shown in Figure 5, which were acquired during a slow drift with a towed array hanging vertically (Figure 3b). These data were acquired as part of a seafloor backscatter experiment unrelated to

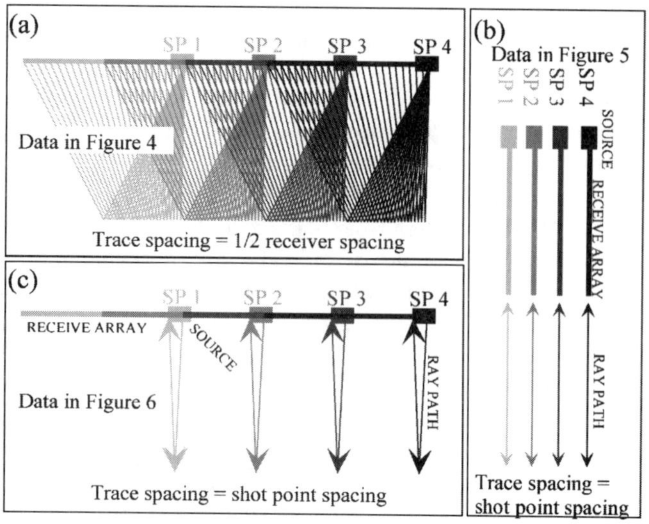

Figure 3. The unconventional imaging techniques used in Figures 4-6 are shown here. If shot spacing is carefully controlled the bounce points corresponding to individual source and receiver pairs form a continuous sequence from shot to shot (a), with trace spacing equal to one half of the receiver spacing, (which is 2.1 m). The data shown in Figure 5 was acquired using a slowly-towed vertical multichannel array (c), which, after stacking, effectively beamforms the signal to enhance signal over noise. A more conventional, single channel imaging technique (c) was used to generate the image in Figure 6.

hydrates, so the array configuration is sub-optimal for determining sediment velocities. Because of this, the depth conversion was performed using the velocity profile of *Holbrook* et al., [1996] for ODP Site 997, acquired just a few hundred meters to the Northwest. The 24 channel gather was corrected using a linear moveout, and each trace was time shifted to ensure that the maximum cross correlation between neighboring traces was at zero lag. The gathers were then stacked, resulting in the 24 fold stacked section seen in Figure 5. This effectively beamforms the array, resulting in better side-lobe and noise suppression than with single channel acquisition. Data from the vertical array are much more strongly affected by heave from rough seas than the data from the horizontal array, so the water bottom in these data has been artificially flattened to generate a coherent image.

Finally, the simplest means of imaging, illustrated in Figure 6, is a near trace gather, with each trace time shifted to correct for varying source depth. These data were acquired from the Southwestern flank of the Blake Ridge [*Rowe* and *Gettrust*, 1993a], approximately 120 km South of the data from the crest. As discussed below, important details are visible in each of Figures 4-6, which exhibit a range of image clarity.

1.4 Velocity Analysis

Because the technique of common midpoint CMP binning allows full angular coverage over a minimal lateral area, providing significant improvements over shot gathers, significant effort has been applied to correctly binning DTAGS data into CMP gathers. One of the obstacles to CMP binning is knowing the exact lateral position of every source and receiver. Fortunately the CMP binning is more sensitive to the shot to shot changes in position, which are known much better than the absolute geographic positions. The changes in lateral position can be estimated from changes in surface position, but absolute positioning requires extension of surface navigation to points hundreds to thousands of meters behind, and below the towing vessel.

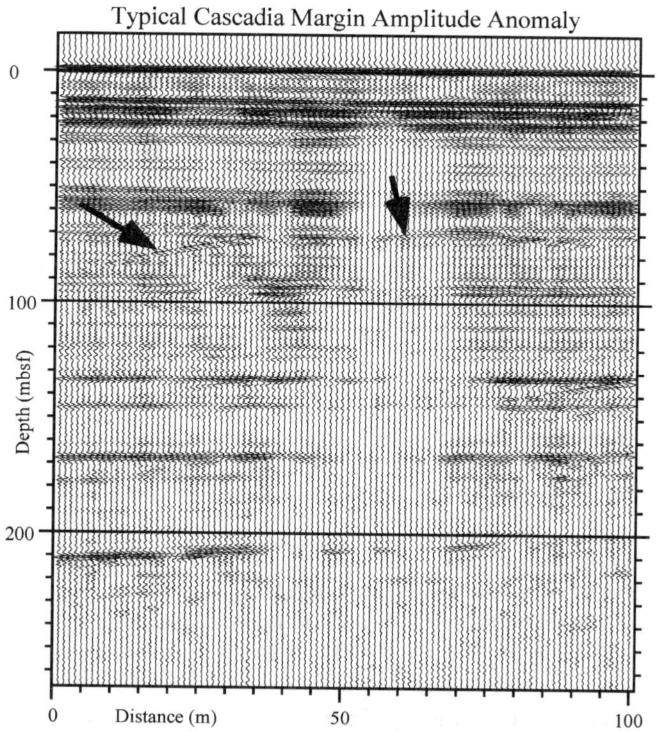

Figure 4. This vertically oriented zone of anomalously low reflectivity is typical of the many seen in DTAGS data acquired on the Cascadia Margin. Arrows mark diffractions emanating from the edge of the anomaly. Data for this image where deconvolved, NMO corrected, binned into 1.05 m wide bins making a single fold data set. The data where corrected for spherical divergence and gained with time variant, but laterally invariant function. The depth conversion assumes a constant velocity of 1600 m/s.

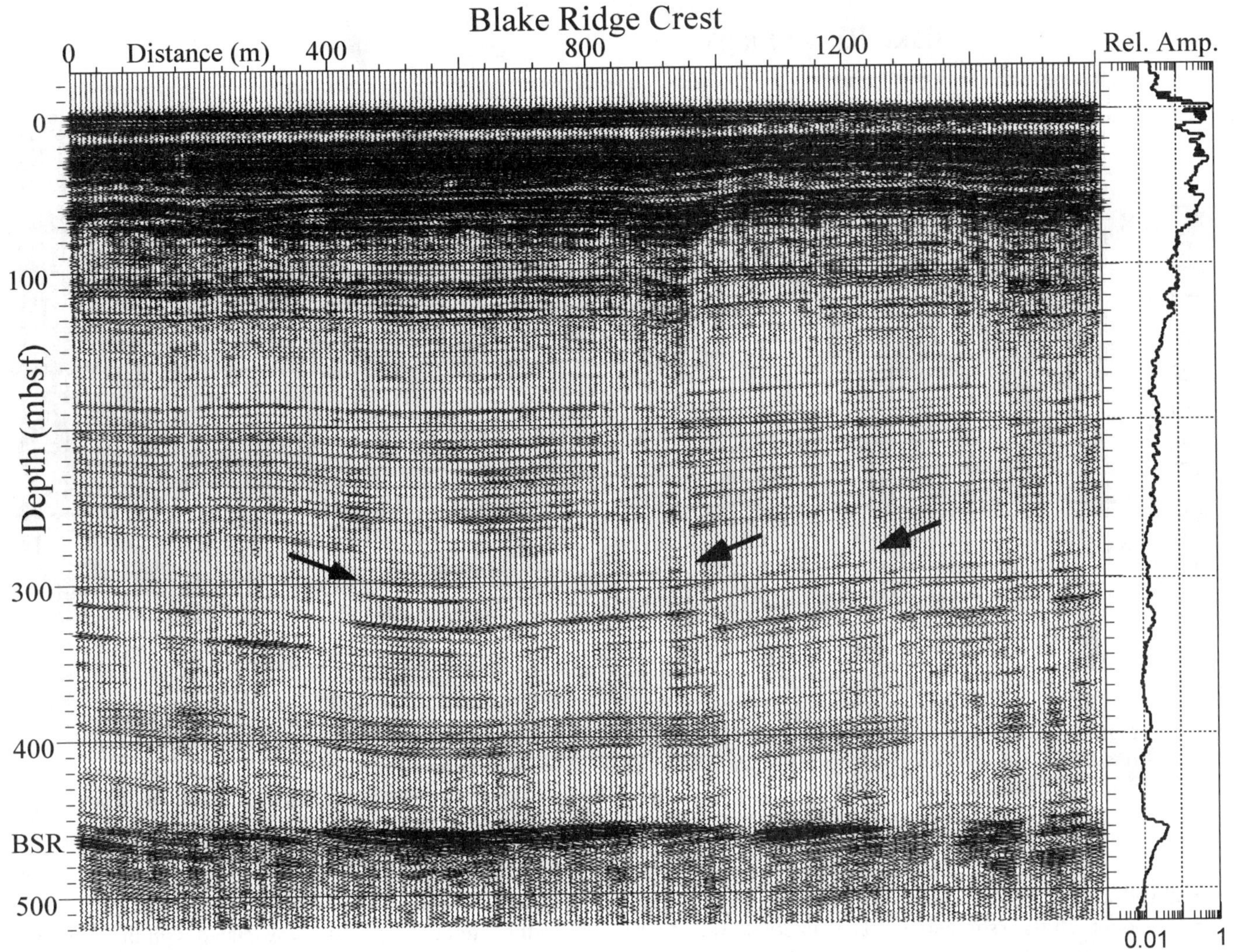

Figure 5. This seismic section from the crest of the Blake Ridge (B in Figure 1) was acquired a few hundred meters Southeast of ODP Site 997. Arrows mark syndepositional faults with offsets of about 15 meters at 400 mbsf but less than 1 meter at 30 mbsf. The BSR here is an intermittent zone of enhanced reflectivity through which individual reflectors can be tracked. The plot to the right shows the average amplitude vs. depth on a log scale, showing the BSR to be about 0.04 as strong as the seafloor reflection. Data for this image were acquired with a 24 channel vertical array, then linearly moved out and stacked. The data where corrected for spherical divergence and gained with time variant, laterally invariant function.

The other obstacle to CMP binning is knowing the vertical changes in position from shot to shot. Efforts to date have focused on altering or "correcting" the DTAGS shot gathers such that reflection event trajectories appear as they would had all the sources and receivers been at the same depth in the water column [*Walia* and *Hannay*, 1999]. The correction results in a coherent trajectory for reflection events between traces acquired with different source-receiver pairs, as in a CMP gather. Corrections can be made using trace dependent static time shifts, as was done for the data in *Rowe* and *Gettrust* [1993a,b], for which tests on synthetic data indicated an interval velocity accuracy of plus or minus 5%, (approximately 1% for each degree of dip in the array). The velocities determined using this technique for data over the Bermuda Rise are consistent with velocities obtained from nearby DSDP holes and OBH data [*Gettrust* et al. 1988]. However, because the angle of arrival of reflection events changes with increasing time, a single static shift will only be exact for a single time in the section. To address this issue, *Walia* and *Hannay* (1999)

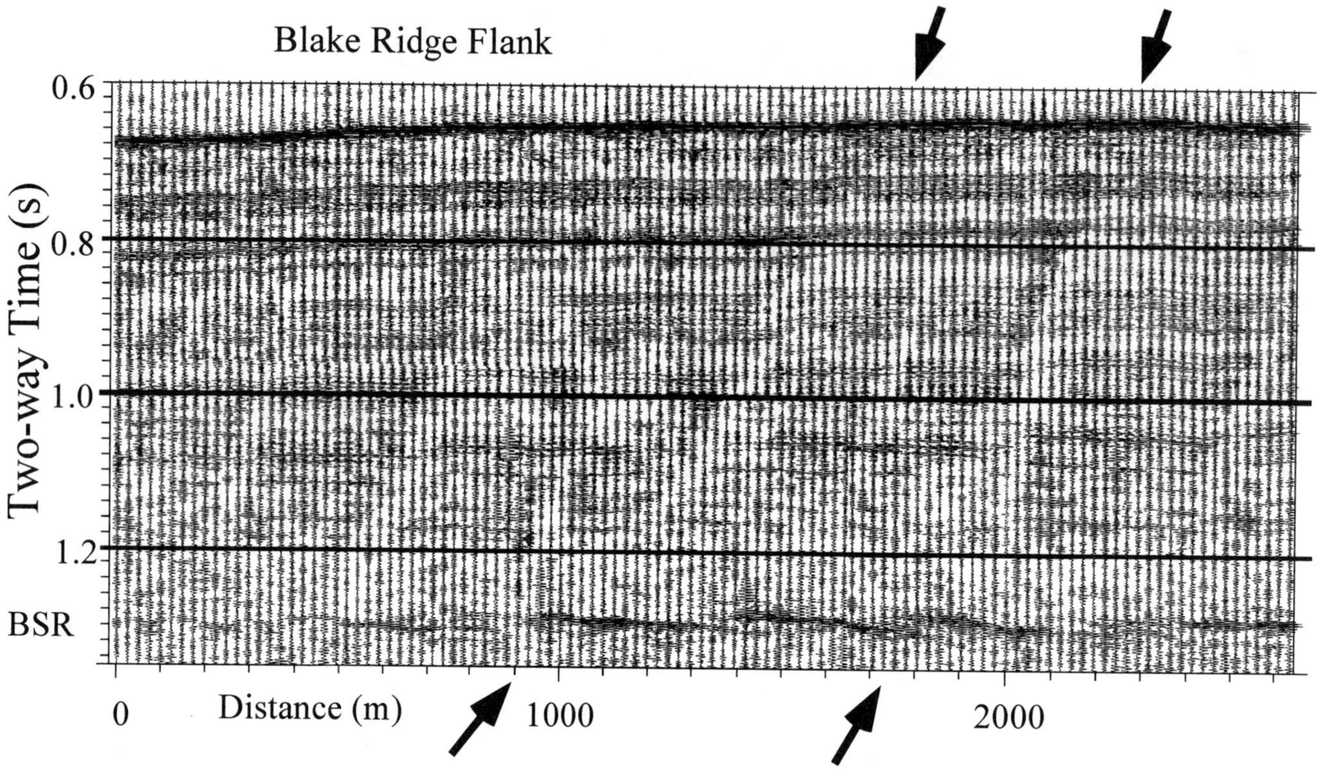

Figure 6. These DTAGS data from the Southwestern flank of the Blake Ridge reveal offset in the BSR where faults (marked by arrows) pass through the hydrate stability zone (*Rowe* and *Gettrust* [1993a]). Note also the intermittent nature of the BSR. The image here is a single channel section gain corrected with a time variant, laterally invariant function.

employed a more sophisticated correction technique using time variant shifts (stretching). The interval velocities determined with this method are consistent with those determined from surface-tow and from sonic logs at a nearby location (Figure 7, rightmost plot).

2. OBSERVATIONS FROM DEEP-TOW DATA

2.1. Cascadia Margin

The first area upon which we focus is the Cascadia Margin. The data shown in Figure 4 were acquired approximately 2 km South of ODP site 889 in an area known to contain methane hydrates [*Westbrook* et al., 1994] and to exhibit a strong BSR in surface-tow data [*Hyndman* and *Spence*, 1992]. Anomalously high seismic velocities [*Hyndman* and *Spence*, 1992] suggest the hydrate is widespread just above the BSR (on the scale of hundreds of meters), with an average concentration of 30% of the pore space. There is also evidence of elevated pore pressure and high permeability in the area [*Westbrook* et al., 1994], suggesting upward fluid advection, as would be expected from the lateral compression associated with the underlying convergent margin.

Several interesting features that appear at the scales resolved with DTAGS data are vertically oriented regions of anomalously low seismic amplitude extending through the HSZ. A typical example of these features is shown in Figure 4. The anomaly is only ~15 m wide at 30 mbsf near it's top, widening to 40-50 m at 210 mbsf, below which few coherent reflectors are visible anywhere in the section. The morphology of this feature is similar to the "columnar disturbances" observed world wide by Hovland and Judd [1988], but is still poorly understood. The anomalies in Figure 4 are much smaller than the VAMPs reported in the Bering Sea by *Scholl* and *Cooper* [1978], (50-100 meters wide, vs. 1-2 km wide) and also do not exhibit the clear velocity push-downs or pull-ups associated with the VAMPs. Many of the major reflectors in Figure 4 can be tracked through the anomaly and appear to exhibit enhanced reflectivities for about 10 m on either side of the anomaly. Diffractions from the edge of the anomaly are

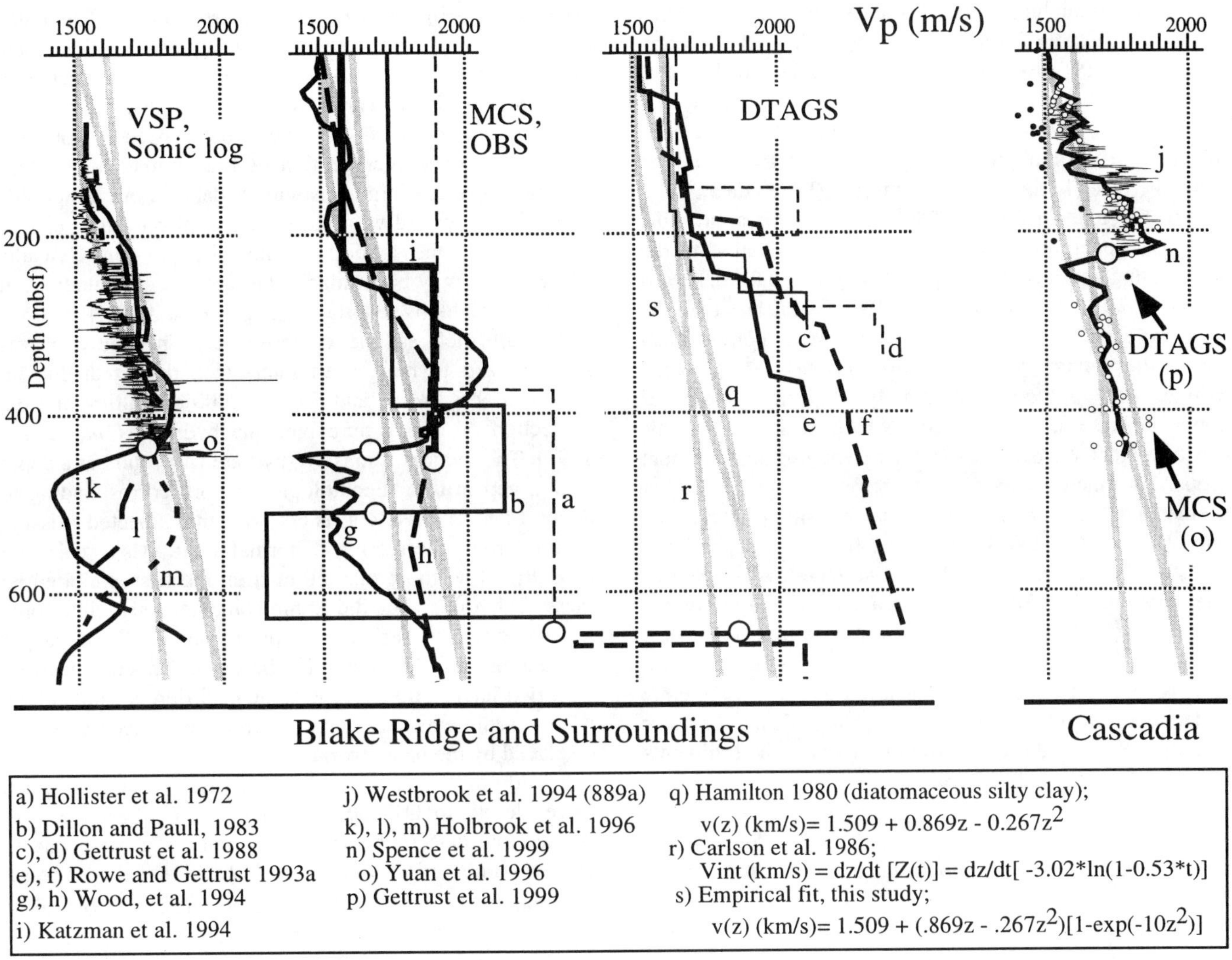

Figure 7. P-wave velocity (Vp) profiles using various techniques from the Blake Ridge and Bermuda Rise (left three plots) and Cascadia margin near ODP sites 889 and 890 (right most plot) exhibit a modest to significant increase in the HSZ. The BSR, usually seen as a sharp Vp decrease, lies at different depths for many of these plots, which are all displayed as a function of depth below the seafloor. Solid white circles indicate the depth to the BSR on profiles where appropriate.

visible on some, but not all of the reflectors (arrows, Figure 4), suggesting a few, very localized velocity or density contrasts at this boundary. Although no velocity pull-up or push-down is visible on the reflectors within the anomaly, it is possible that these events are from out of the plane of the section and actually exist outside the anomaly. The lateral resolution (accuracy in lateral placement of scatterers) is about ±5 meters in this section, estimated from the ability to locate the source of the diffractions.

Several hundred meters to the South, a similar amplitude anomaly is associated with a 10 m deep indentation in the seafloor. Extensive, high-resolution, surface-tow data, indicates that the indentations are spatially limited to less than 100 m, (*Ross Chapman*, U. of Victoria, pers. com.). i.e. they are pockmarks. By analogy it is likely that there is no preferred orientation to the anomaly, and its dimensions across the plane of Figure 4 are similar to its dimensions within the plane. The shape of the anomaly, and the appearance of a surface expression with a some of the anomalies, suggest that they are shoulders of roughly conical, or bell shaped features, many of which may have seafloor expressions out of the plane of Figure 4. We therefore assume that the anomalies are different views of separate, but similarly shaped features. The exact cause is unknown, although the vertical orientation of these features is very suggestive of some association with fluid advection through the HSZ. Because high concentrations of hydrates have been associated with localized fluid advection within the HSZ [*Wood* and *Ruppel*, 2000], we believe methane hydrates and gas are likely responsible for the amplitude anomalies seen in the Cascadia Margin DTAGS data.

Also revealed in the DTAGS data from this area is the very low apparent reflectivity of the BSR relative to the surrounding sediments. The reflectivity of ~0.02 observed with DTAGS data [*Gettrust* et al, 1999] is far less than the ~0.1 reflectivity found for the BSR with conventional surface-tow data [*Hyndman* and *Spence*, 1992] which exhibits an amplitude approximately as strong as the water bottom reflection. *Spence* et al. [1999] have shown that this can be explained by a vertical gradient in bulk sediment properties which occurs over several wavelengths at higher frequencies, thus creating a gradual transition at 450 Hz, but a sharp contrast at 45 Hz. Some frequency dependence is also exhibited in mid-frequency surface-tow, as well as deep-tow data. The BSR in this area is almost certainly due to gas accumulation, which we would assume to have a well defined top, even at DTAGS scales (a meter or two). The reason for a gradual transition is not completely understood, but may be due to vertical gradients, or other heterogeneities in the concentration of gas in the sediments just below the base of the HSZ.

2.2. *The Blake Ridge, Crest and Flank*

The second area upon which we focus is the Blake Ridge, (Figure 1). Two DTAGS surveys were conducted here, one on the crest of the ridge (Figure 5), in 2800 m of water, very near ODP Site 997, and the other on the southwestern flank of the ridge, in 4000 m of water, (Figure 6). When compared to conventional [e.g. *Dillon* et al., 1996], or even high resolution surface-tow [e.g. *Dillon* et al., 1994], the data from the crest in Figure 5, and flank in Figure 6 exhibit strikingly clear images of faults traversing the sediments between the BSR and seafloor (arrows in Figures 5 and 6). All the faults seen are extensional faults, typically listric in nature, ranging from near vertical a few tens of meters below the seafloor, to 45 degrees near the BSR. The throw on the faults increases gradually with depth for the last 5 Ma, indicating relatively constant activity on a million-year time scale, although possibly episodic activity on a shorter time scale. The faults on the crest penetrate the BSR resulting in little or no offset in the BSR itself. Signal penetration below the BSR is limited, obscuring detection of the level at which the faults sole out or stop [*Rowe* and *Gettrust*, 1993a]. The faults appear to be spaced 100 to 200 m apart, and from data acquired during a turn in which a fault was intersected, at least this fault appeared to trend roughly parallel to bathymetric contours, as would be expected from the compaction of a thick section of fine grained mud. Also notable in these high resolution data is an amplitude anomaly associated with the faults, similar to that seen on the sides of the amplitude anomalies in the Cascadia Margin, and may be similar to the upward continuation, or fingers of blanking reported by *Dillon* et al. [1994]

A fault such as one of those seen in Figure 5 was hypothesized to have been intersected during drilling at ODP Site 997 [*Wood* and *Ruppel*, 2000], resulting in a 0.2 m section of nearly pure methane hydrate [*Paull* et al., 1996]. The section of massive hydrate was found at a depth near a gap in the paleontological record corresponding to about 14 m. This is consistent with the expected missing section from intersecting a normal fault. Also, a second hole drilled no more than 20 m from the first exhibited no massive hydrate, but did exhibit anomalously high sonic log velocity and resistivity, approximately 35 m deeper than the massive hydrate in the first hole. This is consistent with two intersections of the same or different faults along which sediments are intermittently enriched with, or displaced by methane hydrate.

In addition to the faults, the BSR itself is imaged much more clearly in DTAGS data than even in good quality surface-tow data (compare for example, Figures 5 and 6, with Figure 1 in *Dillon* et al. [1996], revealing several important features of what we assume here is the coincident top of gas, and base of hydrate stability. First, the BSR itself is not a single reflection but a zone within which individual stratal reflectors exhibit relatively higher amplitude, (for simplicity we continue to use the term BSR to describe this zone). This observation is consistent with observations on the scale of tens to hundreds of meters indicating the BSR is a simple interface or layer which cuts across strata, but also shows the more complex nature of the BSR when viewed on the scale of meters. The polarity reversal observed in surface-tow data is not readily apparent in Figures 5 and 6, but we know from drilling that a strong decrease in acoustic impedance caused by the top of gas lies at 464 mbsf [*Paull* et al., 1996], coincident here with the top of the zone of enhanced reflectivity in Figure 5.

Also consistent with surface-tow data is the very low reflectivity (blanking) observed in the sediments within the HSZ [*Lee* et al. 1993, 1994]. The data in Figure 5 are displayed with a time variant gain so that the top of gas can be clearly seen, but the true amplitudes in this section (mean of the instantaneous amplitude of each trace, corrected only for spherical divergence and not effective attenuation), are shown in the plot to the right in Figure 5. The reflectivity observed in the DTAGS section from 100 mbsf to the BSR, is 10 to 100 times weaker than the water bottom reflection, comparable to the values (e.g. about 30 times weaker) reported by *Lee* et al. [1994] for conventional surface-tow data.

Another issue raised by the data in Figures 5 and 6 is the intermittent nature of the BSR on the scale of meters to tens of meters. The reflection amplitudes change by about a factor of five over just a few traces (trace spacing is about 8 m), suggesting, not unexpectedly, poor lateral communication in these clay rich sediments. The breaks, or holes in the BSR appear to be associated with the faults mentioned above, although the data do not show this to be unequivocally true.

Although quite similar in most respects to the BSR on the ridge crest, the BSR on the ridge flank exhibits significant offsets (up to approximately 20 m) at the faults [Rowe and Gettrust 1993b]. Because the faults are syndepositional, and no 20 m scarp exists at the seafloor, we assume that the offset in the BSR is not related to the offset exhibited in the strata, and that this offset is relatively stable. A stable discontinuity in temperature across a fault would be difficult to explain, but a pressure change from hanging wall to foot wall due to differential stress may be responsible for the discontinuous stability boundary, and subsequent offset in the BSR. *Rowe* and *Gettrust* [1993a] estimate this pressure contrast to be about 200 kPa.

Another difference between the Blake Ridge crest and flank, and possibly also related to the faults, is the somewhat higher P-wave velocity (Vp), and the lateral variability of Vp in the HSZ on the flank. Figure 8 (redrawn from *Rowe* and *Gettrust*, [1994]) shows Vp as a fraction of the baseline reference Vp profile. The reference profile was an average of nearby velocity analyses which showed no BSR, and were assumed to be hydrate free. Figure 8 shows the anomaly beginning at approximately 220 mbsf, consistent with the top of hydrate occurrence from direct sampling (200-240 mbsf from *Paull* et al., [1996]) and inference from velocity anomaly (230 mbsf from *Wood* et al. [1994]), and increasing downward to the BSR (buried here at approximately 700 mbsf as opposed to 464 at the crest). Surprisingly, the velocity field here exhibits vertical features, with some isoanomaly contours perpendicular to the relatively horizontal strata in this area. Reported values of Vp at the base of the HSZ range from 1.8 to 2.5 km/s [*Rowe* and *Gettrust* 1993a, 1994]. The lower of these values is consistent with the downward projection of gas free sediment Vp at the crest (curve s in

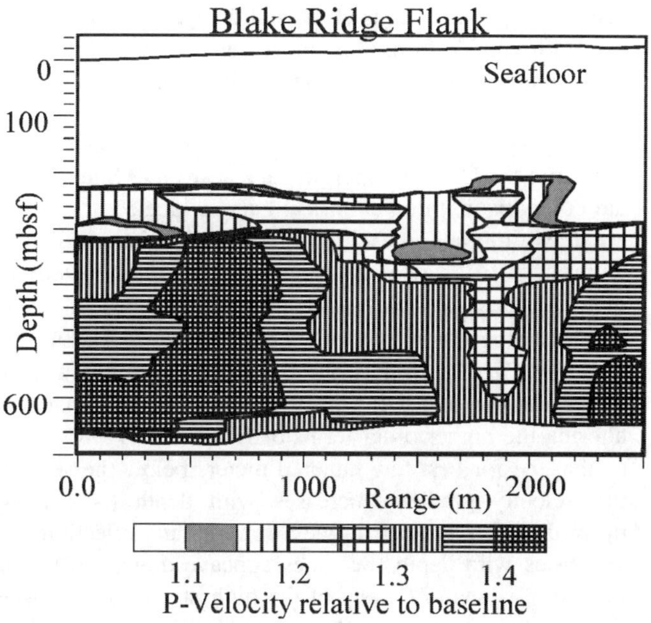

Figure 8. A velocity section based on velocity analyses from the Blake Ridge Flank and plotted in units of fraction deviation from a reference profile, shows significant lateral variability and isoanomaly contours which are vertical in some areas, (redrawn from *Rowe* and *Gettrust*, [1994]).

Figure 7, which is probably close to the hydrate free Vp as well), and the higher is consistent with early measurements by *Hollister* et al. [1972], at DSDP Site 103 (curve a in Figure 7). Although DTAGS velocities are not available from the crest, even the very local VSP measurements suggest a much lower overall Vp, and significantly less lateral variability across the crest.

3. ANALYSES AND INTERPRETATION

3.1. Sediment P-Wave Velocity

Quantitative estimates of hydrate concentrations based on the high velocity anomaly expected from hydrate enriched sediment [*Stoll* et al 1979, and others] have been made using DTAGS data [*Rowe* and *Gettrust* 1994] as well as conventional surface-tow data [*Wood* et al. 1994, *Yuan* et al., 1996]. For both kinds of data, the difficulty in applying this technique is due to two factors; uncertainty in the baseline velocity profile which matches what the sediments would exhibit if no hydrate were present; and uncertainty in the relationship between the hydrate concentration and effective velocity increase. Although neither of these problems are completely resolved here, the importance of accurate Vp estimates is clear, and comparison of Vp from DTAGS with Vp from other measurements provides insight into possible models of hydrate distribution.

Shown in Figure 7 are several (P-wave only) interval velocity profiles from the Blake Ridge and surrounding area, and the Cascadia margin very near ODP Site 889. To aid comparison, three empirical curves, labeled q, r, and s, are drawn in gray on each plot. The curve labeled "q" is an empirical fit of global data made by *Hamilton* [1980] for diatomaceous silty clay, and the curve labeled "r" is an empirical logarithmic fit of globally acquired DSDP data made by *Carlson* et al., [1986], and s is an empirical fit to the sonic log of ODP Site 997. Curve s was made by altering the higher order terms of *Hamilton's* [1980] curve so that the for first few hundred meters below the seafloor the velocity gradient increases with depth (is concave upward), after which it goes through an inflection and decreases with depth (becomes concave downward). This inflection is seen in many of the highest resolution curves in Figure 7, and cannot be fit with the curves of either Hamilton [1980] or Carlson et al. [1986].

The most local measurements of velocity in the Blake Ridge area are the VSP (curves k, l, and m in Figure 7) and sonic log data (curve o) which also exhibit the lowest values for the sediments within the HSZ. These zero offset measurements were both made using energy traveling one way vertically through the sediment. The measurements made using a finite angle range at the sea surface (OBS, curve i, and MCS curves a, b, g, and h) all exhibit a Vp profile somewhat higher than that of the more local measurements. This is not unexpected given the reported P-wave anisotropy in these sediments [*Pecher* et al., 1997]. The finite offset measurements are based on travel times of seismic energy passing through the sediments at some angle with the vertical, and for tens to hundreds of meters laterally through the HSZ. The best resolved profile in this group, g, and the widest angle measurement, i, parallel the zero offset measurements for about the first 240 m below the seafloor after which they diverge, indicating higher Vp just above the BSR than the VSP or log data. A depth of 240 mbsf corresponds quite closely to the shallowest occurrence of both recovered hydrate and hydrate inferred from the chloride anomaly proxy [*Paull*, 1996].

The Vp profiles obtained from DTAGS data (curves c, d, e, f in Figure 7) also show an increase in the downward Vp gradient to approximately 200 ± 50 m, even in the profiles from the Bermuda Rise (curves c and d) where no hydrates are found. The Vp curves from the Blake Ridge flank (curves e and f), especially curve f, exhibit significantly higher velocities than seen on the crest, and may be indicative of greater advection, and greater hydrate formation on the flank than on the crest.

The Vp profiles from the Cascadia margin displayed in Figure 7 exhibit somewhat higher values than on the crest of the Blake Ridge, and are estimated to contain a higher percentage of hydrate, 20-30% [*Hyndman* and *Spence* 1992, *Spence* et al. 1999] vs. about 5% [*Paull* et al. 1996] for the Blake Ridge crest. Estimated rates of pore fluid expulsion in the two areas are similar, 1.3 mm/yr for Cascadia [*Hyndman* et al., 1993], and less than several mm/yr for the Blake Ridge (determined independently with different means by *Egeberg* and *Dickens*, [1999] and *Ruppel*, [2000]). The Vp trend through the HSZ at the Cascadia site is comparable to that of the Blake Ridge flank, but the relatively shallow BSR, 230 vs. 650 mbsf at the Blake Ridge flank make direct comparison difficult. The best resolved Vp data at the Cascadia site, the ODP Site 889 sonic log (curve j), and the waveform inversion result (curve n), both exhibit the same downward decrease in velocity gradient seen in the curves from the Blake Ridge and surroundings, although the overall Cascadia Vp gradient is higher.

The Vp comparison displayed in Figure 7 raises several important issues concerning the use of velocities to quantify hydrate distribution. If the baseline reference Vp for hydrate- and gas-free sediments exhibits a gradient increase (concave upward velocity profile) somewhere within the HSZ (e.g. 240 mbsf), this could potentially decrease the magnitude of the positive Vp anomaly associated with the HSZ, and decrease estimates of hydrate concentration. Straight line trends, or even simple global curves, such as those derived by *Hamilton* [1980] or *Carlson* et al. [1986], may introduce a systematic error in estimates of hydrate concentration. Exacerbating this uncertainty is the fact that the accuracy of both the velocity analysis and reference profile decrease with depth. Where the BSR is deep, and hydrate concentrations possibly the highest, the uncertainty in concentration is also highest.

Perhaps most significant is that all of the seismic reflection methods rely on near horizontal reflections, and inherently average the properties of the medium over a scale of approximately 10 meters for DTAGS, and 100 meters for conventional seismic frequencies. Massive hydrate, such as that found at ODP Site 997, or even interstitial hydrate could exist in significant quantities without affecting reflection velocity analyses if oriented along near vertical fluid advection pathways.

3.2 A Model of Hydrate Distribution

The contributions from DTAGS, combined with many other measurements, have strengthened the notion that upward advection of pore fluids is an important if not dominant factor in the transport of methane, and therefore the distribution of methane hydrate within the HSZ. Although the transfer of methane through the HSZ entirely in the gas phase is possible under certain circumstances, the discontinuity in the BSR is evidence of the difficulty in moving bubbles through sediment, leaving transport by advecting fluids or chemical diffusion [*Xu* and *Ruppel*, 1999], more likely mechanisms.

The advective transport of methane is likely to be distributed between two end member cases, diffuse and channeled flux. The diffuse flux, within which each grain is surrounded by moving fluids, as well as chemical diffusion, is controlled by the permeability and porosity of the frequently near-horizontal strata within the HSZ. This flux is therefore likely to have a large lateral component, and be conformable with these strata. The channeled flux, in which a few preferred pathways transport the majority of the fluid, is controlled by the fluid flux pathways; it is likely to be oriented vertically. The relative amounts of methane transported in each of these ways is not known, and could be quite variable in both space and time.

The methane transport model also affects the distribution of the hydrate within the HSZ [*Xu* and *Ruppel*, 1999]. We assume diffusion and diffuse flux are more likely to result in the evenly distributed, disseminated hydrate, typically invisible in cores and only detected by temperature or chloride anomalies. Channeled flux may result in pathways continually clogging with hydrate, and re-establishing at the nearest weak spot, namely the boundary between the sediment and the just formed hydrate blockage. This dynamic, expanding block may completely fill the pore spaces or even exclude the sediment, and when cored appears as a section of completely cemented or massive hydrate.

The two styles of hydrate accumulation can be thought of as two subreserviors (disseminated and concentrated) of the overall methane hydrate reservoir. Because channeled fluid flux tends also to transport heat, the concentrated subreservoir is not only proximal to fluid pathways but may also be warmer, and therefore less stable. If solubility conditions or methane supply changes, due to sea level or bacterial activity changes, for example, the advecting fluids which were once a source of methane for hydrate formation may relatively quickly become a methane sink for dissociating hydrate. Thus the concentrated subreservoir may be significantly more volatile than the disseminated subreservoir. As alluded to earlier, the concentrated reservoir may also be much more difficult to detect and quantify, using multichannel seismic techniques better suited for near-horizontal features.

In areas of more intense fluid advection, such as the Cascadia Margin, the pore fluids may carry enough heat into the HSZ to significantly distort the base of the HSZ, perhaps so much so that the base of the HSZ intersects the seafloor and hydrate is stable nowhere below this point. The steep-sided bell-shaped hole through the HSZ, which we hypothesize is what is depicted in Figure 4, could be the base of the HSZ, with gas on the inside and below, and hydrate accumulations on the outside and above. The reduced seismic amplitude anomalies seen in the DTAGS data are consistent with the presence of gas within the anomaly, and the occasional diffractions observed at the edges of the anomaly are consistent with highly localized (less than a few meters in scale) gas or hydrate accumulations. A thermal perturbation would also result in a wider anomaly at the BHSZ, where hydrates are less stable, than near the seafloor. This character is also seen in Figure 4. However, as in the case of the frequency dependent BSR discussed above, seismic data alone provide insufficient constraints to unambiguously distinguish between a gas-filled zone which absorbs seismic energy, and a hydrate-filled zone which scatters seismic energy.

3.3. Blanking

The conceptual model of enhanced hydrate accumulation associated with vertical fluid flux is also consistent with at least some of the anomalously low seismic reflection amplitudes (blanking) observed on the Blake Ridge. Along the preferred pathways the hydrate concentrations may be high enough to affect seismic velocity. *Dillon* et al. [1994] describe vertical continuations, or fingers of blanking that extend upward into the HSZ from faults below the BSR. Such blanking zones may continue to even shallower depths in the section along unresolved faults. Cementation from high concentrations of hydrate are more likely in fault zones than in adjacent, unfaulted sediments characterized by presumably lower fluid flux. However, these potentially cemented fault zones are not oriented horizontally, may be quite irregular in shape and impedance, and may thus be more efficient in scattering seismic energy outward rather than directly upward to the receivers. Thus, a "blank" zone in seismic data may imply a paucity of nearly horizontal reflectors, not necessarily a lack of impedance contrasts. The kind of scattering described here typically exhibits strong frequency dependence [e.g. *Clay* and *Medwin*, 1977] and would likely yield different results when studied with seismic systems of different frequencies.

Another phenomenon that may be responsible for blanking at larger scales is the increasing throw with depth on the syndepositional faults imaged in the DTAGS seismic data (Figures 5 and 6). A spherical spreading seismic wave has a finite lateral resolution that decreases with decreasing frequency and increasing distance from the reflector. Inhomogeneities in the surface of the reflector, are averaged when below the limits of resolution, and generally result in lower reflectivity being recorded by the receivers [*Clay* and *Medwin*, 1977]. For wavelengths of 30 to 60 m, (approximately 30 to 60 Hz), a 15 m displacement in a reflector creates destructive interference, and the observed reflectivity is diminished. For data of this frequency we would therefore expect a decrease in observed reflectivity with depth, as the lateral heterogeneity increases due to increasing fault throw. Destructive interference will increase with the distance from the source as more heterogeneity is incorporated into the first Fresnel zone. The lateral heterogeneity of the faulted sediments, combined with the non-horizontal distribution of hydrate, suggests that the reduced reflectivity is more likely due to scattering and destructive interference than to reduced impedance contrasts from sediment cementation. Relative hydrate concentration may still be responsible for seismic blanking, if the pathways containing the hydrate are the faults which degrade the reflection amplitudes.

4. CONCLUSIONS

The high resolution images and Vp profiles prepared from DTAGS data provide information at a lateral and vertical scale which compliments not only the spatially extensive, but lower resolution conventional surface-tow seismic techniques, but also the sub-meter scale well and core data. This broad range of scales allows for a much more thorough understanding of gas hydrate distribution, and has contributed to a scale dependent model of hydrate distribution. On a scale of hundreds to thousands of meters, hydrate is certainly distributed horizontally in that it is limited to the HSZ, but on the scale of tens of meters and smaller, the localized accumulations may be oriented vertically. Seismic images reveal vertically oriented features such as faults in the Blake Ridge and zones of diminished amplitude in the Cascadia Margin which are likely associated with upward fluid advection and hydrate concentration.

Quantification of hydrate by analysis of the expected Vp anomaly is plagued by several key problems, including the vertical orientation of possibly significant quantities of hydrate associated with fluid advection conduits. Seismic velocity analyses are most sensitive to thick, horizontal, layers, not irregular, vertical slabs.

DTAGS data have also allowed newer, more sophisticated interpretations of the BSR, due to the scale dependence of this phenomenon. In the Blake Ridge the BSR is a very distinct zone of enhanced amplitude through which individual reflections can be tracked. Conversely, portions of the BSR in the Cascadia Margin appear as gradational boundaries appearing strong in seismic data only when wavelengths are much longer than the thickness of the gradient layer.

Because of the detailed fine structure revealed at the meter to ten meter scale, there appears to be significant promise in highly coordinated, multidisciplinary studies, in which environments around individual fluid conduits may be studied and placed in the proper chemical and hydrologic context.

Acknowledgments. The authors thank C. Paull and W. Dillon for inviting this contribution, and K. Andreassen and P. Miles for helpful reviews. This work was supported by ONR Program element 0601153N.

REFERENCES

Carlson, R. L., A. F. Gangi, and K. R. Snow, Empirical reflection-travel time/depth and velocity/depth functions for the deep-sea sediment column, *J. Geophys. Res.*, 91, 8249-8266, 1986.

Clay, C. S., and H. Medwin, *Acoustical Oceanography: principles and applications*, John Wiley & Sons, New York, 1977.

Clennel, M. B., M. Hovland, J. S. Booth, P. Henry, and W. J. Winters, Formation of natural gas hydrates in marine sediments. Part 1: Conceptual model of gas hydrate growth conditioned by host sediment properties, *J. Geophys. Res.*, In press, 1999.

Cragg, B. A., R. J. Parkes, J. C. Fry, A. J. Weightman, P. A. Rochelle, and J. R. Maxwell, Bacterial populations and processes in sediments containing gas hydrates (ODP Leg 146 Cascadia Margin), *Earth Planet Sci. Lett.*, 139, 497-507, 1996.

Dickens, G. R., M. M. Castillo, and J. C. G.Walker, A blast of gas in the latest Paleocene: Simulating first-order effects of massive dissociation of oceanic methane hydrate, *Geology*, 25, 259-262, 1997.

Dillon, W. P., M. W. Lee and D. F Coleman, Identification of Marine hydrates in situ and their distribution off the Atlantic coast of the United States. Annals of the New York Academy of Sciences, 17, 364-380, 1994.

Dillon, W. P., D. R. Hutchinson, and, R. M. Drury, Seismic reflection profiles on the Blake Ridge near Sites 994, 995, and 997, edited by C. K. Paull, R. Matsumoto, and P. J. Wallace, et al., Proc. ODP, Init. Repts., 164: College Station, TX (Ocean Drilling Program), 47-56., 1996.

Egeberg, P. K. and G. R. Dickens, Thermodynamic and pore water halogen constraints on gas hydrate distribution at ODP Site 997 (Blake Ridge), *Chemical Geology*, 153, 53-79, 1999.

Gettrust, J. F., M. Grim, S. Madison, and M. Rowe, Results of a deep-tow multi-channel survey on the Bermuda Rise, *Geophys. Res. Lett.*, 15, 12, 1413-1416, 1988.

Gettrust, J. F., J. H. Ross, and M. M. Rowe, Development of a low frequency, deep-tow geoacoustics system, *Sea Technology*, 32, 23-32, 1991.

Gettrust, J. F., W. T. Wood, D. A. Lindwall, R. Chapman, R. Walia, D. Hannay, G. Spence, K. Louden, R. MacDonald, and R. D. Hyndman, New Seismic Study of Deep Sea Gas Hydrates Results in Greatly Improved Resolution, *Eos*, 80, 38, 439-440, 1999.

Hamilton, E. L., Geoacoustic modeling of the sea floor, *J. Acoust. Soc. Am.*, 68, 5, 1313-1340, 1980.

Hollister, C. D., J. I. Ewing, D. Habib, J. C. Hathaway, Y. Lancelot, H. Luterbacher, F. J. Paulus, C. W. Poag, J. A. Wilcoxon, and P. Worstell, Sites 102-103-104 - Blake-Bahama Outer Ridge (northern end), in Initial Reports of the Deep Sea Drilling Project, edited by C. D. Hollister, J. I. Ewing, et al., U.S. Government Printing office, Washington, D. C., 11, 135-143, 1972.

Holbrook, W. S., H. Hoskins, W. T. Wood, R. A. Stephen, D. Lizarralde, and Leg 164 Science Party, Methane Hydrate and Free Gas on the Blake Ridge from Vertical Seismic Profiling, *Science*, 273, 1840-1843, 1996.

Hovland, M., and A. G. Judd, *Seabed pockmarks and seepages*, Graham & Trotman, Boston, 1988.

Hyndman, R. D., and G. D. Spence, A seismic study of methane hydrate marine bottom simulating reflectors, *J. Geophys. Res.*, 97, 6683-6698, 1992.

Hyndman, R. D., K. Wang, T. Yuan, and G. D. Spence, Tectonic Sediment Thickening, Fluid Expulsion, and the Thermal Regime of Subduction Zone Accretioary Prisms: The Cascadia Margin off Vancouver Island, *J. Geophys. Res.*, 98, 12, 21865-21876, 1993.

Katzman, R., W. S. Holbrook, and C. K. Paull, Combined vertical-incidence and wide-angle seismic study of a gas hydrate zone, Blake Ridge. *J. Geophy. Res.*, 99, 17, 975-17, 995, 1994.

Korenaga, J., W. S. Holbrook, S. C. Singh, and T. A. Minshull, Natural gas hydrates on the southeast U. S. Margin: Constraints from full waveform and travel time inversions of wide angle seismic data, *J. Geophys. Res.*, 102, 15345-15365, 1997.

Lee, M. W., D. R. Hutchinson, W. P. Dillon, J. J. Miller, W. F. Agena, and A. B. Swift, Method of estimating the amount of in-situ gas hydrates in deep marine sediments, *Mar. Petr. Geol.*, 10, 493-506, 1993.

Lee, M. W., D. R. Hutchinson, W. F. Agena, W. P. Dillon, J. J. Miller, and A. B. Swift, Seismic character of gas hydrates on the southeastern U. S. continental margin, *Marine Geophys. Res.*, 16, 163-184, 1994.

Markl, R.G., G. M. Bryan, and J. I Ewing, Structure of the Blake-Bahama outer ridge, *J. Geophy. Res.*, 75, 4539-4555, 1970.

Paull, C. K. and W. P. Dillon, Appearance and distribution of the gas hydrate reflection in the Blake ridge region, offshore southeastern United States, Map, U.S. Geological Survey, Reston, VA, 1981.

Paull, C. K., W. Ussler, III, and W. P. Dillon,. Is the extent of glaciation limited by marine gas-hydrates?, *Geophys. Res. Lett.*, 18, 432, 434, 1991.

Paull, C. K., R. Matsumoto, and P. J. Wallace, et al., Proc. ODP, Initial Reports, College Station, TX (Ocean Drilling Program), 164, 1996.

Pecher, I. A., W. S. Holbrook, R. A. Stephen, H. Hoskins, D. Lizeralde, D. R. Hutchinson, and W. T. Wood, Offset-vertical seismic profiling for marine gas hydrate exploration - is it a suitable technique? First results from ODP Leg 164, Offshore Technology Conference, 1997.

Rowe, M. M., and J. F. Gettrust, Faulted structure of the bottom simulating reflector on the Blake Ridge, Western North Atlantic, *Geology*, 21, 833-836, 1993a.

Rowe, M. M., and J. F. Gettrust, Fine structure of Methane hydrate-bearing sediments on the Blake Ridge as determined from deep-tow multi-channel seismic data, *J. Geophys. Res.*, 98, 463-473, 1993b.

Rowe, M. M. and J. F. Gettrust, Methane hydrate content of Blake Outer Ridge sediments, Annals of the New York Academy of Sciences, 17, 491-494, 1994.

Ruppel, C., Fluid flux and thermal regimes in the Blake Ridge gas hydrate province, *Earth Planet. Sci. Lett.*, submitted 2000.

Scholl, D. W. and A. K. Cooper, VAMPs - Possible Hydrocarbon-Bearing, Structures in the Bering Sea Basin, *AAPG Bull.*, 62, 2481-2488, 1978.

Singh, S. C., T. Minschull, and G. Spence, Velocity structure of a gas hydrate reflector, Science, 260, 204-207, 1993.

Sloan, E. D., *Clathrate Hydrates of Natural Gases*, New York (Marcel Deker), 1990.

Spence, G. D., R. D. Hyndman, N. R. Chapman, R. Walia, J. F. Gettrust, and R. N. Edwards, Anals of the New York Academy of Sciences, Third International Conference of Gas Hydrates, Park City, Utah, July 18-22, 1999.

Stoll, R. D., J. Ewing, and G. M. Bryan, Anomalous wave velocities in sediments containing gas hydrates, *J. Geophys. Res.*, 76, 2090-2094, 1971.

Stoll, R. D., G. M. Bryan, PHYSICAL PROPERTIES OF SEDIMENTS CONTAINING GAS HYDRATES, *J. of Geophys. Res.*, 84, B4, April 10, 1979.

Walia, R. and D. Hannay, Source and receiver geometry corrections for deep-towed multi-channel seismic data, *Geophys. Res. Lett.*, 26, 13, 1993-1996, 1999.

Wellsbury, P. K., B. A. Goodman, Cragg, and R. J. Parkes, The geomicrobiology of deep marine sediments from Blake Ridge containing methane hydrate (Sites 994, 995 and 997),edited by C. K.Paull, R. Matsumoto, P. J. Wallace, and W. P. Dillon, 2000 Procedings of the Ocean Drilling Program, Scientific results.

Westbrook, G. K., B. Carson, R. J. Musgrave, et al., Proc. ODP, Init. Rpts., College Station, TX (Ocean Drilling Program), 146 (Pt. 1), 55-125, 1994.

Wood, W. T., P. L. Stoffa, and T. H. Shipley, Quantitative detection of methane hydrate through high resolution seismic velocity analysis. *J. Geophys. Res.*, 99, 9681-9695, 1994.

Wood, W. T., and C. Ruppel, Seismic and thermal investigations of the Blake Ridge hydrate area: A synthesis, edited by C. K. Paull, R. Matsumoto, P. J. Wallace, and W. P. Dillon, Proc. ODP Sci. Results, College Station, TX, (Ocean Drilling Program) 164, 253-264, 2000.

Xu, W., and C. Ruppel, Predicting the occurrence, distribution, and evolution of methane gas hydrate in porous marine sediments, *J. Geophys. Res,* 104, 5081-5096, 1999.

Yuan, T., R. D. Hyndman, G. D. Spence, and B. Desmons, Seismic velocity increase and deep-sea gas hydrate concentration above a bottom-simulating reflector on the northern Cascadia continental slope, *J. Geophys. Res.*, 101, 13, 655-13-671, 1996.

Wood, Warren T. and Joseph F. Gettrust, Naval Research Laboratory, Code 7432, Stennis Space Center, MS 39529, USA

Comparison of Elastic Velocity Models for Gas-Hydrate-Bearing Sediments

Myung W. Lee and Timothy S. Collett

U.S. Geological Survey, Denver Federal Center, Denver Colorado

One of the distinct physical properties of gas-hydrate-bearing sediments is elevated seismic velocities. A number of velocity models and equations have been presented to describe the effect of gas hydrate on the seismic velocities; e.g., pore-filling model, cementation model, effective medium theory, a weighted equation, and time-average equation. The data set from Mallik 2L-38 gas hydrate research well drilled in northern Canada provided us a unique opportunity to test the velocity models for gas-hydrate-bearing sediments. Velocities predicted from an effective medium theory and those from a weighted equation are compared with observed well log velocities. In the case where there is no gas hydrate in the pore space, P-wave velocities predicted from the effective medium theory are lower than those from the weighted equation when porosity is less than about 30% and higher when porosity is higher than about 30%. For S-waves, effective medium theory predicts generally higher velocities than those from the weighted equation. Both theories predict similar increases in P- and S-wave velocities when gas hydrate occupies the pore space. Even though gas hydrate concentration in the pore space is not known accurately, analyses using both P- and S-wave velocities and their ratios enable us to test the validity of velocity models. Considering only P-wave velocities, there is not much difference between the effective medium and weighted equation. However, considering both P- and S-wave velocities and their ratios, the weighted equation is preferred to the effective medium theory in predicting elastic wave velocities for gas-hydrate-bearing sediments at the Mallik 2L-38 well.

INTRODUCTION

Gas hydrates, or clathrates, are ice-like crystalline solids composed of water molecules surrounding gas molecules. In recent years, methane hydrates have been the focus of many investigations because of their widespread occurrence in most of the world oceans and in permafrost regions [*Kvenvolden*, 1993]. Also, large amounts of methane may have escaped from deep-sea gas hydrate-bearing sediments to the atmosphere as a result of dissociation of gas hydrates caused by sea-level and climatic changes, and acted as a negative feedback control on global temperature fluctuations [*Dillon et al.,* 1991; *Paull et al.*, 1991]. Gas hydrate is also of potential importance as a potential energy resource and significance as sea floor stability and safety issues [*Dillon et al.,* 1993]. To use seismic techniques to quantify the amount of gas hydrate or to infer the physical properties of gas-hydrate-bearing sediments, a predictive model that relates the amount of gas hydrate in the pore space of sedimentary sequence to seismic velocities is required.

Gas hydrates exhibit relatively high elastic velocities, both compressional (P-wave) and shear wave (S-wave) veloci-

Natural Gas Hydrates: Occurrence, Distribution, and Detection
Geophysical Monograph 124

Published in 2001 by the American Geophysical Union

ties, compared to the pore-filling fluids, therefore the velocity of gas hydrate-bearing sediments are usually elevated [*Stoll*, 1974; *Tucholke et al.*, 1977]. Earlier studies of the velocity model for gas-hydrate-bearing sediments used various forms of time average equations [*Miller et al.*, 1991; *Bangs et al.*, 1993; *Wood et al.*, 1994]. In order to fit the observed velocity of hydrate-bearing sediments, *Miller et al.* [1991] used a three-component (water, gas hydrate, and non-gas-hydrate-bearing sediment) time-average equation and *Wood et al.* [1994] used a two-component (gas hydrate and non-gas-hydrate-bearing sediment) time-average equation.

Hyndman and Spence [1992] used a velocity model that predicts the velocity of hydrate-bearing sediment by assuming that the porosity reduction by infilling with gas hydrates has the same effect on velocity as the normal compaction porosity reduction with depth.

Dvorkin and Nur [1993] developed a cementation model for the velocity of gas-hydrate-bearing sediments. This theory predicts a dramatic increase of P- and S-wave velocity even with a small amount of gas hydrate, because intergranular gas hydrate cementation significantly increases the elastic moduli of the dry rock frame. *Helgerud et al.* [1999] proposed an effective medium model for elastic-wave velocities in unconsolidated and high porosity sediment and successfully applied this theory to P-wave velocity data from Site 995 drilled by Ocean Drilling Program (ODP) on Leg 164. Also *Jakobsen et al.* [2000] developed a theory to relate seismic velocities to the gas hydrate concentration based on a combination of a self-consistent approximation, a differential effective medium theory, and applied this theory to P- and S-wave velocity data from Site 995 drilled by Ocean Drilling Program (ODP) on Leg 164.

In order to overcome problems associated with the unconsolidated nature of near-surface sediments, *Lee et al.* [1996] proposed a weighted equation using an empirically derived time average equation [*Wyllie et al.*, 1956] and *Wood* [1941] equation. The weighted equation accurately predicted elastic velocities of permafrost samples [*Lee at al.*, 1996] and estimated consistent in situ gas-hydrate concentrations from the P- and S-wave data acquired at Mallik 2L-38 hydrate research well [*Lee and Collett*, 1999].

The purpose of this paper is to compare observed velocities of gas-hydrate-bearing sediments with those predicted from two theories: one theory based on the effective medium model [*Helgerud et al.*, 1999] and the other based on the weighted equation [*Lee et al.*, 1996]. The data used in this study are a suite of logs including P-wave and S-wave velocities obtained from the Mallik 2L-38 gas hydrate research well drilled in the Mackenzie River Delta of northern Canada [*Collett et al.*, 1999]. The bore hole conditions and well logs associated with the Mallik 2L-38 well were excellent in the gas-hydrate-bearing interval and provided a unique and excellent data set for testing velocity models for gas-hydrate-bearing sediments

VELOCITY EQUATIONS

The fundamental equations needed to compare the result of weighted equation (WE) [*Lee et al.*, 1996] and the effective medium theory (EMT) [*Helgerud et al.*, 1999] are presented here.

A) Weighted Equation

A three-phase weighted equation for the compressional-wave (P-wave) velocity is given *by Lee et al.*, [1996]:

$$\frac{1}{V_p} = \frac{W\phi(1-S)^n}{V_{p1}} + \frac{1-W\phi(1-S)^n}{V_{p2}} \tag{1}$$

where

V_p : compressional (P) velocity of hydrate-bearing sediments; V_{p1}: compressional velocity of hydrate-bearing sediments computed from the three-phase Wood equation; V_{p2} : compressional velocity of hydrate-bearing sediments computed from the three-phase time-average equation; W : a weighting factor; ϕ : sediment porosity (as a fraction); S : concentration of hydrate in the pore space (as a fraction); n : a constant simulating the rate of lithification with hydrate concentration.

The shear-wave (S-wave) velocity of hydrate-bearing sediments [*Lee et al.*, 1996] is given by

$$V_s = V_p[\alpha(1-\phi) + \beta\phi S + \gamma\phi(1-S)], \tag{2}$$

with

$$\alpha = V_s/V_p\big|_{martix},\ \beta = V_s/V_p\big|_{hydrate},\ \gamma = V_s/V_p\big|_{fluid}.$$

The weight W can be estimated from the P-wave velocities of non-gas-hydrate-bearing sediments and exponent n can be optimally obtained by comparing gas hydrate concentrations estimated from both P-wave and S-wave velocities [*Lee et al.*, 1996; *Lee and Collett*, 1999]. A value of $W > 1$ favors the Wood equation, which is valid for particles in suspension and a value of $W < 1$ favors the time-average equation, which is appropriate for consolidated sediments. As n increases, the weighted equation approaches the time average equation. The values for the constants used in this study are W = 1.44, n = 1, P-wave matrix velocity = 4.94

Table 1. Elastic constants and some relevant parameters used for velocity models.

Name	Symbol Used	Values Used	Equation Used	Sources
Weight	W	1.44	1	*Lee and Collett* [1999]
Exponent	n	1	1	*Lee and Collett* [1999]
V_s/V_p for non-hydrate-bearing sediment	α	0.56	2	*Lee and Collett* [1999]
V_s/V_p for hydrate	β	0.51	2	*Sloan* [1998]
Volumetric Clay Content	C_v	0.3	1 and 7	From well logs
Shear Modulus of Quartz	$G_s(Gpa)$	45	3 and 7	*Helgerud et al.* [1999]
Bulk Modulus of Quartz	$K_s(Gpa)$	36	3 and 7	*Helgerud et al.* [1999]
Shear Modulus of Clay	$G_c(Gpa)$	6.85	3 and 7	*Helgerud et al.* [1999]
Bulk Modulus of Clay	$K_c(Gpa)$	20.9	3 and 7	*Helgerud et al.* [1999]
Shear Modulus of Hydrate	$G_h(Gpa)$	2.54	3 and 7	**Sloan* [1998]
Bulk Modulus of Hydrate	$K_h(Gpa)$	6.41	3 and 7	**Sloan* [1998]
Density of Quartz	ρ_s (g/cm^3)	2.65	8	*Helgerud et al.* [1999]
Density of Clay	ρ_c (g/cm^3)	2.58	8	*Helgerud et al.* [1999]
Density of Hydrate	ρ_h (g/cm^3)	0.91	8	*Sloan* [1998]
Critical Porosity	ϕ_c	0.38	3, 4, and 5	*Nur et al.* [1998]
Number of contacts per grain	m	9	3	*Dvorkin and Nur* [1996]

* The shear and bulk moduli are computed from the P-wave and S-wave velocities with the density given in *Sloan* [1998)] for structure I type gas hydrate.

km/s, α = 0.56, β = 0.51, and γ = 0. Details of the constants and conditions as estimated at the Mallik 2L-38 drill site are given in *Lee and Collett* [1999].

B) *Effective Medium Equation*

Helgerud et al. [1999] proposed a model to predict the velocity in unconsolidated, high porosity sediments with gas hydrates based on the effective medium theory proposed. This first-principle-based effective medium model computes the moduli of sediments based on Hertz-Mindlin theory [*Mindlin*, 1949], modified Hashin-Shtrikman (H-S) bounds [*Dvorkin and Nur*, 1996], and *Gassmann's* [1951] equation.

This approach starts with a computation of the effective bulk moduli of the dry rock frame at critical porosity ϕ_c [ϕ_c = 36-40 %: *Nur et al.*, 1998]. At critical porosity the effective bulk modulus (K_{hm}) and shear modulus (G_{hm}) of the dry rock frame are computed using the Hertz-Mindlin contact theory [*Mindlin*, 1949].

$$K_{hm} = \left[\frac{m^2(1-\phi_c)^2 G^2 P}{18\pi^2(1-\nu)^2}\right]^{\frac{1}{3}},$$

$$G_{hm} = \frac{5-4\nu}{5(2-\nu)}\left[\frac{3m^2(1-\phi_c)^2 G^2 P}{2\pi^2(1-\nu)^2}\right]^{\frac{1}{3}} \quad (3)$$

where P is the effective pressure; K, G, and ν are the matrix bulk modulus, shear modulus and the Poisson's ratio, respectively; m is the average number of contacts per grain in the sphere pack [$m \approx$ 8-9: *Dvorkin and Nur*, 1996].

Effective pressure is calculated as the difference between the lithostatic and hydrostatic pressure and is given by $P = (\rho_b - \rho_w)gD$, where ρ_b is the bulk density of sediment, ρ_w is water density, g is the gravitational acceleration and D is depth.

The bulk modus (K_{dry}) and shear modulus (G_{dry}) of the dry frame when the porosity of rock frame is below critical porosity are calculated via the modified lower Hashin-Shtrikman (*H-S*) bound [*Dvorkin and Nur*, 1996]:

$$K_{dry} = \left[\frac{\phi/\phi_c}{K_{hm}+4G_{hm}/3} + \frac{1-\phi/\phi_c}{K+4G_{hm}/3}\right]^{-1} - \frac{4G_{hm}}{3}$$

$$G_{dry} = \left[\frac{\phi/\phi_c}{G_{hm}+Z} + \frac{1-\phi/\phi_c}{G+Z}\right]^{-1} - Z,$$

$$Z = \frac{G_{hm}}{6}\left[\frac{9K_{hm}+8G_{hm}}{K_{hm}+2G_{hm}}\right]. \quad (4)$$

For porosity above critical porosity, these moduli are calculated via the modified upper *H-S* bound:

$$K_{dry} = \left[\frac{(1-\phi)/(1-\phi_c)}{K_{hm}+4G_{hm}/3} + \frac{(\phi-\phi_c)/(1-\phi_c)}{4G_{hm}/3}\right]^{-1} - \frac{4G_{hm}}{3}$$

$$G_{dry} = \left[\frac{(1-\phi)/(1-\phi_c)}{G_{hm}+Z} + \frac{(\phi-\phi_c)/(1-\phi_c)}{Z}\right]^{-1} - Z\cdot \quad (5)$$

If the sediment is saturated with pore fluid of bulk modulus K_f, the shear modulus G_{sat} is the same as that of the dry frame and the bulk modulus K_{sat} is calculated from *Gassmann's* [1951] equation as

$$K_{sat} = K\frac{\phi K_{dry} - (1+\phi)K_f K_{dry}/K + K_f}{(1-\phi)K_f + \phi K - K_f K_{dry}/K},$$

$$G_{sat} = G_{dry} \quad (6)$$

In the case that the matrix has mixed mineralogy, the moduli in Equation (3) are calculated from those of the individual mineral constituents using *Hill's* [1952] average formula:

$$K = \frac{1}{2}\left[\sum_{i=1}^{L} f_i K_i + \left(\sum_{i=1}^{L} f_i / K_i\right)^{-1}\right],$$

$$G = \frac{1}{2}\left[\sum_{i=1}^{L} f_i G_i + \left(\sum_{i=1}^{L} f_i / G_i\right)^{-1}\right], \quad (7)$$

where L is the number of mineral constituents and f_i is the volumetric fraction of the *i-th* constituent having K_i and G_i in the solid phase. The solid phase density (ρ) and the bulk density (ρ_b) are computed as

$$\rho = \sum_{i=1}^{L} f_i \rho_i \text{ and}$$

$$\rho_b = (1-\phi)\rho + \phi\rho_w \quad (8)$$

where ρ_i is the density of the *i-th* constituent.

Equation (6) can be used for the elastic velocities of sediment when the pore space is filled with water via the following equation:

$$V_p = \sqrt{(K_{sat} + 4G_{sat}/3)/\rho_b},$$

$$V_s = \sqrt{G_{sat}/\rho_b}. \quad (9)$$

When some of the pore space is filled with gas hydrate, *Helgerud et al.*, [1999] proposed two cases of gas hydrate effect: a) gas hydrate is part of the pore fluid and does not affect the stiffness of the dry frame; or b) gas hydrate is part of the dry frame. Case (a) implies that gas hydrate has negligible effect on the S-wave velocity, which is contrary to the observed S-wave velocity. Therefore, case (b) is considered in this study.

In case (b), gas hydrate is modeled as a load-bearing component of the frame. Therefore gas hydrate is part of the matrix and the porosity is reduced by the gas hydrate saturation. The reduced porosity is defined here as the fluid porosity (ϕ_f), $\phi_f = (1-S)\phi$, with S being the hydrate concentration or saturation. Because the gas hydrate is part of the matrix, the volumetric fraction of constituents, f_i, shown in Equation (7) should be substituted by

$$f_i \Rightarrow \frac{f_i(1-\phi)}{1-\phi_f}, \quad (10)$$

and the hydrate volumetric fraction (f_h) is given by $f_h = S\phi/(1-\phi_f)$.

POROSITY AND VELOCITY DATA

Porosities derived from the density log and P-wave and S-wave velocities in the Mallik 2L-38 gas hydrate research well are shown in Figure 1. The gas-hydrate-bearing interval between 897 m and 1110 m in depth are characterized by high P- and S-wave velocities. The zones with extreme low porosities with high velocities are hard, well cemented intervals, and the high porosities with low velocity zones are coal intervals. As shown in Figure (1), P- and S-wave velocities of non-gas-hydrate-bearing sediments are around 2.2 km/s and 0.9 km/s respectively, but they reach maximum values around 3.8 km/s and 1.9 km/s in gas-hydrate-bearing-sediments. The majority of sediment's porosity ranges from 20% to 40 % and generally the higher porosity sediments contain higher amount of gas hydrates.

The density porosities shown in Figure 1 are derived from the well log measured bulk densities assuming a two components system: fluid (water) and matrix. Since the density of methane hydrate is less than water (0.91 g/cm^3), the porosity derived from the bulk density log using a two component system (fluid and matrix) for highly concentrated gas-hydrate-bearing sediments is in error [*Collett,* 1998]. Therefore, for hydrate-bearing sediments a three component model using fluid, hydrate and matrix is more accurate than

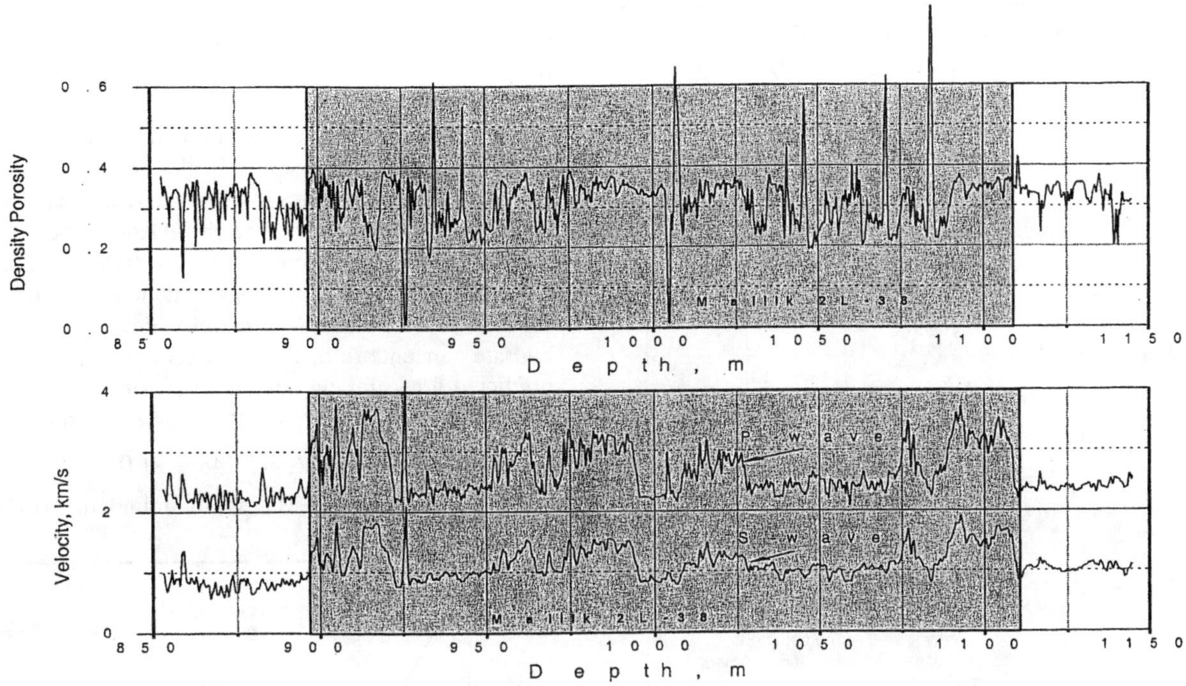

Figure 1. Well logs at Mallik 2L-38 hydrate research well, Mackenzie Delta, Canada. Velocity logs were measured with Schlumberger Dipole Shear Sonic Image Tool (DSI) and porosity was computed from the well log measured bulk densities. The shaded region indicates the hydrated interval (897 –1109 m).

using a two component model for the porosity computation. However, the ranges of porosity correction were less than 2% at the Mallik 2L-38 [*Lee and Collett*, 1999]. The density porosity data shown in Figure 1 were not corrected for the variations in gas hydrate concentrations. However the sediment porosities used later in this paper are based on the average porosity of gas-hydrate-bearing sediments at the Mallik 2L-38 computed after the application of a density-porosity gas hydrate saturation correction.

OBSERVATIONS AND PREDICTIONS

A cross plot of the P-wave velocities and S-wave velocities is shown in Figure 2, where open squares denote velocities for the non-hydrate-bearing sediment and closed circles for velocities within the gas-hydrate-bearing interval. The least squares fitting of a velocity trend for gas- hydrate-bearing sediments is shown in Figure (2) as solid line and is given by $V_p = 1.448V_s + 1.036$.

Figure 2 indicates that the relationship between P-wave and S-wave velocities for non-gas-hydrated-bearing sediments is similar to that of gas-hydrate-bearing sediments. This observation appears to be contrary to a common notion that gas hydrate may have a more pronounced effect on the S-wave velocities than on the P-wave velocities, because the gas hydrate cementation could increase the rigidity of sediments significantly [*Dillon*, 1998]. However, this observation cannot be generalized for all gas-hydrate-bearing sediments, because of the limited data set used here. For example, the S-wave velocities analyzed here are greater than about 800 m/s, which may not be appropriate for unconsolidated marine sediments whose velocities are usually less than 500 m/s [*Gettrust and Rowe,* 1991].

Figure 3 shows the observed P- and S-wave velocities with respect to porosity for non-gas-hydrate-bearing sediments in the Mallik 2L-38 well with velocities predicted from WE and EMT. The parameters for the velocity models are shown in Table 1 and a 1000 m depth was used for all EMT velocities. The predicted velocities from WE follow the observed velocities well. The P-wave velocity from EMT follows the observed P-wave velocity well also for porosity range of 20-40%. As the porosity decreases, the P-wave velocities from EMT are less than those from the WE and at higher porosities, velocities computed from EMT are higher than those from WE. The major difference between two theories is in the computed S-wave velocities. As indicated in Figure 3, S-wave velocities from EMT are much higher than the observed velocities at the higher porosities and converge to those from WE at lower porosities.

Figure 4 shows the observed and predicted velocity ratios of V_p/V_s with respect to S-wave velocities for gas-hydrate-

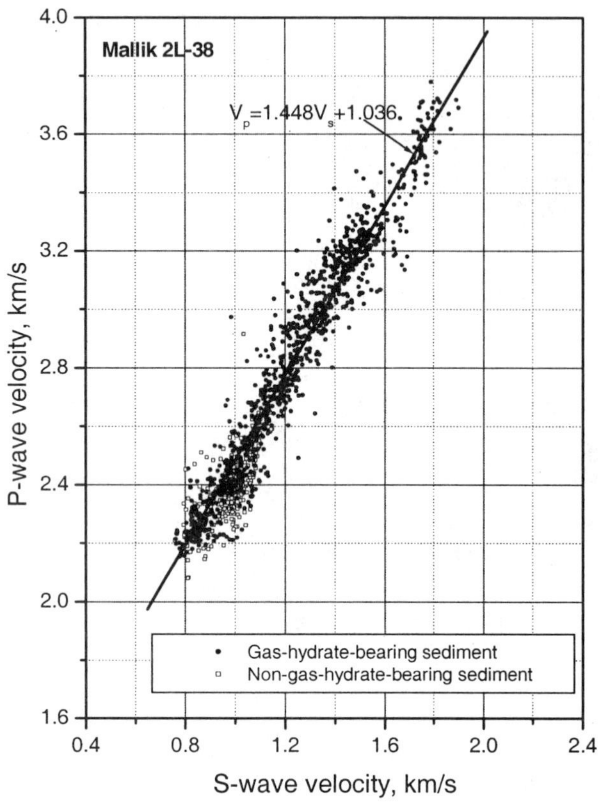

Figure 2. Observed P- and S-wave velocities. Open squares denote the velocities of non-gas-hydrate-bearing sediments inside the gas-hydrate-bearing interval (897-1109 m) and closed circles for velocities in the gas-hydrate-bearing interval (897-1109 m). The least squares fitting line is derived from the velocities for gas-hydrate-bearing sediments.

bearing sediments in the Mallik 2L-38 well. The predicted V_p/V_s ratios were computed by varying the gas hydrate concentration from 0 to 100 % for a sediment having a 33% porosity (33% porosity is the average porosity of gas-hydrate-bearing sediments at Mallik 2L-38). It is obvious that the prediction by WE is more accurate than the prediction by EMT, because parameters for WE were derived from the base line data. The performance of EMT is worse at low S-wave velocities or at low gas hydrate concentrations. If shear modulus is adjusted to match observed V_s in EMT, the overall match in Figure 4 could have been better, but it is difficult to justify based on the first principle.

DISCUSSION

The parameters for the weighted equation (W and α) can be derived or estimated from the observed P-wave and S-wave velocities of non-gas-hydrate-bearing sediments [*Lee and Collett,* 1999]. The weight (W) of the weighted equation can be found from a least squares method by fitting the observed velocities versus porosity to the weighted equation. Therefore, WE fits the observed velocities well within the observed porosity ranges. However, the validity of the equation outside the observed porosity range is questionable and WE is difficult to apply in the case that there are not enough base line samples to estimate W and α.

The effect of gas hydrate on the elastic velocities modeled by WE seems to be accurate, demonstrated by the relationship shown in Figure 4. The result indicates that WE predicts accurate P- and S-wave velocities with respect to gas hydrate concentration. The least squares fitting of velocities predicted for sediment having a porosity of 33 % from WE is given by $V_p = 1.40V_s + 1.06$, which is comparable to the observed relationship $V_p = 1.448V_s + 1.036$.

The advantage of EMT is its foundation to the first-

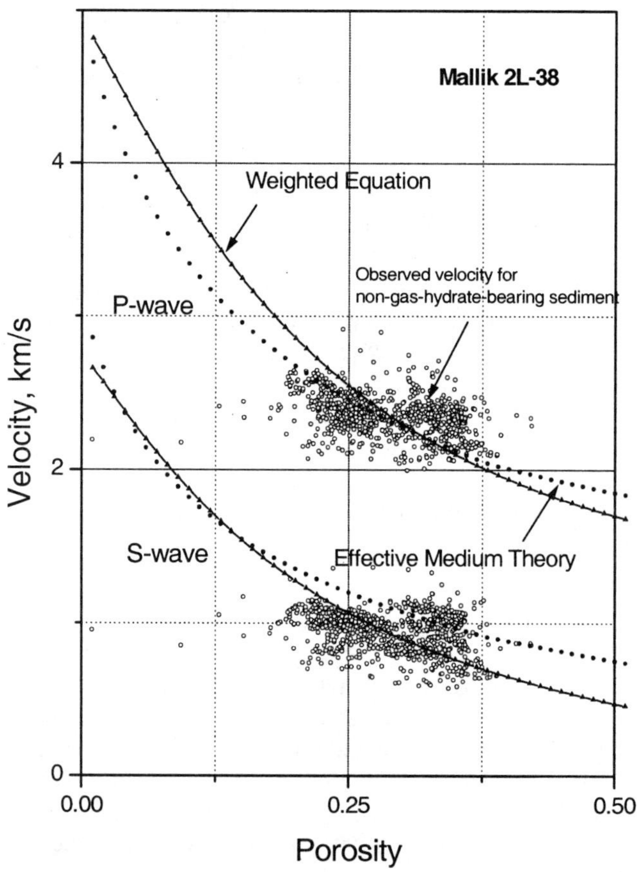

Figure 3. Graph showing velocity versus porosity. Open circles represent the velocities for non-hydrate-bearing sediments at the Mallik 2L-38 well. Two theoretical curves are shown: solid dots from the effective medium theory and line with solid squares from the weighted equation. The elastic constants and parameters are shown in Table 1 and 1000 m depth was used to calculate effective pressure for the effective medium theory.

principle physics [*Dvorkin and Nur*, 1996], so it provides physical insight and rational approach without actual measurements. This theory also explicitly contains the effective pressure term in the formulation, so it can establish the depth dependent velocities. However, the observed velocities do not agree with the predicted velocities. The difficulty with EMT in predicting both P- and S-wave velocities accurately comes from its overestimation of shear modulus for the non-hydrate-bearing sediments (base line), not from the assumption that gas hydrate is a load-bearing component of the frame. Because the predicted S-wave velocity for non-gas-hydrate-bearing sediment is much higher than the observed velocity (Figure 3), any attempt to fit the field measured velocities suffers at the low gas hydrate concentrations. The least squares fitting of velocities predicted for 33 % porosity sediment from EMT is given by $V_p = 1.30V_s + 0.94$, which is not comparable to the observed relationship $V_p = 1.448V_s + 1.036$.

As indicated for EMT shown above, there are only two free parameters at our disposal in order to fit the observed data: critical porosity (ϕ_c) and the average number of contacts per grain in the sphere pack (m). As m decreases, the elastic velocities decrease, which is indicated in Equation (3). The effect of ϕ_c on the velocities is more complex than the effect of m on the velocities as inferred from Equations (4) and (5). In this model, the velocities are maximum around at ϕ_c = 60 % and monotonically decrease away from 60 %. But the permissible ranges of m (about 8-9) and ϕ_c (36-40%) are so narrow that the difference in the computed velocities for various m and ϕ_c is negligible. Also m and ϕ_c affect both P- and S-wave velocity in the same manner, so changes in m and ϕ_c will not significantly change the fit of the observed velocities. As an example, the result using an unrealistic ϕ_c = 30% is also shown in Figure 4; the result of using ϕ_c = 30% matches better than that of using ϕ_c = 38% at low S-wave velocities, but the discrepancy at higher S-wave velocity range is still great.

Figure 5 summarizes velocities computed from the two theories for sediments having a porosity of 33%. For all gas hydrate concentrations, the P-wave velocity calculated from WE is higher than that from EMT. In consequence, the gas hydrate amounts estimated from P-wave velocities using EMT are higher than those estimated using WE. On the other hand, the S-wave velocities from WE are less than those from EMT up to 40 % gas hydrate concentration and similar in the range of 40 to 90% gas hydrate concentration.

Both equations predict moderate increases of P- and S-wave velocities with respect to gas hydrate concentration.

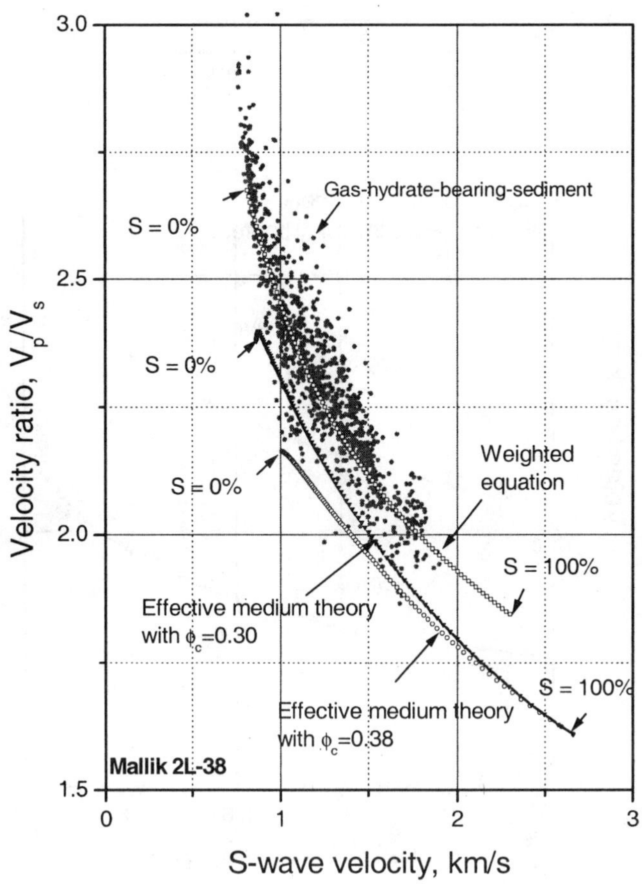

Figure 4. Graph showing the ratio of P-wave velocity to S-wave velocity with respect to S-wave velocity for gas-hydrate-bearing-sediments at the Mallik 2L-38, and comparisons with theoretical predictions. Open circles and open triangles are computed from the effective medium theory and open squares from the weighted equation.

This observation is markedly different from the prediction of the cementation model by *Dvorkin and Nur* [1993], whereby a small amount of gas hydrate in the sediment's pore space significantly increases the elastic velocities of sediment. This implies that the gas hydrates may not significantly cement grain boundaries of the sediment, but they may become part of the frame by just filling the pore space. The low exponent (n = 1) of WE and the assumption of gas hydrate being a load-bearing component of the frame by EMT suggest that the dominant effect of gas hydrate is a pore-filling constituent. However, the higher P-wave velocity from WE relative to EMT suggests that WE predicts a slight lithification of sediment owing to the gas hydrate saturation in the pore space. Similar pore-filling models for the gas-hydrate-bearing sediments are proposed by *Ecker et al.* [1996] based on the amplitude versus offset data at the Blake Ridge and by *Sakai* [1999] based on the elastic ve-

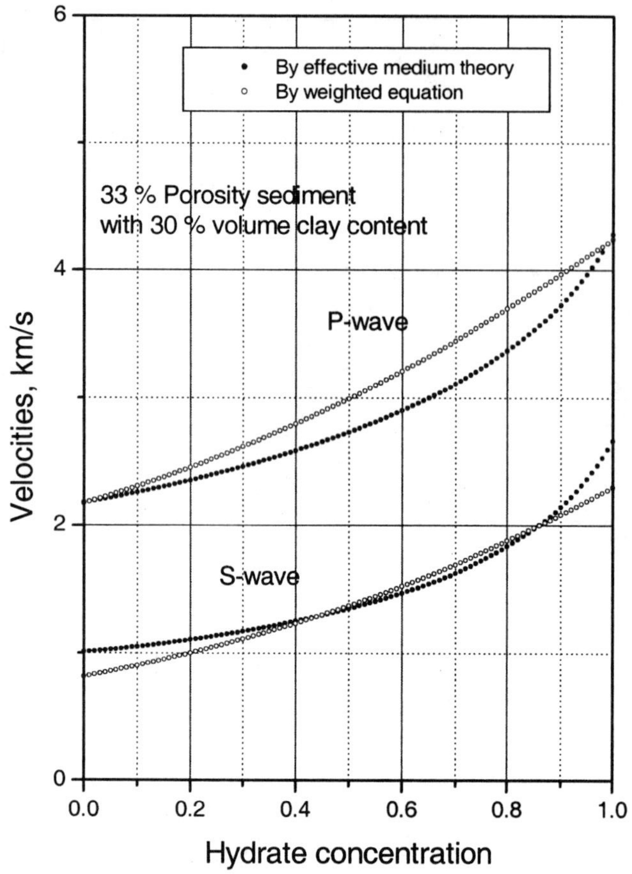

Figure 5. Comparison of predicted velocities for gas-hydrate-bearing sediments with 33 % porosity having 30 % volume clay content. Solid circles represent the predicted velocities from the effective medium theory using two critical porosities (ϕ_c = 30 % and 38 %) and open circles from the weighed equation.

locities at the Mallik 2L-38. On the other hand, based on the velocity data at the Blake Ridge, *Guerin et al.* [1999] proposed that a cementation theory of *Dvorkin and Nur* [1996] agrees well with gas hydrate saturation estimates when gas hydrates are assumed to uniformly coat grains. However, *Guerin et al.* [1999] concluded that additional data in a variety of lithologies and porosity ranges are necessary to determine conclusively the actual hydrate-grain configuration.

Even though the agreement between the observed velocities and predicted velocities from WE (Figure 4) is excellent, this does not prove that WE is accurate in predicting the amount of hydrate in the pore space, because the comparison between observed and predicted values was not performed explicitly with respect to "known" gas hydrate concentrations. To rigorously test any velocity model, a controlled experiment relating elastic velocities with respect to "known" gas hydrate concentrations is required.

CONCLUSIONS

The log data from the Mallik 2L-38 gas hydrate research well provided an excellent opportunity to compare velocity models for gas-hydrate-bearing sediments. Because there was no independent measurement of gas hydrate concentrations, the relationship between elastic wave velocities and gas hydrate concentration could not be completely tested. However, this study indicates that the predictions of P- and S-wave velocities from the weighted equation agree well with the log measured velocity in the Mallik 2L-38 well. The P-wave velocities calculated from the effective medium model agree with the observed velocities for non-gas-hydrate-bearing sediments, but S-wave velocities derived from the effective medium theory are much higher than the observed S-wave velocities.

Even though the parameters for the weighted equation are empirically derived and not based on first-principle physics, the weighted equation is our preferred velocity model because the predictions of elastic velocities of gas-hydrate-bearing sediments from the weighted equation are more accurate than those from effective medium theory by Helgerud *et al.* [1999].

Acknowledgments. Well logs were acquired by Schlumberger Ltd. at the Mallik 2L-38 gas hydrate research well, which was drilled to investigate gas hydrate in a collaborative research project among Japan National Oil Company, Japan Petroleum Exploration Company, Geological Survey of Canada, and United States Geological Survey. We thank John Grow, Christopher Potter, Satish Singh and Ingo Pecher for many helpful comments.

REFERENCES

Bangs, N. L., D. S. Sawyer, and X. Golovchenko, Free gas at the base of the gas hydrate zone in the vicinity of the Chile triple junction, *Geology, 21*, 905-908, 1993.

Collett, T. S., Well log characterization of sediment porosities in gas-hydrate-bearing reservoirs, in *Proceedings of the 1998 Annual Technical Conference and Exhibition of the Society of Petroleum Engineers*, 12 pp., New Orleans, Louisiana, September 27-30, 1998.

Collett, T. S., R. Lewis, S. R. Dallimore, M. W. Lee, T. H. Mroz, and T. Uchida, Mallik 2L-38 downhole well log displays—detailed evaluation of gas hydrate reservoir properties, in Scientific Results from JAPEX/JNOC/GSC Mallik 2L-38 Gas Hydrate Research Well, *Mackenzie Delta, Northwest Territories, Canada,* edited by Dallimore et al., Geological Survey of Canada, Bulletin 544, 1999. Dillon, W. P., Gas hydrate as a potential resource-what do we need to know?, in *Proceedings of the International Symposium on Methane Hydrates Resources in the Near Future?,* pp. 393, Chiba City, Japan, October 20-22, 1998.

Dillon, W. P., J. S. Booth, C. K. Paull, K. Fehlhaber, D. R. Hutchinson, and B. A. Swift, Mapping sub-seafloor reservoirs of green house gas: Methane hydrate, in *Proceedings of Interna-*

tional Symposiums on Marine Positioning, edited by M. Kumar and G. A. Maul, pp. 545-554, Mar. Geod. Comm., Mar. Technol. Soc., Washington, D. C., 1991.

Dillon, W. P., M. W. Lee, K. Fehlhaber, and D. F. Coleman, Gas hydrates on the Atlantic continental margin of the United States-Controls on concentration, in *The future of Energy Gases*, edited by D. G. Howell, *U.S. Geol. Surv. Prof. Pap. 1579*, 313-330, 1993.

Dvorkin, J. and A. Nur, Rock physics for characterization of gas hydrate. in *The future of Energy Gases,* edited by D.G. Howell, *U.S. Geol. Surv. Prof. Pap. 1579*, 293-298, 1993. Dvorkin J., and A. Nur, Elasticity of high-porosity sandstones: Theory for two north sea Datasets, *Geophysics, 61*, 1363-1370, 1996.

Ecker, C., D. Lumlet, J. Dvorkin, and A. Nur, Structure of hydrated sediments from seismic and rock physics, *Proc., 2nd International Conference on Natural Gas Hydrates*, published by PROGEP-NGH conference secretariat, pp. 491-498, Toulouse, France, 1996.

Gassmann, F., Elasticity of porous media, *Vierteljahrsschr der Naturforschenden Gesselschaft, 96*, 1-23, 1951.

Gettrust, J.F., and M.M. Rowe, Constraints on shear velocities in deep-ocean sediments as determined from deep-tow multichannel seismic data, in *Shear waves in Marine sediments*, edited by J.M. Hovem, M.D.Richardson, and R.D. Stoll, pp. 369-378, Kluwer Acad., Norwell, Mass., 1991.

Guerin G., D. Goldberg, and A. Meltser, Characterization of in situ elastic properties of gas hydrate-bearing sediments on the Blake Ridge, *J. Geophy. Res., 104, 17781-17795, 1999.*

Helgerud, M. B., J. Dvorkin, A. Nur, A. Sakai, and T. Collett, Elastic-wave velocity in marine sediments with gas hydrates: Effective medium modeling, *Geophysical Research Letters*, 26, 2021-2024, 1999.

Hill, R., The elastic behavior of crystalline aggregate, *Proc.* Physical Soc., London, A65, 349-354, 1952.

Hyndman, R. D., and G. D. Spence, A seismic study of methane hydrate marine bottom simulating reflectors, *J. Geophy. Res. 97*, 6683-6698, 1992.

Jacobsen, M., J. A. Hudson, T. A., Minshull, and S. C. Singh, Elastic properties of hydrate-bearing sediments using effective medium theory, *J. Geophy. Res. 105*, 561-577, 2000.

Kvenvolden, K. A., A primer of gas hydrates, in *The future of Energy Gases,* edited by D.G. Howell, *U.S. Geol. Surv. Prof. Pap. 1579*, 555-561, 1993.

Lee, M.W., D. R. Hutchinson, T. S. Collett, and W. P. Dillon, Seismic velocities for hydrate-bearing sediments using weighted equation, *J. Geophy. Res.*, 101, 20347-20358, 1996.

Lee, M. W., and T. S. Collett, Gas hydrate amount estimated from compressional - and shear- wave velocities at the JAPEX/JNOC/GSC Mallik 2L-38 gas hydrate research well, in Scientific Results from JAPEX/JNOC/GSC Mallik 2L-38 Gas Hydrate Research Well, Mackenzie Delta, Northwest Territories, Canada, edited by Dallimore et al., Geological *Survey of Canada, Bulletin 544*, 313-322, 1999.

Miller, J. J., , M. W. Lee, and R. von Huene, An analysis of a seismic reflection from the base of a gas hydrate zone, offshore Peru, *AAPG Bull., 75*, 910-924, 1991.

Mindlin R. D., Compliance of elastic bodies in contact, *J. Appl. Mech., Trans. ASME, 71*, 259-268, 1949.

Nur, A., G. Mavko, J. Dvorkin, and D. Galmudi, Critical porosity: A key to relating physical properties to porosity in rocks, *The Leading Edge, 17*, 357-362, 1998.

Paull, C. K., W. III. Ussler, and W. P. Dillon, Is the extent of glaciation limited by marine gas-hydrates?, *Geophys. Res. Lett.*, 18, 432-434, 1991.

Sakai, A, Velocity analysis of vertical seismic profiles (VSP) survey at JAPEX/JNOC/GSC Mallik 2L-38 gas hydrate research well, and related problems for estimating gas hydrate concentration, in *Scientific Results from JAPEX/JNOC/GSC Mallik 2L-38 Gas Hydrate Research Well, Mackenzie Delta, Northwest Territories, Canada*, edited by Dallimore et al., *Geological Survey of Canada, Bulletin 544*, 323-340, 1999.

Sloan, E.D., Jr., *Clathrate Hydrates of Natural Gases,* 705 pp., Marcel Dekker, New York, 1998.

Stoll, R. D., Effect of gas hydrates in sediments, in *Nature of Gases in Marine Sediments,* edited by I. Kaplan, pp. 235-248, Plenum, New York, 1974.

Tucholke, B. E., G. M. Bryan, and J. J. Ewing, Gas hydrate horizons detected in seismic-profile data from the western North Atlantic, *AAPG Bull., 61*, 698-707, 1977.

Wood, A. B., *A Text Book of Sound*, 578 pp., Macmillan, New York, 1941.

Wood, W. T., P. L Stoffa, and T. H. Shipley, Quantitative detection of methane hydrate through high-resolution seismic velocity analysis, *J. Geophys. Res., 99*, 9681-9695, 1994.

Wyllie, M. R. J., A. R. Gregory, and L. W. Gardner, Elastic wave velocities in heterogeneous and porous media, *Geophysics, 21*, 41-70, 1956.

T. S. Collett, U.S. Geological Survey, P.O. Box 25046, MS 939, Denver Federal Center, Denver Colorado 80225 (tcollett@usgs.gov)

M. W. Lee, U.S. Geological Survey, P.O. Box 25046, MS 939, Denver Federal Center, Denver, Colorado 80225 (mlee@usgs.gov)

A Review of Well-Log Analysis Techniques Used to Assess Gas-Hydrate-Bearing Reservoirs

Timothy S. Collett

U.S. Geological Survey, Denver Colorado

Gas volumes that may be attributed to a gas hydrate accumulation within a given geologic setting are dependent on a number of reservoir parameters, two of which, sediment porosity and gas-hydrate saturation, can be assessed with data obtained from downhole well logging devices. The primary objective of this paper is to identify and review existing quantitative well-log evaluation techniques, which permit the calculation of sediment porosities and gas-hydrate saturations in gas-hydrate-bearing reservoirs. The well log response modeling section of this paper includes the assessment of existing gas hydrate well-log evaluation techniques used to characterize sediment porosities and gas-hydrate saturations in gas-hydrate-bearing reservoirs. In the field data application section of this paper, the gas hydrate well-log evaluation techniques described in the response modeling section of this review are tested and used to calculate reservoir porosities and gas-hydrate saturations with the downhole log data from two known gas hydrate accumulations.

INTRODUCTION

Most of the published gas hydrate resource estimates have of necessity been made by broad extrapolation of only general knowledge of local geologic conditions [reviewed by *Kvenvolden*, 1988]. The amount of gas that might be stored in a gas hydrate accumulation is dependent on a number of reservoir parameters, including the areal extent of the gas-hydrate occurrence, reservoir thickness, reservoir porosity, and the degree of gas-hydrate saturation [*Collett*, 1993]. Two of the most difficult reservoir parameters to determine are porosity and the degree of gas-hydrate saturation. Downhole well logs often serve as a source of porosity and hydrocarbon saturation data; however, well-log calculations within gas-hydrate-bearing intervals have been the focus of only a limited number of studies.

Most of the existing gas hydrate well-log evaluation techniques are qualitative in nature and have been developed by the extrapolation of uncalibrated petroleum industry log evaluation procedures [reviewed by *Collett et al.*, 1984; *Mathews*, 1986; *Collett*, 1993; *Lee et al.*, 1993; *Collett*, 1998a, 1998b; *Collett and Ladd*, 2000; and *Collett*, 2000]. To adequately test the utility of standard petroleum log evaluation techniques in gas-hydrate-bearing reservoirs would require numerous laboratory and field measurements. However, only a limited number of gas hydrate occurrences have been sampled and surveyed with open-hole well logging devices (Figure 1). This paper reviews the existing state-of-the art quantitative log analysis techniques used to assess in-situ nature of gas hydrate occurrences. To attain this objective, this paper is subdivided into two main sections: (1) gas hydrate well-log response modeling and (2) field data application. However, this paper first begins with a review of proposed gas hydrate reservoir models at both a macroscopic and microscopic scale. The well-log response modeling section of this paper contains systematic descriptions of existing and proposed quantitative gas hydrate well-log evaluation techniques. The quantitative gas hydrate well-log evaluation techniques described in the response modeling portion of this paper have been used in the field data application section to assess gas hydrate reservoir properties within two known gas hydrate accumulations.

Natural Gas Hydrates: Occurrence, Distribution, and Detection
Geophysical Monograph 124

Published in 2001 by the American Geophysical Union

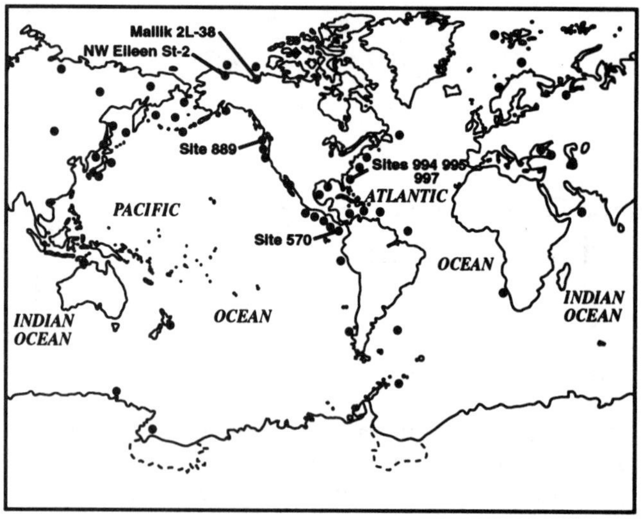

Figure 1. Location of known and inferred gas hydrate occurrences. Also shown are the locations and names of wells with confirmed gas hydrate occurrences that have been wireline logged.

This paper ends with a comprehensive summary of existing quantitative gas hydrate well-log evaluation techniques and a review of the utility of these newly developed techniques in the field.

IN-SITU NATURE OF GAS HYDRATE

Most well log calculations assume that simple mixing rules can be applied to evaluate reservoir conditions; however, little is known about the nature and texture of gas hydrate occurrences. The physical and chemical conditions that result in different forms (disseminated, nodular, layered, massive) and distributions (uniform or heterogeneous) of gas hydrates are not understood [reviewed by *Sloan,* 1990]. The existing and proposed quantitative well-log evaluation techniques assume uniform distribution of gas hydrates as interstitial deposits or cement. Under these conditions, gas-hydrate-bearing reservoirs can be evaluated using standard well-log evaluation techniques developed for mixed multi-component rock-matrix, water, gas, ice, and/or gas hydrate systems. However, a massive gas hydrate such as that encountered at DSDP Site 570 [*Mathews,* 1986] would not be conducive to conventional well-log interpretation procedures since a 3- to 4-m-thick massive gas hydrate occurrence (with no sediment matrix) does not adhere to simple mixing rules. It is necessary, therefore, to systematically review descriptions of known gas hydrate occurrences and evaluate existing gas hydrate reservoir models at both microscopic and macroscopic scales in order to assess the utility of standard well-log evaluation techniques in gas-hydrate-bearing reservoirs. This section begins with a review of published gas hydrate sample descriptions from both marine and permafrost environments, which is followed by an interpretive discussion of the formation microscanner and microimager log data from coreholes in a marine and terrestrial permafrost setting. This section concludes with a review of existing and proposed microscopic and macroscopic gas hydrate reservoir models.

Recovered Gas Hydrate Samples

This review of the nature of gas hydrate occurrences has relied extensively on an offshore gas hydrate sample database published by *Booth et al.* [1996] in which more than 90 recovered marine gas hydrate samples from 15 different geologic regions are described. In general, most recovered gas hydrate samples have been described as cements, nodules, or as laminae and veins, which tend to be characterized by dimensions of a few centimeters or less. In several cases, thick pure gas hydrate layers measuring as much as 3- to 4-m-thick have been sampled [DSDP Site 570; *Shipboard Scientific Party,* 1985]. In both marine and terrestrial permafrost environments, the thickness of identified gas-hydrate-bearing sedimentary sections varies from a few centimeters to as much as 30 m [*Collett,* 1993; *Booth et al.,* 1996]. Most pure hydrate laminae and layers, however, are often characterized by thicknesses of millimeters to centimeters [*Booth et al.,* 1996; *Dallimore and Collett,* 1995]. *Booth et al.* [1996] conclude that gas-hydrate-bearing sedimentary sections tend to be tens of centimeters to tens of meters thick, but thick zones of pure hydrate are relatively rare and only represent a minor constituent.

The *Booth et al.* [1996] data review along with other published gas hydrate sample descriptions from the Mackenzie Delta [*Dallimore and Collett,* 1995; *Dallimore et al.,* 1999] and the Blake Ridge [ODP Leg 164, *Shipboard Scientific Party,* 1996], confirm that gas hydrates are usually uniformly distributed within sediments and in most cases standard well-log evaluation techniques, with suitable gas-hydrate corrections, can be used to evaluate gas-hydrate-bearing reservoirs.

Formation Imaging Logs

Leg 164 of the Ocean Drilling Program [*Shipboard Scientific Party,* 1996] was designed to investigate the occurrence of gas hydrate in the sedimentary section beneath the Blake Ridge at Sites 994, 995, and 997 (ODP

Leg 164 is further discussed later in this paper). One of the downhole logging tools deployed on Leg 164 was Schlumberger's formation microscanner (FMS), which produces high-resolution images of the microresistivity character of the borehole wall that can be used for detailed sedimentological and structural interpretations [*Ekstrom et al.,* 1986]. It is also possible that the FMS can be used to make high-resolution electrical images of gas hydrates in the wellbore; thus yielding information about the nature and texture of gas hydrate occurrence in a rock interval. The FMS makes continuous measurements of detailed vertical and lateral variations in the wellbore, which allows the detection of very thin anomalous features such as fractures which have an opening on the order of a few to tens of microns. The FMS generates a high-resolution gray or color image of the borehole wall, with black and white (darkest or lightest color) indicating low and high microresistivities respectively.

FMS images from a zone of inferred gas hydrate in ODP Hole 995B were reprocessed and examined. Electrical resistivity and acoustic transit-time well logs from Hole 995B indicate the presence of gas hydrate in the depth interval between 193 and 450 mbsf on the Blake Ridge. The deep measuring resistivity log also reveals two conspicuous high electrical resistivity intervals between 218 and 242 mbsf (Figure 2), which have been interpreted to contain about 10% bulk volume gas hydrate [*Collett and Ladd,* 2000]. The FMS image in Figure 2 was generated in dynamic mode, thus the measured microresistivites within the depth interval displayed (217-245 mbsf) have been normalized such that 64 individual shades of gray are depicted within a one-meter sliding depth window. In Figure 2, the standard FMS four-arm display is shown in the left-hand-track, while the (deep) resistivity log from the dual-induction device ran in Hole 995B is plotted in the right-hand-track. The four-arm FMS image is characterized by light (high resistivity) to dark (low resistivity) bands, which in many cases can be traced across the display in all four arms. These continuous bands, likely represent distinct stratigraphic units with thicknesses ranging between 10 and 50 cm. The fine-scale bedding imaged in Figure 2, however, is not visually apparent in the recovered cores [*Shipboard Scientific Party,* 1996]. The apparent bedding in Figure 2 is likely caused by small variations in sediment porosity (on the order of 1 and 2 percent) and water content, which could cause the small microresistivity variations detected by the FMS in Hole 995B. It is also possible that the light (high resistivity) stratigraphic beds may contain gas hydrates, which are characterized by high electrical resistivities, that have been directly measured by the FMS. Since the FMS does not yield calibrated data easily, there is no way to conclusively differentiate gas

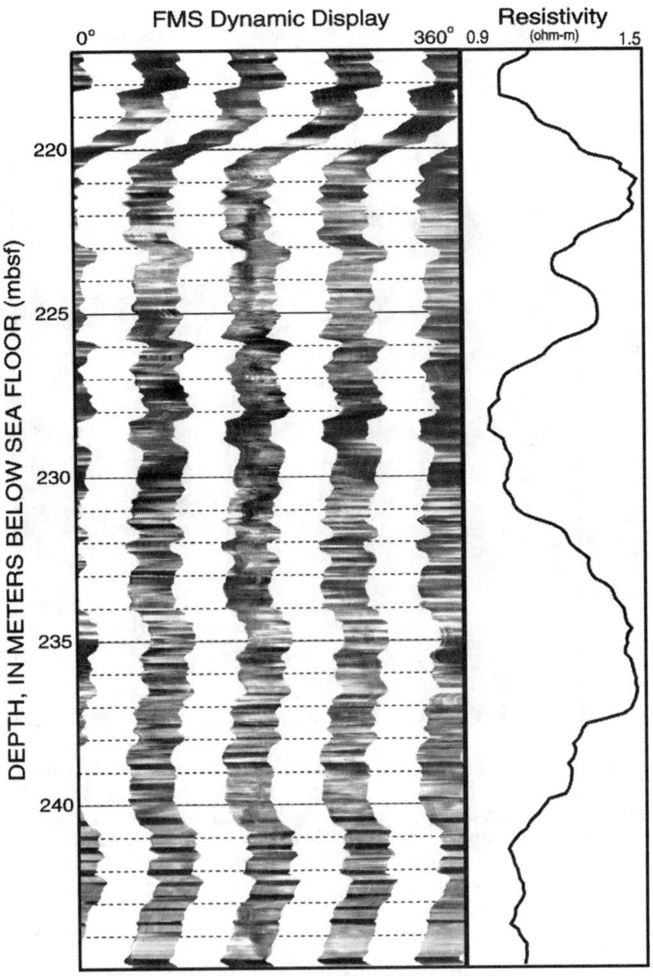

Figure 2. Formation microscanner log from the zone of downhole log inferred gas hydrate occurrence in ODP Hole 995B. Also shown is a log plot of the downhole electrical resistivity log in Hole 995B [modified from *Collett,* 2000].

hydrates from other relatively high resistivity material. However, the FMS image in Figure 2 does suggest that the gas hydrate inferred to occur at Site 995 likely occur as disseminated pore-filling constituents throughout the zone depicted and that small porosity variations between relatively thin stratigraphic beds, measuring between 10-50 cm thick, have a direct effect on the amount of gas hydrate reservoired within the sedimentary section.

An additional observation that supports the interpretation that the FMS log has directly measured gas hydrates at Site 995 comes from the comparison of the four-arm FMS image and deep resistivity log in Figure 2. In general, the FMS image within the two high (deep) resistivity zones (219-227 and 231-240 mbsf) are characterized by relatively lighter colors or higher FMS microresistivites which would collectively contribute to the high (deep) resistivity nature

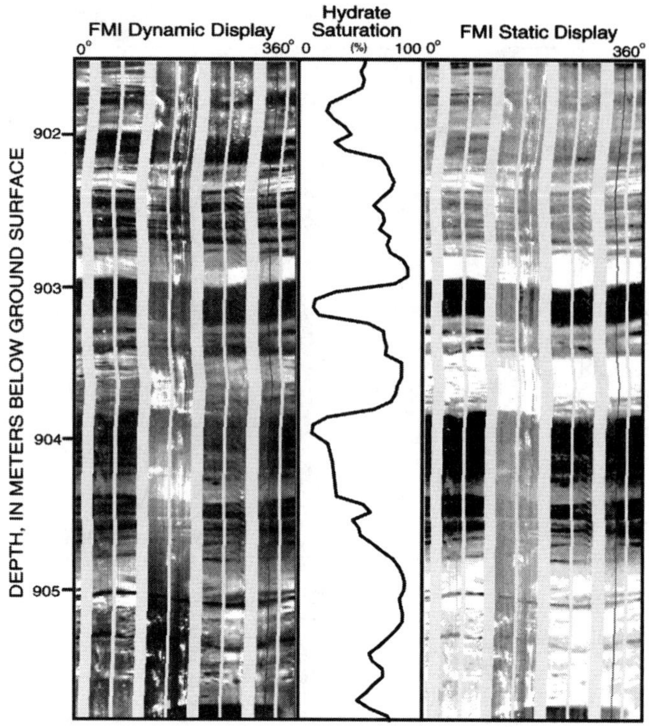

Figure 3. Formation microimager log from the zone of downhole log inferred gas hydrate occurrence in the Mallik 2L-38 well. Also shown is a log plot of gas-hydrate saturations (percent of pore-space occupied by gas hydrate) calculated from the downhole electrical resistivity log in the Mallik 2L-38 well [modified from *Collett,* 2000].

of the two anomalous zones that are interpreted to contain about 10% bulk volume gas hydrate.

Additional borehole microresistivity log data is available from the JAPEX/JNOC/GSC Mallik 2L-38 as hydrate research well drilled in the Mackenzie Delta of Canada [*Collett et al.,* 1999]. Downhole logs and the examination of recovered cores from the Mallik 2L-38 well confirm the occurrence of a thick gas-hydrate-bearing stratigraphic section at the Mallik drill-site within the subsurface depth interval between 889 and 1,101 m. The Schlumberger full-bore formation microimager (FMI) log, which was ran in the Mallik 2L-38 well, operates similarly to the FMS device, but the FMI contains four additional electrode pads; thus it images more of the borehole (about 80% of a 20 cm diameter borehole). The FMI log from the confirmed gas-hydrate-bearing interval in the Mallik well was processed in a manner similar to the processing performed on the FMS log from Site 995. In addition to the normal dynamic processing, a static image was also generated (Figure 3). In a static FMI image the entire log is treated as one unit and the microresistivities are placed into 16 separate shades of gray. Therefore, intervals within the log that have the same shade of gray will have similar resistivity log values, unlike the dynamic normalized presentation; although the static processed FMI display is a less detailed image. Similar to the FMS image from Site 995, the FMI images from Mallik 2L-38 in Figure 3 are characterized by light (high resistivity) and dark (low resistivity) bands, which likely represent distinct stratigraphic units. When compared to the gas-hydrate saturation log (Mallik 2L-38 gas-hydrate saturation calculations are further discussed later in this paper) in the center track of the display in Figure 3, it is apparent that the light (high resistivity) stratigraphic intervals on the FMI displays correlate with stratigraphic units that are inferred to contain high concentrations (saturations) of gas hydrate. Analysis of cores from the high resistivity FMI inferred gas-hydrate-bearing stratigraphic units in the Mallik 2L-38 well confirm the occurrence of disseminated pore-filling gas hydrate.

The FMS data in Figure 2 and the FMI data in Figure 3 indicate that the in-situ gas hydrates at these two drill-sites are mostly pore-filling constituents within thinly layered beds and that standard well-log evaluation techniques, with suitable gas hydrate corrections, can be used to assess the porosity and gas hydrate content of the sediments on the Blake Ridge and in the Mackenzie Delta.

Gas Hydrate Reservoir Models

Most discussions on the nature or texture of gas hydrate occurrences deal with macroscopic issues [reviewed by *Booth et al.,* 1996]. However, information on the occurrence of gas hydrates at the pore-scale is needed since many gas hydrate reservoir physical properties are controlled by microscopic parameters [*Dvorkin and Nur,* 1993]. Of particular concern is the acoustic nature of gas-hydrate-bearing sediments [*Lee et al.,* 1993].

Dvorkin and Nur [1993] along with *Ecker et al.* [1996] have proposed and examined two "micromechanical" models that represent the two extreme cases of gas hydrate occurrence at the pore-scale (Figure 4): (Model-1) gas hydrate cements grain contacts and increases the stiffness of the sediment; and (Model-2) gas hydrate is located away from grain contacts in the bulk pore-volume and it does not affect the stiffness of the sediment frame. *Dvorkin and Nur* [1993] experimentally demonstrated that even small amounts of intergranular cementation, such as proposed by gas hydrate Model-1, can dramatically increase the stiffness of granular material. *Dvorkin and Nur* [1993] used the intergranular gas hydrate cementation model (Model-1) to explain the occurrence of seismic bottom-simulating-reflectors (BSR's), which they attributed to a strong increase of the elastic moduli of the rock due to the occurrence of gas hydrates at the base of the gas hydrate stability zone. *Ecker et al.* [1996] amplitude-versus-offset (AVO) analysis of the BSR on the Blake Ridge, however, concluded that only reservoir Model-2 could qualitatively reproduce the observed BSR and that gas hydrates at the pore-scale are located away from the intergranular contacts, in large pores. *Ecker et al.* [1996] further concluded that the sediment above the BSR is uncemented and mechanically weak. Recently reported analyses of downhole acoustic well log data and vertical seismic profile data from the Mallik 2L-38 gas hydrate research well [*Lee and Collett,* 1999, 2001; *Saki,* 1999] support the micromechanical model of *Ecker et al.* [1996] with the occurrence of gas hydrate limited to the bulk pore-volume away from grain contacts (Model-2; Figure 4). However, *Helgerud et al.* [2000] analysis of the acoustic log data from Site 995 on the Blake Ridge suggests that gas hydrate acts as a load-bearing component of the sediment frame. For now, the micromechanical gas hydrate pore models of *Dvorkin and Nur* [1993] and *Ecker et al.* [1996] should be considered preliminary until additional laboratory and field observations are available.

As previously discussed at the beginning of this section, standard well-log evaluation techniques have been developed for relatively simple, mixed multi-component systems: rock matrix (consisting of quartz, calcite, and/or clay), water (including clay-bound- and free-water), and hydrocarbons (gas and/or oil). In permafrost and relatively deep marine environments, however, other reservoir constituents can include gas hydrates and permafrost ice. Before evaluating well-log responses to gas hydrate in the next section of this paper, it is necessary to define a series of reservoir models for the occurrence of gas hydrates in nature. Models A and B (Figure 5a-b) represent complex gas-hydrate-bearing reservoirs both below (Model A) and above (Model B) the base of ice-bearing permafrost in a terrestrial setting. In both of these models the sediment matrix consists of a simple mixture of quartz, calcite, and a relatively small amount of clay. Gas hydrate reservoir Models A and B assume no free-gas phase (no excess gas): all of the available gas is in the gas hydrate. The only difference between Model A and B is that Model B assumes that all of the free-water and some of the clay-bound-water are frozen. Reservoir Model C (Figure 5c) represents a clay-rich marine gas hydrate reservoir. Reservoir Models C and A are similar, but Model C assumes the clay content of the sediment and associated volume of bound-water are higher. Reservoir Model D (Figure 5d) assumes that a free-gas phase exists and that all of the available water is included in the gas hydrate. Water, being relatively abundant in nature, should not be a gas hydrate-limiting factor in most reservoirs. Therefore, gas hydrate reservoir Model D will not be examined in detail.

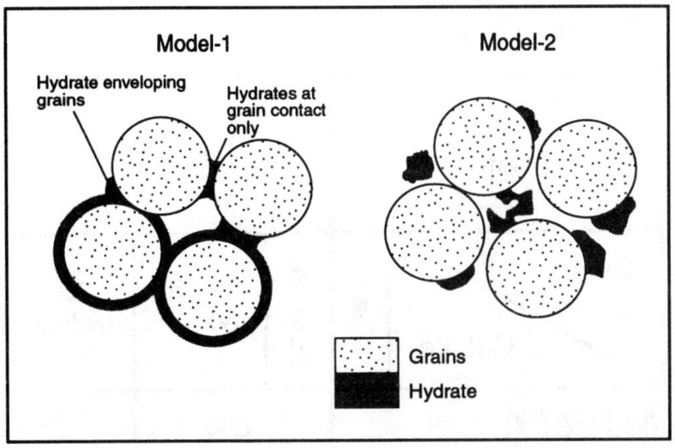

Figure 4. Two gas hydrate reservoir pore-models: (Model 1) gas hydrate cement grain contacts and envelops the grains; and (Model 2) gas hydrate is deposited away from grain contacts [modified from *Ecker et al.,* 1996].

WELL-LOG RESPONSE MODELING

In this paper, a comprehensive set of quantitative gas hydrate well-log evaluation techniques have been described and used to ascertain critical gas hydrate reservoir data. Since gas hydrates are characterized by unique chemical compositions and distinct electrical resistivity and acoustic physical properties, it is possible to obtain gas-hydrate reservoir porosity and hydrocarbon saturation data by characterizing the electrical resistivity and acoustic physical properties of gas hydrates and assessing the chemical composition of the pore-filling constituents within a gas-hydrate-bearing reservoir. The well-logging devices that show the greatest promise of yielding gas hydrate reservoir data, including porosity, are the density and neutron porosity logs (Table 1), which primarily respond to the chemical composition of the pore-filling constituents. The electrical resistivity, acoustic transit-time, and neutron spectroscopy well logs (Table 1) can yield highly accurate gas-hydrate saturation information.

This section of the paper deals with the review of the known and modeled well-log responses attributed to the presence of gas hydrates. This review has relied heavily on the following published gas hydrate well log studies; including *Collett* [1983], *Pearson et al.* [1983], *Collett et al.* [1984], *Mathews* [1986], *Collett* [1993], *Dvorkin and Nur* [1993], *Lee et al.* [1993], *Collett* [1998a, 1998b], *Collett and Ladd* [2000], *Collett and Wendlandt* [2000], *Collett et al.* [1999], *Guerin et al.* [1999], *Helgerud et al.* [1999, 2000], and *Collett* [2000]. Special recognition is also given to several other published gas hydrate well-log summaries authored by *Goldberg and Saito* [1998] and *Goldberg et al.* [2000]. Details of the operating principles

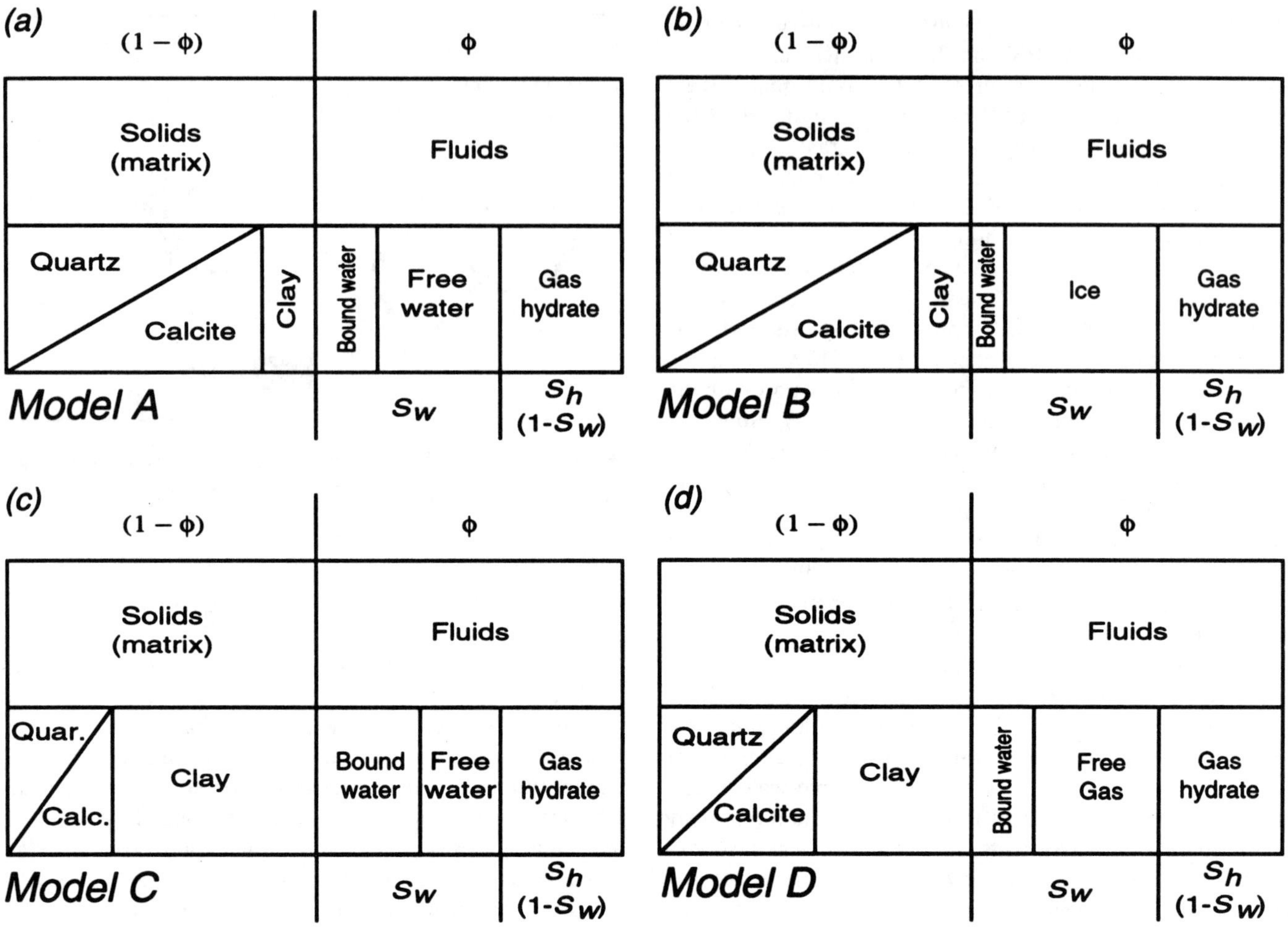

Figure 5. (a) Complex permafrost associated gas-hydrate reservoir model for conditions below the base of ice-bearing permafrost (Model A). (b) Complex permafrost associated gas-hydrate reservoir model for conditions above the base of ice-bearing permafrost (Model B). (c) Complex marine (clay-rich) gas-hydrate reservoir model (Model C). (d) Complex free-gas- and gas-hydrate-bearing reservoir model (Model D).

of the various logging tools can be found in *Serra* [1984], *Goldberg* [1997], and *Hearst et al.* [2000].

Gamma-Gamma Density Logs

In conventional formation analysis, density logs are primarily used to assess in-situ sediment porosities. The basic design of a density sonde consists of a gamma-ray source and detectors placed some distance apart. The emitted gamma rays are scattered within the formation and some of these scattered gamma rays return to the detector. The bulk density of the formation controls the attenuation of the log-emitted gamma rays.

The density of natural gas hydrates has not been directly measured, however, it is possible to calculate gas hydrate density from the unit cell dimensions reviewed by *Pearson et al.* [1983]. The density of a gas hydrate is calculated by adding the mass of the guest molecules to the lattice mass and dividing by the cell volume. The typical bulk-density of a Structure-I methane hydrate is about 0.9 g/cm^3.

In conventional formation density logging, porosities are derived from this standard relation:

$$\varnothing = \frac{\rho_m - \rho_b}{\rho_m - \rho_w} \qquad (1)$$

where ρ_m is the known matrix density (g/cm^3), ρ_w is the fluid (water) density (g/cm^3), and ρ_b is the log-measured formation bulk-density (g/cm^3). To calculate porosities within a gas-hydrate-bearing rock unit a modified density equation has been developed for a three-component system

Table 1. Categories of well logs which yield gas hydrate reservoir data. The operational parameters controlling the well-log responses are given, along with potential gas hydrate applications.

Well log	Gas hydrate application	Well log measurement
Density	Reservoir porosity	Measures electron density of the formation
Neutron porosity	Reservoir porosity	Measures (mostly) the hydrogen content of the formation
Electrical resistivity	Gas hydrate saturation Reservoir porosity	Measures both the electrical resistivity and conductivity of the formation
Acoustic transit-time	Gas hydrate saturation	Measures the attenuation of sound in the formation
Neutron spectroscopy	Gas hydrate saturation	Measures the amount of carbon, oxygen, hydrogen, silica, calcium, and other reservoir matrix and pore-filling constituents

(water, hydrate, matrix) [modified from *Pearson et al.,* 1983]:

$$\rho_b = (1-\varnothing)\rho_m + (1-S_h)\varnothing\rho_w + S_h\varnothing\rho_h \qquad (2)$$

ρ_b = Bulk-density, g/cm^3
ρ_m = Matrix density, g/cm^3
ρ_w = Water density, g/cm^3
ρ_h = Hydrate density, g/cm^3
S_h = Hydrate saturation, fractional %
$\varnothing$ = Porosity, fractional %

Figure 6 [modified from *Collett,* 1998b] is a plot of the three-phase (water-hydrate-matrix; Reservoir Model C, Figure 5c) modified density equation for all possible porosity conditions in a marine sedimentary section. Figure 6 reveals that gas hydrates can cause a small but measurable effect on density-derived porosities. At relatively high porosity (>40%) and high gas-hydrate saturation (>50%) the density log-derived porosities need to be corrected for the presence of gas hydrate. If the density of the gas hydrate is known, it is possible to obtain gas-hydrate-corrected, density-derived porosities with Equation 2.

Neutron Porosity Logs

Neutron logs are used principally for delineation of porous formations and determination of their porosity. Neutron porosity devices emit fast neutrons from either radioactive sources or particle accelerators. As the fast neutrons collide with the nuclei of atoms in the rock medium the neutrons thermalize or slow. The slowed neutrons are eventually captured by atoms and gamma rays of capture are emitted. The slowing of fast neutrons emitted by the logging tool is caused by collisions with mostly hydrogen nuclei, whose mass is almost identical to that of a neutron. Thus, neutron logs that measure the number of slowed neutrons are actually measuring the amount of

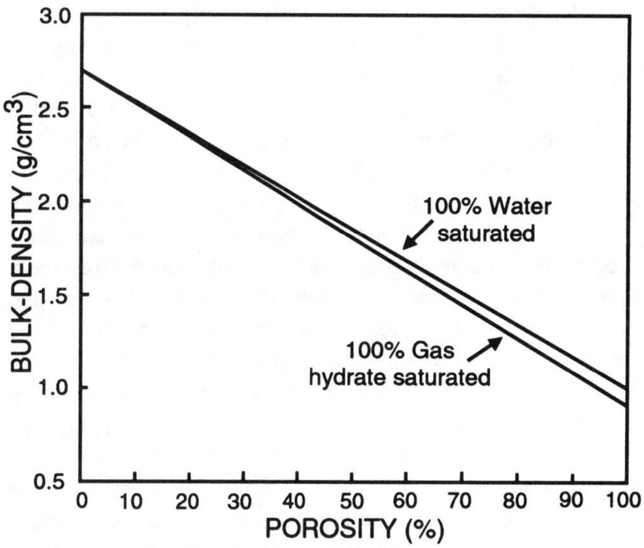

Figure 6. Plot of the modified three-component density equation (Equation 2) for all possible sediment porosity conditions and gas-hydrate saturations in a marine sedimentary section (Reservoir Model C, Figure 5c) [modified from *Collett,* 1998b].

Table 2. Elemental concentration of carbon, hydrogen, oxygen, silicon, and calcium in various reservoir constituents.

Reservoir constituent	Chemical formula	Bulk-density (g/cm^3)	Elemental concentrations ($x10^{22}$ atoms/cm^3)				
			Carbon	Hydrogen	Oxygen	Silicon	Calcium
Gas hydrate Structure-I	$7.598CH_4$+ $46H_2O$	0.9	0.439988	7.084028	2.662037	0	0
Methane gas	CH_4 2.580 MPa 273.15 K	(1.209 mol/dm^3)	0.072807	0.291226	0	0	0
Pure water	H_2O	1.0	0	6.685516	3.342758	0	0
Ice	H_2O	0.92	0	6.016964	3.008482	0	0
Quartz	SiO_2	2.65	0	0	5.311965	2.655983	0
Calcite	$CaCO_3$	2.71	1.630520	0	4.891559	0	1.630520

hydrogen within the pore-space of the rock sequence, which relates to the amount of water and hydrocarbons, including gas hydrates, which are present.

In *Collett* 1998b, two techniques were used to evaluate the effect of gas hydrate on the neutron porosity log. In the first technique, the hydrogen index of a Structure-I methane hydrate was calculated and compared to the hydrogen index of other reservoir constituents. The second technique involved modeling expected neutron porosity well log response to gas hydrates with a numerical multigroup computer simulator [SNUPAR].

Collett et al. [1984] developed a neutron porosity log gas hydrate correction factor based on assessing the hydrogen content of a Structure-I methane hydrate and other potential pore-filling constituents. As discussed in *Collett et al.* [1984], an increase in the hydrogen content would relate to an increase in the number of slowed neutrons and an increase in the rate at which the neutrons are slowed. Thus, a rock saturated with gas hydrate would more effectively slow (thermalize) the log emitted neutrons than a rock saturated with only water, since gas hydrates contain more hydrogen per unit volume [*Collett et al.,* 1984]. In *Collett* 1998b, the amount of hydrogen and other neutron-slowing atoms (carbon, hydrogen, oxygen, silicon, calcium) that are present in a cubic centimeter of pore-volume or matrix of the following substances were calculated: water, methane gas, ice, Structure-I methane hydrate, quartz (sandstone) and calcite (limestone) matrix (Table 2). These calculations assumed standard chemical formulas and bulk densities for each of the reservoir constituents being considered [reviewed by Collett, 1998b].

Hydrogen indexes are often used to compare hydrogen concentrations in different reservoir constituents. The hydrogen index (*HI*) of a substance is defined as the ratio of the concentration of hydrogen atoms per cubic centimeter in the substance, to that of pure water at 24°C ($6.686x10^{22}$ hydrogen atoms/cm^3). Pure water therefore has a *HI* of 1.0. Since a Structure-I methane hydrate contains $7.084x10^{22}$ hydrogen atoms/cm^3 (Table 2), the *HI* of Structure-I methane hydrate is 1.059, which is near the *HI* of pure water. Since Structure-I methane hydrate and pure water have similar hydrogen concentrations and *HI* values it can be generally assumed that neutron porosity logs, which are calibrated to pure water, are not significantly affected by the presence of gas hydrates. It is possible, however, to further quantify the effect of gas hydrates on neutron porosity logs. The following equation, reviewed by *Serra* [1984], was used by *Collett* [1998b] to determine the effect of gas hydrate on measured neutron porosities:

$$\varnothing_N = \varnothing[HI_h S_h + HI_w S_w] \tag{3}$$

$\varnothing_N$ = Measured neutron porosity, fractional %

$\varnothing$ = Actual porosity, fractional %

HI_h = Hydrocarbon hydrogen index

S_h = Hydrocarbon saturation, fractional %

HI_w = Water hydrogen index

S_w = Water saturation, fractional %

The above equation is plotted in Figure 7 [modified from *Collett,* 1998b] for all possible porosity conditions and assuming the reservoired hydrocarbons are gas hydrates with a 1.059 *HI*. As shown in Figure 7, a Structure-I methane hydrate has relatively little effect on neutron porosity measurements within reservoirs with porosities below 40%; however, at relatively high porosities (>40%) and high gas-hydrate saturations (>50%), neutron log calculated porosities need to be corrected for the presence of gas hydrates.

To predict the neutron tool response to complex mixtures of rock minerals having various pore-filling constituents, such as gas hydrates, it is necessary to conduct a more comprehensive assessment of all the physical parameters affecting the neutron porosity measurements. Careful mathematical response modeling has contributed greatly to the development of existing neutron well-log evaluation techniques in complex reservoirs [reviewed by *McKeon and Scott,* 1988]. These modeling studies have been used to characterize the response functions of the individual devices, to augment and check experimental measurements, and in many cases to model conditions that are not possible to simulate in the laboratory. In *Collett* [1998b], the response of the neutron well logs to gas hydrates has been modeled with a numerical multigroup simulator (SNUPAR; Schlumberger Nuclear Parameters, Version 2.C). The primary reason for conducting the SNUPAR modeling efforts in *Collett* [1998b] was to determine the macroscopic cross section (Σ), and thermal decay time for a Structure-I methane hydrate, and to confirm the calculated hydrogen index (*HI*). SNUPAR model results [Table 3; modified from *Collett,* 1998b] indicate that the relative effect of gas hydrate on neutron porosity logs are minimal, since all of the calculated nuclear parameters for Structure-I methane

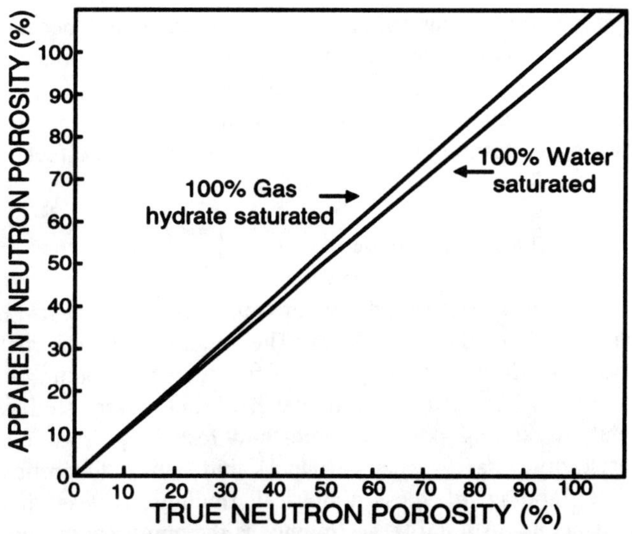

Figure 7. Plot of the hydrogen index (*HI*) neutron porosity equation (Equation 3) for all possible sediment porosity conditions and gas-hydrate saturations [modified from *Collett,* 1998b].

Table 3. SNUPAR calculated nuclear parameters of various reservoir constituents assuming a 4.5 MeV (AmBe) neutron source.

Reservoir constituent	Chemical formula	Bulk-density (g/cm^3)	Neutron slowing-down length (*Ls*, cm)	Neutron migration length (*Lm*, cm)	Thermal neutron decay time (*tau*, μsec)	Hydrogen index (*HI*)	Thermal neutron capture cross section (Σ, cm^{-1}x10^3)
Gas hydrate Structure-I	7.598CH_4+ 46H_2O	0.9	7.12	7.3	195.7	1.05	23.226
Methane gas	CH_4 2.580 MPa 273.15 K	(1.209 mol/dm^3)	194.33	288.99	4812.0	0.04	0.945
Pure water	H_2O	1.0	7.23	7.44	204.8	1.00	22.200
Ice	H_2O	0.92	8.04	8.35	227.5	0.90	19.980
Quartz	SiO_2	2.65	28.81	44.14	998.7	0.00	4.551
Calcite	$CaCO_3$	2.71	24.33	32.70	642.9	0.00	7.070

hydrate are similar to those for pure water and because most neutron porosity logs are calibrated to the response of pure water. Therefore, neutron porosity logs should yield accurate total reservoir porosities in gas-hydrate-bearing rock units, assuming no free-gas or other borehole effect.

Electrical Resistivity Logs

The electrical resistivity of most rock intervals is controlled by the pore-fluids. The relation between rock and pore-fluid resistivity has been studied in numerous laboratory and field experiments. From these studies, relations among porosity, pore-fluid resistivity, and rock resistivity have been found, but the theoretical understanding of these empirical findings is less than satisfactory. Among these findings is the empirical relation established by *Archie* [*Archie,* 1942], which is used to estimate water saturations in a gas-oil-water-matrix systems. In comparison, gas-hydrate- and gas/oil-bearing reservoirs appear to be characterized by relatively similar resistivities, which suggest that the *Archie* relation may be used to calculate gas-hydrate saturations. In *Pearson et al.* [1983], *Collett et al.* [1984], *Mathews* [1986], *Collett* [1993], *Collett and Ladd* [2000], *Collett et al.* [1999], and *Collett* [2000] the usefulness of the *Archie* relation in gas-hydrate-bearing reservoirs has been evaluated. Presented in this section and utilized in the field data application section of this paper are two forms of the *Archie* relation. Two additional electrical resistivity relations that deal with the effect of clay and clay-bound-water on the electrical properties of a clay-bearing reservoir rock are also examined.

Water content and pore-water salinity are the most significant factors controlling the electrical resistivity of a formation. Other factors influencing resistivity of a formation include the concentration of hydrous and metallic minerals, volume of hydrocarbons including gas hydrates, and pore structure geometry. As previously discussed, gas-hydrate-bearing sediments exhibit relatively high electrical-resistivities in comparison to water saturated units, which suggests that a resistivity log could be used to determine gas-hydrate saturations.

There is only one reported laboratory study of resistivities of (tetrahydrofuran) hydrate-bearing sediments [*Pearson et al.,* 1986]; however, there is an extensive body of literature on the electrical properties of partially frozen sediments [reviewed by *Hoyer et al.,* 1975]. Because ice and gas hydrates are electrical insulators, the resistivities of gas hydrate and permafrost deposits are largely controlled by the salinity of the unfrozen pore-waters or brine inclusions. *Archie's* equation [*Archie,* 1942] is an empirical relationship between water content (saturation) and the resistivity of water-saturated sediments:

$$R_t = aR_w \varnothing^{-m} S_w^{-n} \tag{4}$$

R_t = Formation resistivity (from log), ohm-m

a = Empirically-derived parameter

R_w = Resistivity of formation water, ohm-m

$\varnothing$ = Porosity, fractional %

m = Empirically-derived parameter

S_w = Water saturation, fractional %

n = Empirically-derived parameter

Pearson et al. [1983] have shown that as gas hydrates form in the laboratory or as pore-waters freeze, the amount of bulk water decreases, S_w and R_w are reduced; S_w because some of the available pore-space is now filled with a solid nonconductor, and R_w because the dissolved salts are concentrated in the remaining unfrozen water. If the brine is not near saturation, the effect of gas hydrate formation on R_w is relatively easy to quantify because an increase in salt concentration causes a decrease in R_w.

Another resistivity approach for assessing gas-hydrate saturations (S_w) is based on a modification of the so-called *Archie* quick look log analysis technique that compares the resistivity of a water-saturated sediment (R_o) to the resistivity of a hydrocarbon-bearing sediment (R_t) [reviewed by *Collett and Ladd,* 2000]:

$$S_w = \left(\frac{R_o}{R_t}\right)^{\frac{1}{n}} \tag{5}$$

where R_o is the resistivity of the sedimentary section if it contained only water (S_w=1.0), R_t is the resistivity of the hydrocarbon-bearing intervals (log value from gas-hydrate-bearing interval), and n is an empirically derived *Archie* parameter [*Archie,* 1942]. This form of the *Archie* relation is based on the following logic: if the pore-space of a sediment is 100% saturated with water, the deep-reading resistivity device will measure the resistivity of the 100% water-saturated section (R_o). This measured R_o value is considered to be a relative baseline from which hydrocarbon (gas hydrate) saturations can be determined within nearby hydrocarbon-bearing intervals. The logic behind this R_o/R_t method is similar to the supporting logic behind the Pickett crossplot method [reviewed by *Collett,* 1993].

It is well established that the occurrence of shale (clay) in reservoir rocks can result in erroneous values of water saturations as calculated from downhole log data. All of the *Archie* related equations discussed previously in this section apply to non-shaly formations that contain only non-conductive insulating matrix rocks. Shale contained in rocks, however, conducts electricity, which affects the logging tool measurements. Many log analysis models have been developed to account for the affect of shales on resistivity log measurements. Considered in this review [modified from *Collett,* 2000] are only two of the more widely used shaly-sand interpretive models: (1) the Indonesian and (2) the Waxman-Smits.

The Indonesian model [*Poupon and Leveaux,* 1971] was developed to account for over estimation of water saturations in Indonesian shaly-sand reservoirs. It was recognized that the volume of shale (clay) affects the true electrical resistivity of a formation. The empirical Indonesian model which relates downhole log-measured electrical resistivities to water saturations was developed by statistically evaluating frequency crossplots of matrix clay volumes (V_{cl}) estimates and computed water saturations (S_w).

In 1968, *Waxman and Smits* [1968] proposed a saturation-resistivity relationship for shaly formations that related the resistivity contribution of the shale to the cation exchange capacity (CEC) of the shale. Simply stated, the *Waxman-Smits* equation requires the pore-water resistivities to be corrected for the contribution of conductive clay, which is a function of the CEC of the clays. In 1974, *Waxman and Thomas* [1974] extended the *Waxman-Smits* relation to hydrocarbon-bearing shaly-sand reservoirs. The *Waxman-Smits* relation compares the electrical conductivity (C_t) of the hydrocarbon-bearing shaly-sand reservoir (through the formation factor, F) to the reservoir water saturations (S_w), water conductivity (C_w), and the CEC per unit pore-volume (B, Q_v).

Acoustic Transit-Time Logs

Acoustic transit-time logs measure the propagation of elastic vibrations through a medium. There are two main types of elastic waves; compressional and shear. The movement of compressional and shear sound waves through a hydrocarbon-bearing reservoir is controlled by many factors, including the density of the rock matrix, porosity, and hydrocarbon saturations. In most acoustic well log studies, the *Wyllie* time-average equation relates the compressional velocity of the bulk reservoir material to the velocity of the rock matrix and fluid to determine porosity. However, studies of seismic data from the eastern continental margin of the United States have shown that the volume of gas hydrate in marine sediments can be estimated by measuring interval velocities [*Lee et al.,* 1993]. Presumably, acoustic transit-time well logs can also be used to measure the amount of gas hydrate within a drilled sedimentary unit.

Since *Lee et al.* [1993], acoustic modeling of gas hydrate accumulations have followed two general paths dealing with either the development if empirical acoustic relations or the application of multiphase wave scattering theory, grain contact models, or effective medium models to directly calculate gas-hydrate concentrations for various geologic conditions. *Lee et al.* [1996] generally pursued the development of a series of empirical acoustic equations that appear to accurately predict the acoustic velocity of gas-hydrate-bearing sediments. *Lee et al.* [1996] also reintroduced and applied the multiphase wave scattering theory developed by *Kuster and Toksöz* [1974], to assess the acoustic properties of gas-hydrate-bearing sediments. An alternative gas hydrate acoustic model has been proposed by *Dvorkin and Nur* [1993], which is based on estimating acoustic properties of cemented gas-hydrate-bearing sediments from grain contact theory. Subsequently, *Guerin et al.* [1999] used the *Kuster-Toksöz* multiphase wave scattering model and the cementation model of *Dvorkin and Nur* [1993] to predict gas-hydrate saturations on the Blake Ridge. Most recently, an additional acoustic model for gas-hydrate-bearing sediments, based on effective medium theory, has been proposed and tested by *Helgerud et al.* [2000].

To thoroughly review and compare the relatively large number of proposed gas hydrate acoustic models and theories would require more pages of text than any reasonable editor would provide. Therefore, this paper does not contain any detailed examination of the proposed gas hydrate acoustic models. However, in a series of recent publications, by three different groups of researchers, the downhole acoustic log data from the Site 995 borehole on the Blake Ridge has been used to assess in-situ gas hydrate saturations by the *Lee* weighted average equation [*Lee,* 2000], the wave scattering theory developed by *Kuster-Toksöz* [*Guerin et al.,* 1999], the grain contact model developed by *Dvorkin-Nur* [*Guerin et al.,* 1999], and the effective medium theory [*Helgerud et al.,* 2000]. This published series of analyses of the same downhole log data set affords us with the opportunity to compare the results of all the proposed gas hydrate acoustic models and theories. In the field application section of this paper, each of the proposed acoustic methods used to estimate gas-hydrate saturations at Site 995 have been reviewed and compared. It should also be noted that *Lee and Collett* [2001; this volume] have also compared the results of the effective medium theory [*Helgerud et al.,* 2000] and the empirical weighted average equation [*Lee et al.,* 1996] analysis of the

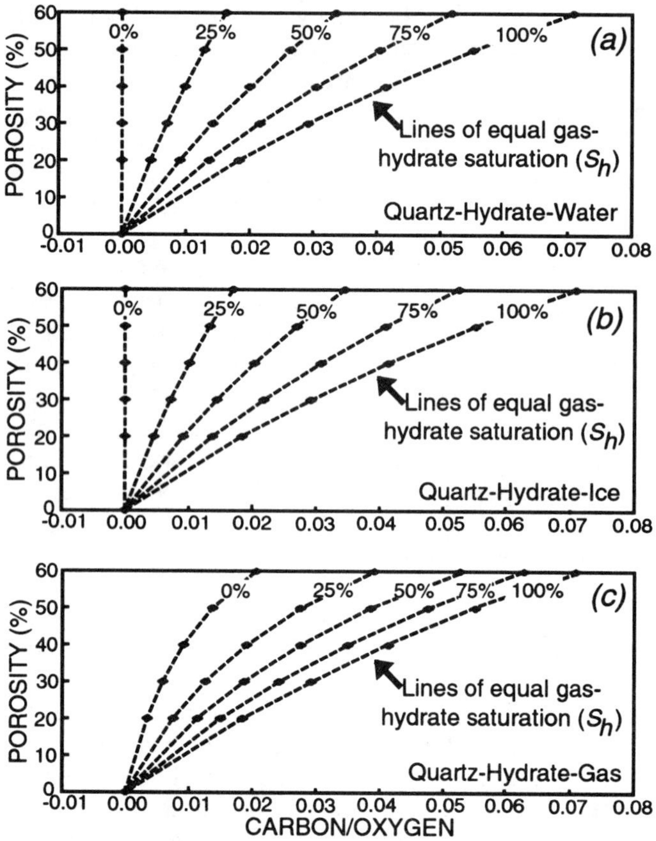

Figure 8. Fan charts for carbon/oxygen interpretation of gas-hydrate saturation in a quartz (sandstone) reservoir with various pore-filling constituents: (a) gas hydrate and water; (b) gas hydrate and ice; and (c) gas hydrate and free-gas [modified from *Collett and Wendlandt,* 2000].

acoustic log data from the Mallik 2L-38 well in northern Canada.

Neutron Spectroscopy Logs

Neutron-induced gamma ray spectroscopic well logging has been established as a means of evaluating the chemical composition of hydrocarbon reservoirs. The Schlumberger geochemical tool string (GLT) provides measurements of most of the elements present in rock-forming minerals, which can be used to construct detailed mineralogic models. In addition, GLT-derived data on the carbon and oxygen content of hydrocarbon reservoirs can be used to determine oil and gas saturation information. In *Collett and Wendlandt* [2000], the theoretical carbon and oxygen content of several hypothetical gas-hydrate occurrences were calculated and a series of carbon/oxygen fan charts that can be used to calculate gas-hydrate saturations were constructed. The effect of complex reservoir conditions (including the presence of clay and dispersed organic carbon) on the GLT gas-hydrate saturation calculations were also assessed in *Collett and Wendlandt* [2000] and reviewed below.

By combining elemental yields from neutron spectroscopy well logs, reservoir parameters including porosities, lithologies, formation fluid salinities, and hydrocarbon saturations (including gas hydrates) can be calculated [*Collett and Wendlandt,* 2000]. The GLT can be operated in two timing modes: inelastic, which mainly measures the neutron reactions in the high energy range (elements quantified: carbon, calcium, iron, oxygen, sulfur, silicon), and capture-tau mode which measures the gamma rays emitted from neutron capture (elements quantified: calcium, chlorine, iron, hydrogen, sulfur, silicon). In ODP boreholes, the GLT is usually run in only capture-tau mode; however, on ODP Leg 164 a selected number of inelastic measurements where made. GLT elemental yields can be calibrated using XRD-XRF measurements and core descriptions to infer downhole mineralogy variations. After post-field processing, the GLT data may be used to make detailed assessments of mineralogical compositions.

The amount of a particular element in a rock is not only controlled by the matrix mineralogy but also by the amount of water and hydrocarbons, including gas hydrates that are present in the rock. The carbon/oxygen elemental ratio shows promise of yielding accurate gas-hydrate saturations. As previously demonstrated in *Collett* [1998b], it is possible to estimate the amount of carbon and oxygen that are present in a cubic centimeter of pore-volume or matrix for the following substances: water, methane gas, ice, structure-I methane hydrate, sandstone (quartz) matrix, and limestone (calcite) matrix (Table 2). By using simple mixing rules and the elemental concentrations in Table 2 it is possible to develop elemental relations (ratios) that yield gas-hydrate saturations. Depicted in Figures 8 and 9 are the results of the carbon/oxygen reservoir modeling efforts published by *Collett and Wendlandt* [2000]. All of the reservoir conditions modeled assume a simple three-component reservoir model, which consists of a quartz or calcite matrix (no clay), gas hydrate, and either water, ice, or methane gas. The carbon/oxygen and reservoir porosity (0-60 %) crossplots in Figures 8 and 9 are similar to conventional oil and gas carbon/oxygen fan charts [*Woodhouse and Kerr,* 1992]. Of concern, however, is the limited range of expected carbon/oxygen ratios in gas-hydrate-bearing sandstone (quartz) reservoirs (Figure 8). For example, the maximum range of the carbon/oxygen ratio for a gas-hydrate- and water-bearing reservoir (Figure 8a) with 40% porosity would only be 0.04, which is near the likely uncertainty of the carbon/oxygen log measurement [*Woodhouse and Kerr,* 1992]. The uncertainty in carbon/oxygen log measurements in gas-hydrate-bearing limestone (calcite) reservoirs should be less of a problem (Figure 9).

Although it is possible to generate a series of carbon/oxygen fan charts for relatively simple reservoir conditions, it is necessary to develop a more complete understanding of the chemistry of the entire borehole environment in complex natural occurring reservoir conditions. The amount of carbon or oxygen measured by the GLT is not only controlled by the chemistry of the pore-fluids within the formation, but also by the chemistry of the rock matrix. An equation relating all of the carbon and oxygen sources associated with a borehole in a gas hydrate accumulation has been developed in *Collett and Wendlandt* [2000]. The carbon-oxygen ratio (COR) gas-hydrate saturation equation, developed by *Collett and Wendlandt* [2000], can be modified to accommodate more complex mixtures of minerals, such as clay and the presence of organic carbon.

FIELD DATA APPLICATION

Gas hydrates have been inferred to occur at about 50 locations throughout the world [reviewed by *Kvenvolden,* 1988]. Gas hydrates have only been sampled and surveyed with open-hole well logging devices at five locations (Figure 1): (1) North Slope of Alaska, (2) Middle-America Trench off the Pacific coast of Guatemala, (3) Cascadia continental margin off the Pacific coast of Canada, (4) Blake Ridge on the southeastern continental margin of the United States, and (5) the Mackenzie River Delta in northern Canada. The next several sections of this paper review the results of completed gas hydrate well log studies on the Blake Ridge and in the Mackenzie River Delta (Figure 1, Table 4). In this review, the Blake Ridge and Mackenzie Delta drill sites were selected because they are characterized by distinctly different geologic conditions. The sediments that make up the Blake Ridge are dominated by high clay contents and the concentration of gas hydrates on the Blake Ridge appears to be low. Conversely, the Mackenzie River Delta gas hydrate accumulation at the Mallik drill site occurs predominantly within thick sandstone sequences (low clay contents) at relatively high concentrations.

Blake Ridge, Atlantic Ocean

Leg 164 of the Ocean Drilling Program (ODP) [*Shipboard Scientific Party,* 1996] was designed to investigate the occurrence of gas hydrate in the sedimentary section beneath the Blake Ridge. In *Collett and Ladd* [2000], the logged interval in the ODP Leg 164 boreholes (Sites 994, 995, and 997; Figure 10) has been divided into three Logging Units and they predicted that disseminated gas hydrates occur throughout Logging Unit 2 at all three sites.

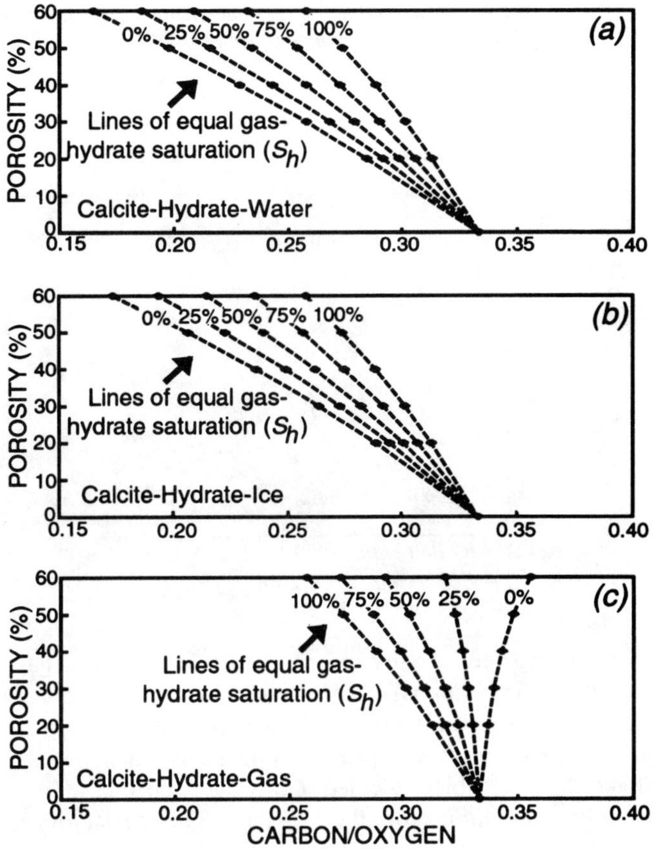

Figure 9. Fan charts for carbon/oxygen interpretation of gas-hydrate saturation in a calcite (limestone) reservoir with various pore-filling constituents: (a) gas hydrate and water; (b) gas hydrate and ice; and (c) gas hydrate and free-gas [modified from *Collett and Wendlandt,* 2000].

Collett and Ladd [2000] have shown that the log-derived porosities (density and neutron porosity logs) from the Blake Ridge boreholes are generally higher than the core porosities. It is likely that the density and neutron porosity log measurements have been degraded by poor borehole conditions and the core-derived sediment porosities are the best available porosity data from the Blake Ridge.

In *Collett and Ladd* [2000] the standard *Archie* relation yielded water saturations (S_w) ranging from about 100% to a minimum of about 80% in the Blake Ridge boreholes (Figure 11). The quick look calculated water-saturations are very similar to the water saturations calculated by the standard *Archie* relation. In *Collett* [2000], the *Waxman-Smits* shaly-sand calculations appear to significantly overestimate the amount of in-situ gas hydrates at Site 995

Table 4. Resistivity and acoustic constants used to calculate water and gas-hydrate saturations within the downhole log-inferred gas hydrate occurrences at ODP Sites 994, 995, and 997, and in the Mallik 2L-38 well.

Site/Well identification	Archie a	Archie m	R_w	R_o	Porosity	Lee r	Lee W
Site 994	1.05	2.56	Ref. 2	Fig. 10	Meas. core por. Av. core por.	1.0	1.1
Site 995	1.03	2.53	Ref. 2	Fig. 10	Meas. core por. Av. core por.	1.0	1.1
Site 997	1.07	2.59	Ref. 2	Fig. 10	Meas. core por. Av. core por.	1.0	1.1
Mallik 2L-38	0.62	2.15	Ref. 1	Ref. 1	Density log	1.0	1.6

Ref. 1 = *Collett et al.*, 1999
Ref. 2 = *Collett and Ladd,* 2000

(when compared to other independent calculations of gas-hydrate concentrations). The likely cause of this problem can be attributed to the high CEC nature (ranging from 32 to 44 meq/100 grams of sediment) of the sediments on the Blake Ridge, which yielded Q_v values outside of the experimental limits of the *Waxman-Smits* relation as described by *Waxman and Thomas* [1974].

For the Blake Ridge boreholes, the *Lee* weighted average equation (compressional-wave version), yielded for the most part gas-hydrate saturations (S_h) ranging from 0% to a maximum of about 20% (Figure 12; modified from *Lee,* 2000], which are compatible to the gas-hydrate saturations calculated from interstitial-water chloride-freshening trends [*Shipboard Scientific Party, 1996*] and electrical resistivity log data.

In order to estimate gas-hydrate saturations with the downhole acoustic log data from Site 995 on the Blake Ridge, *Guerin et al.* [1999] used an inverted *Kuster-Toksöz* approach in which they calculated a series of model derived bulk and shear moduli that were compared to the downhole log derived moduli to calculate gas hydrate concentrations. As shown in Figure 13, the *Kuster-Toksöz* approach as described by *Guerin et al.* [1999] yielded gas-hydrate saturations generally greater than those predicted from

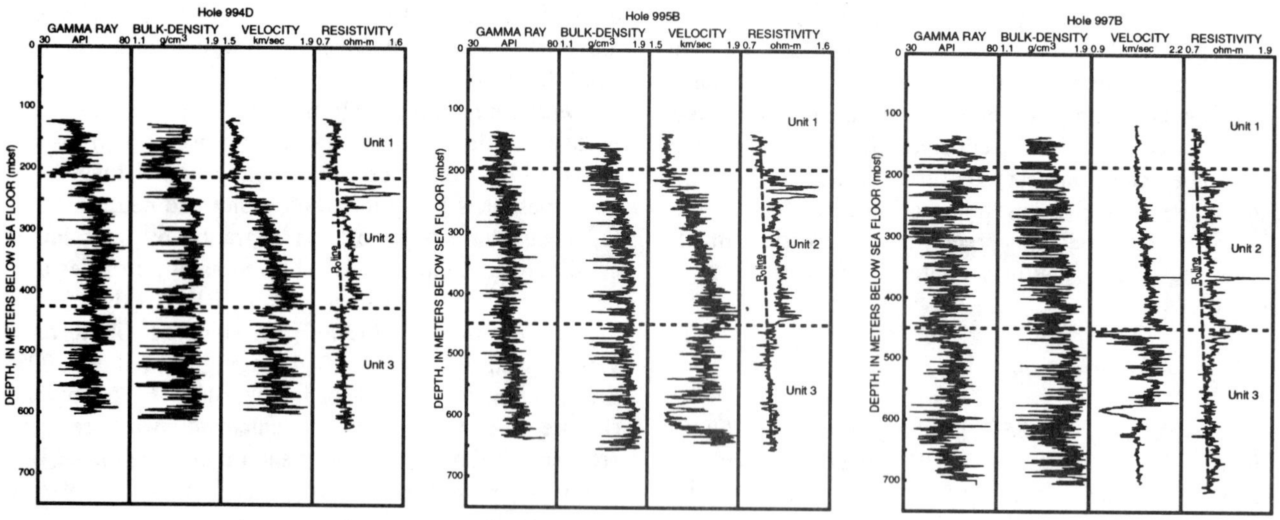

Figure 10. Downhole log data from ODP Holes 994D, 995B, and 997B.

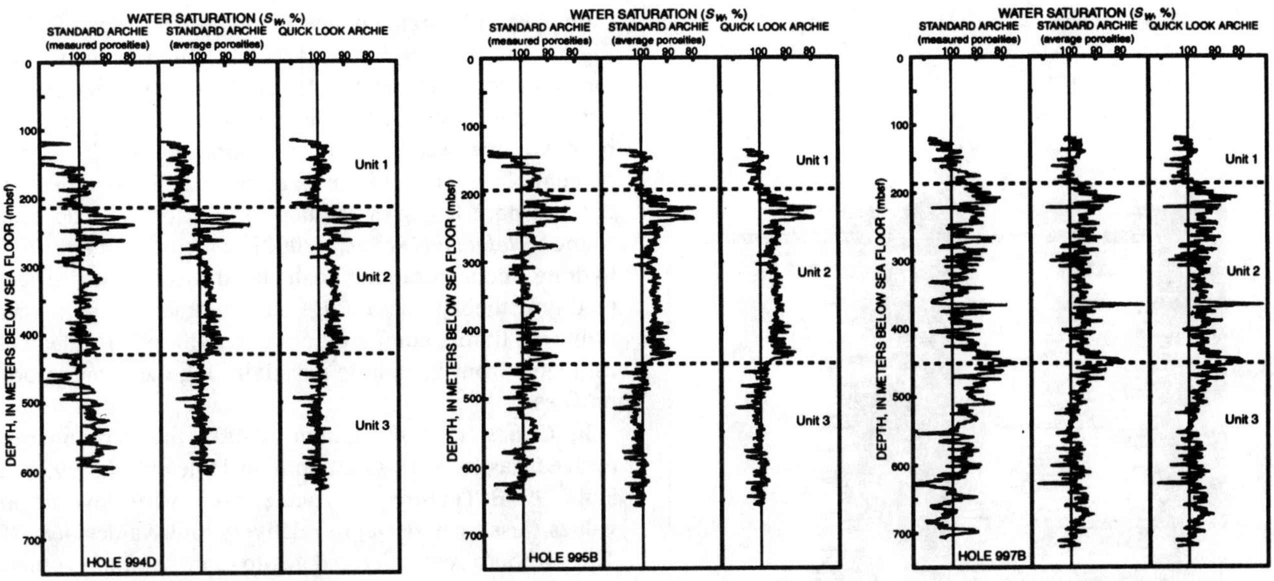

Figure 11. Standard and quick look *Archie* derived water saturations in ODP Holes 994D, 995B, and 997B [modified from *Collett and Ladd,* 2000].

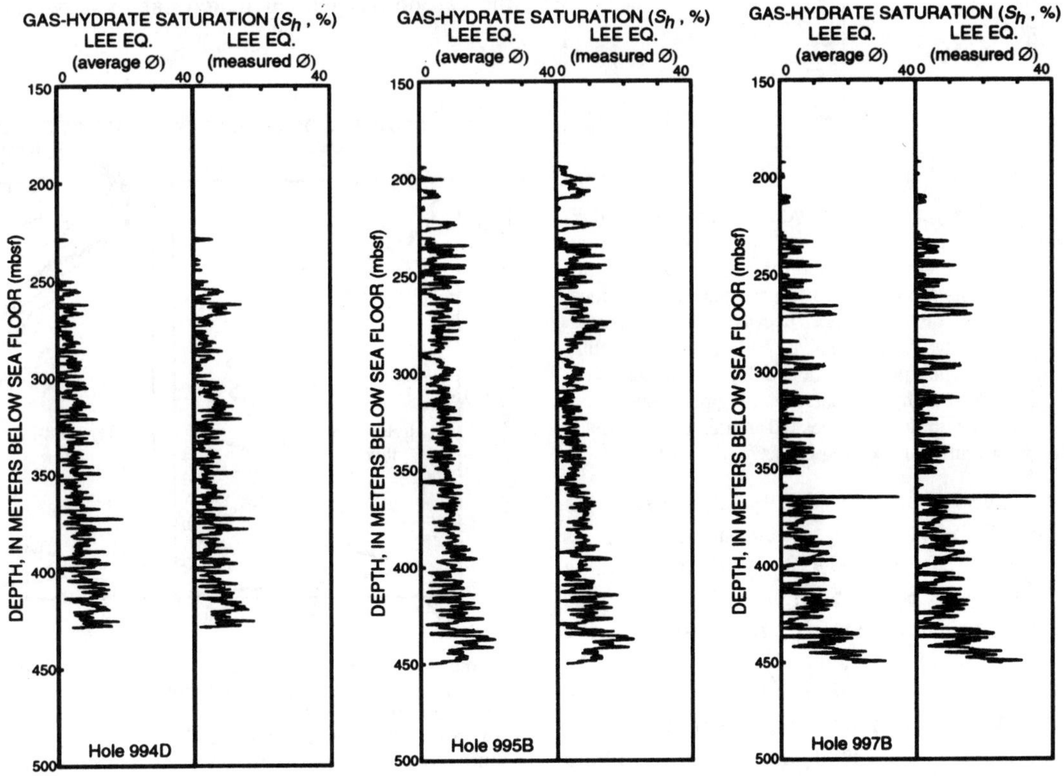

Figure 12. Acoustic velocity derived gas-hydrate saturations in ODP Holes 994D, 995B, and 997B as determined from the Lee weighted average equation [modified from *Lee,* 2000].

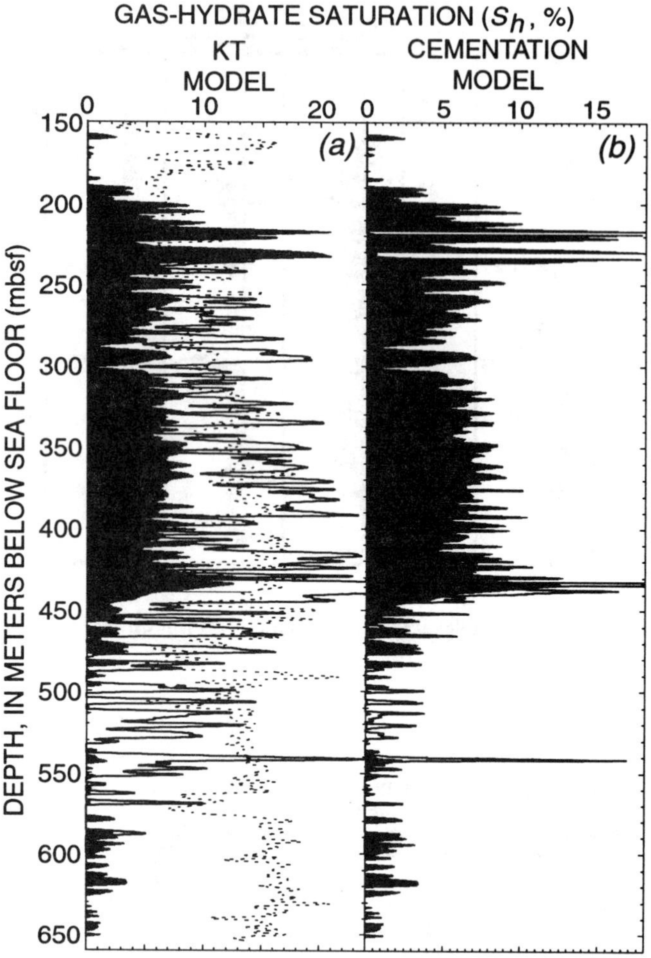

Figure 13. Acoustic velocity derived gas-hydrate saturations in ODP Hole 995B as determined from (a) the *Kuster-Toksöz* wave scattering theory, and (b) the *Dvorkin-Nur* cementation model (modified from *Guerin et al.*, 1999], results are compared to gas-hydrate saturations estimated from resensitivity log (gray area). The *Kuster-Toksöz* derived gas-hydrate saturation log plots have been obtained from either the bulk modulus (solid line) or shear modulus (dotted line) approach. The cementation model derived gas-hydrate saturation log plots have been obtained by assuming gas hydrates are deposited at grain contacts (narrow black solid area) or uniformly coating grains (solid line).

other sources. *Guerin et al.* [1999] also used the cementation model (based on grain contact theory), as proposed by *Dvorkin and Nur* [1993, 1996], to estimate gas-hydrate saturations at Site 995. Figure 13 also contains a plot of gas-hydrate saturations as calculated from the *Dvorkin-Nur* cementation model. At Site 995, the cementation model appears to yield gas-hydrate saturations similar to those calculated by other independent methods.

In *Helgerud et al.* [2000] analysis of the downhole acoustic log data from Site 995, they used the effective medium theory, along with several assumptions about the effect of gas hydrate on sediment elastic moduli, to estimate gas hydrate concentrations on the Blake Ridge. In Figure 14, *Helgerud et al.* [2000] plotted acoustic (sonic) log data together with calculated velocity profiles for a 100% water saturated sediment and three model derived bulk volume gas hydrate concentrations of 2, 4, and 8 percent. In Figure 14, the gas hydrate is assumed to be either a pore filling constituent or part of the load-bearing frame. *Helgerud et al.* [2000] concluded that the gas hydrate concentrations calculated from the effective medium theory, assuming gas hydrate is part of the sediment frame, compare favorably with other independent estimates from downhole resistivity log data and chlorinity profiles.

In *Collett and Wendlandt* [2000], the carbon/oxygen-derived gas-hydrate saturations in Holes 994C, 995B, and Hole 997B (Figure 15) range from very low negative values (less than -10%) to relatively high values near 30%. The carbon/oxygen-derived gas-hydrate saturations that fall below 0% (or 100% water-saturated), which is theoretically impossible, are likely caused by enlarged borehole conditions, which have degraded the inelastic neutron measurements that are not adequately corrected by the procedures developed in *Collett and Wendlandt* [2000]. Within zones of relatively high quality GLT measurements, the carbon/oxygen-calculated gas-hydrate saturations from

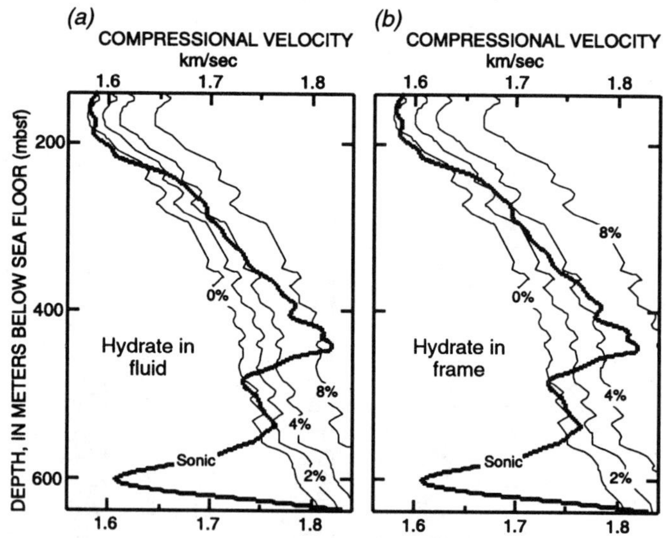

Figure 14. Acoustic velocity derived gas-hydrate saturations in ODP Hole 995B as determined from the effective medium theory [modified from *Helgerud et al.*, 2000]. (a) Comparison of downhole measured compressional-wave acoustic velocities (sonic) with effective medium model results assuming gas hydrate is a pore-filling constituent. (b) Comparison of downhole measured compressional-wave acoustic velocities (sonic) with effective medium model results assuming gas hydrate is part of the sediment frame.

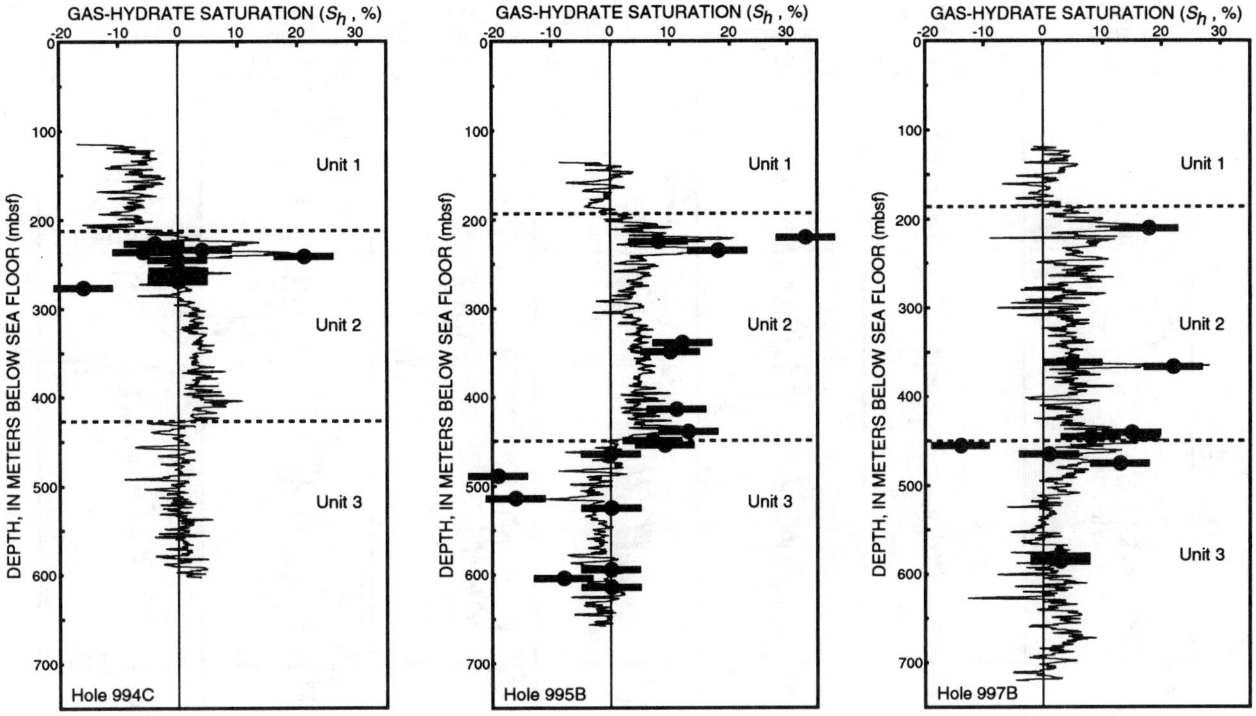

Figure 15. Carbon/oxygen (GST) calculated gas-hydrate saturations (S_h) (shown as discrete point measurements with error bars) in ODP Holes 994C, 995B, and 997B. Also shown (as a continuous line plot) for comparison purposes are the gas-hydrate saturations calculated with the standard *Archie* electrical resistivity method in the same hole (Figure 11) [modified from *Collett and Wendlandt,* 2000].

the Blake Ridge compare favorably to the resistivity-derived gas-hydrate saturations.

Mackenzie River Delta, Canada

The JAPEX/JNOC/GSC Mallik 2L-38 gas hydrate research well was designed to investigate the occurrence of in-situ natural gas hydrates in the Mallik area of the Mackenzie River Delta of Canada (Figure 1) [*Collett et al.,* 1999]. Downhole electrical resistivity and acoustic transit-time logs from the Mallik 2L-38 well confirm the occurrence of in-situ gas hydrates at the Mallik drill-site within the subsurface depth interval between 897.25 and 1,109.5 m (Figure 16).

In *Collett et al.* [1999] and *Collett* [2000], the comparison of core-derived porosities with the shale and hydrate corrected density and neutron log derived porosities from the Mallik 2L-28 well, generally reveals that the corrected density porosities more closely match the core-derived porosities; while, the neutron derived corrected porosities are lower than the core-derived porosities.

The average water saturations (S_w) calculated by *Collett et al.* [1999] with the standard *Archie* relation within the gas-hydrate-bearing units of the Mallik 2L-38 well range from near zero to about 90% (Figure 17). In Figure 17, the results of the quick look *Archie* calculations are also shown as a water saturation (S_w) log trace for the sub-permafrost portion of the well. As shown in Figure 18 [modified from *Collett,* 2000], the Indonesian shaly-sand model-derived water saturations in the shale-free gas-hydrate-bearing sandstone reservoirs in the Mallik 2L-38 well are similar to the water saturations calculated by the standard *Archie* relation. Within the shaly-sand reservoirs, however, the Indonesian model often yielded significantly lower water saturations.

As shown in Figure 19 [modified from *Lee and Collett,* 1999], the *Lee* compressional-wave weighted average equation yielded, for the most part, reasonable gas-hydrate saturations (S_h) ranging from near zero to about 80%, which are comparable but slightly lower than the gas-hydrate saturations (1-S_w) calculated from the electrical resistivity log data. The deployment of the Schlumberger Dipole Shear Sonic Imager (DSI) tool in the Mallik 2L-38 well provided an important data set for accurately estimating the amount of in-situ gas hydrate using both compressional- and shear-wave velocity data. In *Lee and Collett* [1999] the empirical weighted average relationship

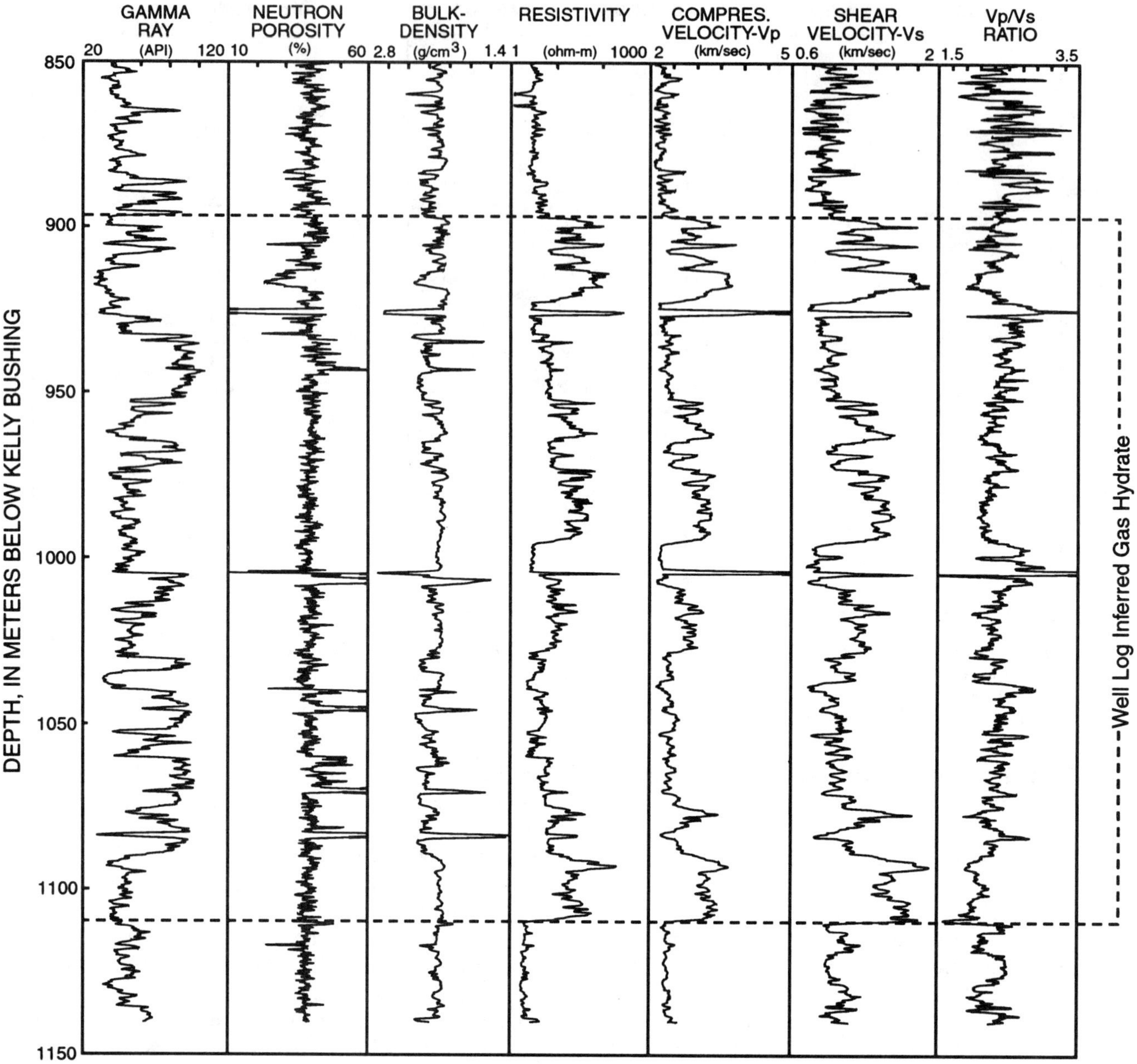

Figure 16. Downhole log data from the Mallik 2L-38 well.

was extended to include shear-wave velocity data. The results of the combined compressional- and shear-wave weighted average acoustic relation assessment of the Mallik 2L-38 well have been included in Figure 19. In the Mallik 2L-38 well, the combined compressional- and shear-wave weighted average acoustic relation yielded similar but slightly lower gas-hydrate saturations than the compressional-wave acoustic velocity calculations.

SUMMARY -- REVIEW OF WELL LOG METHODS FOR GAS HYDRATE RESERVOIR EVALUATION

Porosity Calculations

In the well log response modeling section of this paper, the standard density-porosity relation and a modified three-component (water, hydrate, matrix) density-porosity

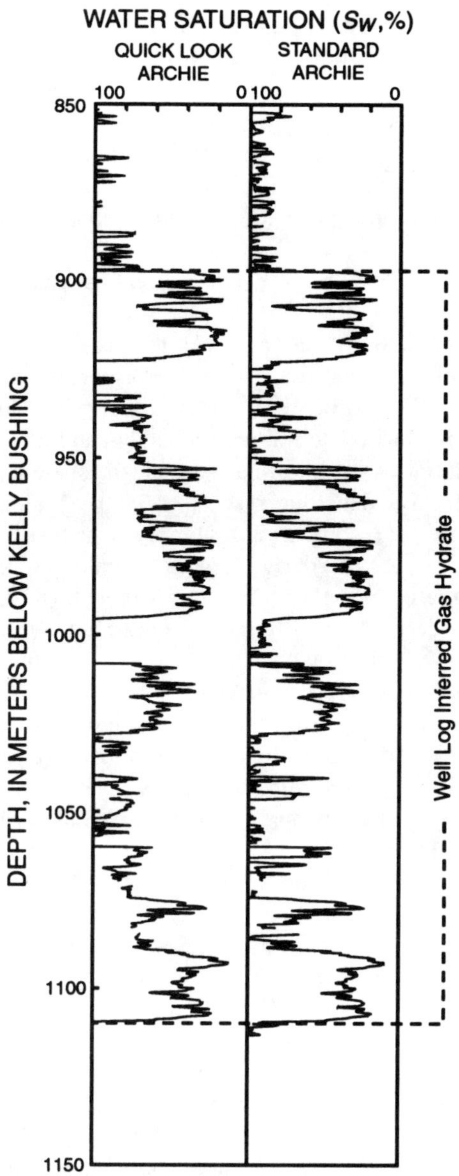

Figure 17. Standard and quick look *Archie* derived water saturations in the Mallik 2L-38 well [modified from *Collett et al.*, 1999].

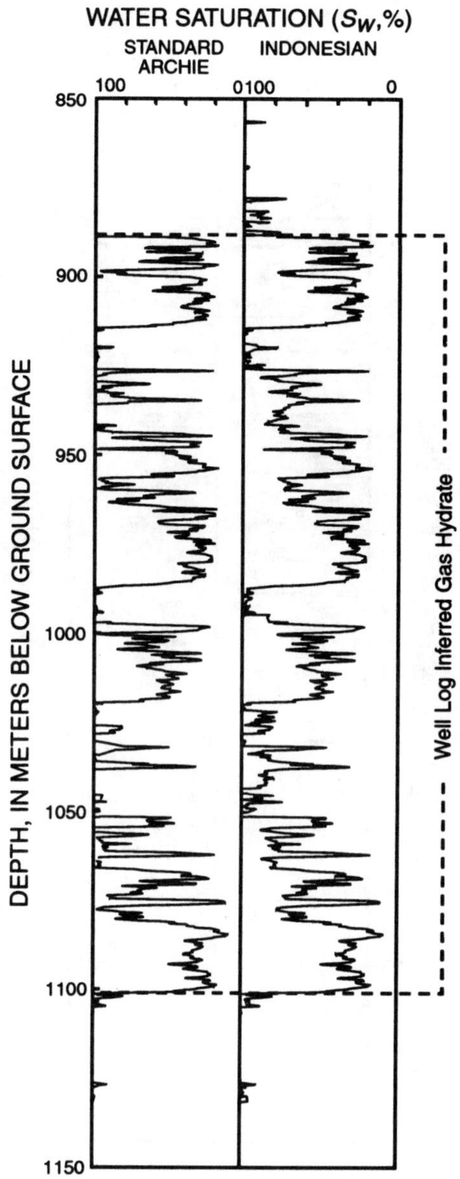

Figure 18. Water saturation log plots for the Mallik 2L-38 well calculated from the standard *Archie* relation and the Indonesian shaly-sand model [modified from *Collett*, 2000].

equation have been reviewed and used to assess the effect of gas hydrate on the density log derived sediment porosities. As previously discussed, at relatively low gas-hydrate saturations (S_h<50%) and with sediment porosities below 40%, the standard density-porosity relation yields relatively accurate results. At higher sediment porosities and gas-hydrate saturations, however, the downhole log-derived bulk-density porosities need to be corrected.

Simple chemical modeling of hydrogen concentrations in gas-hydrate-bearing sediments and complex computer well-log response modeling, as reported in *Collett* 1998b, reveal that gas hydrate has little effect on neutron porosity log responses at relatively low gas-hydrate saturations (S_h<50%) and within reservoirs with porosities below 40%. At relatively high gas-hydrate saturations, however, the neutron porosity log will significantly overestimate porosities.

In summary, assuming good borehole conditions, the density and neutron porosity logs should yield accurate sediment porosities within most gas-hydrate-bearing

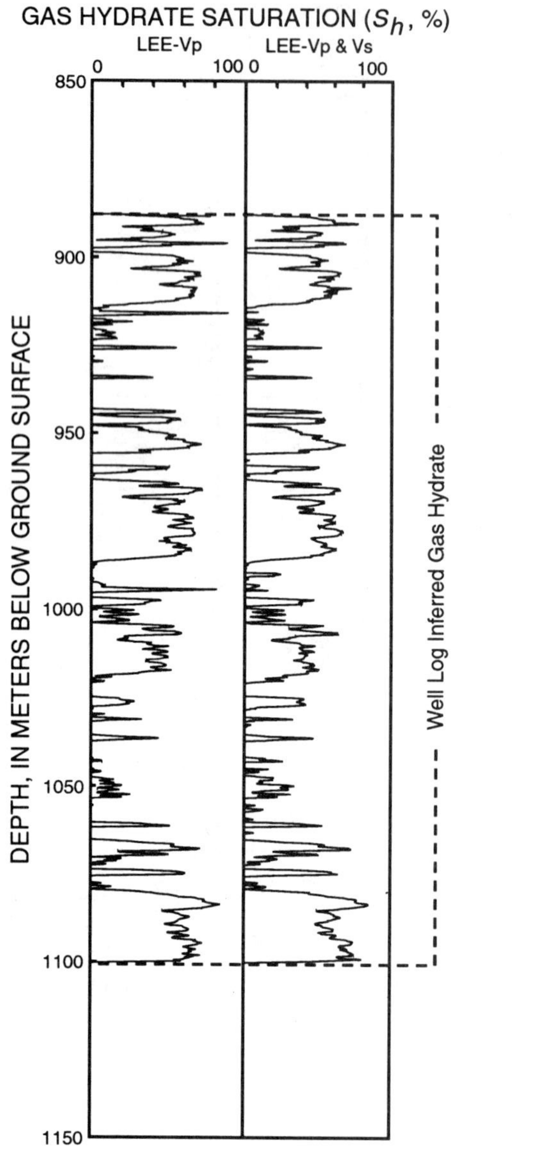

Figure 19. Acoustic velocity derived gas-hydrate saturations in the Mallik 2L-38 well as determined from both the *Lee* compressional- and shear-wave weighted average equations [modified from *Lee and Collett,* 1999].

reservoirs (with appropriate corrections in shale-rich sedimentary sections). However, at relatively high gas-hydrate saturations, density- and neutron-log-derived porosities need to be corrected for the presence of gas hydrate.

Gas-Hydrate Saturations

As shown in the well log response modeling section of this paper, electrical resistivity, acoustic transit-time, and neutron spectroscopy downhole log data can be used to quantify the amount of gas hydrate in a sedimentary section. Two unique forms of the *Archie* relation have been used to calculate water saturations (S_w) from the available electrical resistivity log data at both the Blake Ridge (Figure 11) and Mallik drill sites (Figure 17). The standard *Archie* relation and quick look *Archie* method yielded, for the most part, gas-hydrate saturations that compare favorably with gas-hydrate concentrations calculated by other means.

As documented in *Collett* [2000], the occurrence of shale (clay) in the sediment matrix of a reservoir can affect the calculation of hydrocarbon saturations with downhole resistivity log data. The Indonesian shaly-sand derived gas-hydrate saturations in the relatively clean (shale-free) gas-hydrate-bearing sandstone reservoirs in the Mallik 2L-38 well (Figure 18) compared favorably with the standard (clean-sand) Archie-derived gas-hydrate saturations. Within relatively clay-rich intervals, however, the Indonesian-derived gas-hydrate saturations were often significantly higher than the standard *Archie*-derived saturations. The *Waxman-Smits* derived gas-hydrate saturations for Site 995 on the Blake Ridge [*Collett,* 2000] were significantly higher than the standard (clean-sand) *Archie*-derived gas-hydrate saturations. It appears that the high clay content of the sediments on the Blake Ridge has yielded sediment cation exchange capacities (CECs) beyond the operational limits of the *Waxman-Smits* relation.

In *Lee* [2000] and *Lee and Collett* [1999], the *Lee* weighted average equation has been used along with compressional-wave acoustic log data to calculate gas-hydrate saturations (S_h) in both of the gas hydrate accumulations assessed in this review (Figures 12 and 19). For the most part, the *Lee* weighted average acoustic equation yielded gas-hydrate saturations (S_h) that compared favorably to saturations calculated by other means. Conversely, *Guerin et al.* [1999] determined that the *Kuster-Toksöz* wave scattering theory apparently overestimated gas hydrate saturations at Site 995 on the Blake Ridge. However, *Guerin et al.* [1999] concluded that the grain contact model, proposed by *Dvorkin and Nur* [1993, 1996], appeared to yield reasonable results with the downhole acoustic log data from Site 995. *Helgerud et al.* [2000] also concluded that the proposed effective medium theory yielded gas hydrate saturations at Site 995 that compared favorably with other independent estimates.

As proposed by *Collett and Wendlandt* [2000], carbon/oxygen elemental data from downhole neutron spectroscopy logs can also be used to determine gas-hydrate saturations. Carbon/oxygen elemental data from the geochemical logging tool (GLT) has been used to determine gas-hydrate saturations at all three sites (Sites 994, 995, and 997) drilled on the Blake Ridge during Leg 164. In most cases the carbon/oxygen-calculated gas-hydrate saturations (Figure 15) from the Blake Ridge drill-sites compare favorably with the gas-hydrate concentrations calculated by other methods, however, in several cases the results of the carbon/oxygen calculations were significantly degraded by poor borehole conditions.

In summary, the standard *Archie* relation, the *Lee* weighted average acoustic equation, the grain contact theory, and the effective medium theory can yield gas-hydrate saturations that compare favorably with gas hydrate concentrations calculated by other means.

Acknowledgments. This work was partially funded by the U.S. Department of Energy through the USDOE Interagency Agreement DE-AI21-92MC29214 (Task Number 6). I would like to express my appreciation to Drs. R.F. Wendlandt and E.D. Sloan for all of their technical support and guidance. Special thanks is extended to Drs. C.K. Paull and R. Matsumoto, Co-Chief Scientists of ODP Leg 164, for their support of the an ambitious gas hydrate research cruise. I also wish to thank Dr. D. Goldberg and the entire staff of the Borehole Research Group of the Lamont-Doherty Earth Observatory for the special support they afforded the Leg 164 well logging program. Support for the Mackenzie Delta Gas Hydrate Research Project and the Mallik 2L-38 well logging program was provided by the Japan National Oil Corporation, Japan Petroleum Exploration Company Limited, the Geological Survey of Canada, the U.S. Geological Survey, and the U.S. Department of Energy. I would also like to recognize a long list of groups and individuals, whose collective efforts have been responsible for acquiring and archiving the data and results included within this paper. Most of the historical marine gas hydrate data have been collected by DSDP and ODP crews during Leg 83, 146, 164, and 170; the Mallik data were collected by the Japan Petroleum Exploration Company Limited, the Geological Survey of Canada, the U.S. Geological Survey, and the U.S. Department of Energy.

REFERENCES

Archie, G.E., The electrical resistivity log as an aid in determining some reservoir characteristics. *Journal of Petroleum Technology*, v. 5, p. 1-8, 1942.

Booth, J.S., Rowe, M.M., and Fischer, K.M., Offshore gas hydrate sample database with an overview and preliminary analysis. U.S. Geological Survey Open-File Report 96-272, p. 1-17, 1996.

Collett, T.S., Detection and evaluation of natural gas hydrates from well logs, Prudhoe Bay, Alaska. Proceedings of the Fourth International Conference on Permafrost, July 17-22, Fairbanks, Alaska, National Academy of Sciences Publishers, Washington D.C., p. 169-174, 1983.

Collett, T.S., Natural gas hydrates of the Prudhoe Bay and Kuparuk River area, North Slope, Alaska: American Association of Petroleum Geologists Bulletin, v. 77, no. 5, p. 793-812, 1993.

Collett, T.S., Well log evaluation of gas hydrate saturations. Transactions of the Society of Professional Well Log Analysts, Thirty-Ninth Annual Logging Symposium, May 26-29, 1998, Keystone, Colorado, USA, Paper MM, 1998a.

Collett, T.S., Well log characterization of sediment porosities in gas-hydrate-bearing reservoirs. Proceedings of the 1998 Annual Technical Conference and Exhibition of the Society of Petroleum Engineers, September 27-30, 1998, New Orleans, Louisiana, USA, 12 p. (CD-ROM), 1998b.

Collett, T.S., Quantitative well-log analysis of in-situ natural gas hydrates. Ph.D. Thesis, Colorado School of Mines, Golden, Colorado, p. 1-558, 2000.

Collett, T.S., Godbole, S.P., and Economides, C.E., Quantification of in-situ gas hydrates with well logs. Proceedings of the 35th Annual Technical Meeting of the Petroleum Society of CIM, June 10-13, Calgary, Canada, p. 571-582, 1984.

Collett, T.S. and Ladd, John, Detection of gas hydrate with downhole logs and assessment of gas hydrate concentrations (saturations) and gas volumes on the Blake Ridge with electrical resistivity log data: Proceedings of the Ocean Drilling Program, Scientific Results, v. 164, p. 179-191, 2000.

Collett, T.S., Lewis, R.E., Dallimore, S.R., Lee, M.W., Mroz, T.H., and Uchida, T., Detailed evaluation of gas hydrate reservoir properties using JAPEX/JNOC/GSC Mallik 2L-38 gas hydrate research well downhole well-log displays, *in* Dallimore, S.R., Uchida, T., and Collett, T.S., eds., Scientific Results from JAPEX/JNOC/GSC Mallik 2L-38 Gas Hydrate Research Well, Mackenzie Delta, Northwest Territories, Canada: Geological Survey of Canada Bulletin 544, p. 295-312, 1999.

Collett, T.S. and Wendlandt, R.F., Formation evaluation of gas-hydrate-bearing marine sediments on the Blake Ridge with downhole geochemical log measurements: Proceedings of the Ocean Drilling Program, Scientific Results, v. 164, p. 199-215, 2000.

Dallimore, S.R., and Collett, T.S., Intrapermafrost gas hydrates from a deep core hole in the Mackenzie Delta, Northwest Territories, Canada. *Geology*, v. 23, no. 6, p. 527-530, 1995.

Dallimore, S.R., Uchida, T., and Collett, T.S., Summary. *in* Dallimore, S.R., Uchida, T., and Collett, T.S. eds., Scientific Results from JAPEX/JNOC/GSC Mallik 2L-38 Gas Hydrate Research Well, Mackenzie Delta, Northwest Territories, Canada. Geological Survey of Canada Bulletin 544, p. 1-10, 1999.

Dvorkin, J., and Nur, A., Rock physics for characterization of gas hydrates, *in* Howell, D.G., ed., The Future of Energy Gases. U.S. Geological Survey Professional Paper 1570, p. 293-311, 1993.

Dvorkin, J., and Nur, A., Elasticity of high-porosity sandstones: theory for two North Sea data sets. *Geophysics*, v. 61, p. 1,363-1,370, 1996.

Ecker, C., Lumley, D., Dvorkin, J., and Nur, A., Structure of hydrate sediment from seismic and rock physics. Proceedings of the Second International Conference on Natural Gas Hydrates, June 2-6, Toulouse, France, p. 491-498, 1996.

Ekstrom, M.P., Dahan, C.A., Chen, M.-Y., Loyd, P.M., and Rossi, D.J., Formation imaging with microelectrical scanning arrays. Transactions of the Society of Professional Well Log Analysts 27th Annual Logging Symposium, Paper BB, 1986.

Goldberg, D., The role of downhole measurements in marine geology and geophysics. *Review of Geophysics*, v. 35, no. 3, p. 315-342, 1997.

Goldberg, D., Collett. T.S., and Hyndman, R.D., Ground truth: in-situ properties of hydrate. *in* Max, M.D., ed., *Natural Gas Hydrate in Oceanic and Permafrost Environments*, Kluwer Academic Publishers, The Netherlands, p. 295-310, 2000.

Goldberg, D., and Satio, S., Detection of gas hydrates using downhole logs. *in* Henriet, J.P., and Mienert, J., eds., *Gas Hydrates: Relevance to World Margin Stability and Climate Change*, Geological Society, London, Special Publication, v. 137, p. 129-132, 1998.

Guerin, G., Goldberg, D., and Melster, A., Characterization of in situ elastic properties of gas hydrate-bearing sediments on the Blake Ridge. *Journal of Geophysical Research*, v. 104, 17,781-17,795, 1999.

Hearst, J.R., Nelson, P.H., and Paillett, F.L., *Well logging for physical properties*. Second Edition, John Wiley and Sons, Ltd., New York, p. 1-483, 2000.

Helgerud, M.B., Waite, W.F., Nur, A., Pinkston, J., Stern, L.A., Kirby, S.H., and Durham, W.B., Laboratory measurement of compressional and shear wave speeds through methane hydrate. Proceedings of the Spring Meeting of the American Geophysical Union, June 1-4, Boston, Massachusetts, Abstract T51A-03, 1999.

Helgerud, M.B., Dvorkin, J., and Nur, A., Rock physics characterization for gas hydrate reservoirs, elastic properties. In Holder, G.D., and Bishnoi, P.R., eds., *Gas Hydrates, Challenges for the Future, Annals of the New York Academy of Sciences*, v. 912, p. 116-125, 2000.

Hoyer, W.A., Simmons, S.O., Spann, M.M., and Watson, A.T., Evaluation of permafrost with logs. Transactions of the Society of Professional Well Log Analysts 16th Annual Logging Symposium, Paper 15, 1975.

Kuster, G.T., and Toksöz, M.N., Velocity and attenuation of seismic waves in two-phase media, 1, theoretical formulation. *Geophysics*, v. 39, p. 587-606, 1974.

Kvenvolden, K.A., Methane hydrate--A major reservoir of carbon in the shallow geosphere? *Chemical Geology*, v. 71, p. 41-51, 1988.

Lee, M.W., Gas hydrates amount estimated from acoustic logs at the Blake Ridge, Sites 994, 995, and 997. *in* Paull, C.K., Matsumoto, R., Wallace, P.J., and Dillon, W.P., eds., Proceedings of the Ocean Drilling Program, Scientific Results, v. 164, p. 193-198, 2000.

Lee, M.W., and Collett, T.S., Amount of gas hydrate estimated from compressional- and shear-wave velocities at the JAPEX/JNOC/GSC Mallik 2L-38 gas hydrate research well. *in* Dallimore, S.R., Uchida, T., and Collett, T.S. eds., Scientific Results from JAPEX/JNOC/GSC Mallik 2L-38 Gas Hydrate Research Well, Mackenzie Delta, Northwest Territories, Canada. Geological Survey of Canada Bulletin 544, p. 313-322, 1999.

Lee, M.W., and Collett, T.S., Comparison of elastic velocity models for gas-hydrate-bearing sediments. (this volume), 2001.

Lee, M.W., Hutchinson, D.R., Collett, T.S., and Dillon, W.P., Seismic velocities for hydrate-bearing sediments using weighted equation. *Journal of Geophysical Research*, v. 101, no. B9, p. 20,347-20,358, 1996.

Lee, M.W., Hutchinson, D.R., Dillon, W.P., Miller, J.J., Agena, W.F., and Swift, B.A., Method of estimating the amount of in situ gas hydrates in deep marine sediments. *Marine and Petroleum Geology*, v. 10, p. 493-506, 1993.

Mathews, M., Logging characteristics of methane hydrate. *The Log Analyst*, v. 27, no. 3, p. 26-63, 1986.

McKeon, D.C., and Scott, H.D., SNUPAR--a nuclear parameter code for nuclear geophysics applications. *Nuclear Geophysics*, v. 2, no. 4, p. 215-230, 1988.

Pearson, C.F., Halleck, P.M., McGuire, P.L., Hermes, R., and Mathews, M., Natural gas hydrate deposit: a review of in-situ properties. *Journal of Physical Chemistry*, v. 87, p. 4,180-4,185, 1983.

Pearson, C.F., Murphy, J., and Hermes, R., Acoustic and resistivity measurements on rock samples containing tetrahydrofuran hydrates: laboratory analogs to natural gas hydrate deposits. *Journal of Geophysical Research*, v. 91, no. B14, p. 14,132-14,138, 1986.

Poupon, A., and Leveaux, J., Evaluation of water saturation in shaly formation. Transactions of the Society of Professional Well Log Analysts 11th Annual Logging Symposium, Paper O, 1971.

Saki, A., Velocity analysis of vertical seismic profile (VSP) survey at JAPEX/JNOC/GSC Mallik 2L-38 gas hydrate research well, and related problems for estimating gas hydrate concentration. *in* Dallimore, S.R., Uchida, T., and Collett, T.S. eds., Scientific Results from JAPEX/JNOC/GSC Mallik 2L-38 Gas Hydrate Research Well, Mackenzie Delta, Northwest Territories, Canada. Geological Survey of Canada Bulletin 544, p. 323-340, 1999.

Serra, O., *Fundamentals of well-log interpretation. Developments in Petroleum Science*, v. 1, Elesevier Publishing, Amsterdam, The Netherlands, 1984.

Shipboard Scientific Party, Site 570 (Leg 84). *in* von Huene, R., et al., Proceedings, Deep Sea Drilling Project, Initial Reports, Washington D.C., U.S. Government Printing Office, v. 67, p. 283-336, 1985.

Shipboard Scientific Party, Sites 994, 995, and 997 (Leg 164). *in* Paull, C.K., et al., Proceedings, Ocean Drilling Program, Initial Reports, College Station, Texas, v. 164, p. 99-623, 1996.

Sloan, E.D., *Clathrate hydrates of natural gases*. Marcel Dekker Inc. Publishers, New York, New York p. 1-641, 1990.

Waxman, M.H., and Smits, L.J.M., Electrical conductivities in oil-bearing shaly sands. *Society of Petroleum Engineers Journal*, v. 243, no. 7, p. 107-122, 1968.

Waxman, M.H., and Thomas, E.C., Electrical conductivities in shaly sands, I. The relation between hydrocarbon saturation and resistivity index, II. The temperature coefficient of electrical conductivity. *Journal of Petroleum Technology*, February, p. 213-225, 1974.

Woodhouse, R., and Kerr, S.A., The evaluation of oil saturation through casing using carbon/oxygen logs. *The Log Analyst*, v. 33, no. 1, p. 1-11, 1992.

Timothy S. Collett, U.S. Geological Survey, MS-939, Denver Federal Center, Box 25046, Denver Colorado 80225.

Seafloor Collapse and Methane Venting Associated with Gas Hydrate on the Blake Ridge — Causes and Implications to Seafloor Stability and Methane Release

William P. Dillon, Jeffrey W. Nealon, Michael H. Taylor[1],

U.S. Geological Survey, Woods Hole, MA

Myung W. Lee,

U.S. Geological Survey, Lakewood, CO

Rebecca M. Drury[2], and Christopher H. Anton

U.S. Geological Survey, Woods Hole, MA

A large irregular topographic depression exists at the crest of the Blake Ridge about 280 km off the South Carolina coast, southeastern United States. The feature, which is believed to be a collapse structure, is characterized by normal faulting of strata within the gas hydrate stability zone (GHSZ) on its SW side and by gravity gliding and folding of these subsided sediments in a nappe-like structure on its NE side. Sediments beneath the pre-collapse GHSZ appear to have been chaotically disturbed. The event apparently occurred more than 75 Ka. Subsidence of the seafloor exceeded 100 m over an area of >25 km^2 and the depression represents a volume loss of >13 km^3. This depression is difficult to account for by extension, deep structural movements, landslides, or erosion. We suggest this model: 1. free gas was trapped beneath the GHSZ and overpressures developed; 2. sediments below the GHSZ were mobilized and flowed, causing inflation of the ridge crest; 3. mobilized sediment erupted with significant volume loss; 4. strata within the GHSZ subsided by normal faulting and gravity gliding and folding. If such a sequence of events had occurred on a continental slope/rise, it would probably have triggered a major slide, but on the gentle slopes of the Blake Ridge, the strata above the GHSZ simply collapsed into the evacuated zone. Thus, if this model is correct, this example provides an unusual opportunity to study the process of mobilization of surface layers by gas hydrate-

Natural Gas Hydrates: Occurrence, Distribution, and Detection
Geophysical Monograph 124

Published in 2001 by the American Geophysical Union

[1]Now at: Department of Earth and Space Sciences, University of California, Los Angeles, Los Angeles, CA

[2]Now at: International Faculty Council, University of Washington, Seattle, WA

related processes. The expulsion of sediment may have released large volumes of methane to the ocean/atmosphere system, perhaps 1-4% of the present atmospheric content from the Blake Ridge event, so the process, when also accompanied by landslides, probably has been significant to the global greenhouse.

INTRODUCTION

The Blake Ridge, off the southeastern United States, is marked by a large topographic depression (39 x 23 km) that is located directly on the ridge crest (Plate 1). The depression is considered to be a collapse structure that appears to have been formed by gas hydrate-related processes. The mobilization of sediments that was required to produce this collapse probably would have triggered landslides if it took place on a steeper continental slope/rise, so the study of its formation will provide insight regarding the influence of gas hydrate on seafloor instabilities. The large sediment volume-loss associated with formation of the collapse probably transferred a large volume of methane from the sedimentary reservoir to the atmosphere and the process has probably influenced the global greenhouse. This paper will consider the structure, causes, relationship to gas hydrate processes, and implications of this feature.

SETTING

The Blake Ridge is a sediment drift deposit of Cenozoic age [*Markl et al.,* 1970] and it is perhaps the area of greatest concentration of gas hydrate on the United States Atlantic continental margin [*Dillon and Paull,* 1983, *Dillon and Popenoe,* 1988; *Dillon et al.,* 1994, 1995; *Shipley et al.,* 1979; *Tucholke et al.,* 1977; *Paull and Dillon,* 1981]. The ridge is a broad, accretionary pile of sediment that has accumulated at the site of interaction of major ocean bottom currents [*Bryan,* 1970; *Markl and Bryan,* 1983; *Mountain and Tucholke,* 1985; *McCave and Tucholke,* 1986]. Ridge sedimentation was initiated near the middle of Oligocene time after an episode of ocean floor erosion and the ridge has evolved by accretion on its southwestern flank and erosion on its northeastern flank [*Tucholke and Mountain,* 1986]. The surface of the ridge is generally smooth except for minor differential erosion features on its northeastern (eroding) flank [*EEZ-Scan 87 Scientific Staff,* 1991], and a rough topographic depression at the crest of the ridge at about 32°N [*Dillon et al.,* 1998] that is the topic of this paper. Studies at the Blake Ridge during early Deep Sea Drilling Project (DSDP) work [Leg 11, *Hollister, Ewing et al.,* 1972] provided the initial discovery that gas hydrate exists in marine sediments. Since its initial discovery, other DSDP/Ocean Drilling Program (ODP) drilling efforts have been devoted to sampling gas hydrate at the Blake Ridge during Leg 76 [*Kvenvolden and Barnard,* 1983] and Leg 164 [*Paull, Matsumoto, Wallace, et al.,* 1996]. The Leg 11 work also identified the cause of the Bottom Simulating Reflection (BSR), a seismic reflection that cuts through reflections from strata and parallels the seafloor. The BSR was determined to be a reflection from the top of free-gas-bearing sediments generally overlain by gas-hydrate-bearing sediments and thus it marks the base of the GHSZ in seismic profiles.

METHODS

The Blake Ridge collapse structure was analyzed using bathymetry, seismic reflection profiles and deep-towed sidescan sonar imagery. Bathymetry (Plate 1, Figure 1) was contoured from all available echo-sounder data, primarily 3.5 kHz profiles that were available from U.S. Geological Survey (USGS) and University of North Carolina cruises in the area. Single or 2-channel seismic profiles were collected in 1992 and 1995 with seismic systems designed to image the gas-hydrate zone and the region directly below it (to 1.5-2.0 s two-way reflection time below the seafloor). The profiles are part of a set of 4-km-spaced seismic profiles that cover a 4,800 square-kilometer area on the Blake Ridge. Moderately small, pneumatic seismic sound sources provided adequate energy to penetrate through and below the region of gas-hydrate stability, while still maintaining the best possible resolution. Sources used were a 160-cubic inch (2.62-liter) airgun or a generator/injector (GI) gun, in which the generator chamber (which generates the primary signal) was 105 cubic inches (1.72 liters), and the injector chamber (which controls bubble pulsing) was also 105 cubic inches. One deep-penetration multichannel seismic profile provides imaging to basement and velocity data across the collapse structure; its seismic source was a 1310-cubic inch (21.45 liter) airgun array and a 3600-m, 48-trace streamer was used. Swept frequency (Chirp) sidescan sonar (26 - 33 kHz) obtained from a deep-towed multisensor system, provided complete coverage of the collapse. The deep-towed package also provided subbottom profiles at a swept frequency band of 2 - 7 kHz. Seismic processing of the single-channel profiles was accomplished using Promax 7.2 software. The single channel data were corrected for depth and stacked. The velocity field developed from the multi-channel line was applied to the stacked section using F-K migration. The section then underwent minimum phase predictive deconvolution and a 40-160 HZ Ormsby bandpass filter.

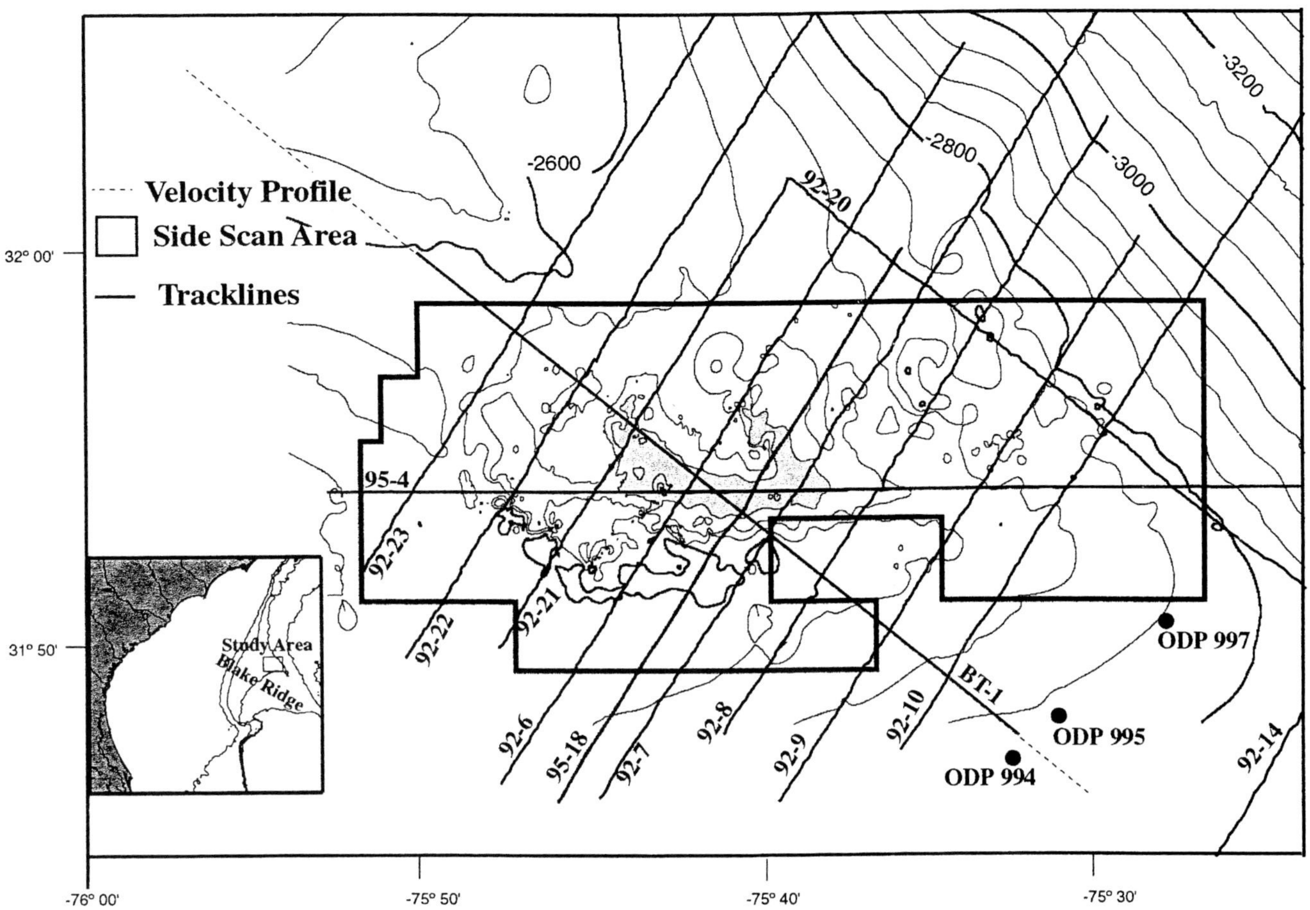

Figure 1. Data locations. Gray shading shows area of inferred vent.

Finally a 300 ms automatic gain control was applied. The sections were plotted with approximately 10 to 1 vertical exaggeration. Faults were mapped using all available seismic profiles, the deep-towed sidescan image, and high resolution seismic profiles from the deep-towed fish. The medium resolution seismic data (airgun and GI gun sources) in SEG-Y format, selected processed profiles, tracklines and the sidescan images are available on a pair of CD-ROMs [*Taylor et al.,* 1999].

SEISMIC PROFILE AND SIDESCAN SONAR DATA ANALYSIS

The inferred collapse feature on the crest of the Blake Ridge will be discussed using a series of profiles that cross the ridge axis starting from the southeast (Figure 1). The collapse, is marked by disturbance of the seafloor, and its limits are indicated on the profiles (Figures 2-6). The broad regional structure of the northwest-trending Blake Ridge outside of the collapse, is characterized by uniformly southwest-dipping strata (Figure 2), so strata on the southwestern flank of the ridge conform to the seafloor. On the northeastern flank of the ridge, strata also dip to the SW, and thus are truncated at the NE-sloping seafloor. Erosional scour of the NE flank of the ridge has left differentially eroded strata, which form rough areas of seafloor that extend along the ridge for >130 km [*EEZ-Scan Scientific Staff,* 1991]. The structure displays the effect of a general southwestward migration of the Blake Ridge by erosion on the northeast side and deposition on the southwest side. The BSR is conspicuous beneath the ridge crest (Figure 2), indicating significant gas accumulation; it is weak, but discernible beneath the collapse, and stronger, but discontinuous to the northeast, where variations in gas content might result from stratigraphically-controlled variations in porosity in the dipping beds.

The southeastern part of the collapse is narrow and variable in character; line 92-10 (Figure 2) shows minor disturbance of the overall ridge structure. Some profiles further down the ridge near the ODP drillsites 994, 995, and 997 [*Dillon et al,* 1996] show a section of the ridge that has not undergone significant collapse. The collapse structures

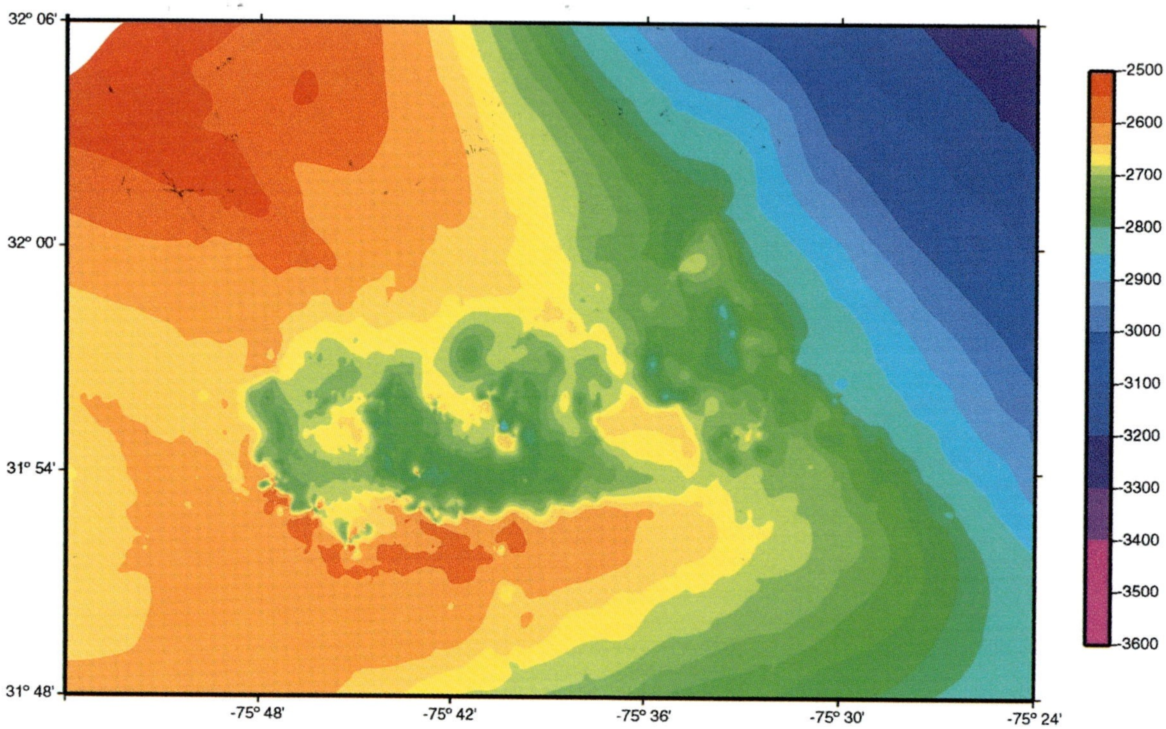

Plate 1. Bathymetric map of the crestal depression on the Blake Ridge.

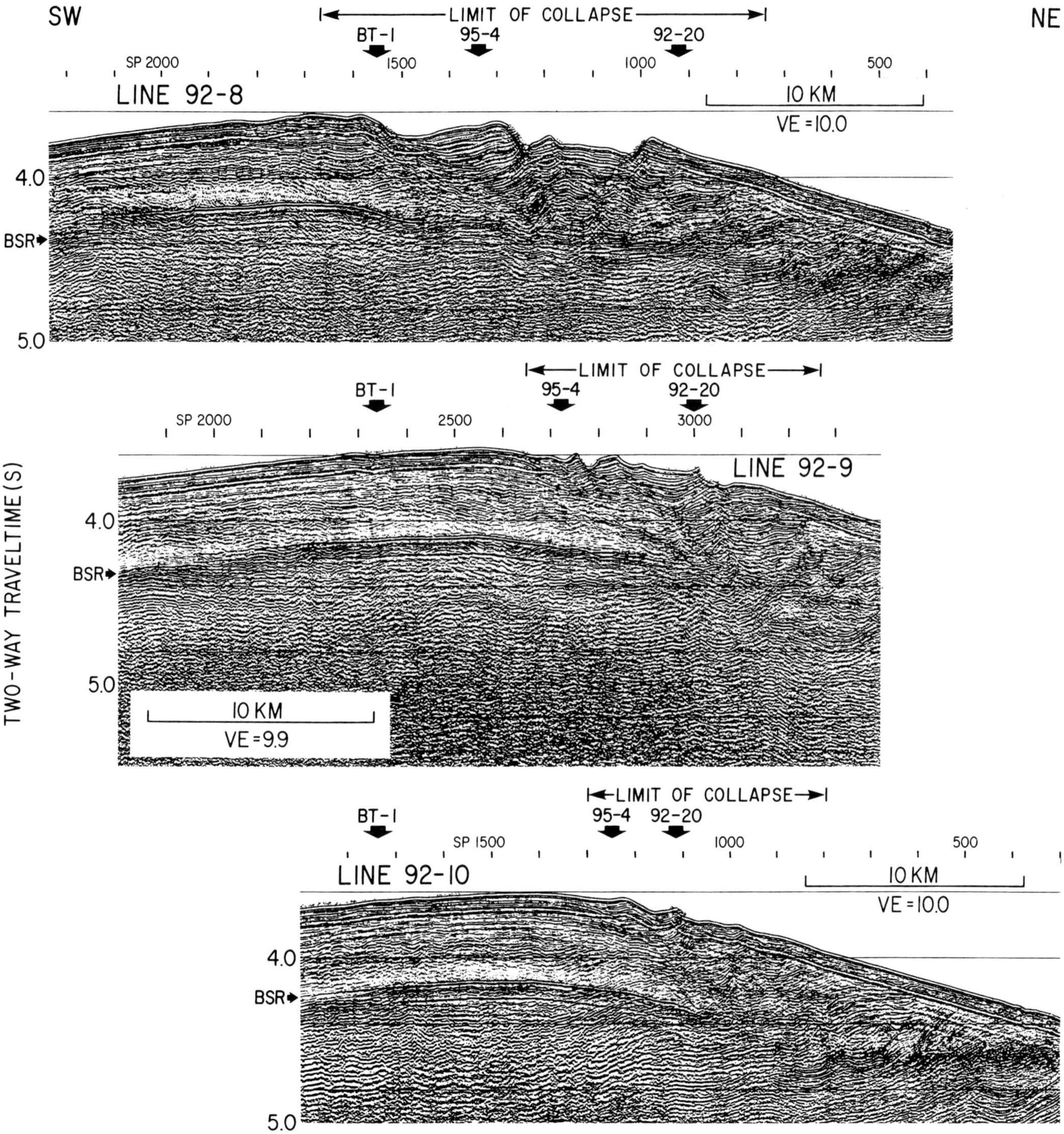

Figure 2. Seismic reflection profiles 92-8, 92-9, and 92-10. Locations shown on Figure 1.

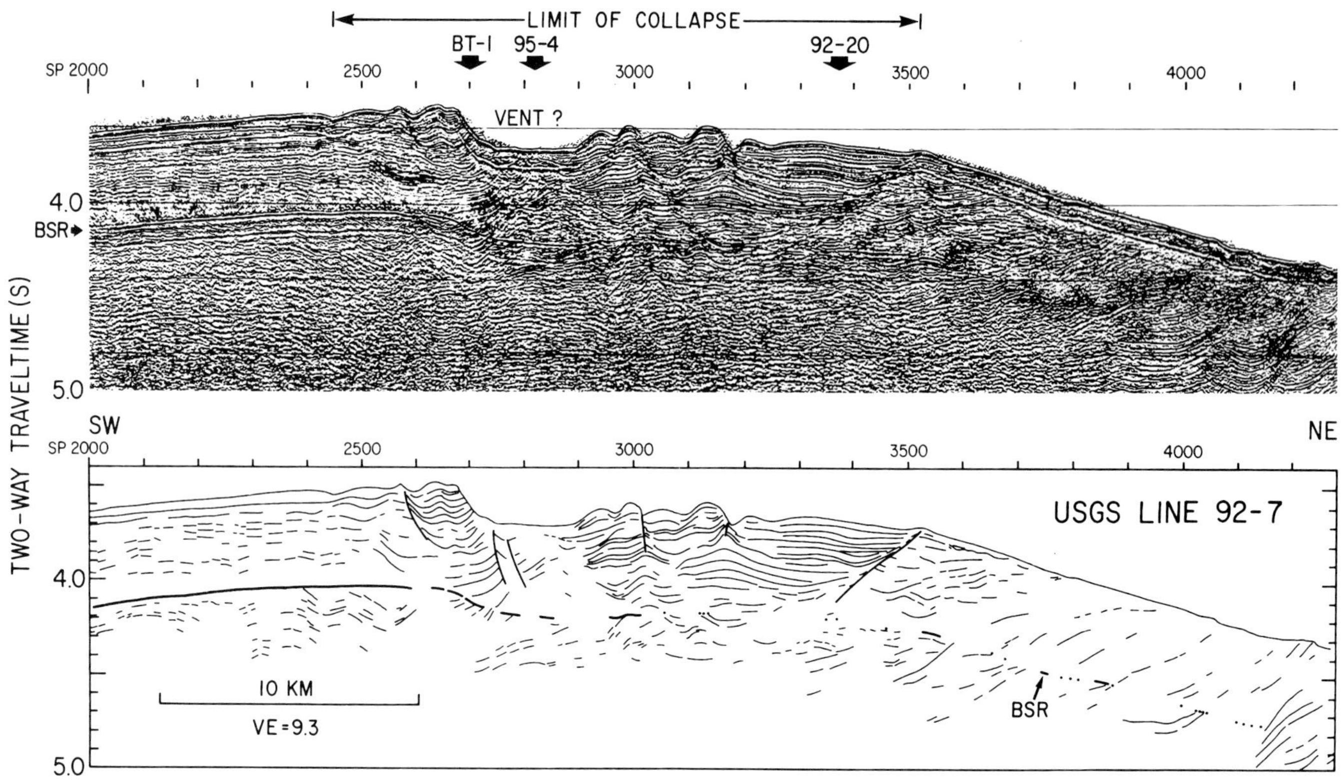

Figure 3. Seismic reflection profile 92-7 and interpretation. Location shown on Figure 1.

in profiles 92-9 and 92-8 (Figure 2), appear to consist primarily of distorted fault blocks bounded by moderately-dipping faults. Note the ~10x vertical exaggeration of most of the figures. For example, true dips of the two faults that bound the central graben on line 92-8 are 10.2° for the southwestern fault (reaches seafloor at shotpoint 1242) and 7.8° for the northeastern fault (reaches seafloor at shotpoint 1019). Significant subsidence of the seafloor appears on line 92-8. At least one older, but comparable, disruption has apparently also affected this region of the Blake Ridge; its structure is visible to the NE and below the recent collapse and is evident in profile 92-7 (Figure 3) between approximately shotpoints 3450 and 3700 at about 4.0-4.5 s time-depth and also appears in the profiles of Figure 2.

The next three profiles to be considered extend across the central zone of disruption of the collapse feature (Figures 3, 4, and 5), where the structures of the northeast and southwest sides are markedly different. On the southwestern side of the collapse, a series of fault blocks appear to have subsided along listric normal faults and truncated strata crop out on fault scarps. Although these blocks contain structures such as drag folds that suggest significant subsidence, parts of the upper portions of the blocks project above the general seafloor surface of the Blake Ridge (Figures 3, 4, and 5) and form the peripheral high that exists along the southern and southwestern boundaries of the depression (Plate 1). Ubiquitous, small-throw, near-vertical growth faults that are spaced ~500 m apart [*Rowe and Gettrust,* 1993] are visible in these profiles and exist over a large area of the southern flank of the Blake Ridge outside of the collapse area. These appear to be structurally unrelated to the major collapse feature.

In contrast to the block faulting on the southern side of the collapse, the sediments on the northern side form an internally-folded and faulted nappe structure. This is underlain by a clearly defined décollement in profiles 92-7 (Figure 3) and 95-18 (Fig 4), whereas the structure in profile 92-6 (Figure 5) is similar, but the detachment at the base is not as clearly apparent. In all profiles, the reflection amplitudes are greater from reflectors within the decollement, and to some extent within the fault blocks, than from reflectors outside the collapse, as shown by the true amplitude profile of Figure 4.

The BSR, representing the base of the GHSZ, occurs at about 0.5 s subbottom (about 400 m) across the entire set of profiles (Figures 3, 4, and 5). As the seafloor has generally subsided in the collapse area, and the base of the GHSZ almost certainly paralleled the seafloor prior to the structural deformation, we conclude that the present BSR in the area is located deeper below the sea surface than before the disruption. Presumably it migrated downward because the seafloor was lowered and cooling of sediments has

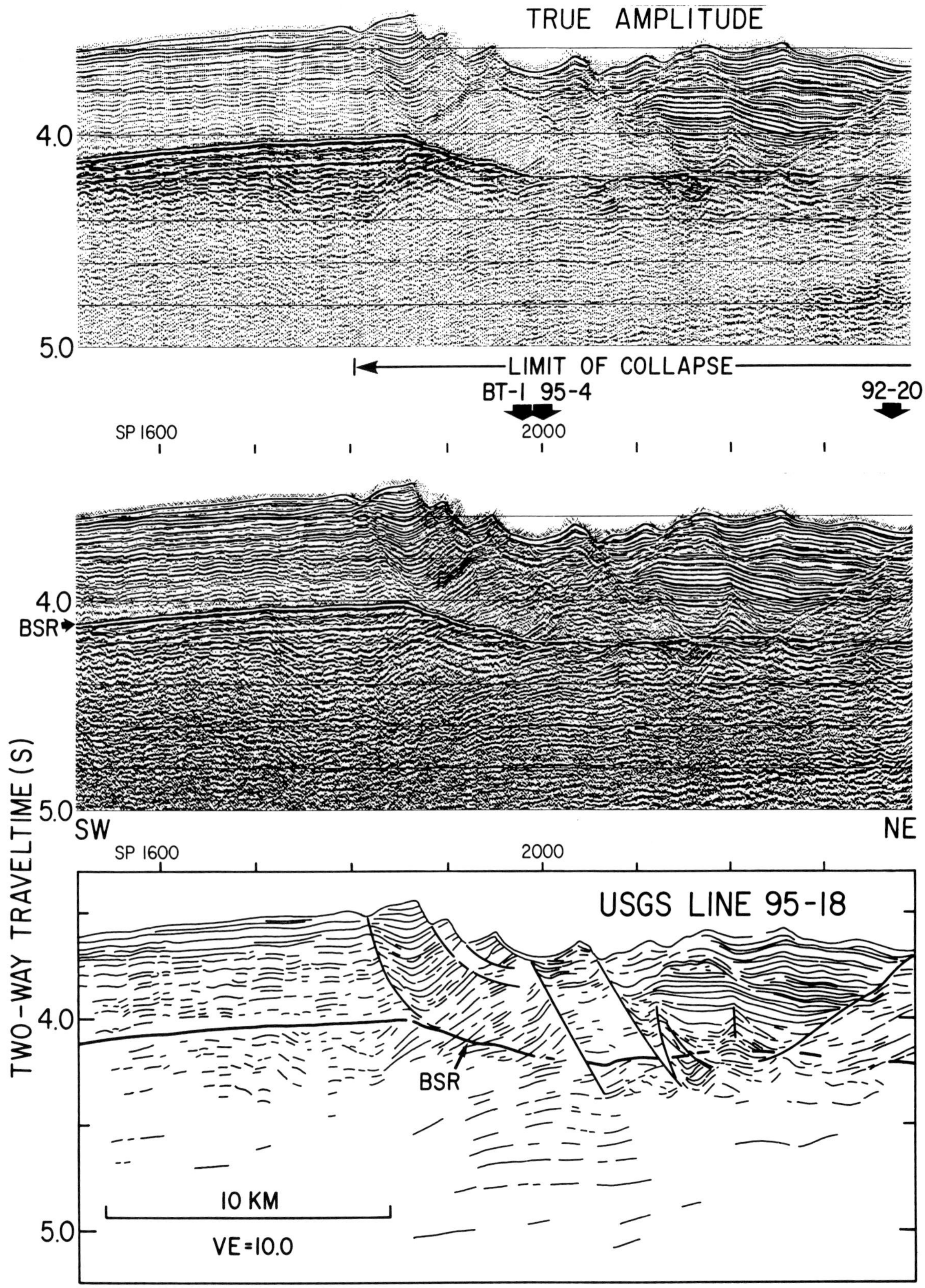

Figure 4. Seismic reflection profile 95-18 and interpretation. Location shown on Figure 1. Top panel shows true-amplitude display of data.

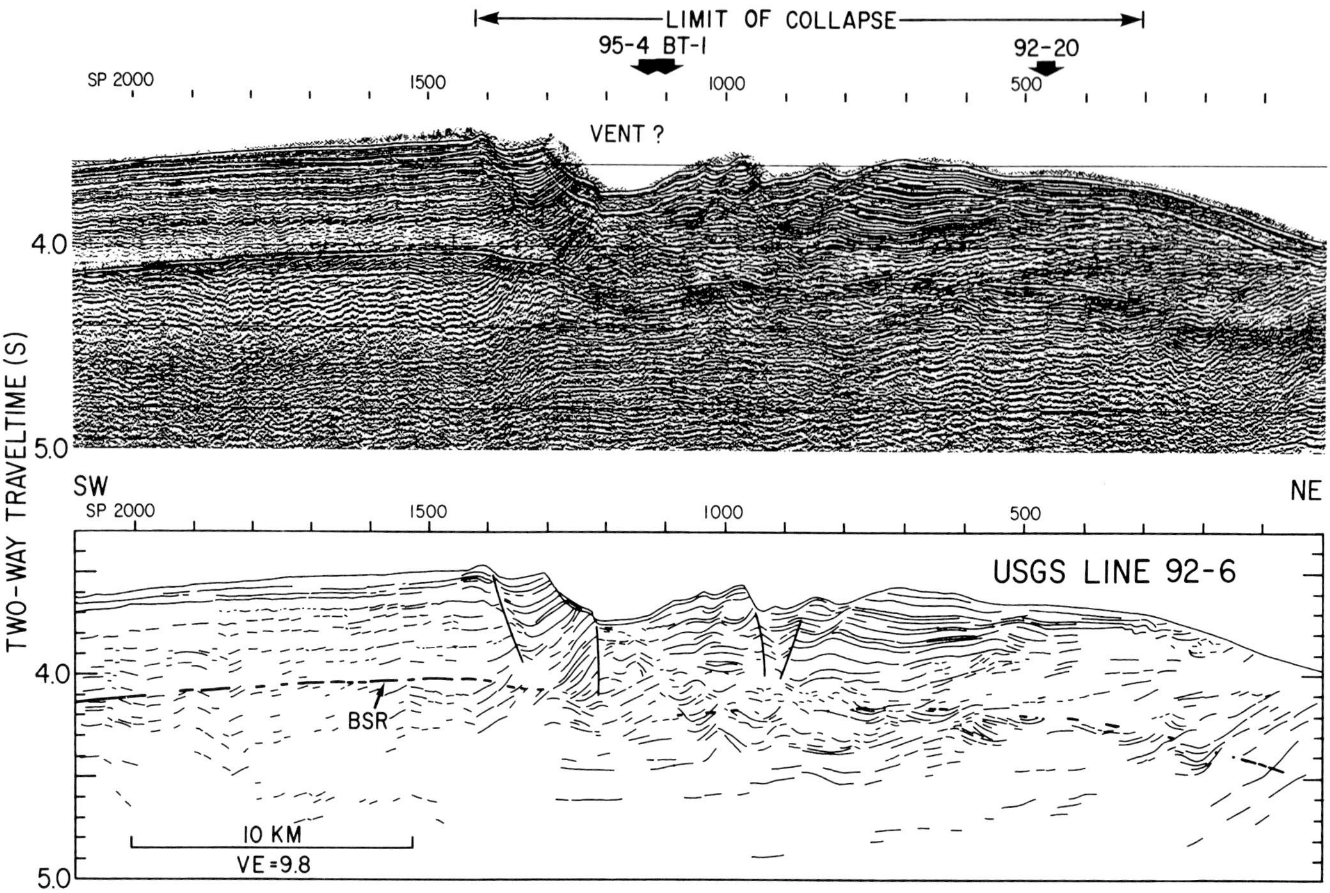

Figure 5. Seismic reflection profile 92-6 and interpretation. Location shown on Figure 1.

reestablished the thermal gradient that controls the depth to the base of the GHSZ.

Two regions of chaotically disturbed strata appear in the profiles. These are clearest in profile 92-6 (Figure 5), where the first of these, a nearly horizontal zone of disturbance extends beneath the collapse structure from about shotpoint 1400 to 700; the zone's top is near 4.0 s and it is about 0.3 s thick. The disturbed zone extends roughly 0.15 s above and 0.15 s below the general depth of the present irregular BSR. However, note that if the trend of the BSR is smoothly extrapolated from the undisturbed regions outside the collapse and through the collapse structure, the extrapolated line will demarcate the top of the disturbed zone. The extrapolated boundary probably also indicates the location of the BSR prior to collapse. Similar highly disturbed regions extending above and below the present BSR are also noted in Figurers 3 and 4.

The second region of chaotic disturbance extends downward from the seafloor beneath the deepest parts of the collapse depression, and appears to connect at its base to the other disturbed region. This region is marked "vent?" in Figures 3, 5, and 6 (line 92-21)

The final three cross-ridge profiles illustrated (Figure 6) define the northwestern part of the collapse structure. Moving northwest, the next profile, 92-21 (Figure 6) is comparable to profile 92-6 (Figure 5), having a block-faulted structure to the south and a mildly deformed nappe to the north, with no well-defined decollement. At the deepest point in the collapse, we observe a convex, apparently accreted pod of sediment, rather than a flat seafloor. Reflections representing the pre-deposition seafloor extend beneath the pod. Beneath that feature is a zone of chaotic reflections, in this case occupying a triangular area in the profile (peak of triangle about shotpoint 576).

Profile 92-21 (Figure 6) shows a distinct peripheral upfolding of strata to the south of the southwesternmost fault, essentially outside of the collapse structure. A comparable, but smaller up-bending of strata also appears in profile 92-6 (Figure 5).

Profiles 92-22 and 92-23 (Figure 6) show that the collapse dies out in the upridge (northwest) direction. Although the sediments on the northeast side of the collapse have subsided and are mildly folded, the reflections seem to

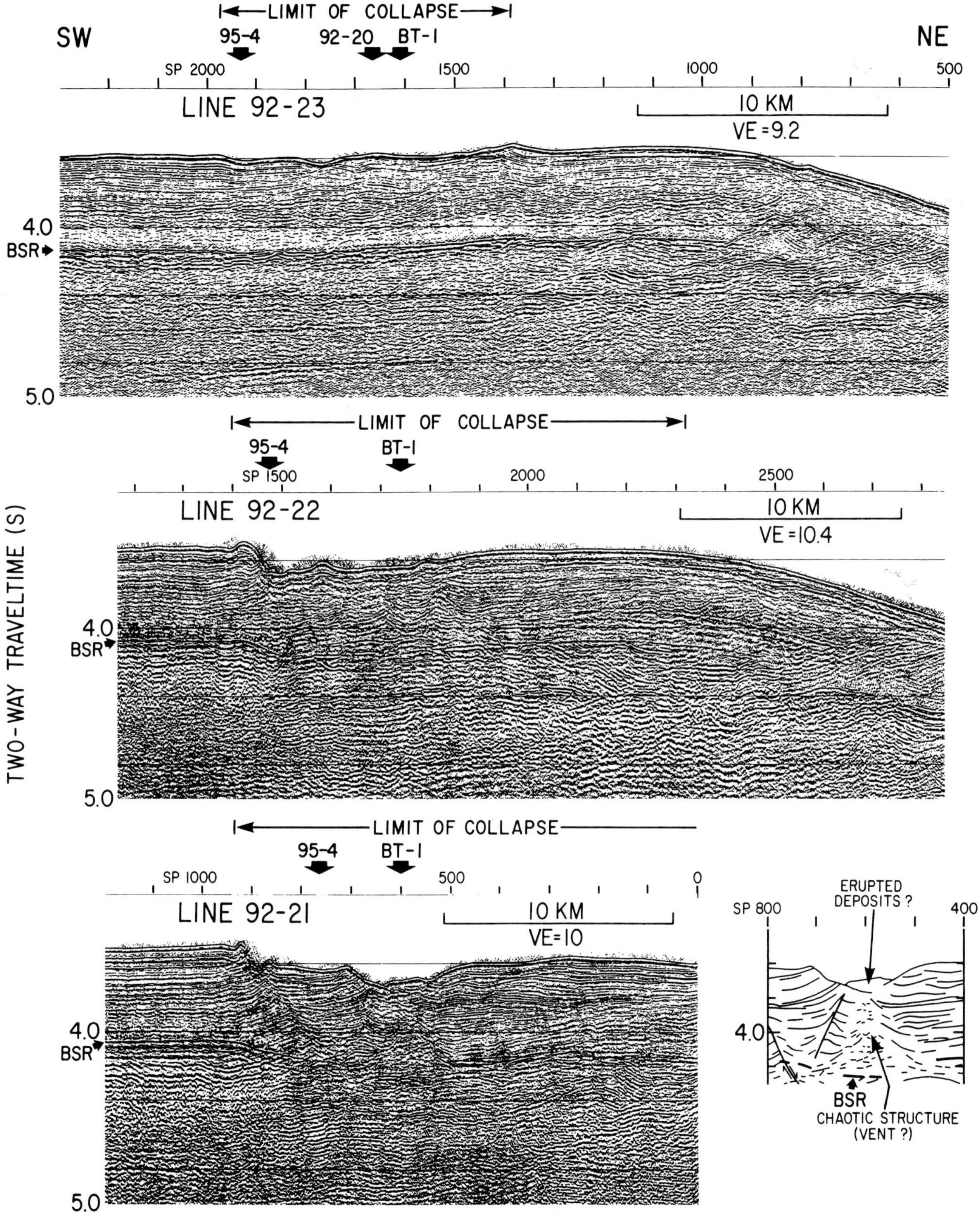

Figure 6. Seismic reflection profiles 92-23, 92-22, and 92-21. Locations shown on Figure 1. Interpretation for central part of collapse structure shown for Line 92-21, shotpoints 800-400.

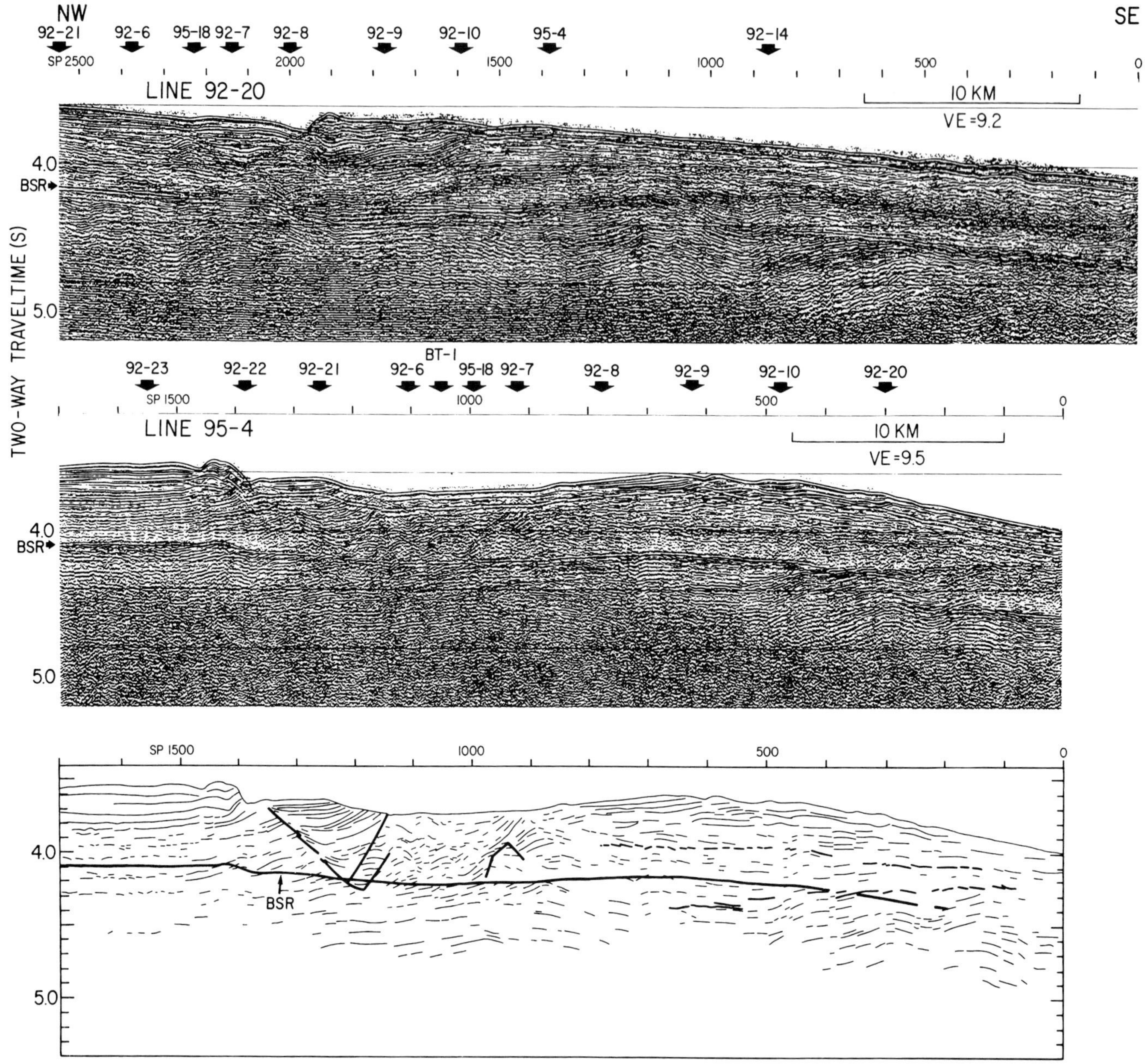

Figure 7. Seismic reflection profiles 92-20 and 95-4 and interpretation of profile 95-4. Locations shown on Figure 1. Profiles are oriented approximately parallel to the Blake Ridge axis. Heavy lines indicate either the BSR (labeled), faults (shotpoints 1400-1100), or the boundary of an inferred fault block (shotpoints 1000-900). Medium weight lines indicate the inferred main décollement surface (upper set) and a deeper set of strong reflections, which may also represent a detachment, although its nature is unclear.

be traceable out of the region of subsidence to undisturbed horizons — no obvious décollement appears. Profiles 92-22 and 92-23 also seem to show a stratigraphically deeper, presumably older zone of disruption, particularly evident in 92-22 between shotpoints 2000 and 2500 at about 4.1-4.5 s depth. Note that a zone of strata directly above the deeper disturbed zone (from about 0.35 to 0.55 s below the seafloor) appears to be undisturbed and separates it from the surface collapse structure. Therefore the deeper disrupted zone appears to be an older collapse, perhaps comparable to the feature observed at depth in profiles of Figures 2 and 3. The magnitude of disruption of the deeper

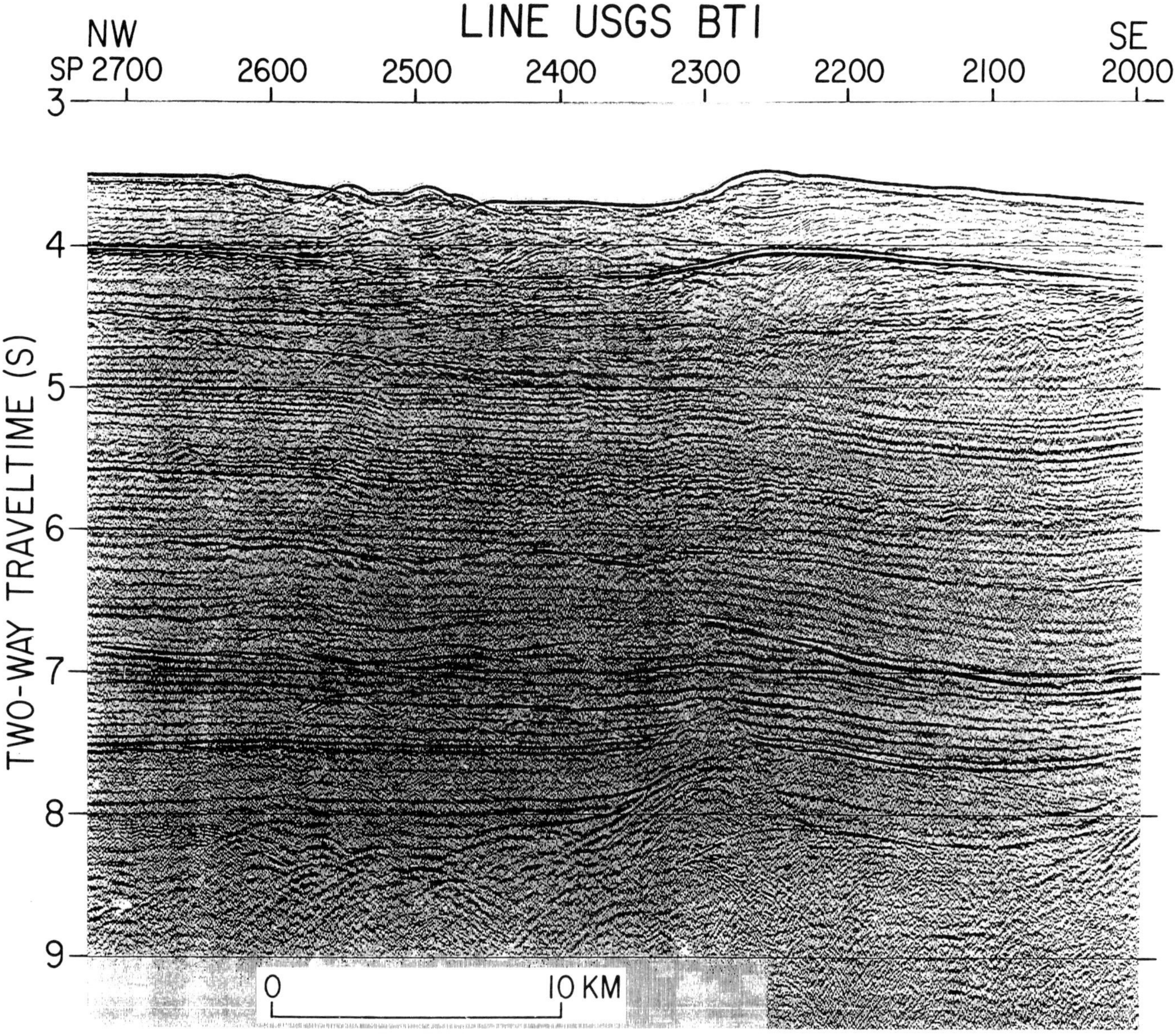

Figure 8. Multichannel seismic reflection profile BT-1. Location shown by solid trackline on Figure 1.

feature seems to increase up ridge on moving from lines 92-21 to 92-23 (Figure 6) whereas the magnitude of the surface collapse becomes less.

Several ridge-parallel lines cross the other profiles and confirm the structure (Figures 7 and 8). Line 92-20 (Figure 7) extends along the northeastern side of the collapse area. At the northwest end of this profile, we observe the base of disturbed sediment at about 3.8 s, deepening southeastward and reaching a maximum depth of about 4.4 s between profiles 92-8 and 92-9 (near shotpoint 1900), then rising, but becoming obscure toward the southeast. The seafloor shows minor undulations to about shotpoint 300, where an inflection occurs, and the seafloor becomes smooth.

Line 95-4 (Figure 7) crosses the center of the collapse structure at an angle of about 55° to mapped fault scarps (Figure 1). The collapse structure extends from roughly shotpoints 1400 to 300. Northwest (left) of the collapse, flat-lying strata are bent up on approaching the feature near shotpoint 1400). Fault subsidence on the northwest side of the collapse takes the form of a folded, downdropped block between shotpoints 1350 and 1150. The deepest part of the depression displays a flat seafloor from shotpoint 1150 to

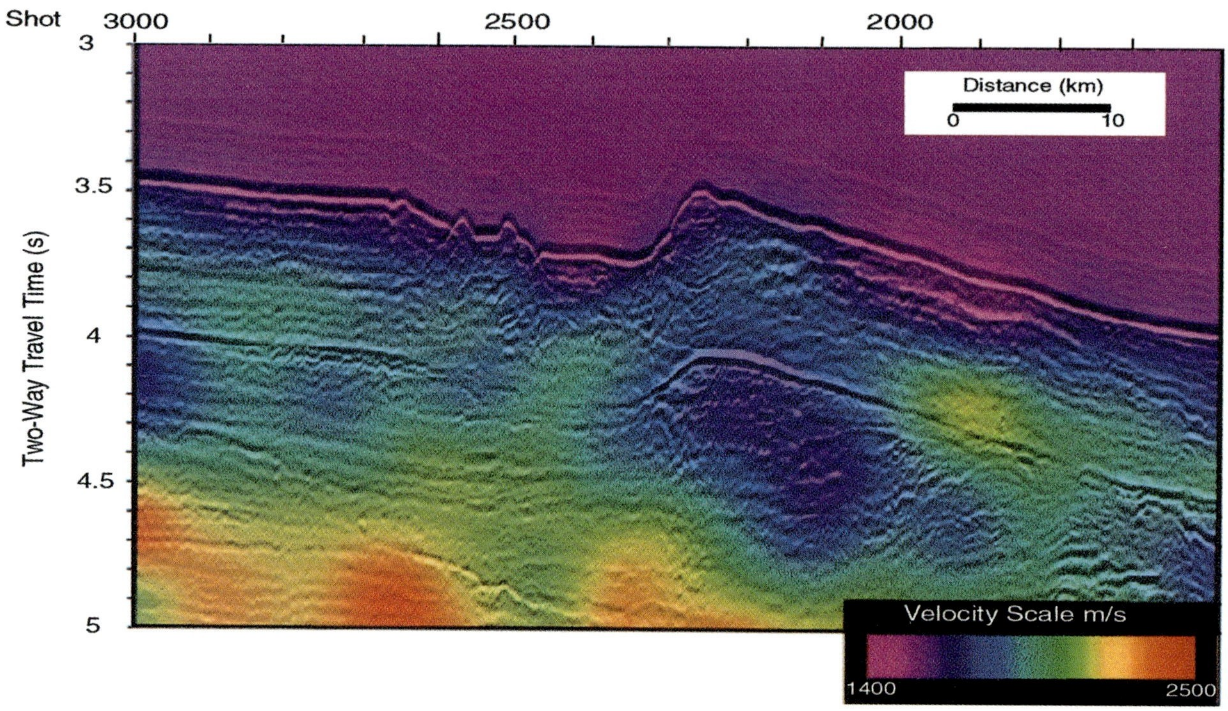

Plate 2. Multichannel seismic reflection profile BT-1 with velocity overlay in color. Location shown by solid trackline plus dashed extensions on Figure 1.

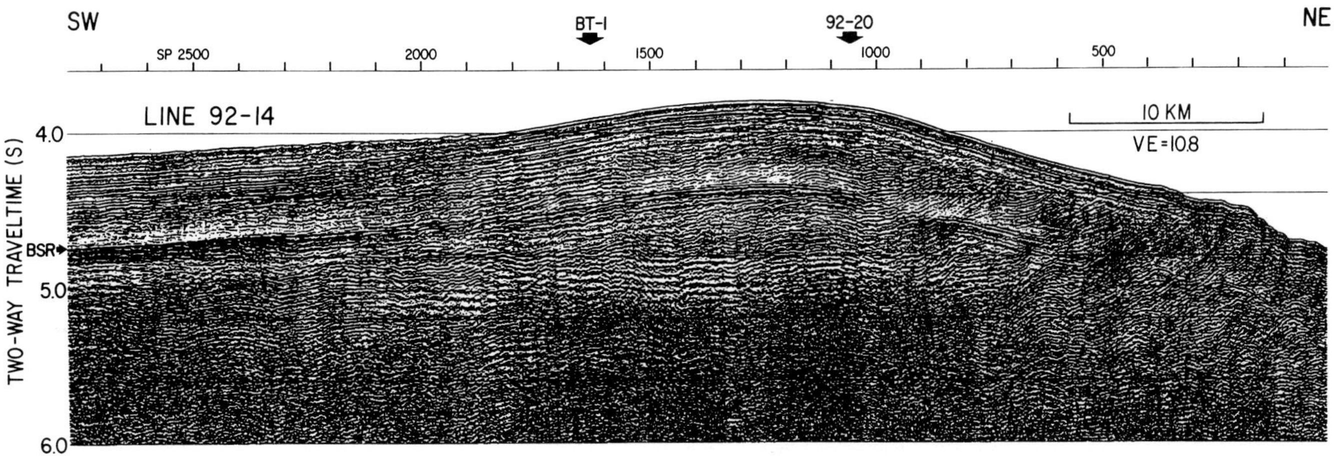

Figure 9. Seismic reflection profile 92-14. Location shown on Figure 1.

900, underlain by chaotic structure, representing a longitudinal crossing of the inferred vent structures noted in Figures 3, 5, and 6. To the right, the profile diagonally crosses the region that has been characterized as a nappe. The décollement may be represented by a set of strong reflections that occur between 3.9 and 4.1 s (marked by medium weight lines in the interpretation).

Profile BT-1 is a deep-penetration multichannel profile that shows the entire sedimentary sequence down to basement (Figure 8). Oceanic basement is a rough surface that intersects the left side of the figure at about 8.4 s. An unconformity at about 6.8 s at the left side of the figure is horizon A^u, initiated by early Oligocene erosion that planed the seafloor [*Mountain and Tucholke,* 1985]. Strata above the A^u unconformity are the hemipelagic clays and silts that formed the Blake Ridge. Disruption of strata in the primary collapse, from about shotpoint 2630 to 2270, extends down near the present BSR (to a total depth of 4.2 s at the central collapse, e.g. shotpoint 2400), but does not extend much deeper. Clearly, reflectors from 4.5 s to basement were not involved in the collapse.

The velocity structure along a somewhat longer section of multichannel seismic line BT-1 is shown in Plate 2. The overall pattern is typical of gas hydrate areas [e.g. *Wood and Ruppel,* 2000]; away from the crater, the profile shows low velocities near the seafloor generally increasing down to the BSR at about 0.55 s subbottom, then a general decrease in velocity, before increasing downward again. This velocity structure would be interpreted as a downward increase in gas hydrate to the base of the GHSZ, and an accumulation of gas, with associated low velocities, below the base of the GHSZ. A culmination of the ridge crest occurs just down-ridge (right) of the collapse. A zone of extremely low velocities under the BSR beneath this bathymetric culmination probably signifies a greater concentration of gas there. Conversely, the relatively high velocities directly below the deepest part of the collapse depression suggests a reduced concentration of gas there.

If we extrapolate the trends of seafloor across the collapse depression on line BT-1 (Plate 2 and Figure 8), we see a discrepancy between the bathymetric trend downridge (right) of the depression and upridge. If the downridge seafloor has been undisturbed since the time of collapse, this suggests that the collapse area actually constituted part of a small bulge on the ridge crest prior to collapse. Profile 92-14 (Figure 9) crosses the Blake Ridge downridge of the collapse (at shotpoint 1770 on line BT-1, Plate 2) and intersects the bulge. Small increases in seafloor slope that form inflections at shotpoints 1800 and 850 on the upper flanks of the ridge (Figure 9) may suggest the limits of the feature.

A sidescan-sonar image of the seafloor in the collapse area (Figure 10) shows the distribution of fault scarps. Almost all of the scarps face northeast to east, as would be expected from the seismic profiles. The striped appearance of the scarps results from backscatter from the truncated, exposed strata on the fault scarp faces. Absence of features like pockmarks and mud volcanoes suggests lack of ongoing gas seepage. This conclusion is reinforced by recent measurements of sulfate gradients in cores. The gradients are steepest (sulfate depletion occurs shallowest) below the crests of structural hills compared to valleys, suggesting that methane flux is greatest at hills *[Paull et al.,* 2000]. This may be consistent with the presence of the strongest BSRs beneath hills, indicating presence of more trapped gas there (e.g. Figure 5). Faults characteristically appear in valleys (e.g. Figures 3, 4, 5). The implication is that gas is primarily migrating laterally to culminations below the base of the GHSZ and then diffusing upward, rather than migrating up faults and venting at the seafloor. No measurements of methane content in near-bottom waters have been made, however.

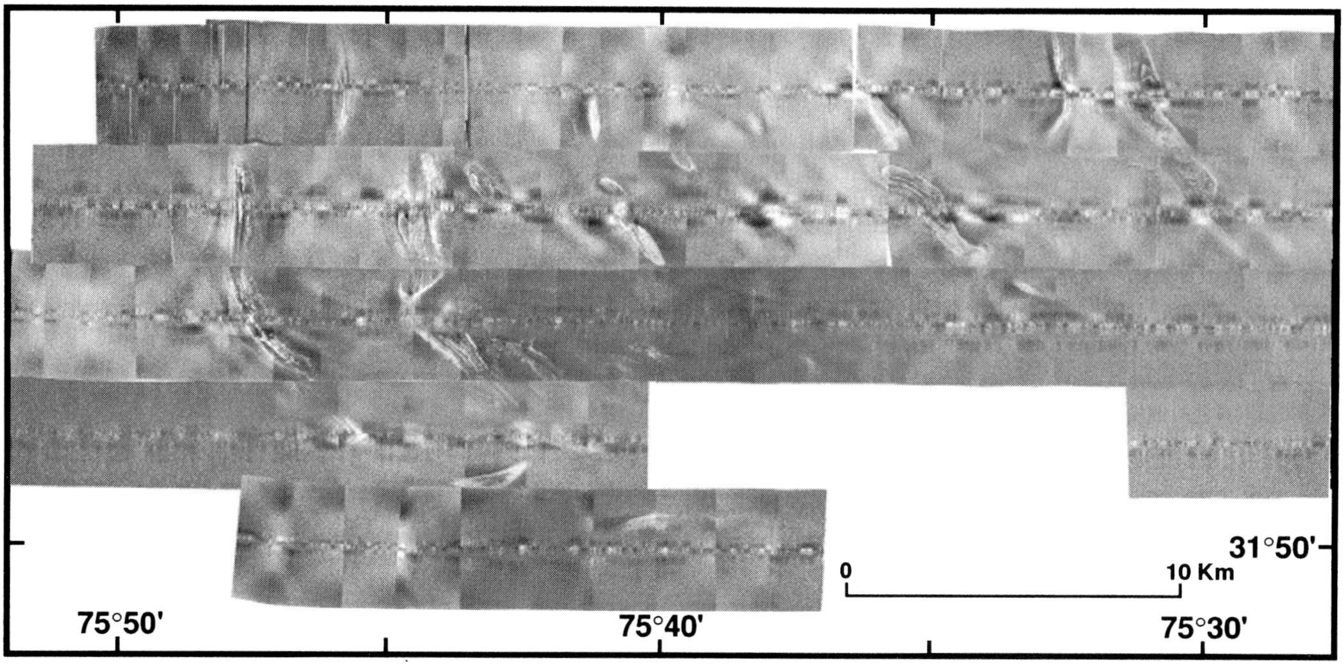

Figure 10. Mosaic of sidescan-sonar data over collapse structure. Location of mosaic shown on Figure 1. A map of fault scarps is presented in Figure 11, which indicates the sense of throw on the faults. The striped appearance of the fault scarps is caused by backscatter from truncated strata on the scarp face.

DISCUSSION

Structure of the Collapse and Adjacent Areas

A restricted area of depression is observed at the crest of the Blake Ridge. Seismic profiles across the depression show that it is formed by two modes of subsidence, block faulting and gravity gliding. Normal faulting of blocks (Figures 3, 4, 5), generally occurs on the southwest side of the collapse area with faults mostly downthrown to the northeast (Figure 11). Faults dominantly trend parallel to the ridge crest (northwest), but swing around to a northerly trend on the western side of the collapse. Subsidence by gravity gliding of a mass of sediment above a décollement is observed over part of the northeast side of the collapse area, but in the upridge (northwest) and downridge (southeast) directions from the central collapse, the décollement disappears and strata simply sag toward the depression without detachment. Where the décollement is present, internal shortening of the gliding mass by complex folding and minor faulting appears We suggest that the difference in structure between the northeast and southwest sides of the collapse may result from the response to the universal southwest dip of strata — where strata crosscut the direction of extension due to subsidence, normal faults were generated, but where the motion approximately paralleled the strata, a décollement and nappe were formed.

In several profiles (e.g., Profile 7, Figure 3), a topographically depressed, relatively flat area exists between the block faulted and gravity gliding terrains. The low area, mapped by a darker pattern in Figure 11, is underlain by very chaotic structure or a seismically transparent zone.

Although the dominant motion clearly is characterized as subsidence, we also see evidence for initial uplift. The structural relationships of the fault blocks to each other and the internal drag folds indicate that the blocks are downdropped progressively northeastward into the depression, but many fault blocks project above the general trend of the Blake Ridge seafloor. This fact and the presence of upturned strata in unfaulted deposits just south and west of the collapse (especially line 92-21, Figure 6, but also lines 92-6, Figure 5, 92-22, Figure 6, and 95-4, Figure 7) seem to require that the region of subsidence was structurally uplifted before it collapsed. The remains of the uplift can be observed as an anomalous shallow band of seafloor around the southern and southwestern limit of the collapse depression (Plate 1). Seafloor depths along profile BT-1 that runs axially down the crest of the ridge (Figure 8; Plate 2) also show that part of the initial uplift still exists and that the extent of the ridge crest involved in the collapse may represent only about half of the extent of the initial uplift. A plot of depth versus shotpoint along a section of BT-1, longer than shown in Figure 8 or Plate 2, (Figure 12)

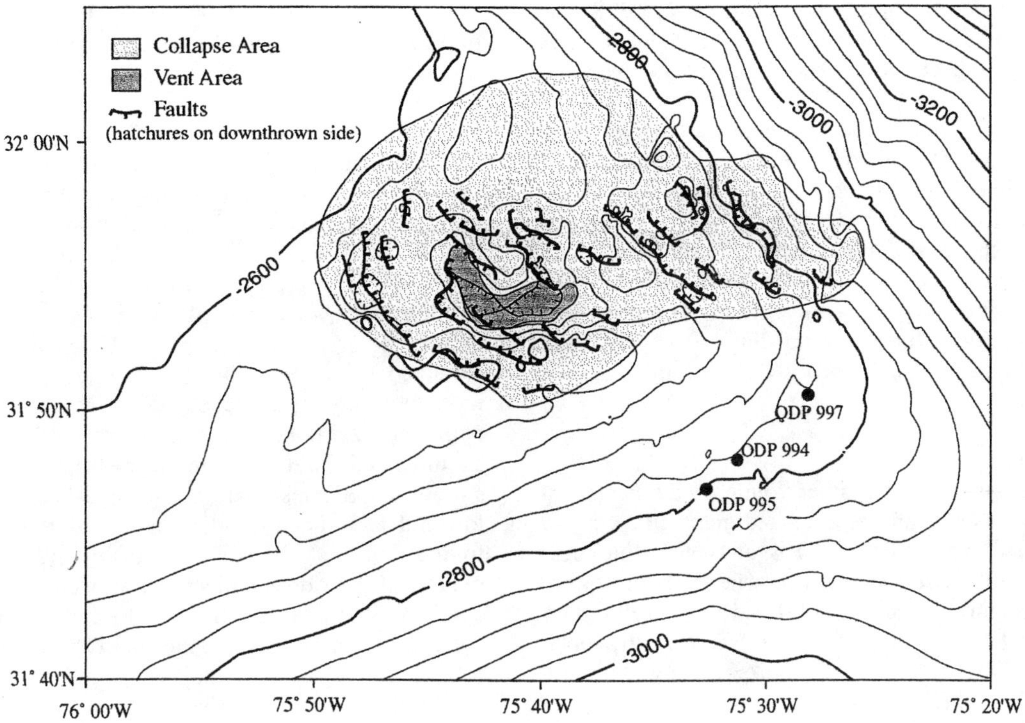

Figure 11. Fault map of collapse structure. Hachures on downthrown side of faults. Light shaded area indicates interpreted area of collapse. Dark shaded area indicates inferred vent area.

indicates that the crestal depth shows its most abrupt discrepancy from a linear trend at the collapse, where we see a well-defined hill downridge from the collapse depression. (It is likely that other, more moderate variations in the plot represent a wandering of the ridge crest relative to the trackline or perhaps variations in the sedimentary construction of the ridge [*Markl and Bryan,* 1983].) Line 92-14 (Figure 9) crosses the uplift at shotpoint 1770 on line BT-1 and may show limits to the inflation of the ridge crest by slight inflections of the seafloor at shotpoints 1800 and 850 on 92-14. The downridge limit of the uplift appears to be the contact of undulating seafloor to smooth seafloor on line 92-20 at shotpoint 300 (Figure 7). In any case, not only is there evidence for an initial uplift prior to collapse, but part of the uplift apparently still exists. The velocity structure (Plate 2) shows that the remnant of the uplift is underlain, beneath the BSR, by a pronounced low-velocity zone that must be interpreted as a region of trapped gas. Prior to the collapse, the entire zone of inflation probably acted as a large gas trap; presumably much gas was vented when the collapse occurred.

In summary, presence of the depression clearly requires that material was removed. Furthermore, profiles show that surface strata several hundred meters thick underwent structural disruption that resulted from subsidence into the depression. The evidence includes listric faults on the southwest side of the structure with backtilted blocks that form fault scarps on which exposed strata crop out. This configuration is observed in seismic profiles (e.g. Figure 3), sidescan images (Figure 10) and high-resolution acoustic

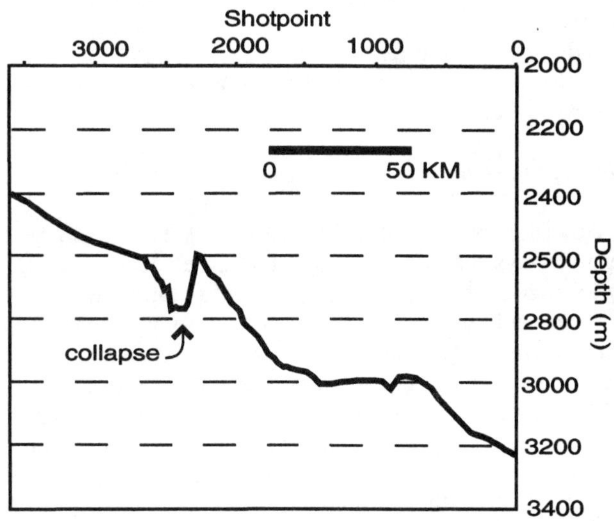

Figure 12. Plot of water depth along line BT-1 which parallels the crest of the Blake Ridge. The part of BT-1 that is plotted here extends beyond the boundaries of Figure 1.

profiles (deep-towed chirp profiles, *Dillon et al.*, 1998, and surface-transducer 3.5 kHz profiles). The nappe on the northeast side of the collapse structure displays structural shortening, with structures that sometimes become more complex and show greater shortening upward (Figure 3). The nappe appears to be continuous along strike to a structure in which the subsidence is a simple sag of the strata, with continuity of beds and no décollement or unconformity at the base. Both patterns, showing block faulting or gravity gliding, are consistent with tectonic subsidence and inconsistent with accumulation of the contorted blocks by sedimentary infilling of a preexistent depression, as suggested by *Holbrook* (this volume).

Inferred Age

The collapse appears to be rather young on the basis that the depression has no appreciable sediment fill, only a sediment drape of zero to a few 10s of meters thickness observable in high resolution profiles. The first cores were taken in the structure in July of 2000 and only preliminary results are available at this time (*D. Clay Kelly, pers. comm.*, 2000). One core taken at approximately shotpoint 570 on line 92-21 (Figure 6) recovered 11 meters of foram/pteropod-bearing mud and apparently nearly penetrated the entire draped section on the basis of comparison to high resolution profiles. This core (PC-9) yielded a magnetic susceptibility record comparable to that of nearby ODP Site 1059 *[Keigwin, Rio, Acton, et al.*, 1998]. Comparison of these records indicates that the distinctive, magnetic susceptibility signature that demarcates the boundary between marine isotope stages 4 and 5 (~75 ka) is preserved at 8 m in this core. Furthermore, micropaleontological investigation of the lowermost sediments in the core reveals the presence of planktonic foraminiferal faunas that typify interglacial conditions, an observation consistent with assigning the lowermost sediments to interglacial stage 5.

Cause of the Collapse

How did the loss of volume occur? There seem to be several ways that a volume loss could be accomplished: 1. by localized erosion; 2. by surface sliding of material away from the collapse region; 3. by outward movement of the flanks of the ridge generating an extensional collapse; 4. by localized basement subsidence at depth; 5. by intermediate depth, slumplike movement that slid a sedimentary mass to the northeast so that the southern and southwestern faults actually represent headwall subsidence and a bulge would have formed on the northeastern side of the ridge; 6. by flow and expulsion of fluidized sediments from some intermediate depth zone below the sediment surface.

In considering each possibility, the last seems to fit the structural evidence best. 1. Surface erosion seems an impossible mechanism, as the surface section (upper ~400-500 m) appears to have been preserved and to have subsided into the depression. 2. Examination of all bathymetric and seismic profiles and the sidescan sonar mosaics indicate that there appear to be no slide scars on the seafloor that would account for the depression by the transport of a volume of material away at the seafloor. The fact that the center of the depression is its deepest part (Plate 1) also argues against a landslide model, which would require some sort of deepening trough extending out of the center of the collapse. A landslide would not be expected to cause faults that extend down to 400 m and more beneath the surface. Furthermore, the volume loss apparently did not result from removal of surface sediments, as indicated in discussion of alternative 1., because the structure indicates that the surface sediments have been lowered tectonically and remain, observed as faulted and folded deposits. 3. All the profiles indicate that outward motion of the ridge flanks has not occurred. The strata are continuous and essentially undisturbed at depth beneath the collapse structure (e.g. Figure 6). The deep structure of the ridge where it has not collapsed (e. g. line 92-14, Figure 9) is no different from the deep structure where it has collapsed (e. g. line 95-18, Figure 4). 4. The continuity of undisturbed reflectors beneath the collapse also eliminates the possibility of an association with basement subsidence, as further demonstrated by the MCS profile (Figure 8) that images basement at about 8 s and shows no faulting between basement and the collapse structure except within about 0.8 s below the seafloor. 5. If a northeastward slump-like movement occurred that displaced volume outward at shallow subbottom depth and allowed crestal collapse, it should have created a bulge on the northeastern flank of the Blake Ridge. It has been suggested that such a bulge could have existed on the northeastern flank and been eroded away. There probably is not time for that to have happened in the last ~100 ky and no bulge is visible (Plate 1, Figure 1). Although the northeast flank is characterized by erosion, which results in the truncated strata at the right side of Figure 9 (shotpoint 100-400), this is a regional pattern as old as the ridge development (order of 30 Ma), and it extends more than 130 km along the ridge [*EEZ-Scan 87 Scientific Staff*, 1991]. The crestal collapse occurs only near the northern 25 km of the erosional patch, so they are not regionally correlated features. In summary, we interpret that neither the surface deposits were removed, nor deep deposits disturbed, and an intermediate-depth slump leaves the problem of where the material went, and therefore we conclude that some intermediate zone was removed by expulsion, which seems to fit with the structures observed (alternative 6.).

Evidence regarding the process of intermediate-depth volume removal may be provided by the two regions of chaotic reflections or acoustic transparency noted previously — one zone forming a layer about 0.3 s thick beneath the faulted and folded strata and the other extending up to the seafloor beneath the bottom of the deepest part of the bathymetric depression. We suggest that the first zone of chaotic reflections marks the disturbed remnants of a zone of sediments that existed just beneath the gas seal in a trap that was formed on the ridge crest. The seal was provided by the base of the GHSZ (and therefore could be unaffected by the ubiquitous small faults) and the structure needed to form the trap was provided by the Blake Ridge on a regional scale and the local bulge on a smaller scale (10's of km's). Such a gas trap would have allowed development of overpressures and we infer that water contents would probably have been high, on the basis of the high porosities (commonly more than 50%) associated with low compaction rates determined by drilling on the Blake Ridge [*Paull Matsumoto and Wallace,* 1996]. Such a region of overpressures and high fluid contents would have had low shear strength and would have flowed easily [*Hovland and Judd,* 1988]. We suggest that the second region of chaotic reflections that extends to the seafloor beneath the deepest part of the collapse, observed in many profiles, probably represents the site of a vent through which mobilized sediments rose and escaped. The general extent of this inferred vent is mapped in Figure 11.

Hypothesized Developmental Stages

We envision the simplified development if the Blake Ridge collapse structure in terms of four steps that are shown as idealized ridge cross-sections in Figure 13. We assume an initial stage as an undisturbed accreted sedimentary drift deposit like much of the rest of the Blake Ridge (like Figure 9) [*Markl, et al.,* 1970; *Markl and Bryan,* 1983]. The base of the GHSZ would have paralleled the seafloor, creating a gas trap at the crest of the ridge. As the ridge accreted on its southwest side, the base of the GHSZ, maintaining its subbottom depth, would have moved upward relative to the sediments, causing dissociation of gas hydrate and releasing gas that would then have migrated upward into the trap. In addition, a drop of sea level would have reduced pressure and caused hydrate to dissociate, contributing still more gas. The Blake Ridge hemipelagic muds have significant clay content, so they would be likely to have low hydraulic conductivity; thus the potential for generating overpressures is high when gas is added to such a system [*Vasseur et al.,* 1996]. Overpressure that developed in such a zone of abundant gas and water can easily have led to mass sediment flow.

The initial effect of the sediment mass flow probably was inflation of the crest of the ridge to form a local bulge. Some subsidence of one or both ridge flanks would be expected ("peripheral subsidence" Figure 13 second panel) as material flowed into the bulge. Thus the beginning of subsidence of the gravity-gliding nappe may have taken place during the inflation stage, prior to fault activity.

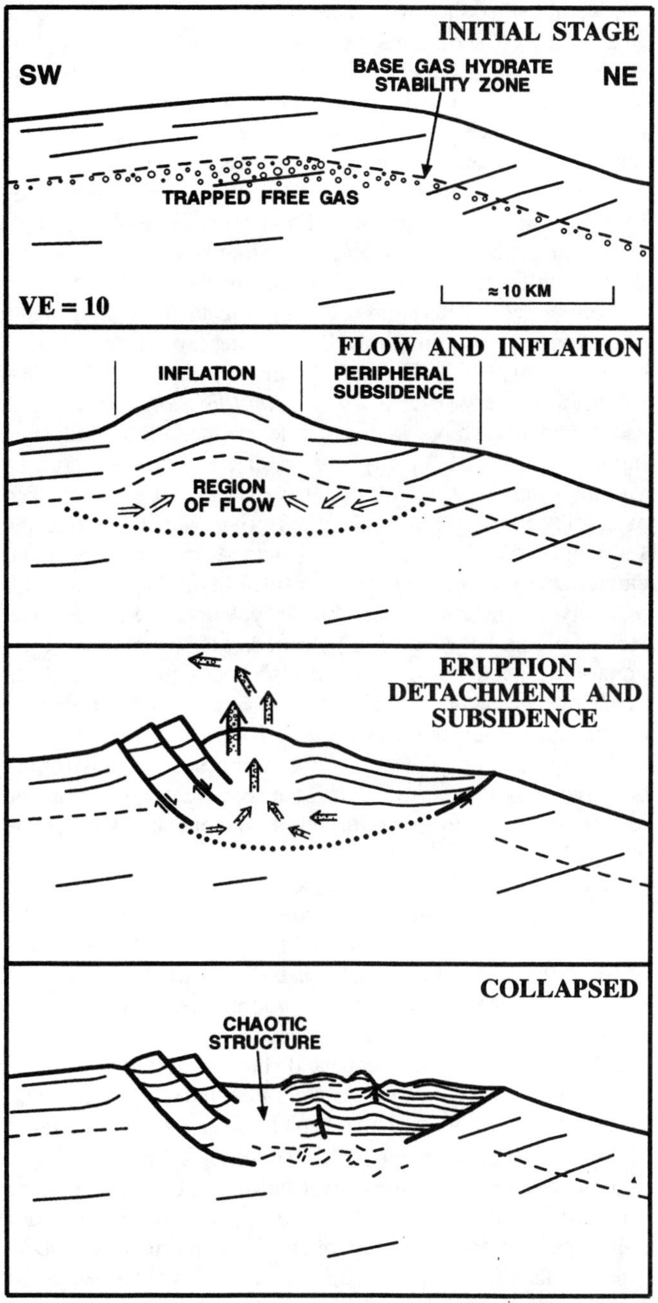

Figure 13. Conceptual model of the collapse process at the crestal region of the Blake Ridge. The region of flow in the second panel is inferred to be a region in which fluid -rich sediments were mobilized. The zone of trapped free gas indicated in the first panel may have included this entire region of flow.

Eventually overpressure built up sufficiently to exceed

the weight of sediment, causing the water- and gas-charged, overpressured sediments beneath the GHSZ to be erupted (Figure 13 third panel). This volume loss accelerated the subsidence of the sediments on the northeast flank of the ridge; the detachment of this mass, forming a décollement, may have taken place at this time in some places. Simultaneously, the southwest flank of the uplift collapsed by block faulting. Expulsion of a large volume of fluid plus sediment probably took place. Unlike a very similar, but much smaller (280 x 400 m) feature in the Gulf of Mexico [*Prior et al.,* 1989] no obvious ejecta field has been observed as yet in the USGS sidescan sonar images collected in 1995 (Figure 10) or 1992 or in GLORIA sidescan imagery [*EEZ Scan 87 Scientific Staff,* 1991]. The sediments may have been so fine grained, included such large quantities of fluid (gas + water), and been ejected so violently that the system might have been best characterized as a diatreme [*Brown,* 1990]. If that was the case, the sediments could have been dispersed in the water and carried away or flowed away as turbidity/debris flows. Or possibly ejecta are present but as yet unrecognized. The lack of ejecta cones at mud volcanoes off Nigeria, and their general absence worldwide, has also been attributed to both injection into the water column and mass transport downslope as debris flows [*Graue,* 2000].

The question of the rate of the eruption — catastrophic or slow — is a most difficult issue, but we suggest that the escape of gas from an overpressured zone is likely to be abrupt. As noted for the development of mud volcanoes [*Graue,* 2000, p.967]: "Because of the compressibility of gas (particularly methane) the seal failure will be accompanied by a gas expansion that provides the explosive power of the system and enables vertical penetration through overburden... The sudden pressure release associated with eruptions will cause an implosion at the root, shaking the undercompacted shale such that the poorly consolidated clay fabric disintegrates into liquefied mud with a complete loss of mechanical strength.... The rapid kinetics of such eruptions are also likely to cause a rapid depletion of pressure to a level below that of the fracture pressure of the seal, which is the point when the roof collapses into the vent and plugs the conduit to surface." The events that lead to eruption of mud volcanoes seem comparable to the Blake Ridge event, except that the seal at the Blake Ridge was the base of the GHSZ, which was a broadly regional phenomenon. A further argument that the release event that formed the Blake Ridge structure occurred over a restricted time period, or at least is not active at present, involves the data of sulfate gradients in cores [*Paull et al.,* 2000], which indicates that degassing through the faults is not occurring presently.

The strata within the GHSZ that collapsed into the volume evacuated by erupted sediments may have been partially cemented by gas hydrate, and, of course, little or no free gas would exist in this zone (it would be in the gas hydrate state), so these strata acted much more cohesively than the mobilized sediments beneath the GHSZ. The system may be conceived as a relatively brittle layer above a fluid one, a model similar to more recent thought regarding some processes of salt diapirism [Jackson 1995]. The Blake Ridge collapse may be thought of as comparable to collapses that occur in regions of salt flow where relatively slow loss of volume at intermediate depth creates thin-skinned extension [*Vendeville and Jackson,* 1992, a and b; *Demercian et al.,* 1993] or the early stages of "raft tectonics" [*Duval et al.,* 1992]. Of course, we infer that the expulsion of gassy sediments would be much more rapid than salt flow.

We note that the reflection amplitudes from stratal reflectors are commonly greater in collapsed sedimentary strata than in undisturbed strata of the GHSZ outside the collapse on the flanks of the ridge. This is clearly apparent in the true-amplitude display of line 95-18 shown in Figure 4. It has been asserted on the basis of many observations in nature that reduction in reflection amplitude, known as "blanking", is associated with gas hydrate accumulations [*Dillon et al.,* 1993,1994; *Lee et al.,* 1993, 1994, 1996; *Lee and Dillon,* in press; but denied by *Holbrook et al.,* 1996]. Blanking is thought to result from preferential accumulation of gas hydrate in the more porous sedimentary strata, where it would increase the velocity of the more porous, initially lower velocity layers by the introduction of high-velocity, gas-hydrate cement. This may create a more uniform, higher velocity, resulting in reduced acoustic impedance contrasts and thus reduced reflection strengths. Why structural disruption should reduce blanking is unclear, unless mechanical stresses at grain level broke down gas hydrate cements, causing the more porous layers to revert to their lower velocity — all unsatisfactorily speculative at this time.

Relation to Sealevel Lowstand

Many authors have pointed out that a sea level drop decreases pressure at the base of the GHSZ, causing gas hydrate dissociation, generating gas overpressures, and weakening sediments, and thus could trigger mass movements [e.g. *Carpenter,* 1981; *Kayen and Lee,* 1991; *Paull et al.,* 1991; *Dillon et al.,* 1993; *Popenoe et al.,* 1993, *Paull et al.,* 1996]. As noted above, there is preliminary evidence that the collapse occurred a relatively short time before the marine isotope stage 4-5 boundary at ~75 ka. Abrupt sealevel lowerings were reported by *Shackleton* [2000] for the periods 98-95 ka (~40 m) and 117-112 ka (~70 m) and similar events were indicated by *Chappell et al.* [1996]. We suggest that the sealevel-change trigger may have weakened sediments on the Blake Ridge and caused collapse of the ridge crest. The major difference between

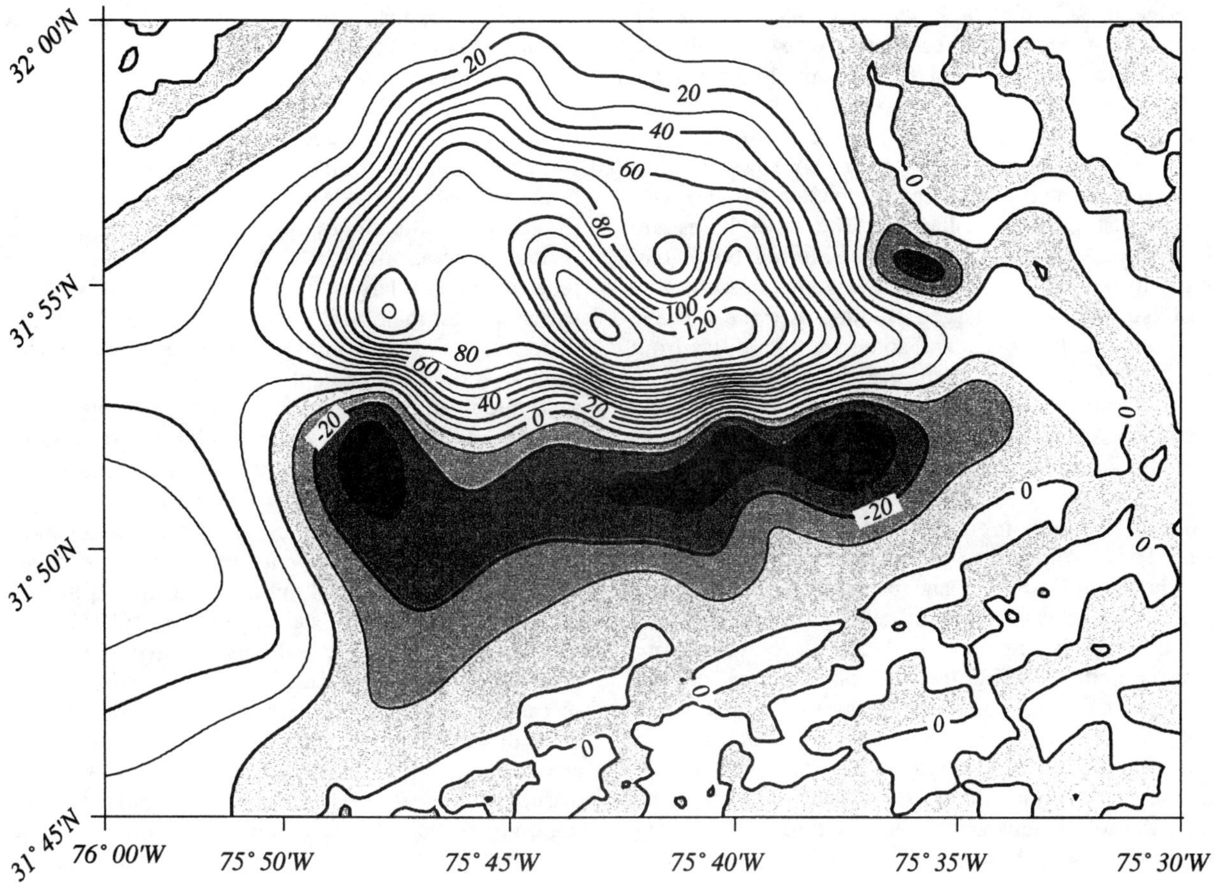

Figure 14. Subsidence pattern of the collapse area. This was calculated by fitting smooth splines along ridge-crossing profiles, measuring the difference between the spline and the seafloor, and extrapolating between these profiles of area-loss to determine volume-loss. Crossings of apparently undisturbed parts of the ridge were used as a template/guide. The isopach contours of volume-loss are shown as positive where the surface has apparently subsided. Shaded areas, shown as negatives, are regions of apparent relative uplift on the basis of the calculation. Actually, all areas that were mobilized during the collapse underwent subsidence, but the presence of a pre-collapse bulge means that values of subsidence estimated by using the splines and represented by contours are smaller than the actual downward movement.

the events on continental slopes and the Blake Ridge may be that seafloor inclinations were lower on the latter so the system erupted in place rather than generating a slide.

Volume Loss and Methane Escape

A large loss of volume was required to produce the collapse depression. We used cross-sections of undisturbed parts of the ridge well away from the collapse as guides to the pre-collapse shape. The volume loss was estimated by placing splines having this inferred pre-collapse shape across the collapse region on all of the crossing profiles. By calculating the area difference at each seismic profile between the spline and the seafloor, and integrating between profiles, we estimated a total volume loss of 13.4 km^3. The isopach map of inferred volume loss (Figure 14) shows a maximum depression of the seafloor by about 130 m, but also indicates a large area of negative subsidence (uplift) on the south side of the collapse, where much of the area exceeds negative values of 20 m. Such negative values were subtracted from the calculated volume-loss. These areas really did subside, but they had been elevated prior to the collapse event, presumably by lateral flow of mobilized sediment below the base of the GHSZ, which formed the elevated bulge. Upbent strata that give evidence for such a structure were noted above.

Using the measured volume loss, we calculated a volume of gas that might have escaped. During ODP Leg 164, determinations of the gas content of sediments beneath the GHSZ were made by several techniques at the ODP holes on the ridge crest (ODP 994, 995, and 997, Figure 1). Perhaps the most dependable estimates were from electric logs [*Collett and Ladd,* 2000] and direct sampling using the Pressure Core Sampler [*Dickens et al.,* 1997a]. The values determined were highest at site 997 at the crest of the ridge, but we suggest that these probably represent low estimates of the gas content in sediments under the GHSZ in the zone of inflation just prior to collapse. The three holes are actually located within the part of the inferred bulge that still exists, in our view. However, we would suppose that the most strongly gas-charged sediments would have escaped in the eruption event and some venting of gas from the area of the ODP holes might also have been possible. Electric log values estimate 4.8% gas in situ directly beneath the base of GHSZ at 997, whereas the Pressure Core Sampler estimated 12%. Multiplying these values by the minimum volume of escaped sediment (13.4 km^3) and converting to STP, this represents $8 \cdot 10^{10}$ or $26 \cdot 10^{10}$ m^3 of methane. These volumes would represent either 1.2% or 3.9% of the amount of methane presently in the Earth's atmosphere — clearly an influence on the global greenhouse from this single event. However, perhaps more important is understanding the process that caused the Blake Ridge collapse. If this process took place on a continental slope, it would probably have generated a slide. If many such collapse/landslide events occurred nearly simultaneously, a major climate shift could have been be triggered A more extensive release of methane from a set of similar events may have triggered the latest Paleocene thermal maximum [*Dickens et al.,* 1997b; *Katz et al.,* 1999]

SUMMARY AND CONCLUSIONS

A large irregular topographic depression (39x23 km) is situated on the crest of the Blake Ridge in the continental rise off South Carolina, USA. The feature is interpreted as a collapse structure caused by processes related to gas hydrate.

The Blake Ridge is a sediment drift deposit, a broad accretionary pile of hemipelagic mud formed by a confluence of bottom currents. The axis of the ridge trends northwest and strata within the ridge dip southwest. The ridge is evolving by accretion on its southwest flank and erosion on its northeast flank.

The collapse is characterized by contrasting structure on its northeast and southwest sides. On the northeast, sediments have slid down into the depression in the form of a broad folded and faulted nappe, which is in part separated from deeper deposits by a décollement at its base. On the southwest side a set of normal faults allowed blocks to sink into the depression, with drag folds within the blocks also indicating subsidence. The presence of upfolded strata outside the collapse to its southwest and the observation that, although the fault blocks are clearly seen to be downdropped, many project above the undisturbed ridge surface, has led to the conclusion that a preexistent bulge was present before subsidence of the strata into the collapse. About half of the bulge appears to remain as a culmination on the ridge crest just southeast (downridge) from the collapse structure. Some of the lowest velocities in the immediate area exist beneath this culmination, which suggests considerable trapping of gas there at present.

The generation of a collapse depression of this size requires that a large volume of sediment was removed, which we estimate at 13.4 km^3 (the size of the remaining depression). This volume loss did not result from either deep structural adjustment or surface sliding, so it must be accounted for by removal of sediments at shallow subbottom depth. Two zones of highly disturbed reflectors or acoustic transparency are inferred to represent regions of former active mobilization of sediments. One zone of apparent disruption extends horizontally under most of the collapse structure area in a zone a few hundred meters thick, and has its upper limit at about the depth of the base of the gas hydrate stability zone (GHSZ) prior to subsidence. This zone is likely to have been a region of horizontal flow of mobilized sediment during collapse. The second zone of disturbed/transparent reflections occurs beneath the deepest part of the collapse depression, extending from the seafloor down to the other zone of disrupted deposits. This zone is suggested to represent a region of vertical flow and venting of mobilized sediment to the ocean.

We suggest a four-stage model (Figure 13) of the evolution of the collapse structure (although it probably developed in a continuous manner). 1. The initial conditions are assumed to be a simple accreted sediment pile like the rest of the Blake Ridge containing gas hydrate with free gas beneath. 2. The second stage would entail flow of overpressured, high fluid-content sediment beneath the GHSZ. This caused inflation of a crestal bulge, where gas became concentrated, and subsidence of a peripheral area, which perhaps represented the incipient subsidence of the northeastern nappe. 3. Overpressures developed to an extent that required release of material from beneath the seal provided by the GHSZ and an eruption caused venting of large amounts of sediments and fluids to the ocean. 4. Finally, in the present collapsed stage, the downdropped fault blocks and nappe occupy most of the collapse structure, with the remnant of a vent structure near the center.

The collapse event is inferred to be rather young, as only a thin (generally 10-20 m) drape of sediment covers the fault blocks. Preliminary results from cores in this drape suggest a faulting event somewhat older than 75 ka. Sealevel lowerings that occurred at ~98-95 ka or ~117-112 ka, by reducing pressure at the seafloor, might have caused increased dissociation of gas hydrate, triggering the event.

At least two other similar structures are interpreted in the deeper strata of the Blake Ridge, so we suggest that this probably is a process that has occurred with some regularity in the history of this continental margin sediment drift. Preservation of the structure may have been aided by its location at the crest of a very broad, flat, sedimentary pile. If similar processes had taken place on steeper parts of the continental slope/rise they would probably have triggered slides,which would remove much of the evidence for the process. Thus the Blake Ridge collapse structure may provide an uncommon opportunity to study a very common process that results in seafloor instability and the generation of slides on continental margins [e.g. *Booth,* 1994].

The model suggests that a large volume of methane probably was concentrated beneath the pre-collapse bulge and released into the ocean/atmosphere system during the collapse event, perhaps in a catastrophic manner. Calculations on the basis of the estimated volume of released material and present concentration beneath the GHSZ (probably representing a minimum value compared to concentration at the time of release) suggest that perhaps 1-4% of the present global atmospheric content of methane was released by this single event. As the processes that triggered the collapse are perhaps world-wide and might have triggered many near-simultaneous events, the global effect may well have been significant.

Acknowledgements. We wish to acknowledge the U.S. Department of Energy, Federal Energy Technology Center, Morgantown, WV for support of 1991, 1992, and 2000 cruises and U.S. Office of Naval Research and Naval Research Laboratory for partial support of our 1995 cruises. We thank the scientists, technicians, and ships officers and crew of the R/V Cape Hatteras during cruises CH 15-91 (1991), CH 12-92 (1992), CH 17-95, CH 18-95 (1995), AND CH-15-00 (2000). William Danforth, Thomas O'Brien, Dwight Coleman, and James Booth of the U.S. Geological Survey and Ingo Pecher of the University of Texas at Austin made valuable contributions to data gathering and scientific interpretation. We thank the reviewers, David Twichell Peter Vogt, Chris Goldfinger and Erich Roth.

REFERENCES

Booth, J.S., W.J. Winters, and W.P. Dillon, Circumstantial evidence of gas hydrate and slope failure associations on the United States Atlantic continental margin, in *International Conference on Natural Gas Hydrates,*, edited by E.D. Sloan, Jr., J. Happel, and M.A. Hnatow, pp. 487-489, *Annals of the New York Academy of Sciences, vol. 715,* 1994.

Brown, K.M., The nature and hydrogeologic significance of mud diapirs and diatremes for accretionary systems, *J. Geophysical Research, 95,* 8969-8982, 1990.

Bryan, G.M, Hydrodynamic model of the Blake Outer Ridge, *J Geophys. Res. 75,* 4530-4537, 1970.

Carpenter, G., Coincident slump/clathrate complexes on the U.S. Atlantic continental slope, *Geo-Marine Letters, 1,* 29-32, 1981.

Chappell, J., A. Omura, T. Esat, M. McCulloch, J. Pandolfi, Y. Ota and B Pillans, Reconciliation of late Quaternary sea levels derived from coral terraces at Huon Peninsula with deep sea oxygen isotope records, *Earth and Planetary Science Letters, 141,* 227-236, 1996.

Collett, T.S., and J. Ladd, Detection of gas hydrate with downhole logs and assessment of gas hydrate concentrations (saturations) and gas volumes on the Blake Ridge with electrical resistivity log data, in *Proc. ODP, Sci. Results 164,* edited by C.K. Paull, R. Matsumoto, P.J. Wallace and W.P. Dillon, College Station TX (Ocean Drilling Program), 179-191, 2000.

Demercian, S., P. Szatmari, and P.R. Cobold, Style and pattern of salt diapirs due to thin-skinned gravitational gliding, Campos and Santos basins, offshore Brazil, *Tectonophysics, 228,* 393-433, 1993

Dickens, G.R., M.M. Castillo, and J.C.G. Walker, A blast of gas in the late Paleocene: Simulating first-order effects of massive dissociation of oceanic methane hydrate, *Geology, 25,* 259-262, 1997b.

Dickens, G.R., C.K. Paull, P. Wallace, & the ODP Leg 164 Scientific Party, Direct measurement of *in situ* methane quantities in a large gas-hydrate reservoir, *Nature, 385,* 426-428, 1997a.

Dillon, W.P., W.W. Danforth, D.R. Hutchinson, R.M. Drury, M.H. Taylor, J.S. Booth, Evidence for faulting related to dissociation of gas hydrate and release of methane off the southeastern United States, *in* Gas Hydrates: Relevance to World Margin Stability and Climate Change, edited by J.P. Henriet, and J. Mienert, *Geological Society, London, Spec. Publication 137,* p.293-30, 1998.

Dillon, W.P., K. Fehlhaber, D.F. Coleman, M.W. Lee, and D.R. Hutchinson, Maps showing gas hydrate distribution off the east coast of the United States, *U.S. Geological Survey, Misc. Field Studies Investigations Map MF2268,* 2 sheets, 1995.

Dillon, W.P., D.R. Hutchinson, and R.M. Drury, Seismic reflection profiles on the Blake Ridge near sites 994,995, and 997, in Paull, C. K., Matsumoto, R., Wallace, P. J., et al., *Proceedings ODP, Initial Reports, v. 164,* College Station TX (Ocean Drilling Program), 47-56, 1996.

Dillon, W.P., M.W. Lee, and D.F. Coleman, Identification of marine hydrates *in situ* and their distribution off the Atlantic coast of the United States, in *International Conference on Natural Gas Hydrates,*, edited by E.D. Sloan, Jr., J. Happel, and M.A. Hnatow, pp. 364-380, *Annals of the New York Academy of Sciences, vol. 715,* 1994.

Dillon, W.P., M.W. Lee, Kristen Fehlhaber, and D.F. Coleman, Gas hydrates on the Atlantic margin of the United States — controls on concentration, in *The Future of Energy Gases* edited by D.G. Howell, *U.S. Geological Survey, Prof. Paper 1570,* 313-330, 1993.

Dillon, W.P., and C.K. Paull, Marine gas hydrates - II: Geophysical Evidence: in, *Natural gas hydrates--properties, occurrences and recovery,* edited by J.L Cox, pp. 73-90, Butterworth Publishers, Boston, MA, 1983.

Dillon, W.P., and P. Popenoe, The Blake Plateau and Carolina Trough, in, *The Atlantic Continental Margin,U.S.*, edited by R.E.Sheridan and J.A. Grow, p. 291-328, The Geology of

North America, v. I-2, Geological Society of America, 1988.

Duval, B., Cramez, C., and Jackson, M.P.A., Raft tectonics in the Kwanza Basin Angola, *Marine and Petroleum Geology, 9,* 389-404, 1992.

EEZ-Scan 87 Scientific Staff, Atlas of the U.S. Exclusive Economic Zone, Atlantic continental margin, *U. S. Geological Survey, Miscellaneous Investigations Series I-2054,* 174 p., 1991

Graue, K., Mud volcanoes in deepwater Nigeria, *Marine and Petroleum Geology, 17,* 959-974, 2000.

Holbrook, W.S., H. Hoskins, W.T. Wood, R.A. Stephen, D. Lizarralde, and the ODP Leg 164 Science Party, Methane hydrate and free gas on the Blake Ridge from vertical seismic profiling, *Science, 273,* 1840-1843, 1996.

Hollister, C.D., J.I. Ewing, et al. *Initial Reports of the Deep Sea Drilling Project, XI,* Washington, U.S. Government Printing Office, 1077 pp. 1972.

Hovland, M., and A.G. Judd, *Seabed pockmarks and seepages, impact on geology, biology and the marine environment,* Graham and Trotman, London, 293 p, 1988.

Jackson, M.P.A., Retrospective salt tectonics, in *Salt tectonics: a global perspective,* edited by M.P.A. Jackson, D.G. Roberts, and S. Snelson, Am. Assoc. Petroleum geologists, Mem. 65, p. 1-28, 1995

Katz, M.E., D.K. Pak, G.R. Dickens, and K.G. Miller, The source and fate of massive carbon input during the latest Paleocene thermal maximum, *Science, 286,* 1531-1533, 1999.

Kayen, R.E., and H.J. Lee, Pleistocene slope instability of gas hydrate-laden sediment on the Beaufort Sea margin, *Marine Geotechnology, 10,* 125-141, 1991.

Keigwin, L.D., D. Rio, and G.D. Acton, et al. Proc. ODP, Int. Repts, 172, College Station, TX (Ocean Drilling Program), 1998.

Kvenvolden, K.A., and L.A. Barnard, Gas hydrates of the Blake Outer Ridge, Site 533, Deep Sea Drilling Project Leg 76, in *Initial Reports of the Deep Sea Drilling Project, 76* , R. E. Sheridan, F. M. Gradstein, et al., p. 353-365, Washington, D.C., U.S. Government Printing Office, 1983.

Lee, M. W., D.R. Hutchinson, W.F. Agena, W.P. Dillon, J. J. Miller, and B. A. Swift, Seismic character of gas hydrates on the southeastern U.S. continental margin, *Marine Geophysical Researches, 16,* p. 163-184, 1994.

Lee, M.W., D.R. Hutchinson, T.S. Collett, and W.P. Dillon, Seismic velocities for hydrate-bearing sediments using weighted equations, *Journal of Geophysical Research, 101,* 20,347-20,358, 1996.

Lee, M.W., D.R. Hutchinson, W.P. Dillon, J.J. Miller, W.F. Agena, and B.A. Swift, A method of estimating the amount of in situ gas hydrates in deep marine sediments, *Marine and Petroleum Geology, 10,* p. 493-506, 1993.

Lee, M.W and W.P. Dillon, Amplitude blanking related to gas hydrate cementation, *Marine Geophysical Researches, in press,* 2000

Markl, R.G., and G.M., Bryan, Stratigraphic evolution of Blake Outer Ridge, *American Association of Petroleum Geologists Bulletin, 67,* p. 666-683, 1983.

Markl, R.G., G.M. Bryan, and J.I.Ewing, Structure of the Blake-Bahama Outer Ridge, *J. Geophys. Res., 75,* 4539-4555, 1970.

McCave, I.N., and B.E. Tucholke, Deep current-controlled sedimentation in the western North Atlantic, in *The Geology of North America, v. M, The Western North Atlantic Region,* edited by P.R. Vogt and B.E. Tucholke, Geological Society of America, pp. 451-468., 1986.

Mountain, G.S., and B.E. Tucholke, Mesozoic and Cenozoic geology of the U.S. Atlantic continental slope and rise, in., *Geologic Evolution of the United States Atlantic Margin,* edited by C.W. Poag, Van Nostrand Reinhold, New York, p. 293-341, 1985.

Paull, C.K., W.J. Buelow, W. Ussler III, and W.S. Borowski, Increased continental margin slumping frequency during sea-level low-stands above gas hydrate-bearing sediments, *Geology, 24,* 143-146, 1996.

Paull, C.K. and W.P. Dillon, Appearance and distribution of the gas hydrate reflection in the Blake Ridge region, offshore southeastern United States, *USGS Misc. Field Studies Map, MF-1252,* 1981.

Paull, C.K., R. Matsumoto, , P.J. Wallace, et al., *Proceedings ODP, Initial Reports, v. 164,* College Station TX (Ocean Drilling Program), 623 pp, 1996.

Paull, C.K., W. Ussler III, and W.P Dillon, Is the extent of glaciation limited by marine gas hydrates?, *Geophysical Research Letters, 18,* 432-434., 1991.

Paull, C., W. Ussler III, T.Naehr, V. Orphan, J. Nealon, W.P Dillon, T. Lorenson, W. Borowski, and W. Gwilliam, Piston cores from the Blake Ridge collapse structure, Geological Society of America Annual Meeting 2000, Abstract 52676.

Popenoe, P., E.A. Schmuck., and W.P Dillon, The Cape Fear landslide: slope failure associated with salt diapirism and gas hydrate decomposition, in *Submarine Landslides: Selected Studies in the Exclusive Economic Zone,* edited by W.C. Schwab, H.J Lee, and D.C.Twichell, *U.S. Geological Survey Bull. 2002,* 40-53, 1993.

Prior, D.B., E.H. Doyle,, and M.J. Kaluza, Evidence for sediment eruptions in the deep sea floor, Gulf of Mexico, *Science, 243,* 517-519, 1989.

Rowe, M.M. and Gettrust, J.F., The fine structure of methane hydrate-bearing sediments on the Blake Outer Ridge determined from deep-tow multichannel seismic data, J. Geophysical Research, *B98,* 463-473, 1993.

Shackleton, N.J., The 100,000-year ice-age cycle identified and found to lag temperature, carbon dioxide and orbital eccentricity, *Science, 289,* 1897-1902, 2000.

Shipley, T.H., M.H. Houston, R.T. Buffler, F.J. Shaub, K.J. McMillen, J.W. Ladd, and J.L Worzel, Seismic evidence for widespread possible gas hydrate horizons on continental slopes and rises: *Am. Assoc. Petroleum Geologists Bull., 63,* 2204-2213, 1979.

Taylor, M., W. Dillon, C. Anton, and W. Danforth, Seismic-reflection surveys of the Blake Ridge, *R/V Cape Hatteras* 1992 and 1995: Data acquisition, navigation and processing, *U.S. Geological Survey, Open File Report 99-372,* 2 CD-ROMs, 1999.

Tucholke, B.E., G.M. Bryan,, and J.I. Ewing,, 1977, Gas-hydrate horizons detected in seismic-profiler data from the western North Atlantic, *Am. Assoc. Petroleum Geologists Bull., 61,* 698-707, 1977.

Tucholke, B.E., and G.S. Mountain, Tertiary paleoceanography of the western North Atlantic Ocean; in, *The Geology of North*

America, v. M, The Western North Atlantic Region edited by P.R. Vogt and B.E.Tucholke, Geological Society of America, p. 631-650, 1986.

Vasseur, G., X. Luo, and P. Jouanna, Generation and dissipation of overpressure in sediments modeling and in-situ measurement of the tightness of sedimentary rocks, *in Fluid Flow through Faults and Fractures in Argillaceous Formations,* pp. 105-119, Proceedings of a Joint Nuclear Energy Agency/European Commission Workshop, Berne Switzerland, 10-12 June, 1996.

Vendeville, B.C., and M.P.A. Jackson, The rise of diapirs during thin-skinned extension, *Marine and Petroleum Geology, 9,* 331-353, 1992a.

Vendeville, B.C., and M.P.A. Jackson, The fall of diapirs during thin-skinned extension, *Marine and Petroleum Geology, 9,* 354-371, 1992b.

Wood, W.T. and C. Ruppel, Seismic and thermal investigations of the Blake Ridge gas hydrate area: A synthesis, in, *Proc. ODP, Sci. Results, 164,* edited by C.K. Paull, R. Matsumoto, P.J. Wallace, and W.P. Dillon, College Station TX (Ocean Drilling Program), 253-264, 2000

Christopher H. Anton, William P. Dillon, Jeffrey W. Nealon, U.S. Geological Survey, 384 Woods Hole Rd. Woods Hole, MA 02543, USA

Rebecca M. Drury, International Faculty Council, Box 352120, University of Washington, Seattle, WA 98195-2120, USA

Myung W. Lee, U.S. Geological Survey, Box 25046, Denver Federal center, MS 960, Lakewood, CO, 80225, USA

Michael H. Taylor, Department of Earth and Space Sciences, 3806 Geology Building, University of California, Los Angeles, Los Angeles, CA 90095-1567, USA

Seismic Studies of the Blake Ridge: Implications for Hydrate Distribution, Methane Expulsion, and Free Gas Dynamics

W. Steven Holbrook

Department of Geology and Geophysics, University of Wyoming, Laramie, WY

I summarize results of active-source seismic investigations, including single-channel seismic reflection, ocean-bottom seismic refraction, and vertical seismic profiles, from the Blake Ridge, offshore South Carolina. The Blake Ridge is a highly dynamic system, in which methane gas is being generated today at depths exceeding 700 m and migrates on the sub-Ma time scale. Seismic velocities indicate that hydrate occupies 3–4% of total rock volume (5–7% of porosity) in a 250-m-thick zone on the crest of the ridge and only about 1–2% of total rock volume 7 km off the crest of the ridge. A significant proportion of methane resides beneath the stability zone, where free gas comprises 1–10% of total rock volume in a zone at least 250 m thick. The impedance contrast responsible for the BSR is caused principally by gas, not hydrate: Site 994 shows that hydrate is not a sufficient condition for generation of a BSR, while Sites 995 and 997 suggest that free gas is a necessary condition. Zones of low reflectance beneath the BSR demonstrate that amplitude "blanking" is not required to explain low reflectance above the BSR on the Blake Ridge. Gas migrates along permeability contrasts at stratal boundaries and can be trapped by faults and at the base of the hydrate stability zone. The Blake Ridge collapse structure, which shows strong circumstantial evidence for expulsion of large quantities of methane into the ocean/atmosphere system, does not appear to have been a single catastrophic event. Rather, the collapse appears to be a relatively long-lived feature filled in the last 2 my at deposition rates of up to 300 m/my. Buried features similar to the collapse beneath the ridge suggest that episodic gas release is a fundamental, recurring part of the hydrate cycle on the Blake Ridge.

INTRODUCTION

The Blake Ridge, offshore South Carolina, provides unique opportunities to understand the geophysical signature of hydrate and free gas in deep marine sediments. Numerous seismic studies have been conducted on the Blake Ridge (Fig. 1), many of which were conducted by the U.S. Geological Survey. The Blake Ridge was also the site of the first ODP drilling leg dedicated entirely to

Natural Gas Hydrates: Occurrence, Distribution, and Detection
Geophysical Monograph 124

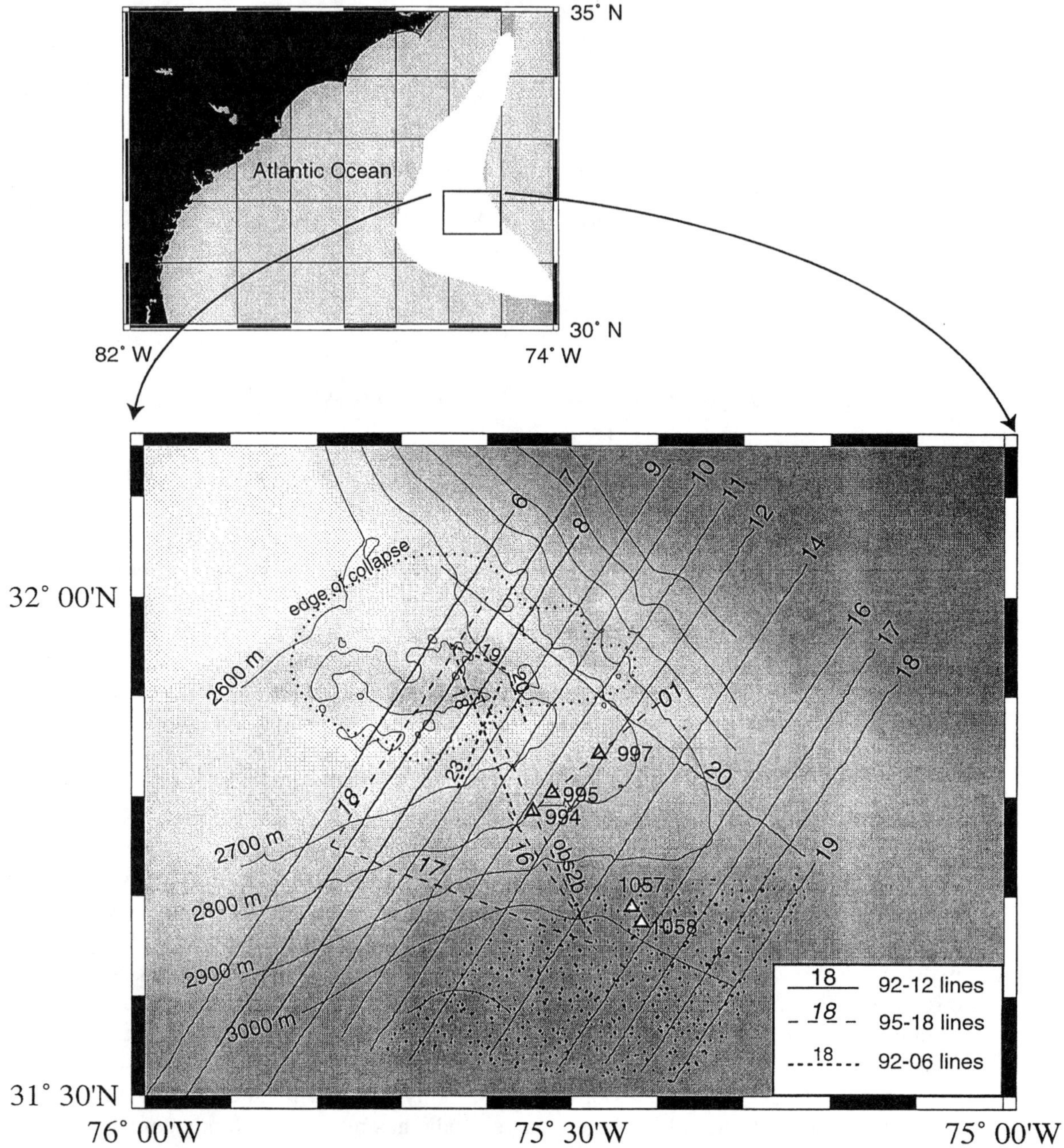

Figure 1. Location of Blake Ridge, single-channel seismic lines used in this paper, and ODP drill sites (triangles). Dotted line marks outline of Blake Ridge collapse structure. Stippled region shows location of young sediment "apron" (e.g., Fig. 3).

researching methane hydrates (Leg 164; Figs 1 and 2), and the data that resulted from that largely successful leg provide invaluable ground truth for interpreting and extrapolating geophysical data [*Paull et al.*, 1996; *Paull et al.*, 2000]. Moreover, Leg 164 demonstrated that the Blake Ridge is particularly well suited to geophysical studies of the hydrate/free gas system: the relatively monotonous lithology [*Paull et al.*, 1996] forms a virtual *tabula rasa* against which geophysical anomalies (e.g., reflectance, velocity, resistivity) can be confidently attributed to hydrate or gas. Bright seismic reflections beneath the BSR, for example, directly locate locally high

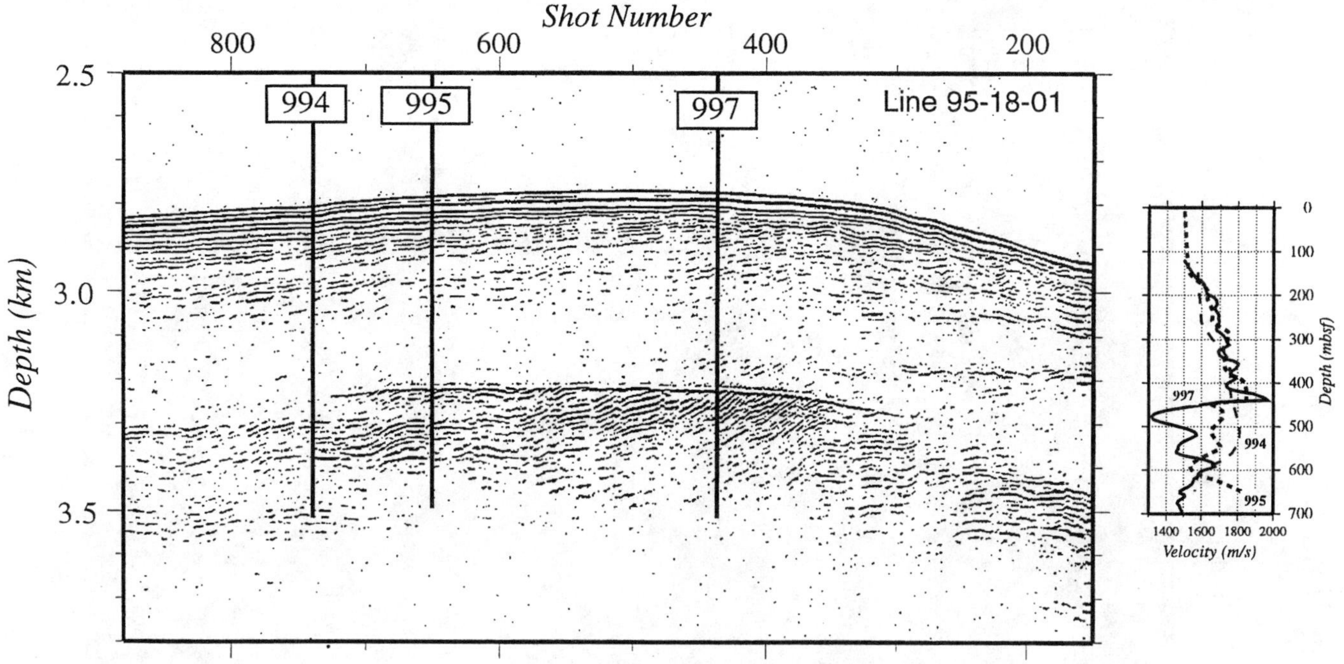

Figure 2. (Left) Seismic line 95-18-01 along ODP Leg 164 drill sites 994, 995, and 997. Notice the disappearance of the bottom-simulating reflection (BSR) at Site 994. (Right) New velocity-depth functions from vertical seismic profiles (VSP) in Sites 994, 995, and 997. Velocities above and below BSR were inverted separately at Sites 995 and 997, resulting in a sharper BSR and more accurate velocities above and below BSR. Low velocities indicative of free gas correspond to reflective zones on the seismic profile, demonstrating that the contrast in reflectance across the BSR is due to enhancement of impedance contrasts by gas beneath the BSR (bright spots).

concentrations of free gas, while zones of high seismic velocities in the hydrate stability zone indicate local concentrations of hydrate.

Despite the relative simplicity of the background lithology, stratigraphy, and tectonics on the Blake Ridge, it is increasingly clear that the distribution of hydrate and free gas is complex and three-dimensional, and that the system as a whole must be regarded as a dynamic one [*Egeberg and Dickens*, 1999; *Wood and Ruppel*, 2000; *Xu and Ruppel*, 1999]. In places, the system appears to be in steady-state equilibrium (e.g., the ridge crest where Leg 164 drilled), while other parts of the system appear to have been perturbed out of equilibrium (e.g., the collapse; [*Dillon et al.*, 1997]). An example of this complexity and dynamism is Line 95-18-16 (Fig. 3), which shows (1) numerous gas-charged zones with complex geometry in a ~300-m-thick zone beneath the hydrate stability zone; (2) apparently gas-free (unreflective) zones below the stability zone; (3) a generally absent BSR reflection, despite the widespread presence of gas; and (4) a bright BSR and gas-charged zone that probably reflects cannibalization of hydrate by a rapidly migrating BSR due to recent, rapid deposition. In order to understand this complex hydrate/gas system, we need to investigate its dynamics. That is, what processes control the formation, migration, and destruction of methane, and at what rates do they occur? Two issues are of prime interest. First, what is the mass balance of the gas/hydrate system where it exists in dynamic equilibrium -- i.e., where, how, and at what rates does methane form, migrate, collect in, and exit hydrate/gas deposits? Second, how does the system respond when perturbed -- in particular, can methane leave the hydrate/gas system, either catastrophically or episodically, to enter the ocean/atmosphere system?

In this paper I analyze existing active-source seismic studies in the vicinity of the Leg 164 drilling transect, focusing on three types of seismic data: single-channel seismic (SCS) data, ocean-bottom seismic (OBS) data, and vertical-seismic profile (VSP) data. I will emphasize the implications of these data sets for the distribution of hydrate, the dynamics of the free gas system, and the origin of the Blake Ridge collapse structure.

HYDRATE, THE BSR AND BLANKING

Various independent measurements made during Leg 164 show that hydrate on the Blake Ridge is present at relatively low average concentrations (2-4% of total sediment volume) in a ~250-m-thick zone overlying the

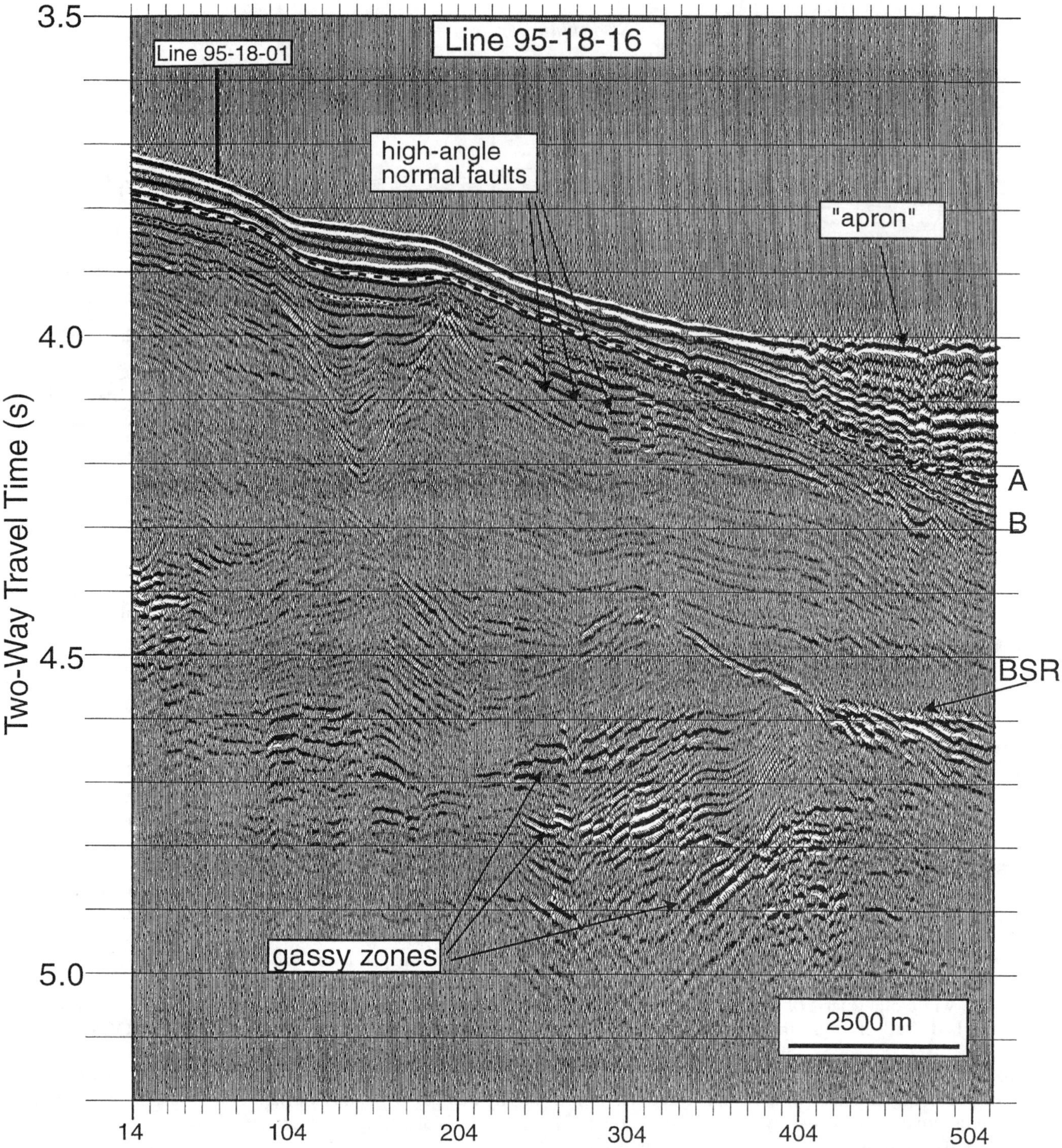

Figure 3. Seismic line 95-18-16, showing complexity of the thick free gas zone. Reflective zones beneath 4.3 s represent gas-charged zones; numerous low-reflectance zones occur between gassy zones, indicating that low reflectance does not require the presence of hydrate. The strong BSR at the southern end of the profile underlies an anomalously thick apron of sediments deposited at a relatively high deposition rate. The age of the apron can be inferred from Reflectors A and B, which can be correlated to ODP Leg 164 drill sites on Line 95-18-01, constraining their ages to be approximately 0.8 and 1.2-1.4 Ma, respectively. Sediments older than about 1.4 Ma are pervasively dissected by high-angle normal faults, while younger sediments appear unfaulted.

BSR, although locally higher concentrations (up to 25%) exist [*Paull et al.*, 1996]. P-wave velocities provide important estimates of mean hydrate and gas concentrations that are generally in accord with estimates from other data (e.g., pore water chemistry, direct sampling of gas, and resistivity logs; [*Collett and Ladd*, 2000; *Dickens et al.*, 1997; *Egeberg and Dickens*, 1999; *Paull and Matsumoto*, 2000; *Paull et al.*, 1996]). On the Blake Ridge, P-velocity information comes from several sources: (1) downhole sonic logs in the Leg 164 holes [*Collett and Ladd*, 2000; *Paull et al.*, 1996], (2) VSPs conducted during Leg 164 [*Holbrook et al.*, 1996; *Paull et al.*, 1996], (3) wide-angle reflection data recorded on ocean-bottom hydrophones (OBH) or seismometers [*Katzman et al.*, 1994; *Korenaga et al.*, 1997], (4) deep-towed seismic data [*Rowe and Gettrust*, 1993; *Wood and Ruppel*, 2000] and (5) interval velocities from a few multichannel seismic reflection lines [*Ecker et al.*, 2000; *Tinivella and Accaino*, 2000; *Wood et al.*, 1994].

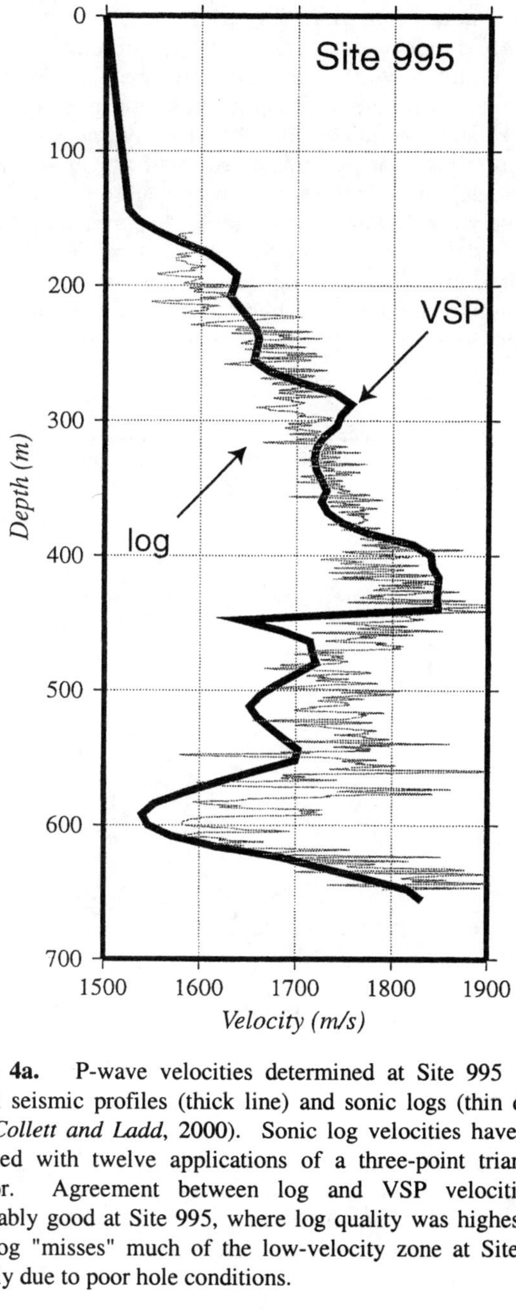

Figure 4a. P-wave velocities determined at Site 995 from vertical seismic profiles (thick line) and sonic logs (thin dotted lines; *Collett and Ladd*, 2000). Sonic log velocities have been smoothed with twelve applications of a three-point triangular operator. Agreement between log and VSP velocities is remarkably good at Site 995, where log quality was highest; the sonic log "misses" much of the low-velocity zone at Site 997, probably due to poor hole conditions.

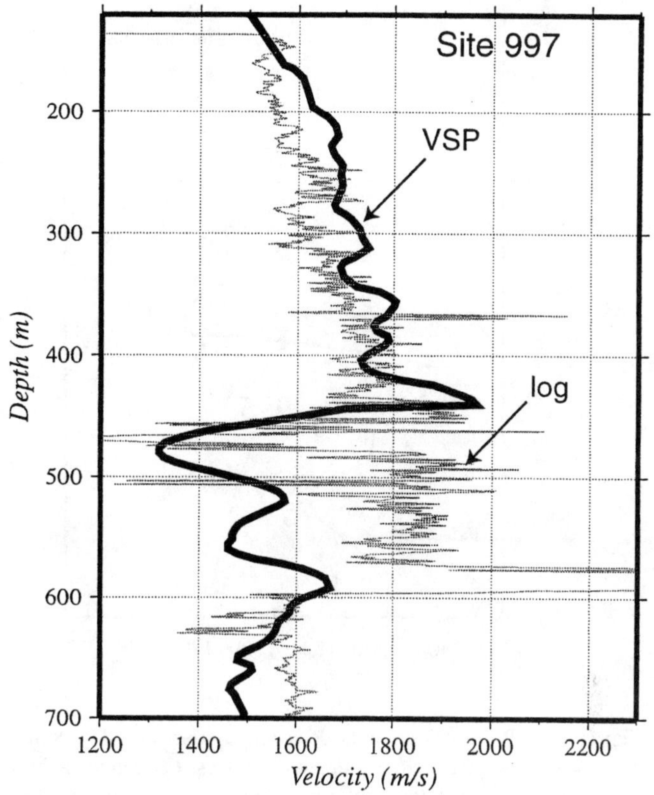

Figure 4b. (b) P-wave velocities at Site 997 from vertical seismic profiles and sonic logs, plotted as in Figure 4a.

Downhole measurements from logs and VSPs provide the most accurate, direct measurements of Vp on the Blake Ridge, with the sole drawback that their geographical distribution is limited to a few holes. In this paper, the VSP velocity profiles from Sites 995 and 997 represent new inversions, differing from those in previous publications [*Holbrook et al.*, 1996; *Paull et al.*, 1996] in that data above and below the BSR were inverted separately, using the method of *Lizarralde and Swift* [1999]. This eliminates the smoothing of the velocity profile across the BSR, resulting in a more accurate determination of BSR depth and of velocities immediately above and below the BSR. In general, Vp profiles produced by sonic logs and VSPs on the Blake Ridge agree well (Fig. 4). The sonic logs provide a higher-resolution record and are thus superior for identifying small-scale

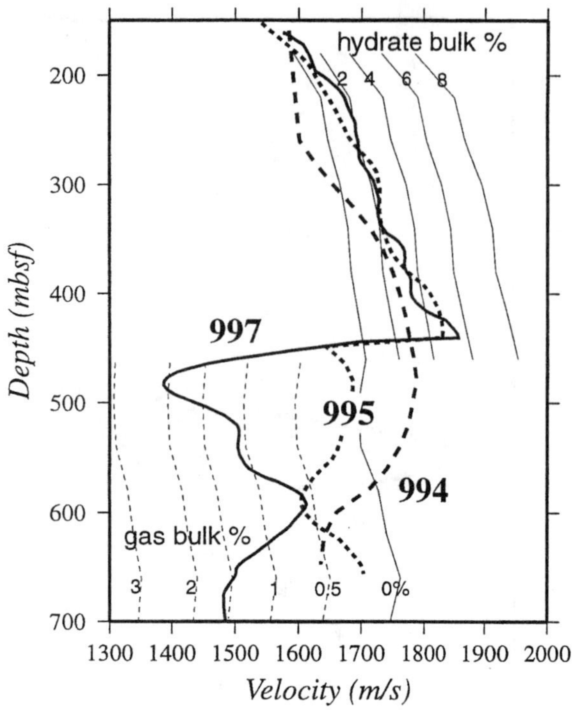

Figure 5a. P-wave velocities determined from Leg 164 VSPs, plotted on predicted velocity-depth curves as a function of bulk hydrate concentration (thin solid lines) and bulk gas concentration (thin dashed lines). Velocities of hydrate- and gas-bearing sediments are calculated from the rock physics model of *Helgerud et al.* [1999], using the hydrate-in-frame and "homogeneous gas" models, corresponding to minimum concentrations.

variations in Vp. VSPs, however, are conducted at surface-seismic frequencies and thus provide more realistic links to seismic reflection data; their longer wavelengths make them less susceptible to degradation due to near-hole conditions, which is especially important in the free gas zone. At Site 995 there is remarkable agreement between the two profiles: both show the velocity drop at the BSR at 448 mbsf, and the long-wavelength variations in Vp beneath the BSR are virtually identical, with the caveat that mean Vp from VSPs is slightly lower than that in the logs. At Site 997, the agreement is less impressive: VSP velocities are generally higher in the hydrate zone and significantly lower in the free gas zone; the sonic log did not detect the free gas zone over > 100 m of its extent. The large excursions in sonic Vp at Site 997 attest to relatively poor hole conditions, which likely accounts for the log/VSP discrepancy there.

To interpret VSP velocities for hydrate and gas concentration, I use the effective-medium model of Helgerud et al. [*Helgerud et al.*, 1999] and a smoothed version of the Vp profiles (Fig. 5). The smoothed profiles emphasize mean hydrate concentrations at each site; although sharp variations are suppressed (except at the BSR), the smoothed profiles show robust average values in the hydrate and gas zone. Bulk hydrate concentrations are near zero at 200 mbsf and increase with depth toward the BSR, reaching about 4-5% for the hydrate-cement model and 14% for the hydrate-as-fluid model (Figure 5).

Uncertainty over which rock physics model applies to the Blake Ridge causes substantial uncertainty in determining absolute concentrations of hydrate and gas from seismic velocities. If hydrate acts as a cement between grains, a little hydrate goes a long way: its effect on sediment elastic properties is much greater at a given concentration than if it acts as pore fill. The hydrate-cement model thus predicts a minimum hydrate concentration, while the pore-fill model produces a maximum estimate. Despite this uncertainty, relative lateral and vertical changes in hydrate (and gas) concentration can be detected with accurate P-wave velocities, even for surprisingly subtle variations at relatively low bulk concentrations. Approaching the crest of the Blake Ridge from Site 994 toward Site 997, both hydrate and free gas concentrations increase. Using minimum estimates, bulk hydrate increases from about 2% at Site 994 to about 3% at Sites 995 and 997; using

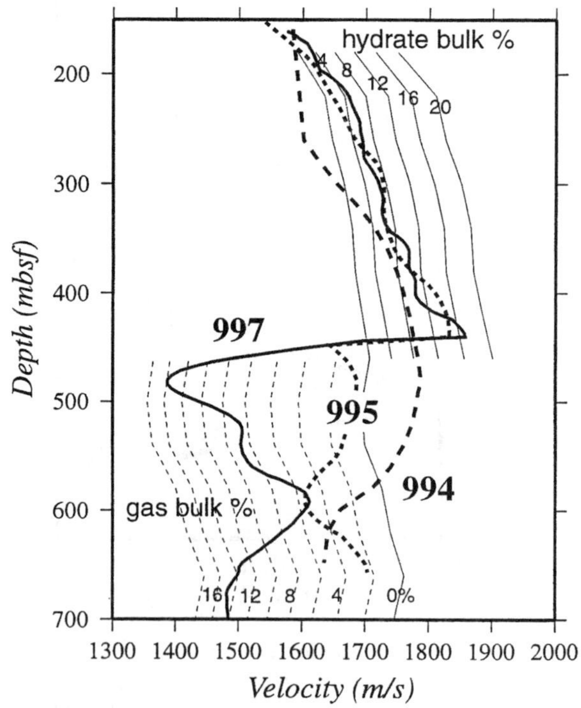

Figure 5b. P-wave velocities determined from Leg 164 VSPs, plotted as in Figure 5a, but with theoretical velocities calculated from *Helgerud et al.* [1999] using the hydrate-in-pore-fluid and the patchy gas models, corresponding to maximum concentrations.

maximum estimates, this lateral increase goes from 4% at Site 994 to 7% at Sites 995 and 997.

These estimates compare well to those determined from other data, such as resistivity logs [*Collett and Ladd*, 2000], pore-water chemistry [*Egeberg and Dickens*, 1999; *Paull et al.*, 1996], and interval velocities from MCS data [*Ecker et al.*, 2000]. Indeed, the minimum (hydrate-cement) estimate of 3% bulk hydrate from the VSP velocities agrees well with the estimate of about 2.5% bulk hydrate (5.3% of porosity) from pore-water chlorinity [*Egeberg and Dickens*, 1999]. This suggests that hydrate acts as a cement in Blake Ridge sediments. Several older estimates of hydrate concentrations on the Blake Ridge, in contrast, are anomalously high, perhaps due either to less accurate (and significantly higher) estimates of Vp [*Rowe and Gettrust*, 1993] or to interpretation of less diagnostic properties (e.g., sediment reflectance; [*Lee et al.*, 1993, 1994]). Hydrate is restricted to the region above the BSR; although *Guerin et al* [1999] inferred from logging data that hydrate and gas coexist in an 80-m-thick zone beneath the BSR, the lack of chloride anomalies in that zone is strong evidence against the presence of hydrate there [*Egeberg and Dickens*, 1999; *Paull et al.*, 1996].

A central question relating to the dynamics of methane recycling in hydrate provinces is whether the BSR is immediately overlain by a zone of concentrated hydrate (a "hydrate wedge"). Upward migration of the hydrate phase boundary in response to continual sedimentation will cause dissociation of hydrate at the base of the stability zone; if the liberated free gas is able to migrate upward into the hydrate stability zone, it will re-freeze into hydrate, resulting in a concentrated hydrate layer at or just above the phase boundary [*Hyndman and Davis*, 1992; *Hyndman and Spence*, 1992]. On the Blake Ridge, two pieces of seismic evidence indicate a concentrated zone of hydrate above BSR: waveform inversion of wide-angle, ocean-bottom seismic data, and detailed VSP results at Site 997.

Waveform inversion provides a detailed model of the short-wavelength velocity structure of an area, by iteratively adjusting an input velocity-depth function in order to match synthetic and observed seismograms wiggle-for-wiggle. An important caveat in waveform inversion is that an accurate starting model of the long-wavelength velocity structure must be provided (for example, from traveltime inversion of wide-angle reflections). Korenaga et al. [1997] applied waveform inversion to OBH data from the Blake Ridge, at a location coincident with Site 997, and found a strong hydrate wedge at the base of the stability zone (Fig. 6). According to this result, about one-third of the impedance contrast at the BSR comes from the high-velocity hydrate wedge, and about two-thirds from the underlying low-velocity gas zone. However, after the Korenaga et al. work was conducted (though published in 1997, the inversion preceded the drilling leg), it became clear that the BSR at the Blake Ridge was underlain by a much stronger and

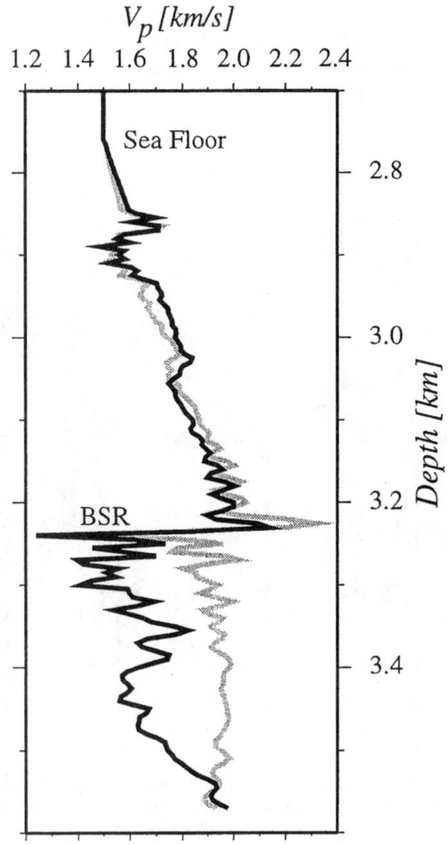

Figure 6. P-wave velocities at Site 997 from waveform inversion of wide-angle, ocean-bottom seismic data by *Korenaga et al.* [1997] (gray line) and using a new starting velocity model (solid line). The new result yields a still-evident, but less prominent, hydrate wedge just above the BSR at 3.22 km depth.

thicker free gas zone than was included in the Korenaga et al. starting velocity model.

I therefore repeated the waveform inversion of Korenaga et al. [1997], using a new starting velocity model that included the thick low-velocity zone imaged by VSPs at Site 997. The resulting synthetic seismograms (Fig. 7) provide a better fit to the observed data than the Korenaga et al. result (an 8% reduction in misfit), thus reaffirming the importance of using an accurate starting velocity model in waveform inversion. In the new model, the hydrate wedge is present but less prominent, with a peak velocity of only 2.15 km/s, compared to the 2.32 km/s of the Korenaga model. In this new model, then, only about 15% of the impedance contrast at the BSR is attributable to the hydrate wedge.

A second, independent line of evidence for a high-velocity wedge at Site 997 comes from the unusually detailed VSP conducted there. In order to determine errors in the VSP-derived velocity function, an unusual "repeat VSP" was conducted at Site 997 during Leg 164. The

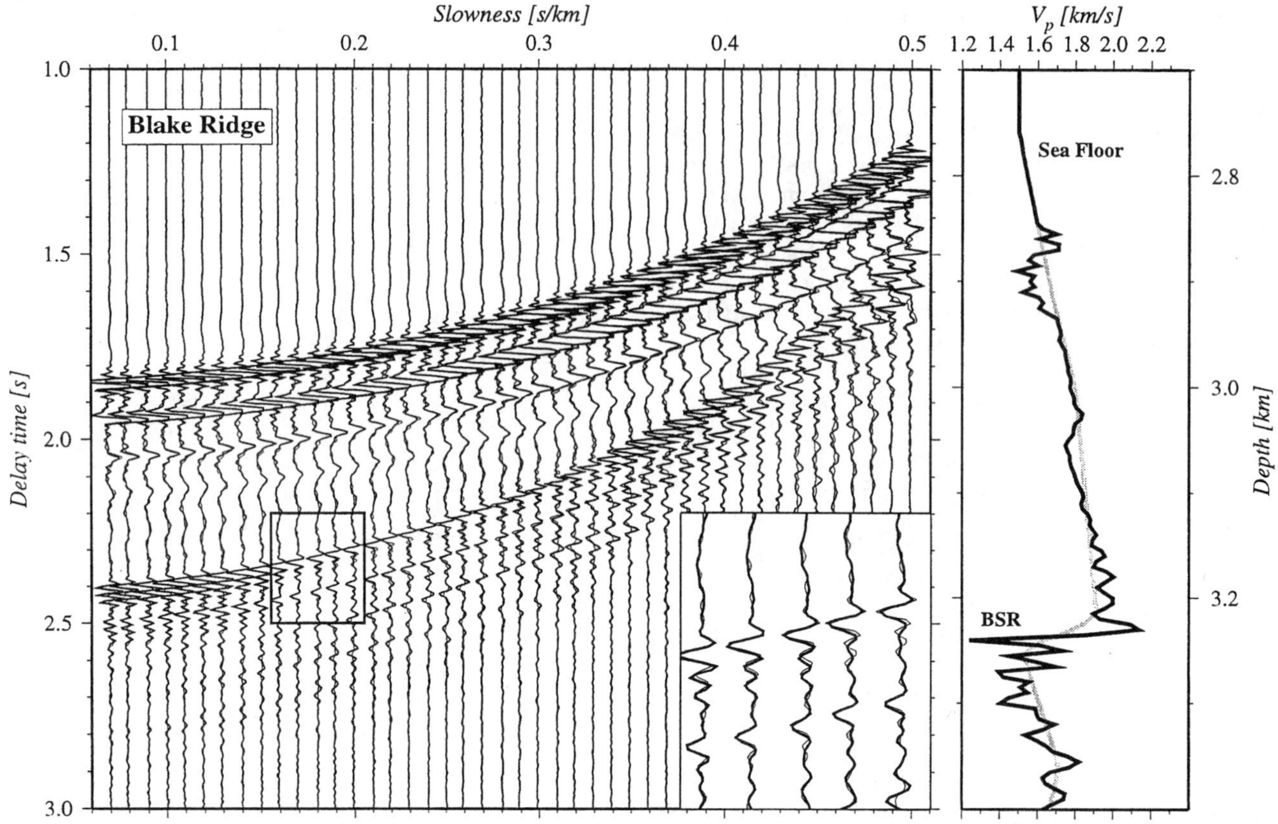

Figure 7. Observed seismic data (dark lines) from an ocean-bottom hydrophone at Site 997 overlaid on synthetic seismograms (light lines) from the velocity-depth model (right) determined by waveform inversion. Inset shows blow-up of data fit over a portion of the BSR reflection. Starting velocity-depth model (light gray line) includes the thick low-velocity zone imaged by VSPs. The new result produces an 8% improvement in data fit over the *Korenaga et al.* [1997] result (Fig. 6).

borehole seismometer was clamped at 8 m intervals in both VSP runs, but the second run occupied positions between the original VSP positions, so that the two VSP runs can be combined into one profile with 4 m spacing. Velocity-depth profiles interpreted independently from the two runs agree remarkably well in all major features (Fig. 8), including the sharp drop in velocity at the BSR and the persistence of velocities of 1500 ± 100 m/s from the BSR to the base of the hole. In general, the two velocity-depth functions lie within ±50 m/s of each other, but local discrepancies of up to 100 m/s exist. The velocity-depth function derived from the combined VSP at 4 m intervals shows clear evidence for a hydrate wedge above the BSR (Fig. 4). This feature represents an excursion of ~150 m/s above the background trend, so it is resolvable given the errors suggested by the comparison of the independent VSPs. This result is also supported by the waveform inversion results shown above, as well as interpretations of sonic and resitivity logs [*Collett and Ladd*, 2000] and chlorinity data [*Egeberg and Dickens*, 1999], all of which suggest that a peak of hydrate concentration occurs just above the base of the stability zone.

Despite the strong evidence for a hydrate wedge at Site 997, waveform inversions at other locations on the Blake Ridge [*Kloska*, 1999] and on the Carolina Rise [*Korenaga et al.*, 1997] show that a layer of concentrated hydrate is not a ubiquitous feature on the southeastern U.S. margin. We currently lack an understanding of distribution of hydrate wedges and of the processes that control their formation and localization, but sedimentation rate and the availability of gas migration pathways (e.g., high-angle faults, dipping strata) are potential candidates. Waveform inversion of long-offset seismic data offers the best hope for determining the regional distribution of hydrate wedges in a particular hydrate province.

The VSP and OBH data clearly show that the strong BSR amplitude on the Blake Ridge is principally due to free gas, though a hydrate lens may contribute to BSR amplitude in places. Importantly, however, Site 994 shows that the BSR is absent where no free gas is present, even though hydrate is present. Thus the presence of hydrate is not a sufficient condition for generating a BSR, whereas the presence of free gas is certainly a sufficient, and probably a necessary, condition. Other data from the area show that abundant free gas can exist beneath the hydrate stability zone without an identifiable BSR: Lines 95-18-17 and -16 show 300-m-thick, strongly reflective zones that indicate gas-charged zones at depths of greater than 450 mbsf, even though a BSR (defined as a bright reflection that cross-cuts strata) is lacking over most of the profile (Figs. 3, 9). The implications of this observation for gas migration will be discussed in a later section.

The contrast in velocities between the VSP and OBH data (Fig. 10) demonstrates that significant (~8%) velocity anisotropy occurs in the Blake Ridge hydrate deposit. The anisotropy is a robust, well-resolved feature of the data: using the VSP-measured velocities to model the OBH data produces large misfits, and vice-versa. This form of anisotropy, transverse isotropy, is caused by horizontal layering: sound waves propagate faster along the layering (as in refraction data) than against it (as in VSP or sonic log data). Although transverse isotropy is common in marine sediments, its implications for interpreting hydrate concentrations may be underappreciated. For example, at Site 997, VSP data show a mean Vp of 1.80 km/s at 400 m depth, corresponding to a minimum (hydrate-in-frame) bulk concentration of 4% hydrate. In contrast, the OBH data show a Vp of 1.95 km/s at the same depth, corresponding to 8% bulk hydrate. The contrast at the peak velocity of the "wedge" is similar: 9% bulk hydrate for the VSP velocity (1.9 km/s), versus 14% for the OBH velocity (2.15 km/s). The correct velocity to use for purposes of determining hydrate concentrations is the equivalent isotropic velocity, which lies between the maximum and minimum velocities. Thus velocities from VSPs predict a minimum concentration of hydrate and a maximum concentration of gas, while velocities from wide-angle measurements may slightly overestimate hydrate concentrations but underestimate gas concentrations.

Reflection amplitudes on the Blake Ridge are generally much stronger beneath the hydrate stability zone than above it. Amplitude "blanking" was originally conceived to explain that contrast, by proposing that hydrate cements grains or fills porosity so as to suppress the impedance contrasts between layers [*Lee et al.*, 1993, 1994; *Shipley et al.*, 1979]. Much of the motivation for the blanking concept can be traced to a classic seismic section from Shipley et al. [1979], often reproduced in later publications (e.g., [*Katzman et al.*, 1994]), that shows a sharp, first-order difference in reflection amplitudes above and beneath the BSR. Two often-overlooked points are worth making in reference to this figure. First, the spectacular abruptness of the contrast, where reflectance is uniformly low above the BSR and uniformly high beneath the BSR, is unique among Blake Ridge seismic data. Other seismic

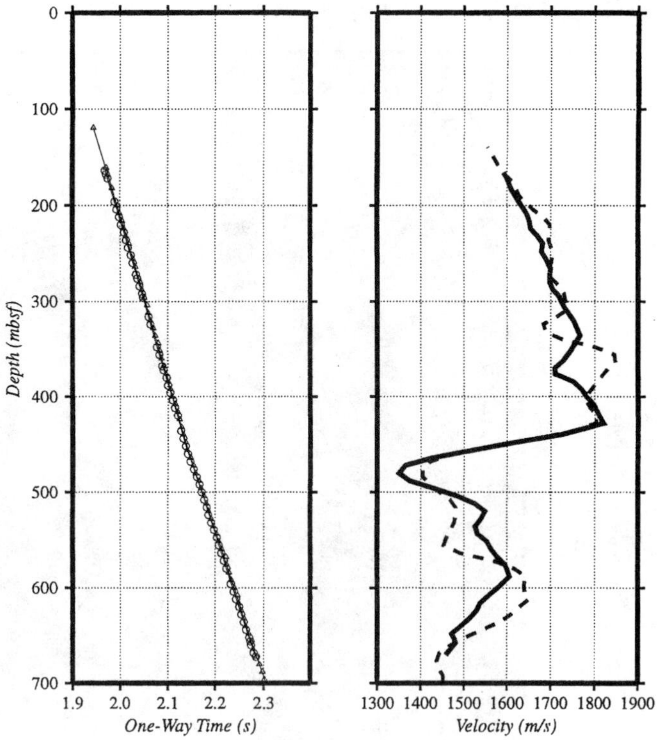

Figure 8. (Left) Travel-time data as a function of depth from two independent runs of a VSP at ODP Leg 164 Site 997 (open circles and triangles). (Right) Independent velocity-depth functions inverted from the travel-time data at left. Inversion results generally agree to within 50 m/s, with rare discrepancies of 100 m/s.

sections on the Blake Ridge typically show more complexly varying reflectance on length scales of 10's to 100's of m (e.g., Figs. 3, 9, 11). Second, the data were "collected with programmed gain increasing with time" [*Shipley et al.*, 1979]. The specific gain function is unknown and cannot be removed from the data (T. Shipley, pers. comm.), so that amplitudes in this figure are highly suspect.

There are, in fact, several significant problems with applying the blanking hypothesis to the Blake Ridge. First, Leg 164 showed that the contrast in reflection amplitudes above and below the BSR are due to enhancement of impedance contrasts by gas-charged zones (the well-known phenomenon of "bright spots"), not due to suppression of impedance contrasts above the BSR. The weak reflectance above the BSR is the natural consequence of a remarkably homogeneous sedimentary section [*Holbrook et al.*, 1996]. The contrast in reflectance that Shipley et al. [1979] described as "a significant decrease in amplitude... above the [BSR]" is thus better described as an increase in amplitude below the BSR. This new view of reflectance on the Blake Ridge removes the original rationale for invoking a blanking mechanism. Second, physical models

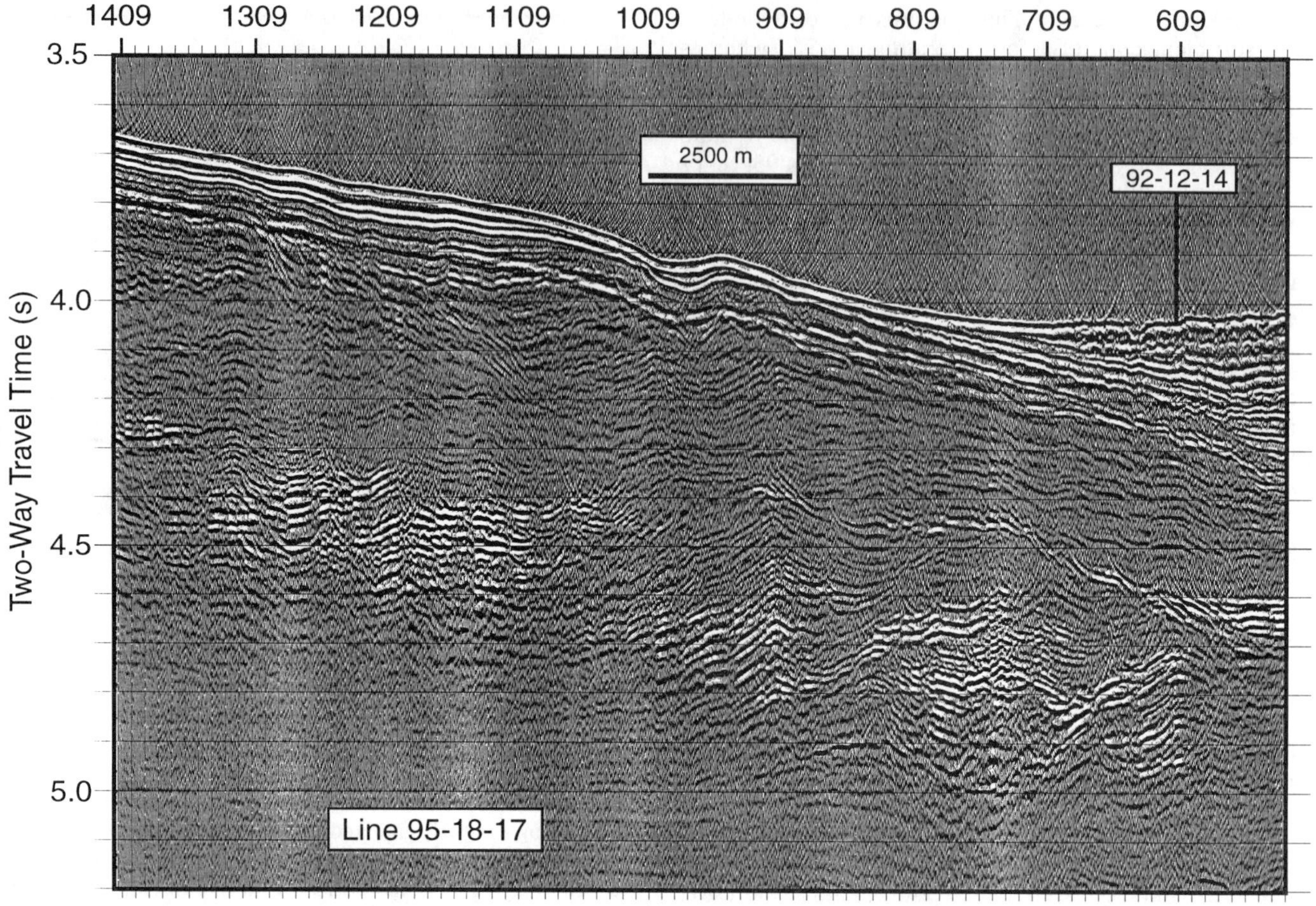

Figure 9. Seismic line 95-18-17, showing complexity of the thick free gas zone. Despite the presence of numerous gas-charged zones, most of the profile lacks a BSR. Location of cross-line 92-12-14 in the young wedge of sediments is shown.

of blanking (e.g., by grain cementation or preferential filling of porous layers) have failed to achieve a significant degree of amplitude suppression without high concentrations of hydrate, which predicts much higher P-wave velocities than are observed on the Blake Ridge [*Lee et al.*, 1993]. Blanking is a viable mechanism at high hydrate concentrations (>30%), but it is not possible to strongly affect impedance contrasts with the small concentrations (<10%) of hydrate present on the Blake Ridge. Third, there are numerous zones beneath the BSR, where hydrate cannot exist, whose reflectivity is as low as the so-called "blanked" zones above the BSR (e.g., Figs. 3, 9). This observation is incontrovertible evidence that low reflectance on the Blake Ridge does not require the presence of hydrate.

I prefer to take a viewpoint dictated by Occam's razor. Data from the Leg 164 transect [*Holbrook et al.*, 1996] showed that strong reflectivity on the Blake Ridge can be completely attributed to two causes: lithological contrasts in the upper 150 m and gas-charged bright spots beneath the BSR. Invoking a third cause, amplitude "blanking," adds unnecessary complexity to our model of reflectivity on the Blake Ridge. The arguments presented above, especially the observation of low reflectivity beneath the BSR, prove that amplitude blanking is not *required* to explain low reflectivity in the hydrate-bearing zone. However, those arguments do not (indeed cannot) disprove that some small amount of amplitude blanking occurs. But in the absence of definitive, quantitative tests of the contribution of "blanking" to low reflectance on the Blake Ridge, amplitude blanking should be considered an unproven, unnecessary mechanism there.

THE DYNAMIC FREE GAS SYSTEM

One of the principal insights gained from Leg 164 is that free gas is distributed in a much broader zone on the Blake Ridge than previously realized. The free gas zone may be thicker than the hydrate-bearing zone on the Blake Ridge, especially considering that the uppermost 200 m of sediments are likely devoid of hydrate [*Paull et al.*, 1996]. This has important implications for both resource considerations and potential release of methane into the ocean/atmosphere system, since gas, unlike hydrate, is mobile. Rather than thinking strictly in terms of "hydrate" deposits, then, we should consider the entire hydrate/gas system.

The amount of free gas present beneath the Blake Ridge can be constrained from the VSP velocities using the model of Helgerud et al. [1999]. P-velocities are anomalously low in a ~300-m-thick zone beneath the hydrate stability zone (Fig. 4). As was the case for estimating hydrate concentrations, the concentration of free gas required to satisfy the reduced Vp depends on which of two rock-physics models applies [*Helgerud et al.*, 1999]. If free gas is homogeneously distributed in the pore space, a given Vp reduction can be achieved with relatively little gas; if free gas is "patchy," more gas is required for a given Vp anomaly. Mean bulk gas concentrations in the thick free gas zone at Site 997 are about 1% for the homogeneous-gas model and 10% for the patchy-gas model. The strongest low-velocity zone is observed beneath Site 997, where P-wave velocity determined by both the VSP and the waveform inversion reaches of minimum of about 1.3 km/s beneath the BSR (Fig. 10). This requires a bulk gas concentration of 3% (5% of porosity) for the homogeneous case and 22% (34% of porosity) for the patchy case. Direct sampling by pressure-core-sampler (PCS) technology suggests that gas occupies 12% of pore space just beneath the BSR [*Dickens et al.*, 1997], a value intermediate between the VSP minimum and maximum estimates. Due to the paucity of PCS samples from beneath the BSR, it is not possible to determine which rock-physics model is appropriate for the free gas zone on the Blake Ridge.

A full understanding of the hydrate/gas system requires knowledge of the rates and locations of methanogenesis, the pathways and time scales of gas migration, and the mechanisms of gas trapping. The prevalence of bright spots that follow strata shows that gas is trapped by permeability contrasts between strata and migrates updip along strata. Abrupt updip terminations of bright spots show that gas is also trapped by faults, which presumably juxtapose permeability contrasts. However, faults may also act as conduits for gas migration and escape, as has apparently happened during formation of the Blake Ridge collapse structure (discussed below), and high-angle normal faults are often invoked elsewhere to provide conduits for gas migration (e.g., [*Hovland et al.*, 1997]). The conditions under which a fault acts as a seal or a conduit in hydrate/gas deposits are not clear. On a larger scale, free gas migrates into the crest of ridge, as evidenced by the higher gas content at Site 997 than at Sites 994 and 995 (Fig. 5), as well as a general pattern of higher BSR

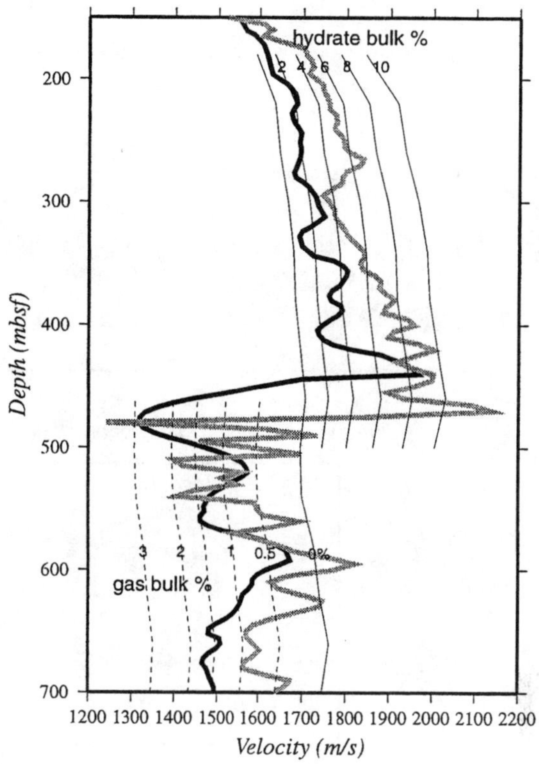

Figure 10. Comparison of P-wave velocities at Site 997 determined by zero-offset VSPs (dark line) and waveform inversion of wide-angle, ocean-bottom hydrophone data (gray line). Wide-angle velocities are uniformly higher than downhole velocities, and the BSR depth predicted by OBH data is deeper than that determined by VSPs, due to transverse isotropy of horizontally layered sediments. This 6% velocity anisotropy can significantly affect estimates of hydrate and gas concentration, as shown by predicted velocity-depth curves for varying bulk hydrate concentration (thin solid lines) and bulk gas concentration (thin dashed lines) [*Helgerud et al.*, 1999].

amplitudes beneath the crest of the ridge than off the ridge (Fig. 11). The BGHZ at the crest of the ridge thus appears to act as a large-scale trap for gas.

Direct constraints on time scales of gas migration are lacking, but some inferences can be drawn from the seismic data. In particular, an apron of young sediments on the southeast end of Line 95-16 (Fig. 3) provides strong evidence for updip gas migration on a sub-Ma time scale. Two clear strata near the base of this apron, which I refer to as reflectors A and B, can be traced to the drilling transect, where biostratigraphy constrains their ages to be about 0.8 Ma and 1.2-1.4 Ma, respectively. The sedimentation rate above reflector A (0.8 Ma) at the end of the Line 16 is about 22 cm/ky (180 m / 0.8 m.y.), which is comparable to the Miocene sedimentation rate of 26 cm/ky on the ridge crest but much higher than the Quaternary rate of ~6 cm/ky [*Okada*, 2000]. The apron was drilled at Sites

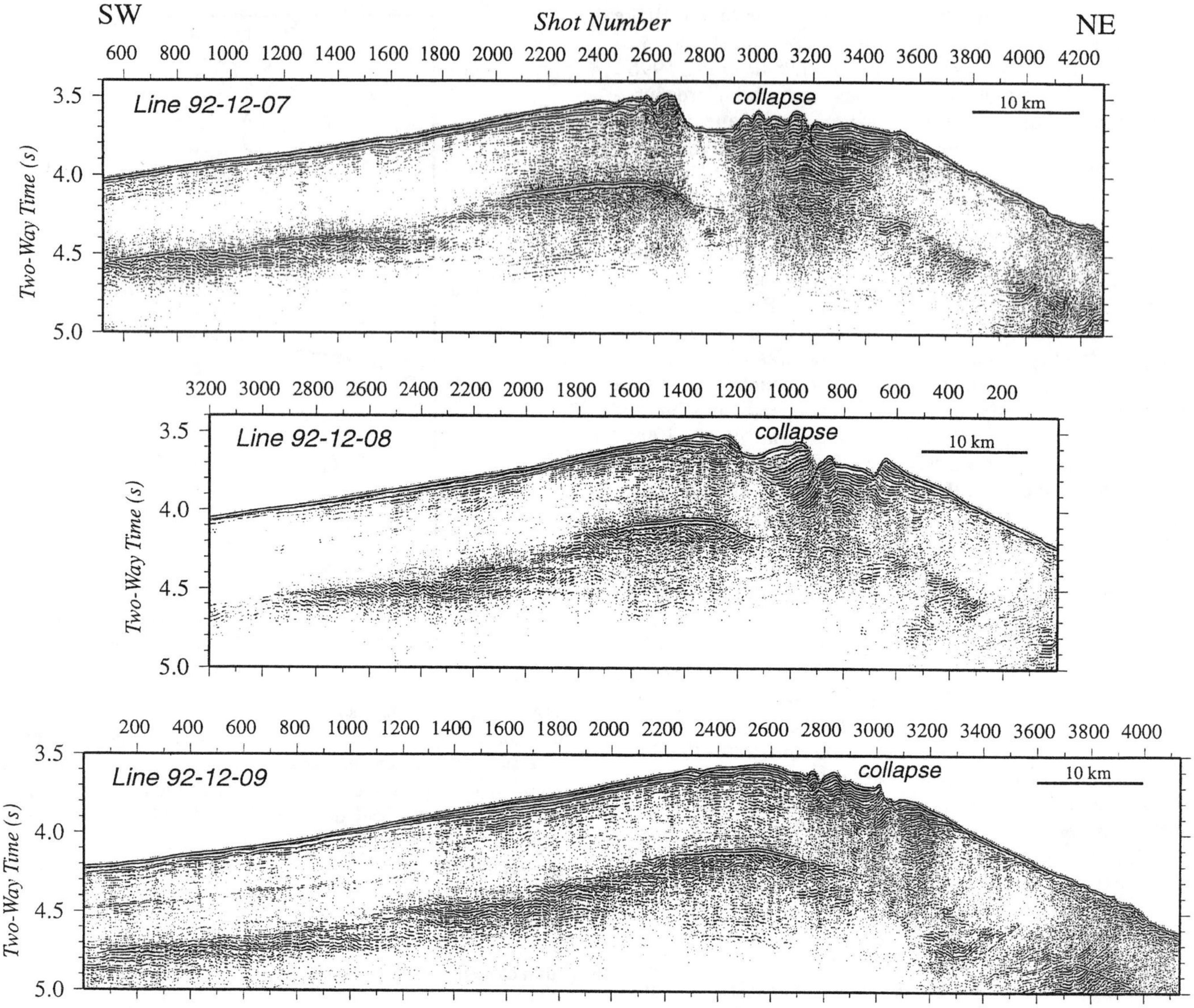

Figure 11. Single-channel seismic lines across the Blake Ridge, from the 92-12 cruise [*Taylor et al.*, 1999]. The Blake Ridge collapse structure is visible on Lines 07, 08, 09, 10, and 11. The young sediment "wedge" is visible on Lines 12, 14, 16, 17, 18, and 19; the dark solid line marks the interpreted position of reflector A (e.g., Fig. 3). A possible buried collapse feature is seen on Line 20 and is visible on crossing lines 14 and 16; similar features are visible on Line 09 (SP 3100-3400), Line 10 (SP 600-900), and Line 11 (SP 2600-3000).

1058 and 1059 on ODP Leg 172, which confirmed that these sediments are young (<1.4 Ma) and deposited at high deposition rates [*Keigwin et al.*, 1998]. Although no tie lines directly link the Leg 172 sites to the USGS network, character correlation suggests that the Reflector A (0.8 Ma) is equivalent to the base of Unit I at Site 1058, which is indeed dated at 0.8 Ma [*Keigwin et al.*, 1998] and that Reflector B corresponds to the base of the apron, which lies just beneath the bottom of Site 1059 (1.4 Ma). Beneath the apron, a strong BSR exists (the BSR is conspicuously lacking elsewhere on the profile), presumably due to breakdown of hydrate as the base of the stability zone migrated upward in response to the high sedimentation rate. The BSR currently resides at the expected depth, indicating that thermal re-equilibration has kept pace with the rapid sedimentation. Significantly, the free gas zone beneath the strong BSR is thinner than the thickness of recently deposited sediments, indicating upward migration and focusing of free gas on sub-1.0 m.y. time scale. (If gas did not migrate upward, the thickness of

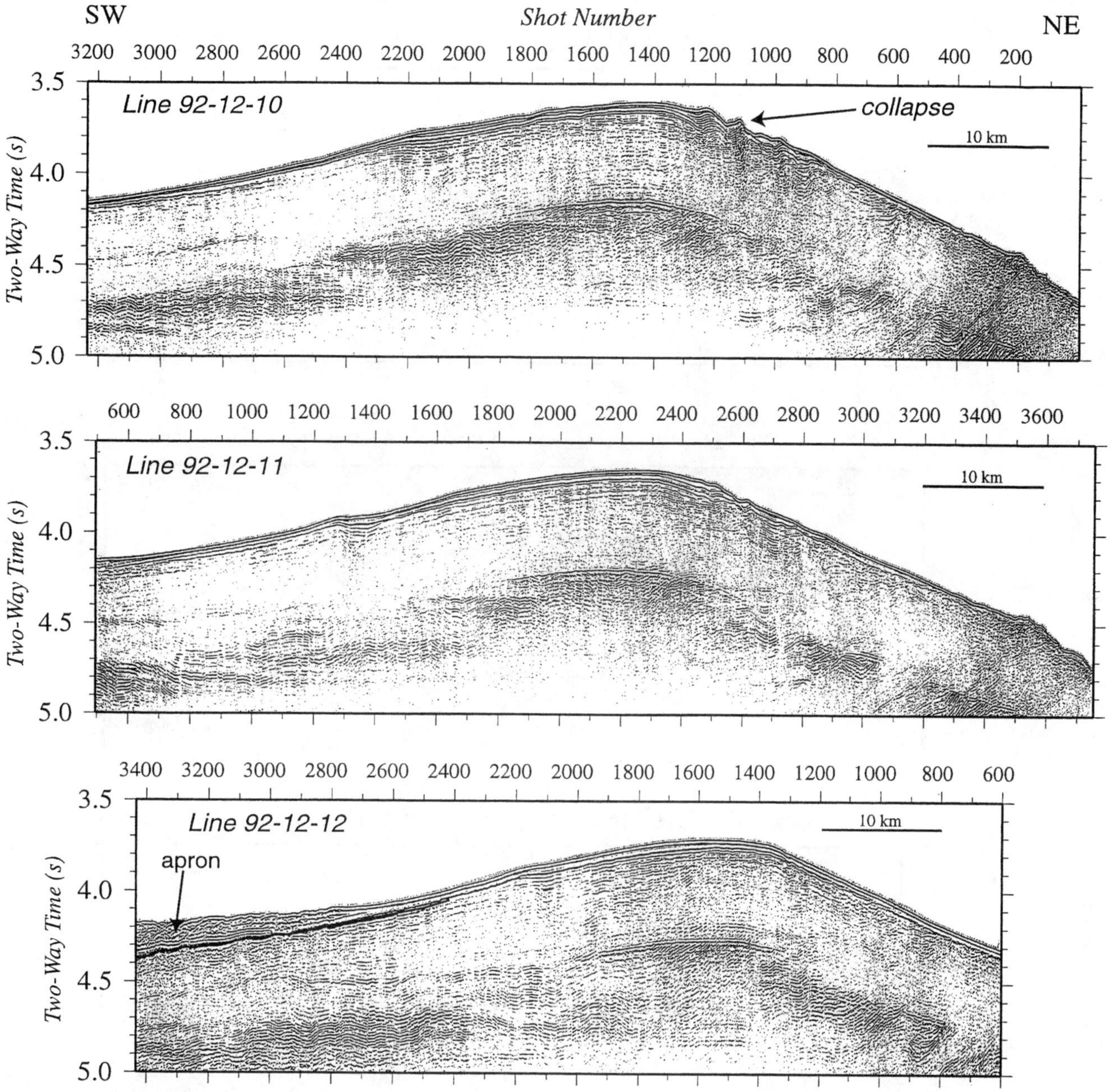

the free gas zone liberated by upward migration of the stability zone would be equal to the thickness of the apron sediments.) If gas does migrate upward on this time scale, then gas currently residing well beneath the BHSZ elsewhere on Line 16 must be produced at those depths (or deeper), rather than having been generated by breakdown of hydrate as the phase boundary passed upward. If this line of reasoning is correct, deeper gas is likely younger than 1 m.y., implying that methanogenesis is likely active at depth in the system [*Wellsbury et al.*, 1997]. The seismic data thus suggest a vital bacterial biosphere in deep marine sediments.

Lines 95-17-16 and 95-17-17 show that a BSR does not necessarily form even when free gas is widespread at depth (Figs. 3, 9). This observation has at least three possible explanations. First, a lack of gas migration pathways may prevent gas from accumulating at the BGHZ. However, the section is dissected by numerous high-angle normal faults, which ought to provide ample migration pathways, and dipping beds provide clear updip migration paths toward the BGHZ. A second possibility is that a relatively low sedimentation rate does not generate sufficient "hydrate recycling" to enhance gas concentrations at the BGHZ. However, the sedimentation rate at these sites is

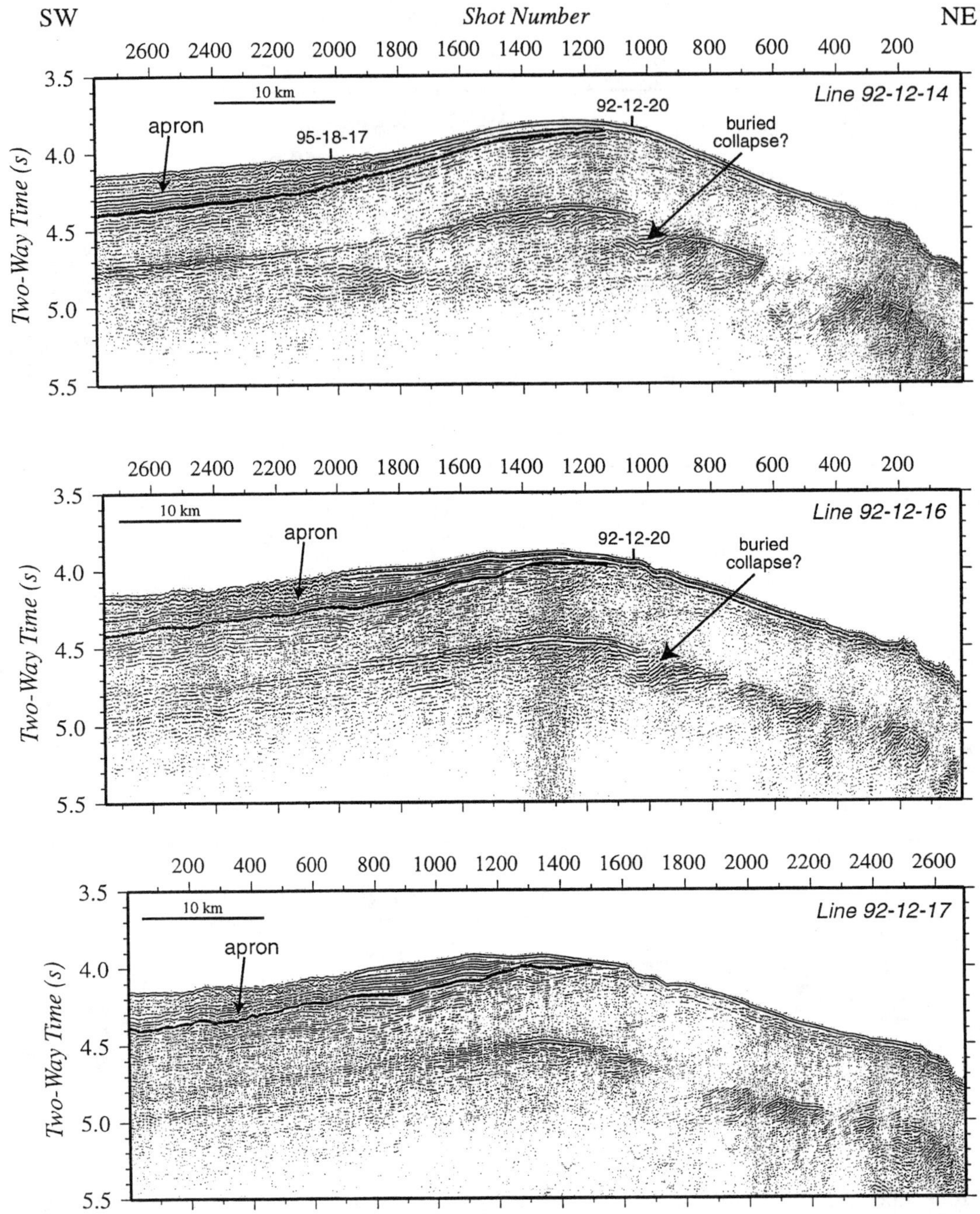

not significantly different from those at Sites 995 and 997, which show a strong BSR. Finally, the gas present at depth may have been produced relatively recently by microbial processes and not yet reached the base of the stability zone. It is quite likely that the true explanation involves a complex interaction of all three of these processes.

THE BLAKE RIDGE COLLAPSE: THE VALVE ON A PRESSURE COOKER?

The Blake Ridge collapse is a large depression on the crest of the Blake Ridge about 10 km northwest of the Leg 164 drilling transect [*Dillon et al.*, 1997]. The structure resembles a large sinkhole: if the collapse is bounded by

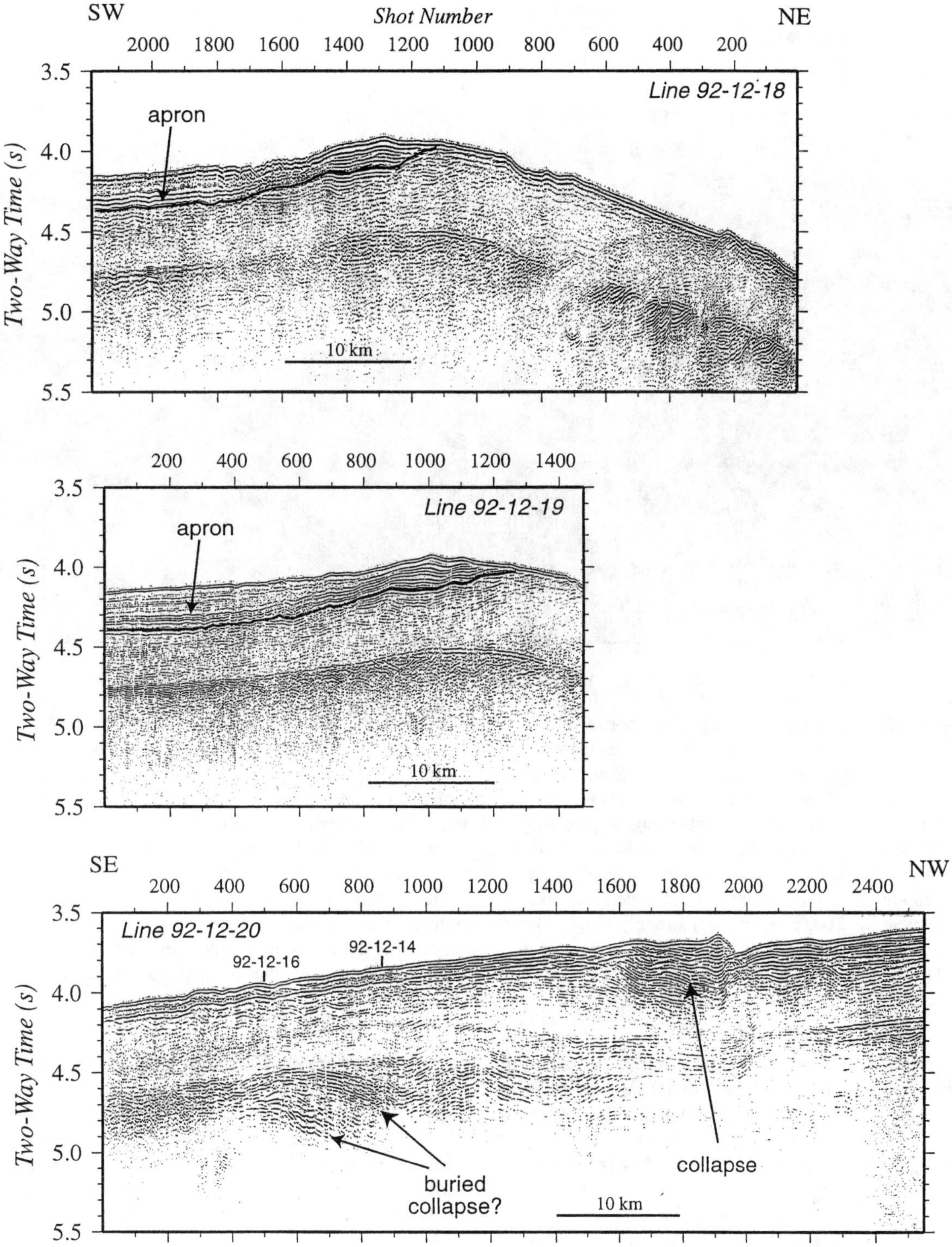

large normal faults, its formation must be accommodated by a significant volume loss at depth, because the Blake Ridge has not experienced regional extension. Two observations suggest a relationship between the collapse structure and the hydrate/gas system: (1) the feature occurs on the crest of the ridge, where gas and hydrate concentrations are greatest, and (2) the bright BSR and sub-BSR reflectivity characteristic of the ridge crest disappears beneath the collapse (Figs. 11, 12; [*Taylor et al.*, 1999]). This latter observation constitutes strong

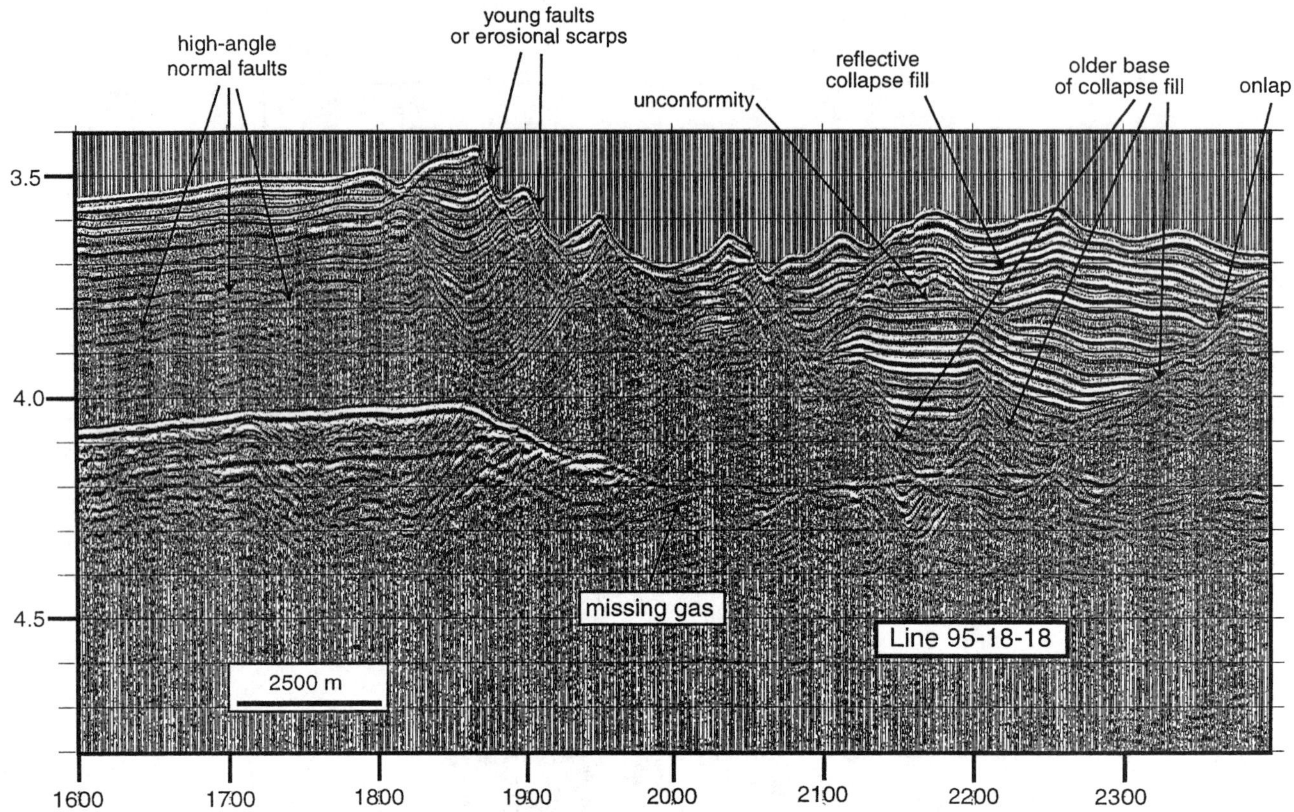

Figure 12. Data along the northeastern portion of Line 95-18-18, across the Blake Ridge collapse structure. The bright BSR and sub-BSR gas reflectivity becomes markedly weaker beneath the collapse, suggesting that methane has been released through the collapse structure. Some faults cut the youngest strata on the ridge, indicating recent activity in the collapse. However, reflective sediments within much of the collapse do not resemble sediments outside the collapse: their reflectance is much higher, and they lack the pervasive high-angle normal faults that dissect Blake Ridge sediments older than about 1.4 Ma. These observations, together with onlapping stratal geometries on the northeast bounding structure, a clear unconformity within the reflective sediments, an apparently faulted surface separating reflective sediments from underlying less-reflective units, and pinching out of strata against the underlying base (near SP 2100 at 3.9 s time) strongly suggest that the brightly reflective strata represent young collapse fill, rather than down-faulted equivalents of the older sediments present at similar sub-seafloor depths elsewhere on the Blake Ridge. This inference is supported by Line obs2b (Fig. 13), which demonstrates that the collapse fill is younger than about 1.4 Ma.

circumstantial evidence for the escape of methane gas from beneath the collapse: the lack of strong reflectivity indicates that little gas is present, and it is reasonable to assume that the gas-charged zone seen elsewhere on the ridge crest also existed here prior to the formation of the collapse. If the missing methane escaped into the water column, the Blake Ridge collapse may be a "smoking gun" of methane exchange between a large hydrate/gas reservoir and the ocean/atmosphere system. *Dillon et al.* [this volume] review seismic data across the collapse and suggest that it formed by a recent, catastrophic event involving gas overpressure, sediment expulsion, normal faulting, and escape of methane into the ocean/atmosphere system.

In this section I present a somewhat different view of the collapse structure. Although I agree with Dillon et al. that the collapse structure is intimately linked to the hydrate/gas system and that methane has escaped into the ocean/atmosphere system, I interpret the seismic structure of the collapse and surrounding ridge as evidence that the collapse structure is a long-lived, periodically reactivated structure.

The sediments inside the collapse are usually portrayed (often in line drawings) as down-faulted equivalents of strata outside the collapse, in a typical normal-fault geometry. The southwest bounding structure of the collapse intersects the BSR at the point that free gas disappears, suggesting that this structure provided the

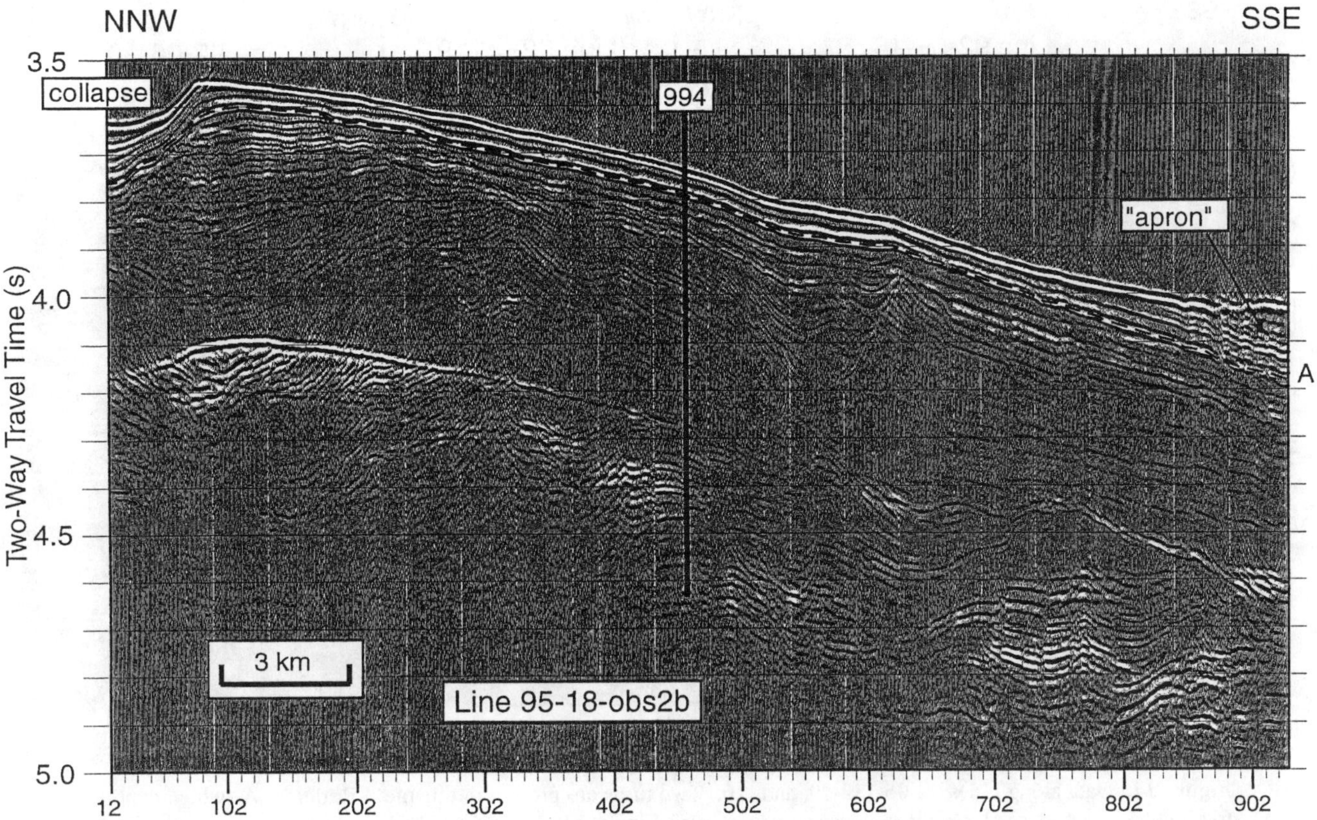

Figure 13. Data along Line 95-18-obs2b, which extends from the "wedge" northward into the collapse. Reflector A (0.8 Ma; see Fig. 3) can be traced continuously from beneath the wedge, through Site 994, and into the highly reflective sediments of the collapse structure. This demonstrates that the reflective strata within the collapse (e.g., Fig. 12) are time-correlative to the young wedge sediments and were deposited at a similarly high deposition rate, filling a collapse depression that was formed during an earlier (~1.4 Ma?) event.

pathway for the escape of gas (Fig. 12). It is important to note, however, this structure differs from typical normal faults in several key respects. The bounding structure on the SW side of the collapse has a dip of only 8-9° (e.g., Fig. 12), though it appears much greater in the highly vertically exaggerated sections typically shown. True normal faults outside the collapse appear nearly vertical in the sections (e.g., Figs. 3, 12). This surface is thus unlikely to have formed by Mohr-Coulomb brittle failure but rather probably marks either (1) the detachment surface of a slide, slump, or fluidized decollement, (2) an unconformity, or (3) the bounding surface of a sedimentary feature. Moreover, strata inside the main part of the collapse do not resemble strata at similar depths outside the collapse, so that it is difficult to correlate strata across the bounding structure of the collapse. This suggests that, contrary to the model of *Dillon et al.* [this volume], these high-amplitude strata are not the "down-faulted" equivalents of strata outside the collapse, but rather represent young sediments deposited at a high deposition rate. Four lines of evidence support this inference.

First, on some seismic lines the prominent 800-kyr Reflector A can be traced continuously into the collapse structure (Figs. 13, 14), and sediments above Reflector A thicken substantially within the collapse, indicating an expanded sedimentary section there. The sediments within the collapse are thus time-correlative to the apron of young sediments south of the collapse (Fig. 3, 9, 13) and were deposited at a similarly high deposition rate. At its deepest point in Figure 14, reflector A implies a deposition rate of about 300 m/my.

Second, the Pliocene-Miocene section on the Blake Ridge outside the collapse is pervasively faulted ([*Wood and Ruppel*, 2000]; Fig. 3). Strata younger than about 800 kyr, in contrast, are generally not cut by high-angle normal faults. Brightly reflective strata inside the collapse are also unfaulted, consistent with the similarly young age proposed here.

Third, the collapse-fill sediments often thin toward and show onlapping stratal geometries onto the bounding structure of the collapse (e.g., Figs. 12, 14, 15). The combination of faultward-thinning and onlap indicates

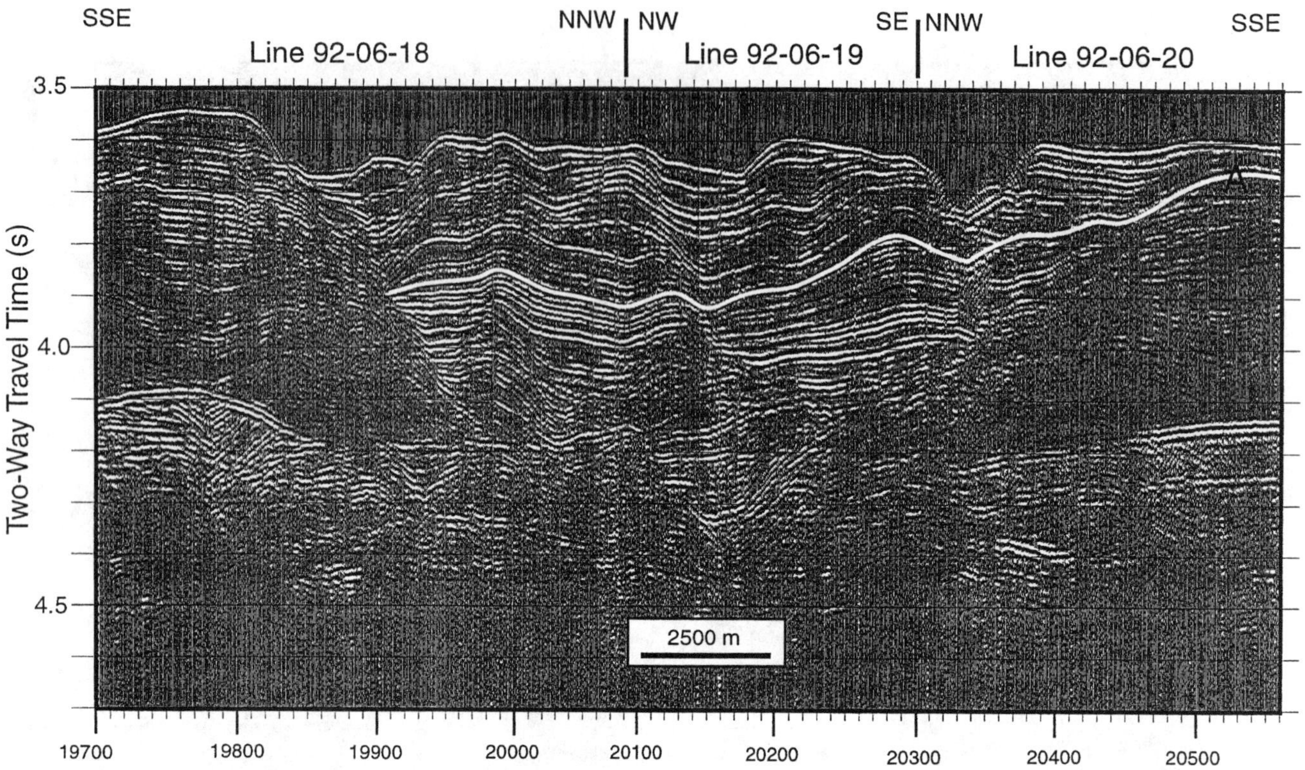

Figure 14. Data along Lines 92-06-18, -19, and -20. Two turns are present in this plot. Reflector A (which continues to Site 995) can be traced into the collapse structure, as on Line obs2b (Fig. 13). Much of the reflective collapse fill on these sections lies beneath Reflector A. Note the pinch-out of strata against the bounding structure at 3.9 s, SP 19900.

clearly that these stratal geometries are not drag folds. Similar faultward-thinning stratal geometries are also seen well within the collapse (e.g., Fig. 12). Onlap occurs on both sides of the collapse (e.g., Figs. 12, 15), indicating that the bounding structures predate the strata. Line 92-12-09 shows that strata thicken into the collapse at both edges, again indicating that much of the stratigraphic section post-dates the formation of the collapse (Fig. 16). Local thickening of strata within the collapse structure (e.g., between SP's 2900 and 3000 of Line 92-12-09; Fig. 16) is also consistent with post-collapse infilling.

Fourth, the high reflectivity of sediments within the collapse is unlike the generally lower reflectivity of Pliocene and Miocene sediments at similar subseafloor depths outside the collapse -- with the notable exception of sediments younger than ~800 kyr. This is a common observation on all lines crossing the collapse structure and is shown particularly clearly on Line 95-18-18 (Fig. 12).

The high reflectance is particularly difficult to explain if the bright intra-collapse sediments are not considered young collapse fill. There is no plausible mechanism by which formation of a collapse would uniformly enhance the reflectance of its hanging-wall section. *Dillon et al.* [this volume] propose "unblanking" of impedance contrasts during formation of the collapse. However, this theory is implausible on two grounds: Not only is blanking unlikely to occur on the Blake Ridge in the first place (as discussed above), but the mechanism by which sediments would later be pervasively and uniformly "unblanked" is even more unclear. The "structural disruption" proposed by Dillon et al. to destroy blanking is unsupported by the seismic data: the collapse-fill sediments are intact, while the surrounding sediments are disrupted by pervasive high-angle normal faults. Other scenarios to explain the high reflectance can be imagined but are equally implausible: (1) Metastable free gas in the sediments, possibly injected into shallower sediments during outgassing event, might enhance impedance contrasts. This is highly unlikely, however, since free gas would not be stable in the shallow sediments above the BSR. (2) Layers with locally high hydrate concentrations interbedded with low-hydrate layers might be imagined. However, nowhere has a reflection from a hydrate-rich layer above the BSR been conclusively observed; nor would enhancement of impedance contrasts due to concentrated hydrate be expected to form such a uniform reflectance over such a large area. (3) Finally, a serendipitous presence of pre-existing reflective units in the location of the future collapse (which to my knowledge no one has proposed) would strain credulity. The implausibility of these alternative explanations for the high reflectance, coupled with the onlapping, outward-thinning geometries of the strata and continuity of a clear basal reflector to thinner, reflective sediments outside the collapse, leaves little doubt that these strata are relatively young collapse-fill sediments. The high reflectance of these strata is then easily explained as the consequence of the lithologic variability that characterizes the < 1 Ma sediments of the Blake Ridge [*Holbrook et al.*, 1996].

Other evidence suggests a long-lived history of collapse, possible episodic reactivation, and the potential existence of older, similar structures elsewhere on the Blake Ridge. First, differential tilting of strata is observed adjacent to the main bounding structure on several lines (e.g., Line 92-12-06, Fig. 16), suggesting a long-lived history of tilting. Second, apparent unconformities within reflective strata inside collapse suggest discrete events (e.g., Fig. 12). Third, the base of the reflective sediments filling the collapse is itself a discrete surface, against which the overlying strata show onlapping geometries, implying that a collapse-forming event predated the deposition of those strata. This is particularly clear on Line 95-18-18 (Fig. 12). Finally, there is evidence elsewhere on the Blake Ridge for structures similar in appearance to the collapse (e.g., Fig. 11; *Dillon et al.*, this volume). These features are characterized by relatively steeply dipping beds, rapid lateral changes in dip, and upward shallowing of dips — all features that are found, perhaps uniquely, in the present-day collapse structure. Although the existing seismic coverage is too sparse to definitively determine the three-dimensional structure of those features, they may represent earlier, now-buried collapse structures that formed elsewhere on the ridge. If so, the Blake Ridge collapse structure currently exposed on the seafloor may not be a unique feature, but rather the latest in a long history of similar structures that form on the southwestward-migrating ridge crest.

The stratigraphic evidence thus overwhelmingly indicates that the BRCS cannot have formed in a single, recent catastrophic event, but rather represents a long-lived structure with multiple outgassing events and a long history of sedimentation. As *Dillon et al.* [this volume] point out, truncated strata at the edge of the collapse's boundary suggest a very recent faulting episode, but this interpretation must be tempered by two caveats: (1) if the truncated strata indicate faults, then the recent faulting episode was not the only episode, and (2) given the quality of the present data, it is equally possible to interpret the truncated strata as due to erosional scours. In any case, it seems possible that the collapse structure, and perhaps other, now-buried structures like it, is an integral part of the dynamic hydrate/gas sytem on the Blake Ridge, episodically releasing gas when overpressures develop, much like the valve on a pressure cooker.

The existing evidence thus suggests that the Blake Ridge collapse structure has provided a link between a large hydrate/gas reservoir and the ocean/atmosphere system over a much longer period of time than previously recognized. Moreover, the structure itself is likely not unique: other collapses probably existed on the crest of the Blake Ridge in the past. A complete understanding of the origin and evolution of this important structure requires better, more densely spaced reflection data, as well as cores and drill holes in the collapse itself.

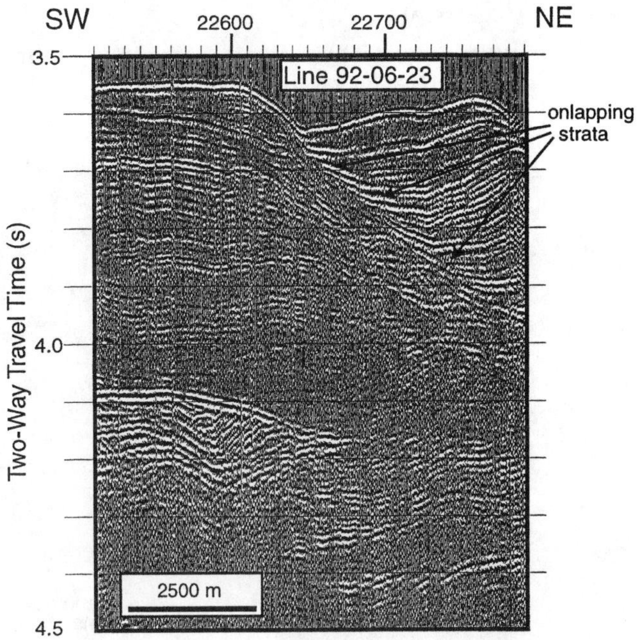

Figure 15. Data along Line 92-06-23, crossing the boundary of the collapse structure. Onlapping geometries and thinning of strata toward the bounding fault are clearly visible, indicating that the collapse-fill strata post-date and were deposited onto the bounding structure.

CONCLUSIONS

1. Both hydrate and gas concentrations increase toward the crest of the ridge from Site 994 to 995/997. Estimates of the concentration of hydrate from VSP velocities agree well with those of other proxies: 2-3% bulk hydrate (5-6% of porosity) exists in the 250 m above the BSR, assuming the hydrate-as-cement rock physics model. Estimates using the hydrate-as-pore-fill model produce significantly higher concentrations (4-7% bulk hydrate) than those from resistivity logs and pore water chemistry, suggesting that hydrate acts to cement grains in the fine-grained sediments of the Blake Ridge.

2. Free gas concentrations at Site 997 average about 1% bulk (assuming homogeneous gas distribution) or 10% bulk (assuming patchy gas distribution). Insufficient independent evidence exists to determine which of these models is correct.

3. A wedge of concentrated hydrate just above the BSR

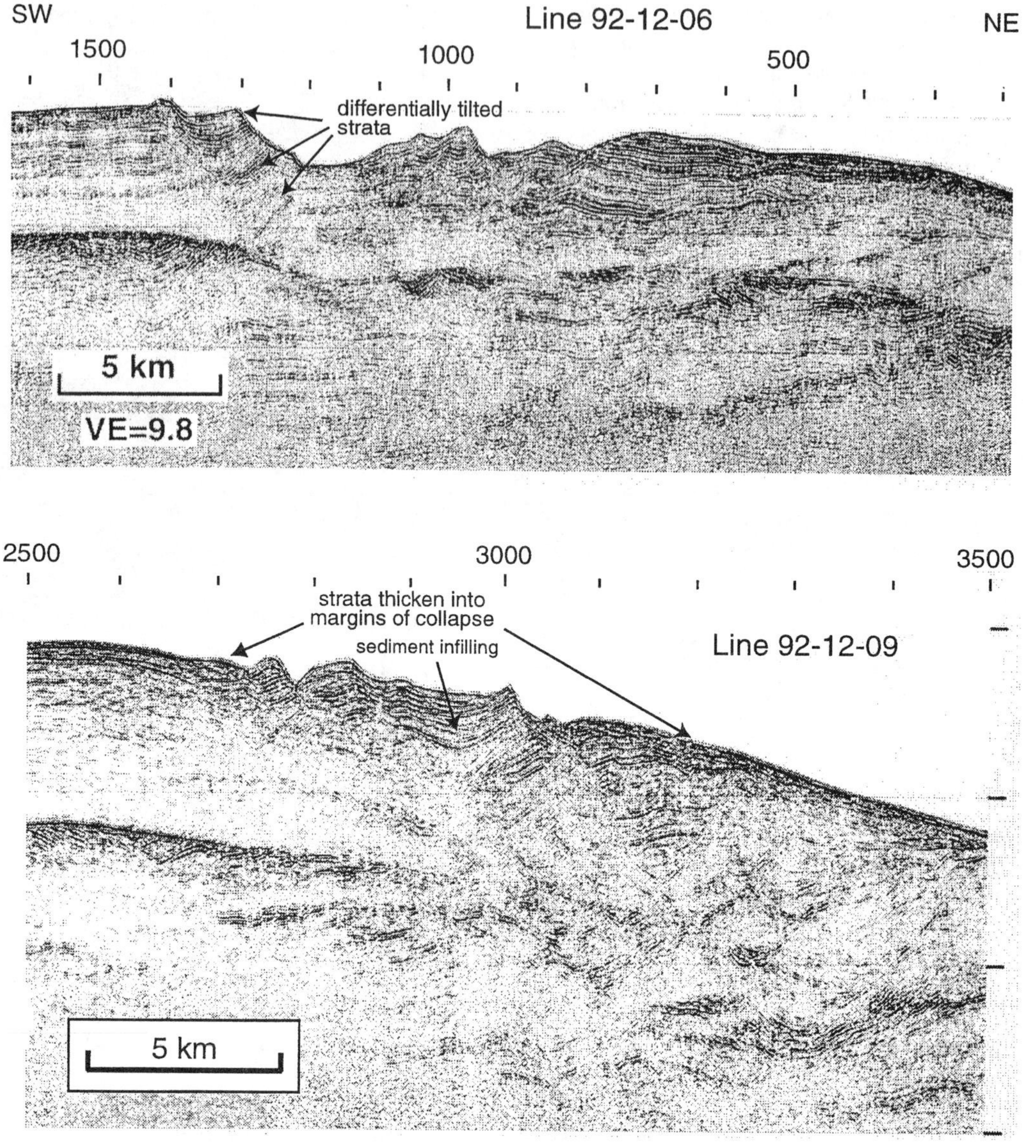

Figure 16. Processed data from *Taylor et al.* [1999] on Lines 92-12-06 and -09. Differential tilts of strata against the southwestern bounding structure on Line 06 are clearly visible, suggesting a long-lived history of offset on that fault. Line 09 shows clearly that young, reflective strata thicken into the margins of the collapse on both sides, consistent with their interpreted origin as young collapse fill.

at Site 997 is independently indicated by detailed VSP velocities and waveform inversion of OBH data. The wedge probably reflects recent (or current) active upwelling of gas into the hydrate stability zone. The wedge is not universally present, however, on the Blake Ridge.

4. Anisotropy (8%) of marine sediments causes significant mismatch between velocities (and BSR depth) determined by VSPs and wide-angle seismic data and must be considered when interpreting hydrate concentrations from seismic velocity data.

5. Amplitude blanking is not required to explain low reflectance above the BSR on the Blake Ridge. Indeed, zones of equally low reflectance below the BSR demonstrate conclusively that hydrate is not required for low reflectance on the Blake Ridge.

6. The Blake Ridge is characterized by a dynamic, thick (~300 m) free gas system underlying the hydrate stability zone. Free gas is much more widespread than the BSR on the Blake Ridge; even areas lacking a BSR typically are underlain a thick, complex gas-charged zone. A strong BSR beneath an apron of young sediments demonstrates

upward mobility of free gas on the sub-m.y. time scale. Free gas is trapped by permeability contrasts between strata, by hydrate at the BSR, and locally by high-angle faults.

7. Strong circumstantial evidence suggest that methane gas has been expelled from beneath the Blake Ridge collapse structure. Contrary to the "catastrophic collapse" model, stratigraphic evidence strongly suggests that the collapse structure is a long-lived, periodically reactivated structure. Brightly reflective strata within the collapse are young (< ~1.4 Ma) sediments deposited at high deposition rate (200-300 m/my) onto an older surface.

Acknowledgments. Although the interpretations presented in this paper are my own, many of the data and ideas owe much to the hard work and generosity of my colleagues. I thank Bill Dillon for stimulating my interest in hydrates back in 1989 and nurturing it ever since, and Charlie Paull for valuable support and leadership on cruises to the Blake Ridge in 1992 and 1995. The VSP data are due to the hard work of Hartley Hoskins, Warren Wood, Ralph Stephen, and the crew of the *Joides Resolution.* I thank Carolyn Ruppel and Jerry Dickens for numerous conversations regarding the dynamics and mass balance of hydrate systems, and for their many contributions to a drilling proposal that stimulated many of the ideas here. Mike Taylor and Bill Dillon provided much of the single-channel seismic data presented here. Dan Lizarralde provided the VSP inversion script and data plotting routines, and Jun Korenaga provided the waveform inversion script. I thank Paul Heller, Barbara John, Harm van Avendonk, and Lloyd Keigwin for important conversations. This work was supported by NSF grant OCE-9504610 and by JOI/USSAC Subaward 7-95.

REFERENCES

Collett, T.S., and J. Ladd, Detection of gas hydrate with downhole logs and assessment of gas hydrate concentrations (saturations) and gas volumes on the Blake Ridge with electrical resistivity log data, in *Proc. ODP, Sci. Results, 164*, edited by C. K. Paull, R. Matsumoto, P. J. Wallace and W. P. Dillon, 179–191, Ocean Drilling Program, College Station, TX, 2000.

Dickens, G.R., C.K. Paull, P. Wallace, and ODP Leg 164 Scientific Party, Direct measurement of in situ methane quantities in a large gas-hydrate reservoir, *Nature*, 385, 426–428, 1997.

Dillon, W.P., W.W. Danforth, D.R. Hutchinson, R.M. Drury, M.H. Taylor, and J.S. Booth, Evidence for faulting related to dissociation of gas hydrate and release of methane off the southeastern United States, *J. Geol. Soc. London*, 1997.

Ecker, C., J. Dvorkin, and A. Nur, Estimating the amount of gas hydrate and free gas from marine seismic data, *Geophysics*, 65, 565–573, 2000.

Egeberg, P.K., and G.R. Dickens, Thermodynamic and pore water halogen constraints on gas hydrate distribution at ODP Site 997 (Blake Ridge), *Chemical Geology*, 153, 53–79, 1999.

Guerin, G., D. Goldberg, and A. Meltser, Characterization of in situ elastic properties of gas hydrate-bearing sediments on the Blake Ridge, *J. Geophys. Res.*, 104, 17,781–17,795, 1999.

Helgerud, M.B., J. Dvorkin, A. Nur, A. Sakai, and T. Collett, Elastic-wave velocity in marine sediments with gas hydrates: Effective medium modeling, *Geophysical Research Letters*, 26, 2021–2024, 1999.

Holbrook, W.S., H. Hoskins, W.T. Wood, R.A. Stephen, D. Lizarralde, and Leg 164 Science Party, Methane hydrate and free gas on the Blake Ridge from vertical seismic profiling, *Science*, 273, 1840–1843, 1996.

Hovland, M., J.W. Gallagher, M.B. Clennell, and K. Lekvam, Gas hydrate and free gas volumes in marine sediments; examples from the Niger Delta front, Marine and Petroleum *Geology*, 14, 245–255, 1997.

Hyndman, R.D., and E.E. Davis, A mechanism for the formation of methane hydrate and sea floor bottom simulating reflectors by vertical fluid expulsion, *J. Geophys. Res.*, 97, 7025–7041, 1992.

Hyndman, R.D., and G.D. Spence, A seismic study of methane hydrate marine bottom simulating reflectors, *J. Geophys. Res.*, 97, 6683–6698, 1992.

Katzman, R., W.S. Holbrook, and C.K. Paull, Combined vertical-incidence and wide-angle seismic study of a gas hydrate zone, Blake Ridge, *J. Geophys. Res.*, 99, 17,975–17,995, 1994.

Keigwin, L.D., D. Rio, G.D. Acton, et al., *Proceedings of the Ocean Drilling Program, Initial Reports*, vol. 172, 949 pp., Ocean Drilling Program, College Station, TX, 1998.

Kloska, M.H., Quantitative measurements of methane hydrate and free gas from full waveform inversion and seismic inversions, Blake Ridge, offshore South Carolina, M.S. Thesis, University of Wyoming, City, 1999.

Korenaga, J., W.S. Holbrook, S.C. Singh, and T.A. Minshull, Natural gas hydrates on the southeast U.S. margin: Constraints from full waveform and traveltime inversions of wide-angle seismic data, *J. Geophys. Res.*, 102, 15,345–15,365, 1997.

Lee, M.W., D.R. Hutchinson, W.F. Agena, W.P. Dillon, J.J. Miller, and B.A. Swift, Seismic character of gas hydrates on the southeastern U.S. continental margin, *Marine Geophysical Researches*, 16, 163–184, 1994.

Lee, M.W., D.R. Hutchinson, W.P. Dillon, J.J. Miller, W.F. Agena, and B.A. Swift, Method of estimating the amount of in situ gas hydrates in deep marine sediments, *Marine and Petroleum Geology*, 10, 493–506, 1993.

Lizarralde, D., and S. Swift, Smooth inversion of VSP traveltime data, *Geophysics*, 64, 659–661, 1999.

Okada, H., Neogene and Quaternary calcareous nannofossils from the Blake Ridge, Sites 994, 995, and 997, in *Proc. ODP, Sci. Results*, 164, edited by C. K. Paull, R. Matsumoto, P. J. Wallace and W. P. Dillon, 331–341, Ocean Drilling Program, College Station, TX, 2000.

Paull, C.K., and R. Matsumoto, Leg 164 Overview, in *Proc. ODP, Sci. Results*, 164, edited by C. K. Paull, R. Matsumoto, P. J. Wallace and W. P. Dillon, 3–10, Ocean Drilling Program, College Station, TX, 2000.

Paull, C.K., R. Matsumoto, P. Wallace, et al., *Proceedings of the Ocean Drilling Program, Initial Reports*, vol. 164, 623 pp., Ocean Drilling Program, College Station, TX, 1996.

Paull, C.K., R. Matsumoto, P.J. Wallace, and W.P. Dillon, in *Proc. ODP, Sci. Results*, 164, pp., Ocean Drilling Program, College Station, TX, 2000.

Rowe, M.M., and J.F. Gettrust, Fine structure of methane hydrate-bearing sediments on the Blake Outer Ridge as determined from deep-tow multichannel seismic data, *J. Geophys. Res.*, 98, 463–473, 1993.

Shipley, T.H., M.H. Houston, R.T. Buffler, F.J. Shaub, K.J. McMillen, J.W. Ladd, and J.L. Worzel, Seismic reflection evidence for widespread occurrence of possible gas-hydrate horizons on continental slopes and rises, *Am. Assoc. Petroleum Geol. Bull.*, 63, 2204–2213, 1979.

Taylor, M., W. Dillon, C. Anton, and W. Danforth, Seismic-reflection surveys of the Blake Ridge, R/V Cape Hatteras 1992 and 1995: Data acquisition, navigation and processing, U.S. Geological Survey Open-File Report, 99-372, 1999.

Tinivella, U., and F. Accaino, Compressional velocity structure and Poisson's ratio in marine sediments with gas hydrate and free gas by inversion of reflected and refracted seismic data (South Shetland Islands, Antarctica), *Marine Geology*, 164, 13–27, 2000.

Wellsbury, P., K. Goodman, T. Barth, B. Cragg, S. Barnes, and R.J. Parkes, A deep bacterial biosphere fuelled by low-temperature acetate generation during burial, *Nature*, 388, 573–576, 1997.

Wood, W.T., and C. Ruppel, Seismic and thermal investigations of the Blake Ridge gas hydrate area: A synthesis, in *Proc. ODP, Sci. Results*, 164, edited by C. K. Paull, R. Matsumoto, P. Wallace and W. P. Dillon, 253–264, Ocean Drilling Program, College Station, Texas, 2000.

Wood, W.T., P.L. Stoffa, and T.H. Shipley, Quantitative detection of methane hydrate through high-resolution seismic velocity analysis, *J. Geophys. Res.*, 99, 9681–9695, 1994.

Xu, W., and C. Ruppel, Predicting the occurrence, distribution, and evolution of methane gas hydrate in porous marine sediments, *J. Geophys. Res.*, 104, 5081–5096, 1999.

W. Steven Holbrook, Deptartment of Geology and Geophysics, University of Wyoming, Laramie, WY 82071-3006

Gas Hydrates Along the Peru and Middle America Trench Systems

Ingo A. Pecher

Institute for Geophysics, University of Texas, Austin, TX

Nina Kukowski

GeoForschungsZentrum Potsdam, Potsdam, Germany

César R. Ranero

GEOMAR Forschungszentrum, Universität Kiel, Kiel, Germany

Roland von Huene

GEOMAR Forschungszentrum, Universität Kiel, Kiel, Germany

The convergent margins of Peru and Middle America are dominantly shaped by subduction erosion, which causes tectonic subsidence of large areas. Both margins are largely non-accretionary and most of the incoming trench sediment is being subducted. Gas hydrates are ubiquitous across the Peru margin with its high organic carbon content. Strong bottom simulating reflections (BSRs) are present on the lower slope. A regional lack of BSRs further upslope in Lima Basin may be explained by rapid subsidence leading to a downward movement of the base of gas hydrate stability and absorption of free gas into the gas hydrate stability zone. BSRs in small areas in Lima Basin are perhaps supported by high rates of methane supply offsetting the absorption of gas. Off Costa Rica, seamount subduction promotes slope failure. A general lack of BSRs in sediment sections that were destroyed by slumping suggests no or a very slow re-formation of BSRs after slumping. Above a subducted seamount, sediments containing a BSR remain intact despite inferred strong uplift. The associated pressure decrease should lead to dissociation of gas hydrates, generation of gas, and overpressuring. We speculate that the gas may escape through faults above the BSR. Thermo-hydraulic modeling based mainly on BSR heat flow indicates high frictional heating in the Peru subduction zone causing a landward increase of the

Natural Gas Hydrates: Occurrence, Distribution, and Detection
Geophysical Monograph 124

thermal gradient. Off Costa Rica, a shallowing of BSR depth reflects locally elevated heat flow coinciding with expected expulsion of fluids at the landward boundary of the frontal prism.

INTRODUCTION

The tectonic environment affects several key parameters for the evolution of gas hydrate systems. The Blake Ridge offshore the U.S. east coast, as a prime example of a gas hydrate bearing passive margin, constitutes a stable environment with relatively low fluid flow [*Egeberg and Dickens*, 1999; *Paull et al.*, 1996]. Rapid tectonism along convergent margins leads to high fluid flow rates and introduces other variables of significance for gas hydrate formation, such as vertical movement of the seafloor, mass wasting, and potentially high rates of terrigenous sediment input from adjacent mountain ranges. The Cascadia margin, containing one of the best studied convergent margin gas hydrate systems, is flooded by sediments from a large continental source area, has a large accretionary wedge, and has a relatively low convergence rate. The margins off Peru and at the Middle America trench system (Figure 1) are generally non-accretionary margins undergoing subduction erosion. They are locally affected by subducting ridges and seamounts, experience rapid convergence, and have a smaller terrigeneous sediment source than for example Cascadia. Most of the sediment on the lower plate is being subducted leading to a relatively small frontal sedimentary prism. The sediment in the frontal prism is mostly debris from the slope carried into the trench or intercepted as it crosses the base of the slope. An important difference between the Peru and the Middle America margins is their organic carbon content, which, as a source for bacteriological methanogenesis is an obvious key factor for gas hydrate occurrences. Organic carbon content is exceptionally high off Peru [*Suess et al.*, 1988] and relatively low off Costa Rica [*Kimura et al.*, 1997]. In this paper, we will compare gas hydrate occurrences off Peru and Middle America as examples for erosional convergent margins. We will in particular focus on the role of vertical tectonism, the link between high fluid flow and gas hydrates, and the effect of slope failure on gas hydrate systems.

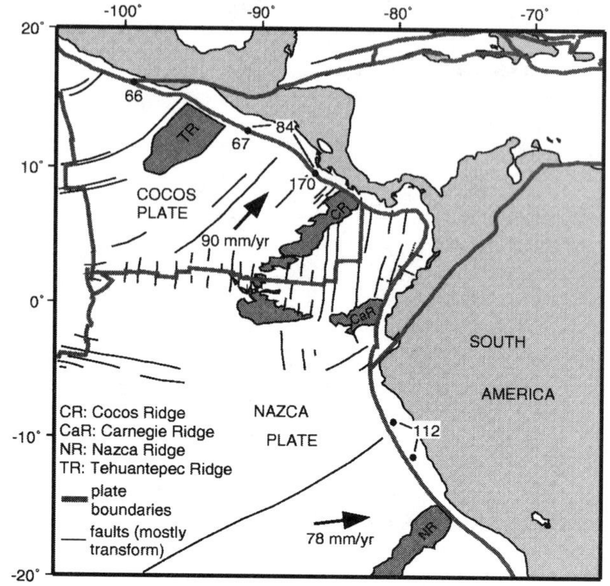

Figure 1. Tectonic setting, Pacific margin of South and Middle America, compiled after *Coffin et al.* [1998]. Arrows with numbers: convergence directions with rates. 67, 66, 84: DSDP legs; 112, 170: ODP legs.

THE PERU MARGIN

Geologic Setting

The Nazca Plate is subducted at the Peru margin beneath the South American continent at a rate of about 78 mm/yr [*DeMets et al.*, 1990]. The post-Miocene tectonic history of this margin between 8° and 15° S is dominated by an oblique subduction of the Nazca Ridge. The northeastern edge of the ridge entered the subduction zone at about 8° S some 8 Ma [*Cande*, 1985]. Its crest swept north to south along the margin - currently it is located at 15° S. Seismic profiles across the margin from 15° S to the north therefore display different stages of the margin's development after subduction of the Nazca Ridge [*von Huene et al.*, 1996]. Ridge subduction causes uplift of the margin and erosion of the accretionary prism. After subduction of the ridge crest, the margin subsides and most of the incoming material is accreted to build a new frontal sedimentary prism. The proportion of subducted compared to accreted sediment gradually increases until the margin enters a steady-state equilibrium where all of the incoming sediment is ultimately subducted, although some of it may be accreted first. The Peru margin was studied during Ocean Drilling Program (ODP) Leg 112 along two corridors at 12° S and 9° S. A new accretionary prism is still being formed at 12° S, where the crest of Nazca Ridge was subducted about 4 Ma. At 9° S, the crest of the ridge was subducted 6.5 Ma and by far most of the incoming sediment seems to ultimately enter the subduction zone. The volume of fluids expelled from subducted sediments is therefore predicted to be higher than at 12° S [*von Huene et al.*, 1996].

The Peru margin is in an oceanic upwelling region with high biologic productivity, which is the cause for the high organic carbon content in the post-Miocene sediments of ODP Leg 112 cores [*Suess et al.*, 1988]. Total organic carbon content typically is over 5% in the upslope Lima Basin at 12° S and 6-8% in Yaquina Basin at 9° S. On the lower slope, it is generally 6-8% (up to 12%) at 12° S and 4% at 9° S. Only in the accretionary prism, which may in part consist of hemipelagic material transported along the trench, it is estimated at less than 1% at 12° S and 2-4% at 9° S [*von Huene and Pecher*, 1999]. The generally high organic carbon content favors methane generation and hence, gas hydrate formation.

Gas Hydrates During Leg ODP 112

Strong bottom simulating reflections (BSRs) are present along much of the Peruvian lower slope (Figure 2). Penetration of the BSR was considered a drilling hazard during Leg 112 and the boreholes were therefore located away from BSRs. Nevertheless, solid gas hydrates were recovered at Sites 685 and 688. Low chlorinity as indirect evidence for gas hydrate occurrences was observed at two other sites. Methane constitutes about 99% of hydrocarbons in both pore waters and gas hydrates, indicating a mainly microbial origin of methane [*Kvenvolden et al.*, 1990; *Kvenvolden and Kastner*, 1990]. Low $\delta\,^{13}C$ values of -65.0 and -59.6‰ PDB in gas hydrate samples at Sites 685 and 688, respectively, support this conclusion [*Kvenvolden and Kastner*, 1990].

Nature of BSRs

The BSR along Peru Line 1018 (Figure 2) displays a mostly strong negative reflectivity suggesting a drop in compressional wave velocity (V_p). Conventional velocity analysis during pre-stack depth migration did not indicate a thick (i.e., for the given data, over ~50 m) low-velocity zone. A thinner layer causes interferences of reflections from its top and base and seismic waveforms therefore contain information about its velocity structure. Both post-stack waveform forward-modeling [*Miller et al.*, 1991] and pre-stack waveform inversion suggest that the strong BSR along this line is caused by a thin (6-18 m; *Pecher et al.*, [1996a]) but pronounced low-velocity layer beneath the base of gas hydrate stability (BGHS). This indicates that the main contribution for the BSR comes from a thin free-gas layer causing low V_p [*Domenico*, 1977] rather than from elevated V_p caused by gas hydrates above the BGHS. The BSR weakens significantly close to its landward termination. Waveform inversion indicates that the BSR in this area is mainly caused by elevated V_p above the BGHS [*Pecher et al.*, 1996b]. This suggests the presence of gas hydrates above the BGHS with only very little free gas, if any, trapped beneath it (Figure 2).

We speculate that gas at this location perhaps escapes through normal faults that appear to cross the BGHS. It is not clear how methane migrates along faults through the gas hydrate stability zone (GHSZ) without being transformed to gas hydrate. Suggested mechanisms include local thermal anomalies within fault zones (too warm to form gas hydrates in the faults) and gas hydrate seals on the fault walls that shut off water supply for gas hydrate formation (for a discussion, see e.g., *Dillon et al.* [1997]). Gas venting at the seafloor within the GHSZ has been observed close to faults at other locations, in particular in the Gulf of Mexico [*MacDonald et al.*, 1994].

The gas zone beneath BSRs off Peru is considerably thinner than the over 250-m thick zone on the Blake Ridge [*Holbrook et al.*, 1996]. We caution that waveform studies depend on accurate large-scale velocity models. Conventional velocity analysis, however, was based on relatively dense and continuous reflectors such that a thick low-velocity layer should have been detected. The gas zone beneath the Blake Ridge appears to be marked by a highly reflective zone probably caused by slight variations of gas content across strata at a generally low gas concentration (where V_p is very sensitive to gas content). Such a high-reflectivity band is not observed off Peru although it may be obscured by reflections from stratigraphic boundaries. It is not clear what may cause the apparent difference between the Blake Ridge and the Peru margin in the thickness of the gas zones beneath the BGHS. For this, we still need to improve our understanding of the "plumbing system" of gas and fluids beneath gas hydrates. Key parameters that we think may affect the thickness of the gas zone include sediment permeability, fluid flux, fluid velocity, and in-situ generation of methane. We caution that little is known to date about the mobility of gas and the mechanisms of gas migration in sediments. The permeability of sediments for free gas, e.g., may be significantly different from that for fluids [*Honarpour et al.*, 1986].

Distribution of BSRs

BSRs are arguably the most widely used proxies for gas hydrates in marine sediments. Gas hydrates have been encountered, however, in regions without BSRs. A better understanding of the factors that control BSR formation in addition to a presence of gas hydrates is therefore essential for improved global quantification of hydrates. The Peru margin is a good location to study BSR occurrences because gas hydrates probably are ubiquitous, however, BSRs are absent over large parts of the margin.

BSRs are absent in the accretionary prism at 12° S. This is explained by the predicted low organic carbon content of

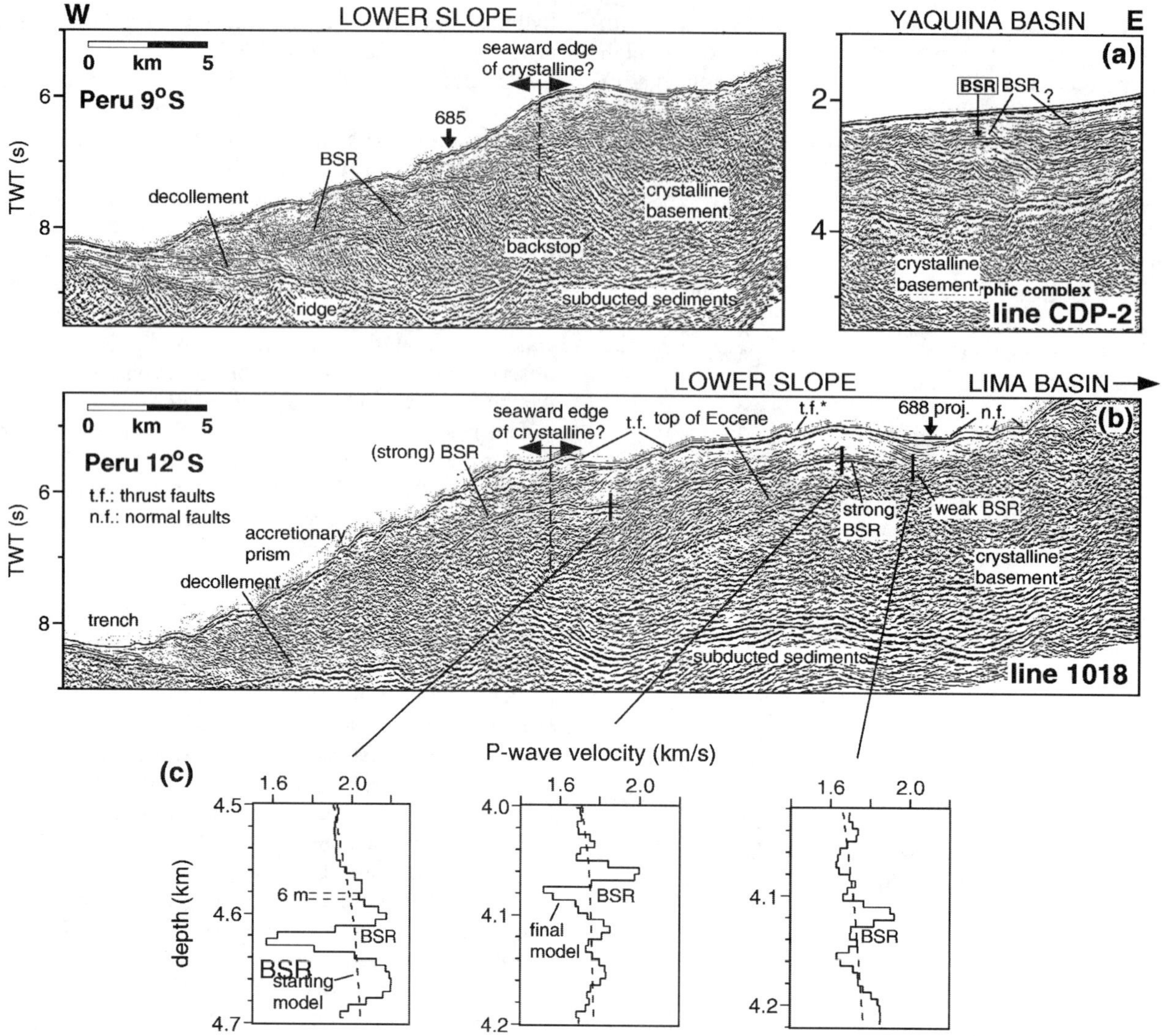

Figure 2. (a) Pre-stack depth migrated section of CDP-2, re-stretched to time to enhance continuity of BSR. Processing includes automatic gain control (AGC). TWT: two-way traveltime. Structural interpretation after *von Huene et al.* [1996]. (b) Section of line 1018, processing as above. The crystalline basement is covered by a thin section of Eocene sediments. The seaward termination of the top of Eocene is interpreted to coincide with the edge of the basement. Structure after *von Huene et al.* [1996]. Weakening of BSR across normal faults close to its landward termination. t.f.*: fault system causing disturbances in thermal gradient in Figure 5. No BSRs are present along this line in Lima Basin further upslope (not shown here). (c) High-resolution velocity profiles across the BSR in (b) from full waveform inversion [*Pecher et al.*, 1996b]. The relatively weak BSR at the easternmost location close to a system of normal faults appears to be mainly caused by elevated velocities above the BSR. This suggests presence of gas hydrate over no or only very little free gas.

the accreted hemipelagic sediments of less than 1% probably resulting in a lack of gas hydrates [*von Huene and Pecher*, 1999]. BSRs are present in the prism at 9° S (Figure 2). There, organic carbon content is estimated at 2-4% in the accreted trench fill, which should lead to higher production rates of biogenic methane than at 12° S [*von Huene and Pecher*, 1999]. *Von Huene and Pecher* [1999] also suggest that the rate of fluid expulsion from the subducted sediments may be higher at 9° S because a larger fraction of the sediment is believed to be subducted than at 12° S. A subducted ridge beneath the accretionary wedge may lead to additional focusing of fluid flow and local

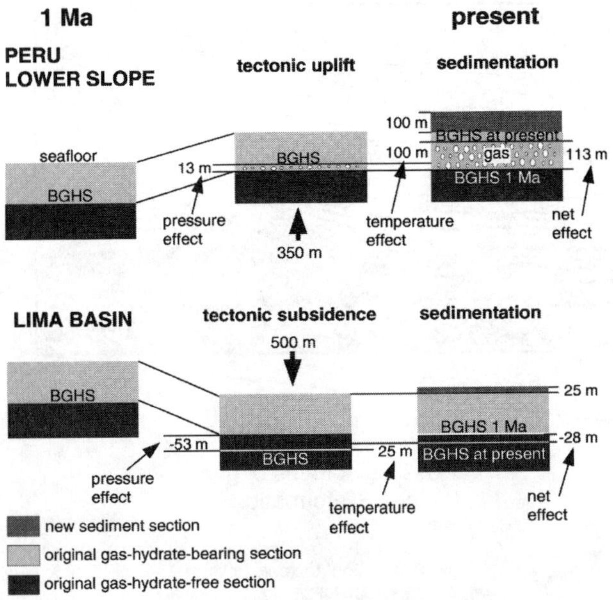

Figure 3. Vertical tectonism and BSRs, after *von Huene and Pecher* [1999]. Uplift on the Peru lower slope causes an upward movement of the BGHS with respect to the sediment column because the associated pressure decrease moves gas hydrate stability towards lower temperatures ("pressure effect"). Sedimentation causes an upward movement of isotherms and hence the BGHS ("temperature effect"). An upward movement of the BGHS in the sediment column is predicted to lead to dissociation of gas hydrates and generation of free gas at the BGHS causing BSRs. Lima Basin is subsiding rapidly. The net movement of the BGHS with respect to the sediment column in Lima Basin is predicted to be downward, absorbing eventual free gas into the GHSZ. See *von Huene and Pecher* [1999] for detailed information on simplifications and input values.

uplift [*Pecher*, 1995]. All three factors –higher organic carbon content, elevated fluid flow, and uplift– are commonly thought to favor BSR occurrence.

BSRs are ubiquitous on the lower slope at 12° S. Moderate uplift due to the formation of the accretionary prism is predicted for the lower slope. Together with high sedimentation rates, this uplift causes significant upward movement of the BGHS and should lead to the generation of free gas, which is consistent with the occurrence of strong BSRs.

No BSRs are observed in Lima Basin along the ODP Leg 112 transect despite the high organic carbon content in its sediments. Lima Basin is still subsiding after subduction of the Nazca Ridge. This subsidence may cause a tectonic suppression of BSRs [*von Huene and Pecher*, 1999] (Figure 3). Tectonic subsidence leads to an increase of pressure, which moves gas hydrate stability towards higher temperatures corresponding to greater depths beneath the seafloor. Free gas at the BGHS is absorbed into the GHSZ and transformed to gas hydrates –the BSR disappears. This process has to offset the effect of sedimentation which leads to an upward movement of isotherms and hence of the BGHS with respect to the sediments. A quantification of both the sedimentation and subsidence effects demonstrates that the BGHS is indeed predicted to move downward with respect to the sediment column in Lima Basin. A BSR has been observed along line CDP-2 at 9° S in Yaquina Basin, which has a similar geologic history as Lima Basin. Subsidence of Yaquina Basin, however, has essentially stopped and hence, BSR occurrence in this basin is consistent with the hypothesis of a tectonic suppression of BSRs [*von Huene and Pecher*, 1999].

Some localized BSRs have been identified in Lima Basin north and south of the ODP Leg 112 transect [*Kukowski and Pecher*, 1999a] apparently beneath morphologically elevated areas (Figure 4). The seismic data from these areas may indicate seafloor erosion, which, together with subsidence, should lead to a downward movement of the BGHS with respect to the sediments. If this observation is correct, methane gas for BSR formation would have to be supplied from below, rather than from dissociation of gas hydrates at the BGHS. We speculate that the morphologic highs may be associated with focusing of fluid flow and hence, an increase of methane supply, such that enough gas is supplied to offset its absorption into the GHSZ. The data shown in Figure 4, however, do not have a sufficiently high resolution to rule out recent sedimentation. We have therefore acquired seismic high-resolution data from this area during a campaign of the R/V *Sonne* in 2000 [*Bialas and Kukowski*, 2000], which should lead to a better understanding of BSR formation in Lima Basin.

Thermal Field of the Peruvian Margin

The depth of BSRs can in theory be used to estimate thermal gradients and heat flow [*Yamano et al.*, 1982]. This method is currently disputed mainly because temperatures at the BGHS inferred from Leg 164 borehole measurements were 0.5-2.9° C lower than theoretical values [*Ruppel*, 1997]. However, it is commonly believed that BSR depth is still a reliable indicator for lateral variations of heat flow. Below, we summarize results from modeling of the thermal field on the Peru margin based on thermal gradients from BSR depth [*Kukowski and Pecher*, 1999b].

The thermal gradient above margin wedges and in accretionary prisms is mostly found to decrease with distance from the deformation front (e.g., *Hyndman et al.*, [1993]). This pattern is attributed to sealing by sediments with a low thermal conductivity and to expulsion of cold pore fluids due to tectonic porosity loss. However, at the Peru margin at 12° S, the thermal gradient computed from the

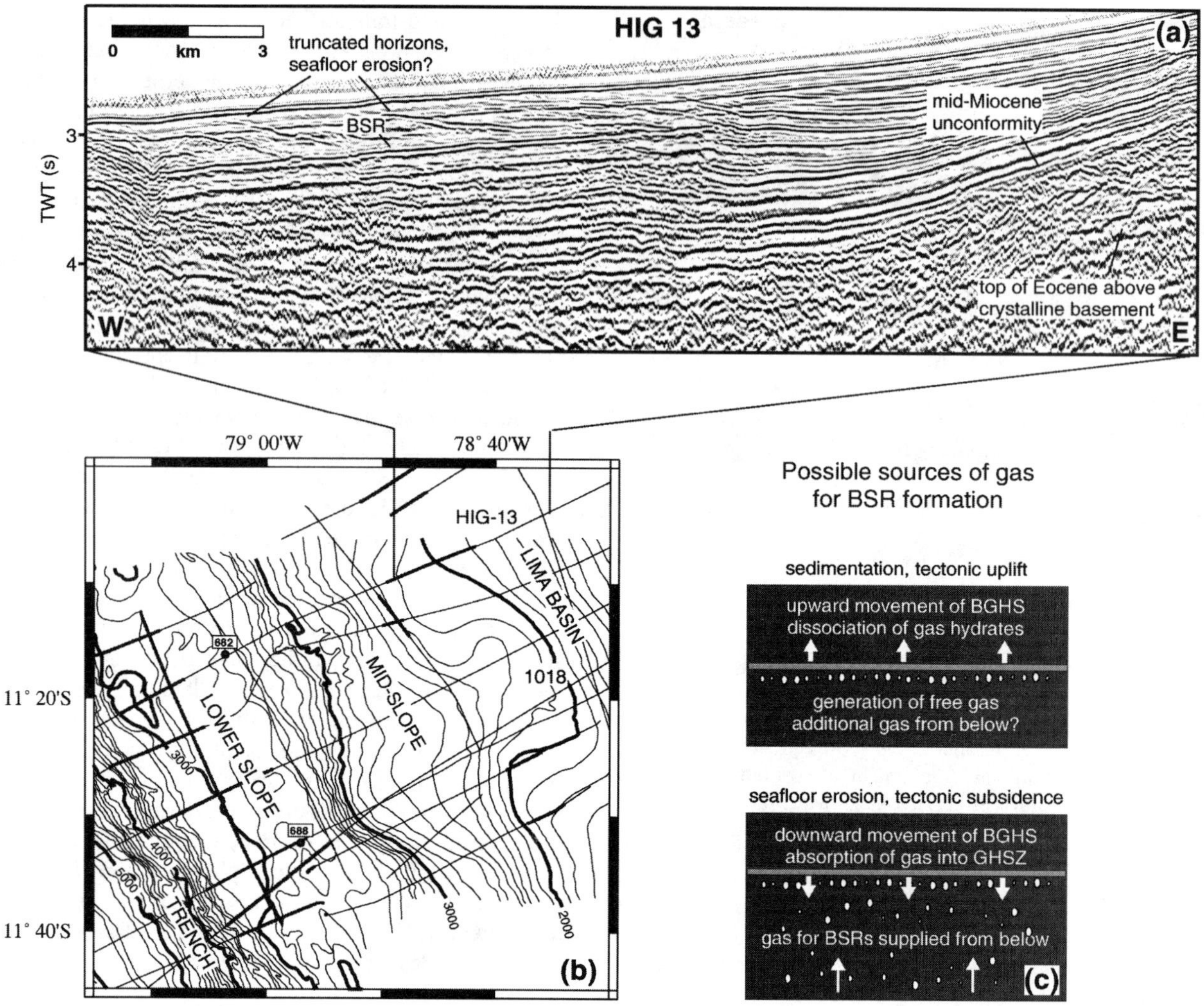

Figure 4. (a) A BSR in Lima Basin, post-stack time migrated section of line HIG-13. Sediment layers above the BSR appear to be truncated at the seafloor, perhaps indicating seafloor erosion. Processing includes AGC. Stratigraphic interpretation after *Ballesteros et al.* [1988]. (b) Bathymetry and BSR occurrence after *Kukowski and Pecher* [1999a]. Bathymetry after *Hussong et al.* [1988], contour interval: 100 m. Thin lines: tracks of seismic reflection data. Thick lines: BSRs. BSRs seem to be confined to two morphologically elevated areas. 685 and 688 are Leg 112 sites. (c) Possible sources of gas at BSRs. Both subsidence and seafloor erosion lead to a downward movement of the BGHS absorbing free gas into the GHSZ. We suggest that BSRs, if present, in an area with a downward migrating BGHS must be maintained by gas from below.

depth of the BSR increases continuously towards the continent (Figure 5). Superimposed to this trend are strong small-scale anomalies, which correlate with major fault zones. Based on these observations together with some results from conventional heat flow measurements [*Yamano and Uyeda*, 1990], *Kukowski and Pecher* [1999b] studied the thermal field at high lateral resolution by applying coupled hydrothermal modeling (Figure 5). This technique allowed a quantitative estimate of the contributions from fluid advection, sediment advection, and frictional heating to the thermal budget.

On a regional basis, fluid advection never contributes more than 10% to the thermal budget within the accretionary prism and significantly less at the slope. The contribution from sediment advection to the thermal field is also very small. Considerable frictional heating along the plate interface, increasing towards the continent, is required to match the thermal constraints. Fluid advection rates are predicted to be in the range of several 10^{-11} m/s to several 10^{-10} m/s in the prism and slightly lower at the slope. In the fault zones, however, fluid advection is the dominant mode of heat transport. Fluid flow is predicted to be at least three orders of magnitude higher.

The thermal structure of the Peru forearc shows significant variations along strike. The depth of BSRs along two seismic profiles at 12° S (1018, Figure 2 and 1017 further south) indicates an increase of the thermal gradient towards the continent. A higher contribution from frictional heating is required along the southern profile in order to match the thermal constraints. At 9° S, where the margin has largely recovered from subduction of the Nazca Ridge, the thermal gradient decreases towards the continent. This thermal structure is more common and indicates lower frictional heating.

MIDDLE AMERICA TRENCH SYSTEM

Geologic Setting

Cocos Plate is subducted beneath the Caribbean Plate between Costa Rica and southern Mexico. The margin off southern Mexico west of the Tehuantepec Ridge was drilled during Deep Sea Drilling Project (DSDP) Leg 66 [*Watkins et al.*, 1982] and studied with seismic and magnetic surveys. Continental crust extends to only 35 km away from the trench [*Karig et al.*, 1978; *Shipley et al.*, 1980]. The margin's history is characterized by tectonic erosion and by formation of a juvenile accretionary prism [*Watkins*, 1989], probably similar to the Peru margin. The outer margin apparently was eroded between ~10-24 Ma. A new accretionary prism started to form ~10-12 Ma. The lower slope now experiences slight uplift caused by recent accretion [*Shipley*, 1982].

From Guatemala to Costa Rica, the frontal ~50 km of the margin appear to have the same basic structure. The bulk of the margin is formed by a wedge-shaped high-velocity body, the margin wedge (Figure 6) [*Ranero and von Huene*, 2000]. Drilling on the continental slope of Guatemala during DSDP Legs 67 [*Aubouin et al.*, 1982] and 84 [*von Huene et al.*, 1985] documented a margin wedge made of igneous rocks of oceanic origin. Igneous breccia was recovered at the base of ODP Leg 170 Site 1042 off Costa Rica close to or from the top of the margin wedge [*Kimura et al.*, 1997]. Seismic reflection data from the Costa Rica margin show a small sediment prism along the front of the margin [*Hinz et al.*, 1996; *Shipley et al.*, 1992], some of which might locally be accreted sediment [*Shipley et al.*, 1992; *von Huene et al.*, 2000], but in the area of ODP Leg 170 it seems to be tectonically kneaded slope sediment [*Kimura et al.*, 1997]. Drilling at the toe of the continental slope penetrated the upper plate and across the subducting sediment to the ocean basement, showing that the entire sediment section underthrusts the front of the margin [*Kimura et al.*, 1997].

Multibeam bathymetry along the continental slope and the facing ocean plate indicates a control on tectonic processes by the topography of the lower plate. Offshore central and southern Costa Rica, subduction of the buoyant Cocos Ridge and large flanking seamounts is eroding the margin [*Ranero and von Huene*, 2000; *von Huene et al.*, 2000]. Several large embayments indicate retreat of the continental shelf (Figure 7). Seamount subduction causes failure and sliding of the oversteepened slope. From the central Nicoya Peninsula to Nicaragua, flexural normal faulting creates a rough terrain in the ocean plate but subduction erosion rates are probably lower [*Ranero and von Huene*, 2000]. A similarly rough topography is subducted off Guatemala, and an effect on lower slope topography has been reported [*von Huene*, 1989].

Organic carbon content in the sediments of the lower slope offshore Mexico is between 1.1 and 2.8%. Offshore Guatemala at DSDP Sites 568 and 570 it is relatively high, typically 3-4%, favoring biogenic production of methane [*Claypool et al.*, 1985]. The organic carbon content of sediments offshore Costa Rica is significantly lower, mostly 0.5-2% [*Kimura et al.*, 1997; *von Huene et al.*, 1985].

Gas Hydrates During DSDP and ODP Legs

Evidence for gas hydrates offshore southern Mexico was found at three DSDP Leg 66 sites [*Shipley and Didyk*, 1982]. The ratio of methane to higher-order hydrocarbons indicated a mainly biogenic origin of the hydrate-forming gas. *Shipley and Didyk* [1982] distinguished between "ice inclusions" that predominantly occurred in unconsolidated mud and "frozen sediments" in porous volcanic ash and fine sands. This was, to our knowledge, the first-ever observation, that the microscopic distribution of gas hydrates in sediment may depend on grain size. *Clennell et al.* [1999] postulated in a thermodynamic study that gas hydrates should occur as nodules in fine-grained sediments, "pushing away" grain material, while they should cement coarser-grained sands.

Two gas hydrate samples were recovered during ODP Leg 67 off Guatemala [*Aubouin et al.*, 1982; *Harrison and Curiale*, 1982]. One sample was a 2-cm long solid gas hydrate [*Harrison and Curiale*, 1982] in a sediment section that consisted primarily of mud and mudstone [*Aubouin et al.*, 1982]. The other sample was a cemented sand. After hydrate decomposition, the sample collapsed [*Harrison and Curiale*, 1982]. This observation is probably still one

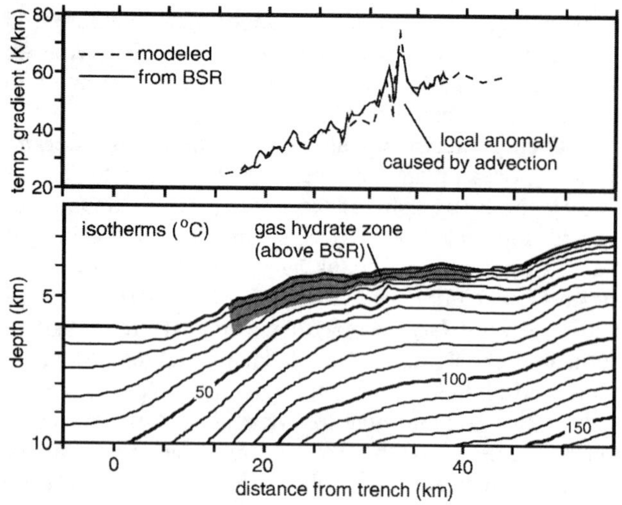

Figure 5. Thermal gradient from BSR depth and from thermo-hydraulic modeling [*Kukowski and Pecher*, 1999b]. Isotherms from modeling. The local anomaly reflects steps in BSR depth across a thrust fault, t.f.* in Figure 2 (b).

of the best demonstrations that gas hydrates may act as cement and that their dissociation may significantly weaken the sediment frame.

During Leg 84 off Guatemala and Costa Rica, drilling had to be terminated out of safety considerations at two sites because of an encounter of gas-hydrate-cemented volcanic ash [*von Huene et al.*, 1985]. Gas hydrates were sampled at three sites [*Kvenvolden and McDonald*, 1985] including a 1.05 m core of massive gas hydrates at Site 570. Measurement of V_p with a Hamilton Frame on 14 samples from this core yielded values between 3.5 and 4 km/s [*von Huene et al.*, 1985]. V_p, density, and resistivity logs indicated that the massive hydrate extended over 4 m in situ [*Mathews and von Huene*, 1985]. Such an encounter of massive gas hydrates may be a "sweet spot" [*Dillon et al.*, 1999] for gas hydrates as a possible energy source. The high permeability expected in faults would facilitate production of methane from gas hydrates. Chemical and isotopic composition of gases indicated a primarily biogenic origin [*Brooks et al.*, 1985; *Kvenvolden and McDonald*, 1985]. Surprisingly, hydrocarbon gases were found at Sites 566 and 570 in fractures at the top of the igneous basement. At Site 566, even gas hydrates were suggested to be present in these fractures [*Hesse et al.*, 1985]. Hydrocarbon gases in the basement were found to be primarily of biogenic origin. Some higher-order hydrocarbon gases, however, appear to be thermally produced from low-temperature chemical diagenesis [*Jeffrey et al.*, 1985]. The sources and migration paths of these gases have not yet been determined. In particular, it is not clear if some hydrocarbon gases are migrating upward from the subduction zone through faults in the basement.

Gas hydrates were also encountered at ODP Leg 170 Site 1040 on the lower slope [*Kimura et al.*, 1997] (Figure 7 for location of Leg 170 study area). Based on chlorinity, there was evidence for an uneven gas hydrate distribution with somewhat higher concentrations in ash layers. Gas hydrates were encountered in the pore spaces of a carbonate accretion. At Site 1041 on the mid-slope, there was visible evidence of gas hydrate in a long sediment section

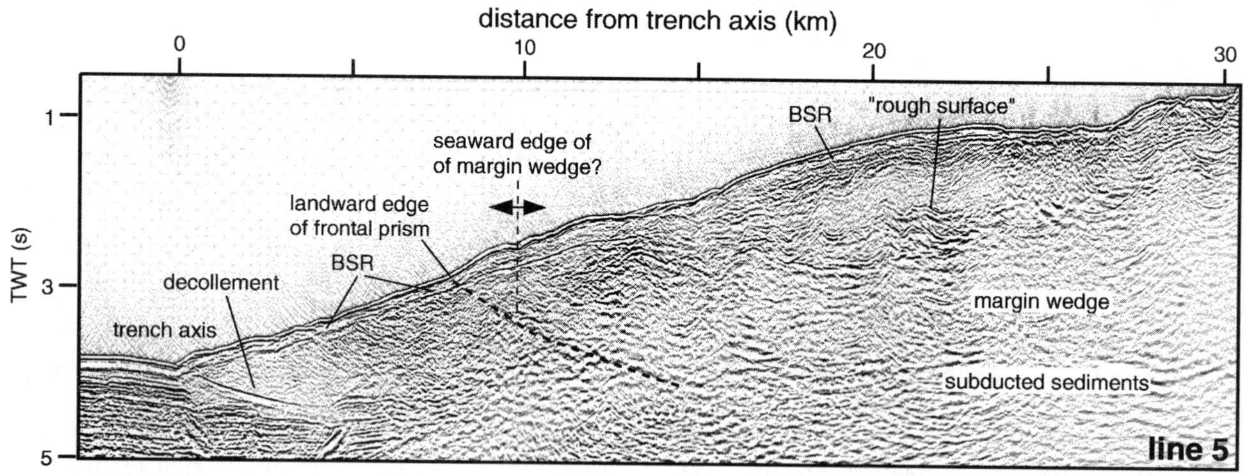

Figure 6. Costa Rica line 5 (see Figure 7 for location), post-stack time migrated, AGC. Structural interpretation after *Hinz et al.* [1996]. The "rough surface" marks the top of the margin wedge, which most likely consists of oceanic igneous rock [*Ranero et al.*, 2000]. The base of the margin wedge is marked by reflections from subducted sediments. The BSR starts at the décollement, and stretches over most of the slope before it pinches out. The minimum of BSR depth (in TWT) over the prism translates into a maximum of heat flow (Figure 10).

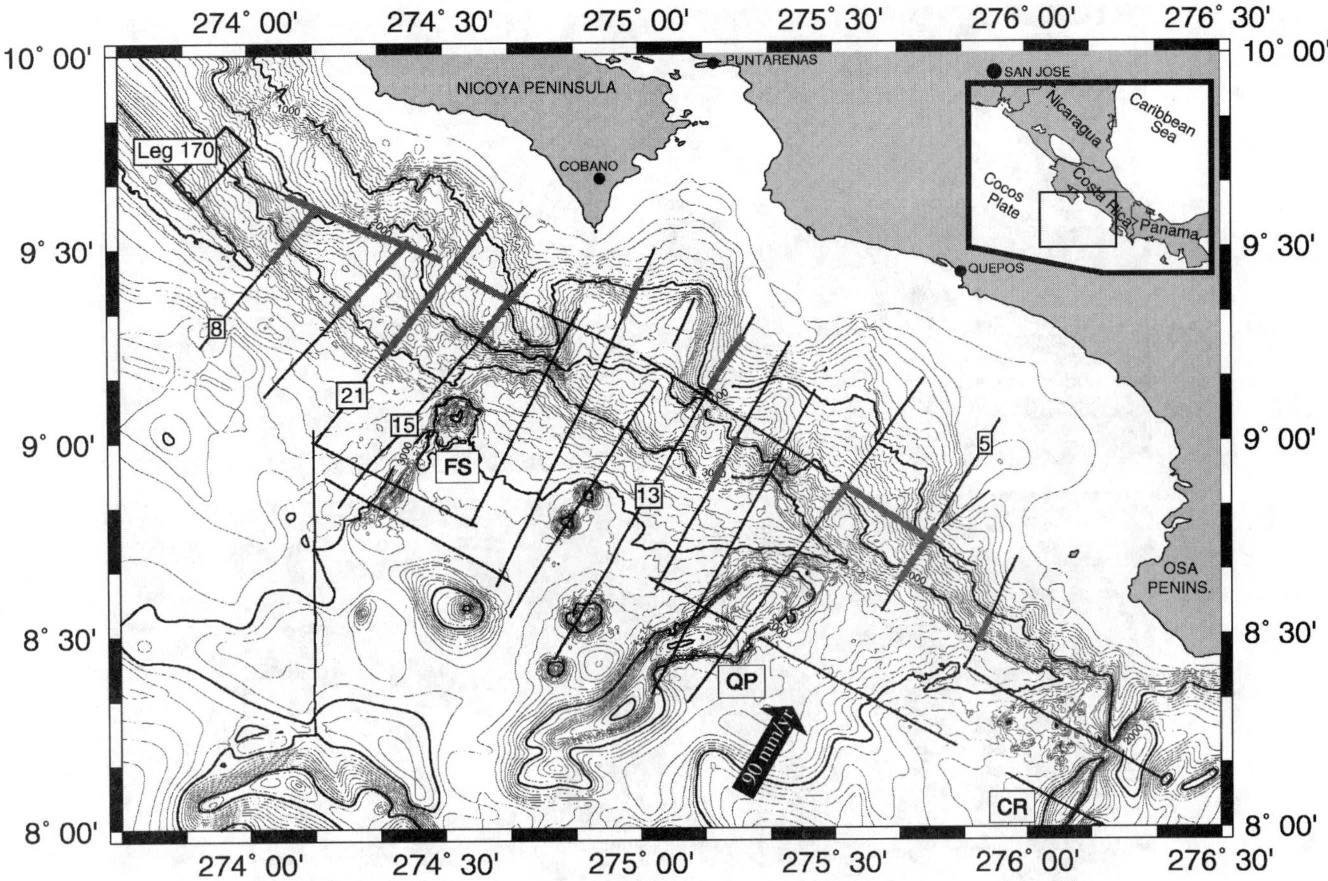

Figure 7. Tracks of seismic data on top of bathymetry [*Hinz et al.*, 1996]. Contour intervals: 100 m. Thick gray lines mark BSR occurrences [*Pecher et al.*, 1998]. Lines 8, 13, 15, and 21 are referred to in the text. The area of the relatively densely spaced ODP Leg 170 sites is marked by a box (indicating coverage with 3-D seismic site survey data [*Shipley et al.*, 1992]). Arrow: Direction and rate of convergence. FS: Fisher Seamount, QP: Quepos Plateau, CR: Cocos Ridge.

between 120 and 280 mbsf. Gas hydrates were mostly disseminated in the pore space or they formed as thin sheets of pure gas hydrates filling microfractures. A massive nodular gas hydrate sample was recovered in silty claystone. Gas hydrates also were detected in two cores as cements in ash layers. Indirect evidence for gas hydrates mainly based on chlorinity was found at other locations. The use of chlorinity profiles for gas hydrate quantification in Leg 170 cores, however, is made difficult because of a possible fresh-water source deeper in the sediment, either water of meteoritic origin or from dehydration processes [*Ruppel and Kinoshita*, 2000]. The gas hydrate stability zone may stretch into the basement because of the extremely low heat flow in this area [*Langseth and Silver*, 1996]. The, compared to sediments, very low porosity of the probably igneous basement makes a presence of free gas at the BGHS unlikely, which may readily explain the lack of BSRs in this area.

Distribution of BSRs at the Costa Rica Margin

BSRs are wide-spread across the Costa Rica margin (Figure 7). Waveform inversion again indicates that BSRs are mainly caused by a thin free-gas layer [*Pecher et al.*, 1998]. BSRs are relatively weak compared to BSRs off Peru, which intuitively is consistent with the low organic carbon content that would suggest a lower supply of methane for BSR and gas hydrate formation. Where the margin is undisturbed, BSRs appear to be present almost continuously from the décollement until they intersect the seafloor at about 700 m water depth.

Most BSRs off Costa Rica occur at relatively shallow depths between 100 and 400 mbsf. *Pecher et al.* [1998] suggested that methane for hydrate formation is probably mainly supplied from below by advected fluids. Sedimentation appears to occur at most areas that are not affected by slope failure (at DSDP Site 565 in the ODP Leg 170

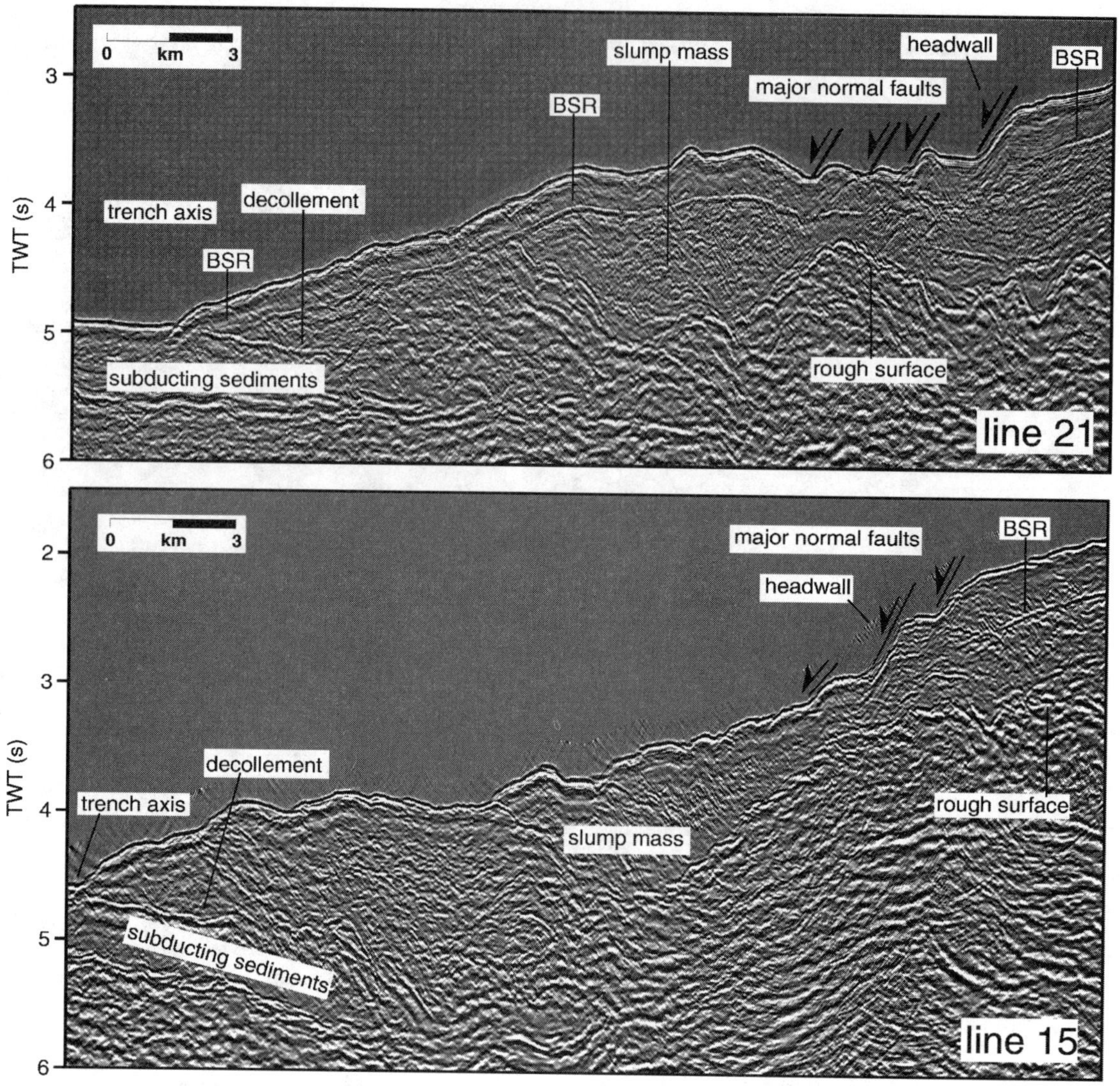

Figure 8. BSRs and slumping, post-stack time migrated sections of lines 15 and 21. The BSR has "survived" slumping or it has re-formed along line 21. Several coherent sediment blocks appear to have slumped. Within these blocks, the sediment stratification probably has remained largely intact. No BSRs are visible across the slump mass in line 15 where the sediment stratification has been destroyed [*Pecher et al.*, 1998].

study area, the Pleistocene sediment accumulation rate was 77 m in 0.47 Myr [*von Huene et al.*, 1985]; also see *Kimura et al.* [1997] for sediment accumulation rates at the ODP Leg 170 sites). The shallower the BGHS, the shorter the time that sediments reside within the GHSZ, depending on the sedimentation rate. Biogenic methane production above a shallow BGHS in areas with sedimentation would have to be very high in order to allow gas hydrate formation from in-situ methane before the sediments are buried beneath the GHSZ. Any quantification, however, requires a more detailed knowledge of sedimentation rates, carbon content, and in particular, biologic activity. Also, we caution that our understanding of the microbiology associated with gas hydrates and free gas is still very limited.

BSRs and Slope Failure

A BSR is present in a slump block where the sediment stratification appears to have remained largely undisturbed, whereas no BSR occurs in an adjacent line across the same slump where sediment stratification has been destroyed (Figure 8). A destruction of BSRs during slope failure can be explained by an escape of free gas into the water column. BSRs seem to be absent everywhere where sediment stratification has been destroyed due to slumping after the subduction of seamounts and they have not re-formed since. It is unlikely that slope failure took place recently everywhere along the seismic lines used to study the BSR distribution (Figure 7). Hence, it appears that BSR formation after slumping takes a geologically significant time, if they form again at all. A lack of BSRs may be caused by thermal disturbances after slumping leading to a discontinuous BGHS in the sediments. Alternatively, methane supply may be too low for BSR and perhaps gas hydrate formation, e.g., because the sediment is too depleted of organic material [*Pecher et al.*, 1998]. We speculate that as another alternative a lowering of permeability after slumping may decrease methane supply for BSR formation. The "survival" of a BSR in a slump block where the original stratification is intact may be explained by gas not being able to escape during slumping. Alternatively, the relatively undisturbed temperature field in the slump block [*Pecher et al.*, 1998] and/or higher sediment permeability may have allowed the BSR to form again as a continuous reflection along a largely undisrupted BGHS.

BSRs are often considered zones of sediment weakness. Off Costa Rica, however, we did not find any seismic evidence that BSRs act as detachment zones. A good candidate for a "weak" layer would be a BSR along line 13 (Figures 7 and 9). A seamount appears to be subducted beneath the slope. This seamount subduction causes strong uplift of a small area on the mid-slope landward of the inferred location of the top of the seamount. From the change of the slope dip, we predict an uplift of ~390 m in roughly 87,000 years (4.5 km/Myr; see Figure 9 for an explanation how we estimated the rate of uplift). We are aware that our approach to determine uplift rates is a simplification, however, for the discussion below we only need very rough estimates. The pressure decrease associated with an uplift of 390 m has caused an upward movement of the BGHS by over 45 m. An additional upward movement of the BGHS is expected from sedimentation. Sediment accumulation rates on the mid-slope at ODP Site 1042 in the upper 225 mbsf are 55-62 m/Myr [*Kimura et al.*, 1997]. If we therefore arbitrarily assume a sedimentation rate of 58 m/Myr, ~5 m of sediments would have been added in 87,000 years. The total upward movement of the BGHS therefore is about 50 m in a sediment column that is only 170-m thick (Figure 9). This should lead to dissociation of gas hydrates, generation of gas, overpressuring, and potentially slope failure at the BSR. However, the sediment column appears to have survived the overpressuring. There is no indication for sediment collapse above the BGHS. Slope failure appears to take place seaward of the crest of the seamount. This location is roughly where slumping would be expected regardless of a presence of gas hydrates.

Dissociation of gas hydrates to free gas should lead to volume expansion and hence, overpressuring. However at greater depths, volume expansion is not nearly as significant as at atmospheric conditions under which the volume expands by a factor of about 164 [*Sloan*, 1990]. At a pressure of 100 bar (approximately 1000 m depth) and a temperature of 13° C (approximately the phase boundary for methane hydrate at 100 bar), the density of methane gas is 0.082 g/cm^3 [*Duan et al.*, 1992]. At full cage occupancy, 1 cm^3 of gas hydrates generates 0.7904 g of water and 0.1224 g of methane (see *Sloan* [1990] for details). If we assume 90% cage occupancy (which is somewhat arbitrary but often used for natural methane hydrates), only 0.1102 g of methane are produced. 1 cm^3 of methane hydrate then yields 0.79 cm^3 of water and 1.34 cm^3 of methane gas. Hence, dissociation of gas hydrates should lead to a total volume expansion by a factor of 2.1. Assuming that 20% of the pore space is filled with hydrates, which is a high estimate compared to the Blake Ridge [*Paull et al.*, 1996], the volume of the pore fill expands only by a factor of about 1.2. A number of densely spaced faults are present in the sediments above the seamount [*Ranero and von Huene*, 2000]. These faults seem to cross the BSR. We therefore speculate that the faults may be escape paths for excess free gas beneath the BGHS alleviating overpressuring. As mentioned for the Peru BSRs, we do not understand, yet, how gas migrates along faults through the BGHS without being transformed to gas hydrates, but gas venting close to faults has been observed at the seafloor in the GHSZ elsewhere. During a 1999 R/V *Sonne* campaign, tubeworms and –possibly chemoherm– carbonate outcrops were detected in the area of faulting, which may indicate venting of fluids and/or gas [*Bialas et al.*, 1999]. This area is the focus of ongoing research that should add significantly to our knowledge of the interaction between gas hydrates and slope failure.

Heat Flow Patterns

BSR heat flow off Costa Rica still needs calibration with conventional heat flow measurements. Preliminary studies [*Pecher*, 1995; *Ranero et al.*, 1997], however, reveal some characteristic patterns (Figure 10). Heat flow estimated from the BSR depth at the western-most line across the margin, line 8 [*Pecher*, 1995], is considerably higher than observed in the Leg 170 study area [*Langseth and Silver*, 1996] (Figure 7). The difference is too large to be explained by inadequate calibration of BSR heat flow. It probably reflects a fundamental change of the heat flow regime that is most likely caused by processes in the subducted oceanic plate. The change occurs near the boundary of oceanic crust generated at Cocos Ridge and East Pacific Rise [*von Huene et al.*, 2000]. Conventional heat

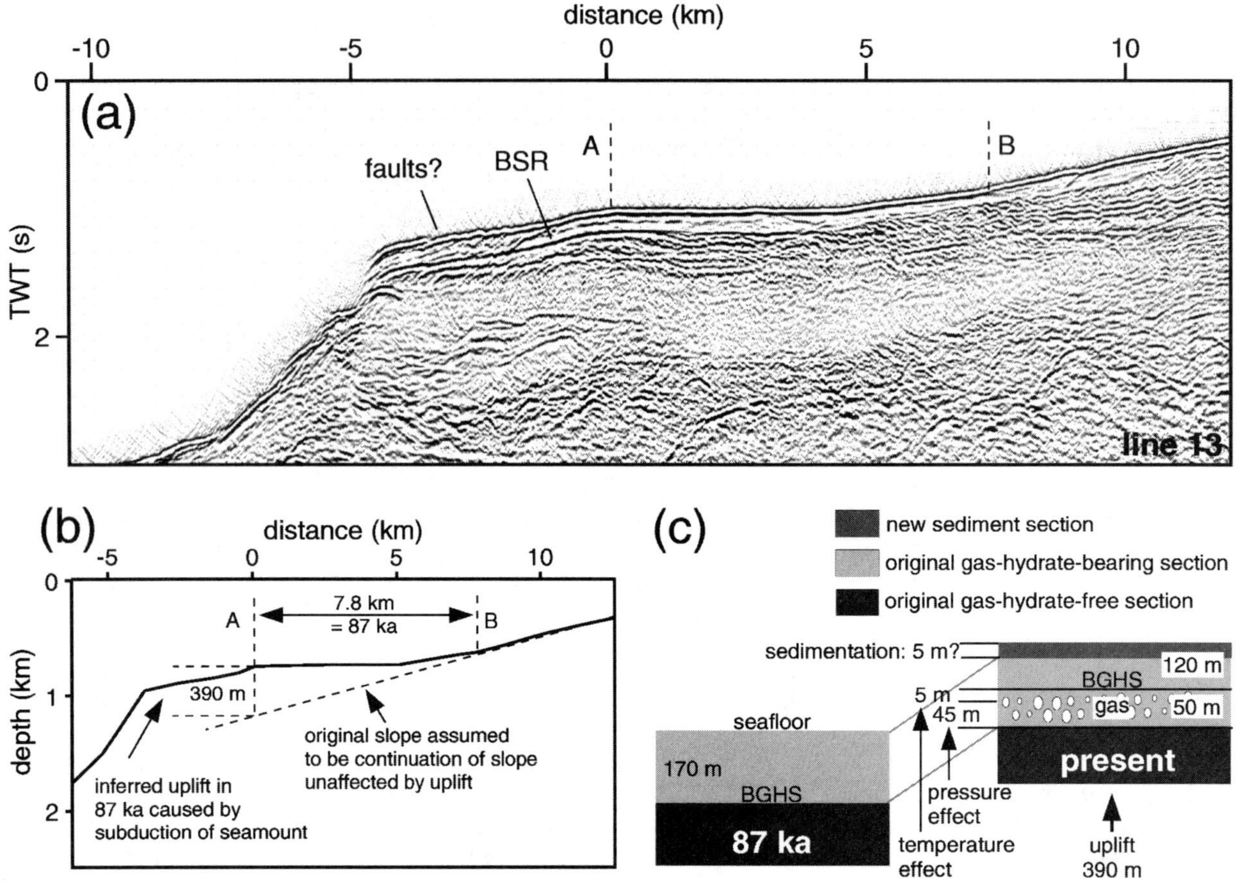

Figure 9. (a) BSR above a subducting seamount, processing includes multiple-suppression and AGC. (b) Estimating the rate of uplift. We estimate the depth of the seafloor at "A" before subduction of the seamount by linearly extrapolating the slope from "B" towards "A". Under this assumption, the slope at "A" was uplifted by about 390 m. Based on the plate convergence rate, we estimate that the leading edge of the seamount migrated from A to B in 87 kyr. These assumptions are grossly simplified (e.g., not accounting for the geometry of the seamount and the dip of the subducting plate) and these numbers are only meant to be rough constraints. (c) Resulting upward movement of the BGHS. Sedimentation rate estimated from the Leg 170 (see text), which is somewhat arbitrary. Using the above assumptions, we predict that gas hydrates have dissociated in a relatively thick layer (50 m in an originally 170-m thick section) generating free gas and possibly overpressure. Sediment stratification, however, appears to have remained intact. Faults may act as conduits allowing gas to escape alleviating overpressure.

flow measurements on the oceanic plate immediately seaward of the trench and west of this boundary suggest low heat flow of about 14 mW/m^2 [*Langseth and Silver*, 1996]. Further to the southwest, earlier measurements typically yielded values between 25 and 60 mW/m^2, whereas heat flow values on the oceanic plate east of this boundary were found to be mostly in the range of 100 to 150 mW/m^2 [*Ladd and Alvarez*, 1985]. The oceanic plate entering the subduction zone is of similar age on both sides of the boundary, about 20 Myr, and would be expected to have a heat flow of roughly 100 mW/m^2 [*Klitgord and Mammerickx*, 1982]. Hence, values to the west of this boundary are anomalously low. *Langseth and Silver* [1996] propose a "seawater heat exchange" mechanism, circulation of seawater through fractures into the upper oceanic crust, as cause for the low heat flow. Further discussion of this dramatic change in heat flow, however, is beyond the scope of this paper.

Across the margin, heat flow increases from the continent towards the deformation front along line 8 reflected by a shallowing of the BSR, up to about 6 km away from the trench, beyond which the BSR cannot be traced. The BSR extends across the frontal prism at line 5. Here, a minimum of BSR depth about 3-8 km landward of the trench, reflects a maximum of heat flow. This maximum is located close to the seaward edge of the margin wedge. Evidence for venting of fluids along faults was detected from submersible dives and seismic data slightly seaward

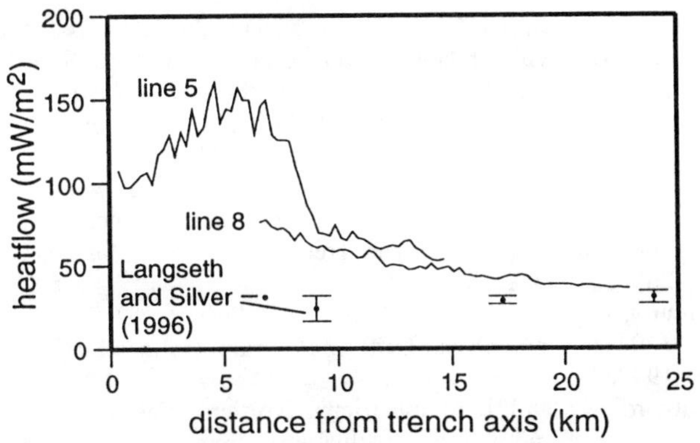

Figure 10. Preliminary heat flow from BSR depth along lines 8 and 5 [*Pecher*, 1995], and from conventional measurements in the Leg 170 study area [*Langseth and Silver*, 1996] about 25 km to the northwest of line 8 (see Figure 7 for locations). Heat flow along line 8 is considerably higher than in the Leg 170 study area. A maximum of heat flow seaward of the margin wedge along line 5 is reflected by a shallowing of BSR depth (Figure 6). For reliable absolute values, BSR heat flow still needs calibration.

of the boundary between margin wedge and the frontal prism in the Leg 170 study area [*McAdoo et al.*, 1996; *McIntosh and Silver*, 1996]. Therefore, *Ranero et al.* [1997] in a preliminary study suggest the maximum of heat flow seaward of the margin wedge to be caused by focused fluid expulsion from the subduction zone.

SUMMARY

We summarize our review of BSR and gas hydrate occurrences on the erosional convergent margins of Peru and Middle America as follows:

- Tectonic subsidence may cause the regional lack of BSRs in Lima Basin. Free gas at the BGHS generating strong BSRs elsewhere on the slope is being absorbed into a downward moving GHSZ.
- Some BSRs occur in Lima Basin beneath small, morphologically elevated areas. BSRs are perhaps supported by upward migrating methane, offsetting absorption into the GHSZ.
- A correlation between microscopic distribution of gas hydrates in sediments and grain size has been documented in several boreholes offshore Middle America. Gas hydrates occur as nodules or are disseminated in fine-grained sediments. In sandy sediments, they act as cements.
- The shallow depth of BSRs off much of Costa Rica suggests a supply of (biogenic) methane for gas hydrate formation from sources below.
- Re-forming BSRs off Costa Rica in sediments after slumping is suggested to take a geologically significant time. Alternatively, methane supply may be too low for BSR renewal and perhaps gas hydrate formation.
- BSRs and hence, gas hydrates, are present above a subducted seamount. The sediment column remains intact despite the inferred strong uplift. The associated pressure decrease is expected to cause generation of free gas from hydrates potentially leading to overpressuring. This process, however, does not appear to cause slope failure. Numerous faults cutting through the BSR perhaps allow gas to escape and alleviate overpressuring.
- Two studies based on lateral variations of BSR heat flow demonstrate the potential of this method. Frictional heating in the Peru subduction zone is inferred from observed increase of heat flow towards the continent. Elevated heat flow over a fault zone may indicate focused fluid flow. Off Costa Rica, a shallowing of BSRs reflects a local maximum of heat flow seaward to the boundary between frontal prism and margin wedge perhaps associated with elevated fluid expulsion.

Acknowledgements. Most plots in this paper have been generated with *GMT* [*Wessel and Smith*, 1991] or *SEISUNIX* (Colorado School of Mines) free software. We thank W. S. Holbrook and an anonymous reviewer for reviewing an earlier version of this manuscript as well as S. Gulick for an internal review. We wish to thank L. Gahagan for compilation of Figure 1. IAP and CRR have been supported by various U.S. and German research grants. This is UTIG publication no. 1539.

REFERENCES

Aubouin, J., R. von Huene, et al., *Init. Repts. DSDP, 67*, 799 pp., U.S. Printing Office, Washington, D.C., 1982.

Ballesteros, M.W., G.F. Moore, B. Taylor, and S. Ruppert, Seismic stratigraphic framework of the Lima and Yaquina forearc basins, Peru, in *Proc. ODP, Init. Repts., 112*, edited by E. Suess, R. von Huene, et al., pp. 77-90, Ocean Drilling Program, College Station, TX, 1988.

Bialas, J., E.R. Flueh, and G. Bohrmann, FS Sonne, cruise report SO144/1&2, pp. 437, Geomar, Kiel, 1999.

Bialas, J., and N. Kukowski, Cruise Report SO 146, GEOPECO, pp. 499, Geomar, Kiel, 2000.

Brooks, J.M., A.W.A. Jeffrey, T.J. McDonald, R.C. Pflaum, and K.A. Kvenvolden, Geochemistry of hydrate gas and water from Site 570, Deep Sea Drilling Project Leg 84, in *Init. Repts. DSDP, 84*, edited by R. von Huene, J. Aubouin, et al., pp. 699-703, U.S. Printing Office, Washington, D.C., 1985.

Cande, S.C., Nazca-South American plate interactions since 50 my B.P., in *Peru continental margin*, edited by D.M. Hussong, S.P. Dang, L.D. Kulm, R.W. Couch, and T.W.C. Hilde, Marine Sciences International, Woods Hole, MA, 1985.

Claypool, G.E., C.N. Threlkeld, P.N. Mankiewicz, M.A. Arthur, and T.F. Anderson, Isotopic composition of interstitial fluids and origin of methane in slope sediments of the Middle America Trench, Deep Sea Drilling Project Leg 84, in *Init. Repts.*

DSDP, 84, edited by R. von Huene, J. Aubuoin, et al., pp. 683-691, U.S. Printing Office, Washington, D.C., 1985.

Clennell, M.B., M. Hovland, J.S. Booth, P. Henry, and W.S. Winters, Formation of natural gas hydrates in marine sediments. Part 1: Conceptual model of gas hydrate growth conditioned by host sediment properties., *J. Geophys. Res., 104*, 22985-23003, 1999.

Coffin, M.F., L.M. Gahagan, and L.A. Lawver, Present-day plate boundary digital data compilation, pp. 5, Inst. for Geophysics, U. Texas, Austin, TX, 1998.

DeMets, C., R.G. Gordon, D.F. Argus, and S. Steins, Current plate motions, *Geophys. J. Int., 101*, 425-478, 1990.

Dillon, W.P., W.S. Holbrook, R. Drury, J. Gettrust, D.R. Hutchinson, J. Booth, and M. Taylor, Faulting of gas hydrate-bearing marine sediments - contribution to permeability, in *Proc. 29th Offshore Technology Conference*, pp. 201-209, 1997.

Dillon, W.P., M.H. Taylor, and C.H. Anton, Gas hydrate in a natural setting - where are the sweet spots?, in *AAPG Annual Convention*, pp. A33, 1999.

Domenico, S.N., Elastic properties of unconsolidated porous sand reservoirs, *Geophysics, 42*, 1339-1368, 1977.

Duan, Z., N. Moeller, and J.H. Weare, An equation of state for the CH4-CO2-H2O system. I. Pure systems for 0 to 1000 C, *Geochim. Cosmochim. Acta, 56*, 2605-2617, 1992.

Egeberg, P.K., and G.R. Dickens, Thermodynamic and pore water halogen constraints on hydrate distribution at ODP Site 997 (Blake Ridge), *Chem. Geol., 153*, 53-79, 1999.

Harrison, W.E., and J.A. Curiale, Gas hydrates in sediments of Holes 497 and 498A, Deep Sea Drilling Project Leg 67, in *Init. Repts. DSDP, 67*, edited by J. Aubouin, R. von Huene, et al, pp. 591-594, U.S. Printing Office, Washington, D.C., 1982.

Hesse, R., J. Lebel, and J.N. Gieskes, Interstitial water chemistry of gas-hydrate-bearing sections on the Middle America trench slope, Deep Sea Drilling Project Leg 84, in *Init. Repts. DSDP, 84*, edited by R. von Huene, J. Aubouin, et al., pp. 727-737, U.S. Printing Office, Washington, D.C., 1985.

Hinz, K., R. von Huene, C.R. Ranero, and the PACOMAR Working Group, Tectonic structure of the convergent Pacific margin offshore Costa Rica from multichannel seismic reflection data, *Tectonics, 15*, 54-66, 1996.

Holbrook, W.S., H. Hoskins, W.T. Wood, R.A. Stephen, D. Lizarralde, and the Leg 164 Scientific Party, Methane hydrate and free gas on the Blake Ridge from vertical seismic profiling, *Science, 273*, 1840-1843, 1996.

Honarpour, M., L. Koederitz, and A.H. Harvey, *Relative permeability of petroleum reservoirs*, 143 pp., CRC Press, Boca Raton, FL, 1986.

Hussong, D.M., T.B. Reed, and W.A. Bartlett, SeaMARC II sonar imagery and bathymetry of the Nazca Plate and Peru Forearc, ODP Leg 112, in *Proc. ODP, Init. Repts., 112*, edited by E. Suess, R. von Huene, et al., pp. 125-130, Ocean Drilling Program, College Station, TX, 1988.

Hyndman, R.D., K. Wang, T. Yuan, and G.D. Spence, Tectonic sediment thickening, fluid expulsion, and the thermal regime of subduction zone accretionary prisms: the Cascadia margin off Vancouver Island, *J. Geophys. Res., 98*, 21865-21876, 1993.

Jeffrey, A.W.A., R.C. Pflaum, T.J. McDonald, J.M. Brooks, and K.A. Kvenvolden, Isotopic analysis of core gases at Sites 565-570, Deep Sea Drlling Project Leg 84, in *Init. Repts. DSDP, 84*, edited by R. von Huene, J. Aubouin, et al., pp. 717-726, U.S. Printing Office, Washington, D.C., 1985.

Karig, D.E., R.K. Cardwell, G.F. Moore, and D.G. Moore, Late Cenzoic subduction and continental margin truncation along norhtern Middle American Trench, *Geol. Soc. Am. Bull., 89*, 265-276, 1978.

Kimura, G., E. Silver, P. Blum, et al., *Proc. ODP, Init. Repts., 170*, 1015 pp., Ocean Drilling Program, College Station, TX, 1997.

Klitgord, K., and J. Mammerickx, Northern East Pacific Rise: Magnetic anomaly and bathymetric framework, *J. Geophys. Res., 87*, 6725-6750, 1982.

Kukowski, N., and I.A. Pecher, Is BSR occurrence in Lima Basin offshore central Peru an indication for methane production below the gas hydrate stability zone?, in *4th International Symposium on Andean Geodynamics*, pp. 404-406, 1999a.

Kukowski, N., and I.A. Pecher, Thermo-hydraulic modelling of the accretionary complex off Peru at 12°S, *J. Geodyn., 27*, 373-402, 1999b.

Kvenvolden, K.A., T.J. Frank, and M. Golan-Bac, Hydrocarbon gases in tertiary and quaternary sediments offshore Peru -- results and comparisons, in *Proc. ODP, Sci. Results, 112*, edited by E. Suess, and R. von Huene, pp. 505-516, Ocean Drilling Program, College Station, TX, 1990.

Kvenvolden, K.A., and M. Kastner, Gas hydrates of the Peruvian outer continental margin, in *Proc. ODP, Sci. Results, 112*, edited by E. Suess, and R. von Huene, pp. 517-526, Ocean Drilling Program, College Station, TX, 1990.

Kvenvolden, K.A., and T.J. McDonald, Gas hydrates of the Middle America trench -- Deep Sea Drilling Project Leg 84, in *Init. Repts. DSDP, 84*, edited by R. von Huene, J. Aubouin, et al., pp. 667-682, U.S. Printing Office, Washington, D.C., 1985.

Ladd, J.W., and A.M. Alvarez, Heatflow, southern Mexico to Costa Rica margin, in *Middle America Trench off western Central America*, edited by J.W. Ladd, and R.T. Buffler, Marine Sciences International, Woods Hole, MA, 1985.

Langseth, M.G., and E. Silver, The Nicoya convergent margin -- a region with exceptionally low heat flow, *Geophys. Res. Lett., 23*, 891-894, 1996.

MacDonald, I.R., N.L. Guinasso, R. Sassen, J.M. Brooks, L. Lee, and K.T. Scott, Gas hydrate that breaches the sea floor on the continental slope of the Gulf of Mexico, *Geology, 22*, 699-702, 1994.

Mathews, M.A., and R. von Huene, Site 570 methane hydrate zone, in *Init. Repts. DSDP, 84*, edited by R. von Huene, J. Aubouin, et al., pp. 773-790, U.S. Printing Office, Washington, D.C., 1985.

McAdoo, B.G., D.L. Orange, E.A. Silver, K. McIntosh, A. L., J. Galewsky, L. Kahn, and M. Protti, Seafloor structural observations, Costa Rica accretionary prism, *Geophys. Res. Lett., 23*, 883-886, 1996.

McIntosh, K.D., and E.A. Silver, Using 3D seismic reflection data to find fluid seeps from the Costa Rica accretionary prism, *Geophys. Res. Lett., 23*, 895-898, 1996.

Miller, J.J., M.W. Lee, and R. von Huene, An analysis of a reflection from the base of a gas hydrate zone off Peru, *Am. Assoc. Pet. Geol. Bull.*, *75*, 910-924, 1991.

Paull, C.K., R. Matsumoto, P.J. Wallace, et al., *Proc. ODP, Init. Repts., 164*, 623 pp., Ocean Drilling Program, College Station, TX, 1996.

Pecher, I.A., Seismic Studies of bottom simulating reflectors at the convergent margins offshore Peru and Costa Rica, Ph.D. thesis thesis, Univ. Kiel, 1995.

Pecher, I.A., T.A. Minshull, S.C. Singh, and R. von Huene, Velocity structure of a bottom simulating reflector offshore Peru: Results from full waveform inversion, *Earth Planet. Sci. Lett.*, *139*, 459-469, 1996a.

Pecher, I.A., C.R. Ranero, R. von Huene, T.A. Minshull, and S.C. Singh, The nature and distribution of bottom simulating reflectors at the Costa Rican convergent margin, *Geophys. J. Int.*, *133*, 219-229, 1998.

Pecher, I.A., R. von Huene , C. Ranero, N. Kukowski, T.A. Minshull, and S.C. Singh, Formation mechanisms of free gas beneath the hydrate stability zone at convergent margins - geophysical evidence from bottom simulating reflectors at the Peruvian and Pacific Costa Rican margins, in *Proc. International conference on natural gas hydrates*, pp. 593-600, 1996b.

Ranero, C.R., and R. von Huene, Subduction erosion along the Middle America convergent margin, *Nature*, *404*, 748-755, 2000.

Ranero, C.R., R. von Huene, E. Flueh, M. Duarte, D. Baca, and K. McIntosh, A cross section of the convergent Pacific Margin of Nicaragua, *Tectonics*, *19*, 335-357, 2000.

Ranero, C.R., R. von Huene, and I.A. Pecher, Tectonic structure and heat flux of the Pacific convergent margin of Costa Rica, *Annales Geophysicae, EGS XXII Assembly*, *15*, 168, 1997.

Ruppel, C., Anomalously cold temperatures observed at the base of the gas hydrate stability zone on the U.S. Atlantic passive margin, *Geology*, *25*, 699-702, 1997.

Ruppel, C., and M. Kinoshita, Heat, methane, and energy flux in an active margin gas hydrate province, offshore Costa Rica, *Earth Planet. Sci. Lett.*, *179*, 153-165, 2000.

Shipley, T.H., Seismic facies and structural framework of the southern Mexico continental margin, in *Init. Repts. DSDP, 66*, edited by J.S. Watkins, J.C. Moore, et al., pp. 775-790, U.S. Printing Office, Washington, D.C., 1982.

Shipley, T.H., and B.M. Didyk, Occurrence of methane hydrates offshore southern Msexico, in *Init. Repts. DSDP, 66*, edited by J.S. Watkins, J.C. Moore, et al., pp. 547-555, U.S. Printing Office, Washington, D.C., 1982.

Shipley, T.H., K.D. McIntosh, E.A. Silver, and P.L. Stoffa, Three-dimensional seismic imaging of the Costa Rica accretionary prism: Structural diversity in a small volume of the lower slope, *J. Geophys. Res.*, *97*, 4439-4459, 1992.

Shipley, T.H., K.J. McMilllen, J.S. Watkins, J.C. Moore, J.H. Sandoval-Ochoa, and J.L. Worzel, Continental margin and lower slope structures of the Middle America Trench near Acapulco (Mexico), *Mar. Geol.*, *35*, 65-82, 1980.

Sloan, E.D., *Clathrate hydrates of natural gases*, 641 pp., Marcel Bekker, New York, 1990.

Suess, E., R. von Huene, et al., *Proc. ODP, Init. Repts, 112,* 1015 pp., Ocean Drilling Program, College Station, TX, 1988.

von Huene, R., The Middle America convergent plate boundary, Guatemala, in *The Eastern Pacific Ocean and Hawaii*, edited by E.L. Winterer, D.M. Hussong, and R.W. Decker, pp. 535-550, Geol. Soc. Am., Boulder, CO, 1989.

von Huene, R., J. Aubouin, et al., *Init. Repts. DSDP, 84*, 967 pp., U.S. Printing Office, Washington, D.C., 1985.

von Huene, R., and I.A. Pecher, Neotectonics and the origins of BSRs along the Peru margin, *Earth Planet. Sci. Lett.*, *166*, 47-55, 1999.

von Huene, R., I.A. Pecher, and M.-A. Gutscher, Development of the accretionary prism along Peru and material flux after subduction of Nazca Ridge, *Tectonics*, *15*, 19-33, 1996.

von Huene, R., C.R. Ranero, W. Weinrebe, and K. Hinz, Quaternary convergent margin tectonics of Costa Rica, segementation of the Cocos Plate and the Central American volcanism, *Tectonics*, *19*, 314-334, 2000.

Watkins, J.S., The Middle America Trench off southern Mexico, in *The Eastern Pacific Ocean and Hawaii*, edited by E.L. Winterer, D.M. Hussong, and R.W. Decker, pp. 523-533, Geol. Soc. Am., Boulder, CO, 1989.

Watkins, J.S., J.C. Moore, et al., *Init. Repts. DSDP, 66*, 864 pp., U.S. Printing Office, Washington, D.C., 1982.

Wessel, P., and W.H.F. Smith, Free software helps map and display data, *Eos Trans. AGU*, *72*, 331, 1991.

Yamano, M., and S. Uyeda, Heat flow studies in the Peru Trench subduction zone, in *Proc. ODP, Sci. Results, 112*, edited by E. Suess, and R. von Huene, pp. 667-682, Ocean Drilling Program, College Station, TX, 1990.

Yamano, M., S. Uyeda, Y. Aoki, and T.H. Shipley, Estimates of heat flow derived from gas hydrates, *Geology*, *10*, 339-343, 1982.

Ingo A. Pecher, Institute for Geophysics, University of Texas, 4412 Spicewood Springs Road, Austin, TX 78759

Nina Kukowski, GeoForschungsZentrum, Potsdam, Germany

César R. Ranero and Roland von Huene, GEOMAR Forschungszentrum, Universität Kiel, Kiel, Germany

Geophysical Studies of Marine Gas Hydrate in Northern Cascadia

R.D. Hyndman[1,2], G.D. Spence[2], R. Chapman[2], M. Riedel[2], R.N. Edwards[3]

This paper reviews the extensive geophysical studies and Ocean Drilling Project (ODP) results that have provided constraints on the occurrence, distribution, and concentration of gas hydrate and underlying free gas beneath the continental slope of the northern Cascadia subduction zone. On this margin there is a large clastic accretionary sedimentary prism, the most common environment for high concentrations of marine gas hydrates. Most information on the gas hydrate has come from a wide range of seismic surveys, including mapping the area of the characteristic bottom-simulating reflector (BSR), determining the depth distribution of hydrate and underlying free gas, and the geological controls of hydrate formation. BSRs are evident beneath about half of the mid-continental slope. Special seismic studies include BSR reflection coefficients, the frequency dependence of the BSR amplitude, BSR amplitude-versus-offset (AVO), and full waveform inversions. Additional information on hydrate concentration is provided by electrical resistivity profiling, and measurements of seafloor compliance. Detailed heat flow surveys have mapped the thermal regime that controls the depth to which gas hydrate is stable. Ocean Drilling Program Leg 146 drilled through the gas hydrate on the mid-continental slope and a reference hole in the adjacent deep sea Cascadia basin. Downhole geophysical logs and core analyses have constrained the concentrations of hydrate and free gas at the drill sites, and provided calibration of the regional geophysical data. In the region of the ODP site, the estimated concentration of hydrate from both geophysical and borehole data is 20-30% of the pore space (10-15% of total volume) over 100 m above the BSR, with an underlying 10-20 m layer of less than 1% free gas.

INTRODUCTION

This article reviews the geophysical surveys, sea floor measurements, and Ocean Drilling Program (ODP) data that constrain the occurrence, distribution and concentration of gas hydrate and underlying free gas on the northern Cascadia subduction zone. The Cascadia subduction zone extends from northern California to Vancouver Island; the focus of this summary is the continental slope off Vancouver Island. In this region, the Juan de Fuca plate is underthrusting the margin approximately orthogonally at about 40 mm/yr. The Juan de Fuca spreading center is located only several 100 km offshore and the oceanic plate is very young, 6-8 Ma, at the margin. The buoyancy of the young hot oceanic plate and the thick incoming sediment section result in a very subdued and shallow trench with a depth of about 2,500 m. Most of the 2-3 km thickness of incoming hemipelagic and turbidite

Natural Gas Hydrates: Occurrence, Distribution, and Detection
Geophysical Monograph 124
Published in 2001 by the American Geophysical Union

[1]Pacific Geoscience Centre, Geological Survey of Canada, Sidney, B.C., Canada

[2]School of Earth and Ocean Sciences, University of Victoria, Victoria, B.C., Canada

[3]Department of Physics, University of Toronto, Toronto, Ontario, Canada

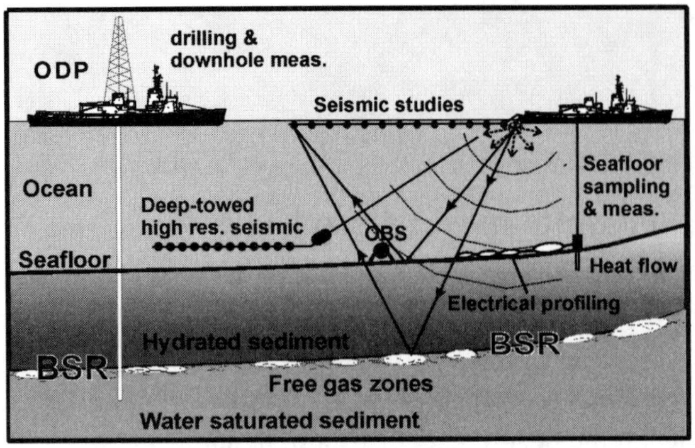

Figure 1. Marine gas hydrate studies.

sediment is scraped off at the deformation front, and accretion since the Eocene has formed a large clastic sedimentary prism. The prism exhibits active folding, thrusting, tectonic consolidation, and pore fluid expulsion. Further landward where the sediment section grows rapidly in thickness due to tectonic shortening, sediments undergo severe tectonic compaction and distributed small-scale deformation. Most of the stratigraphic reflectors are disrupted. Slope basin sediments are deposited on the continental slope, ponded between the anticlinal and faulted structures, and then progressively deformed.

The extensive geophysical and geological data analysed and integrated (Figure 1), include numerous seismic reflection surveys with a variety of systems, seismic refraction, magnetics, gravity, swath bathymetry, seafloor acoustic imagery, heat flow, seismicity and piston coring. Reviews of the geophysical and geological data, the structure, and the margin tectonics have been provided by Davis and Hyndman [1989], Hyndman et al. [1994], and Hyndman [1995]. Ocean Drilling Program (ODP) Leg 146 drilled several wells through the gas hydrate on the mid-continental slope (Figure 2), and a reference no-hydrate, no-gas well in the adjacent deep sea Cascadia basin [Westbrook et al., 1994; Carson et al., 1995]. The ODP downhole measurements and core sample data are an important complement to the geophysical data, especially providing a calibration of the velocity and electrical resistivity estimates obtained from surface measurements. A widespread gas hydrate bottom-simulating reflector (BSR) has been mapped beneath the mid-continental slope at a sub-bottom depth of about 250 m. The shallow BSR depth is a consequence of the relatively shallow water depth and of the high heat flow associated with the young age of the underthrusting oceanic lithosphere. The shallow water depth facilitates high spatial resolution data from the surface geophysical surveys.

SEISMIC DATA

A wide variety of seismic surveys have been used to map and characterize the gas hydrate and underlying free gas on the margin. The seismic systems have frequencies from 20 Hz to 650 Hz; they include: (1) conventional multichannel survey lines, (2) 3-D high resolution multichannel surveys, (3) the very high resolution Deep-Tow Acoustics/Geophysics System (DTAGS) multichannel system that is towed near the seafloor, (4) detailed single channel reflection mapping with several different frequency, small airgun sources, and (5) studies using ocean bottom seismographs. The locations of several conventional multichannel survey lines (shown in later figures) and the region of hydrate occurrence are shown in Figure 2 and the detailed study lines in Figure 3. In addition to the seismic data, 3.5 kHz and 12 kHz profiling, 12 KHz SeaMARC acoustic imaging, and piston coring have provided data on the upper 1-50 m beneath the seafloor. These surveys and sampling demonstrated the occurrence of carbonate deposited on the seafloor from rising methane, and allowed the mapping of fluid expulsion channels.

The seismic data have allowed mapping the regional distribution of the bottom-simulating reflection (BSR) which is characteristic phenomenon of many marine hydrate areas. A BSR occurs beneath much of the middle continental slope off Vancouver Island (Figure 2). An example of a conventional multichannel section across the margin where there is a strong BSR is shown in Figure 4. No BSR has been observed in Cascadia basin, and ODP drilling data indicate that hydrate is not present. Hydrate is not stable at the P-T conditions of the continental shelf sediments at this latitude. The BSR in most low-frequency seismic sections is a single negative pulse that is almost the mirror image of the sea-floor reflection. The BSR reflection amplitude represents a decrease in seismic impedance of 20-30%. There is no continuous reflection above the BSR that indicates the top of a high-velocity hydrate layer, and no continuous reflection below it that represents the base of a low-velocity free gas layer (Figure 5). To produce the observed waveforms the top of the hydrate layer must be gradational upward, and the base of the free gas layer beneath the BSR must be gradational downward (Figure 6).

The surveys and analyses of the seismic data that are described below include mapping the BSR and estimating the BSR reflection coefficient [Clowes et al., 1987; Spence et al., 1991a; 1991b; 1995; Hyndman et al., 1994; Yuan et al., 1994; Fink, 1995], modeling the BSR reflection waveform [Hyndman and Spence, 1992; Fink and Spence, 1999], determining and modeling the frequency dependence of the BSR [Chapman et al., 2001; Riedel et al., 2001], determining detailed interval velocity-depth profiles [Yuan et al., 1996;

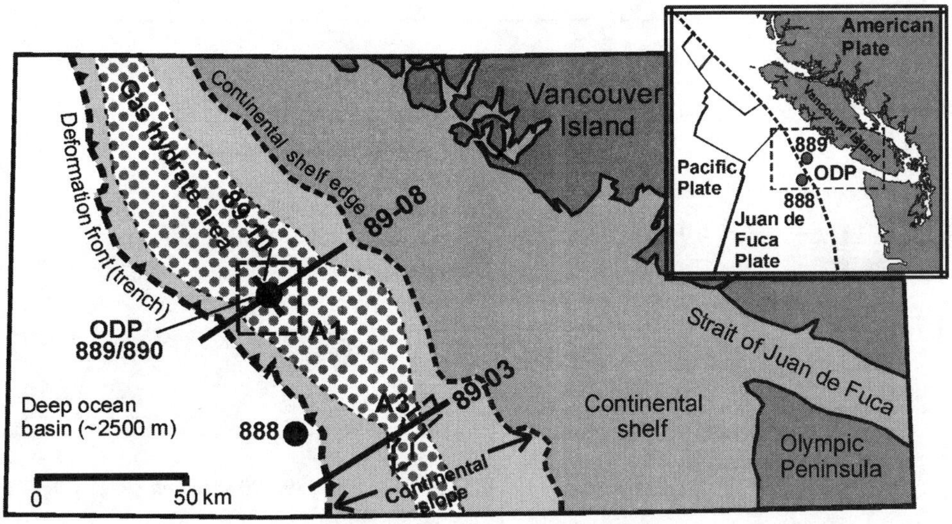

Figure 2. The northern Cascadia margin showing the region where gas hydrate is found. Most detailed seismic data is in region A1 with other detail in A3. The locations of three conventional multichannel lines and ODP Sites 889/890 and reference 888 are also shown.

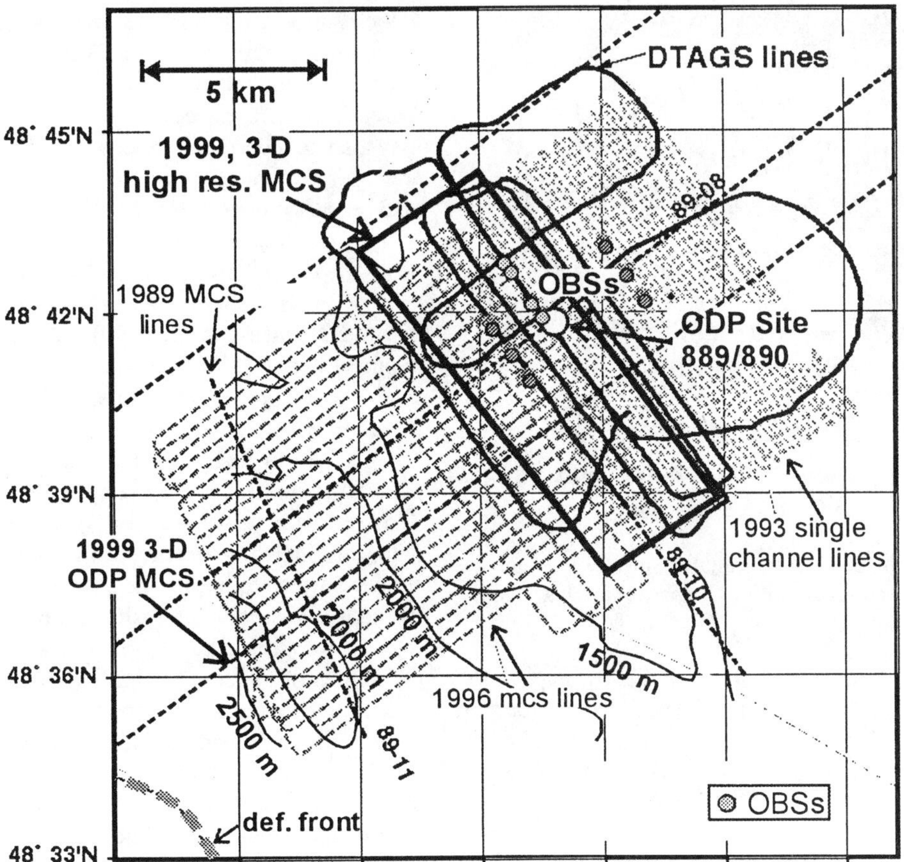

Figure 3. Detail of lines for different seismic reflection systems and for ocean bottom seismographs (OBS) in the area of ODP Site 889/890.

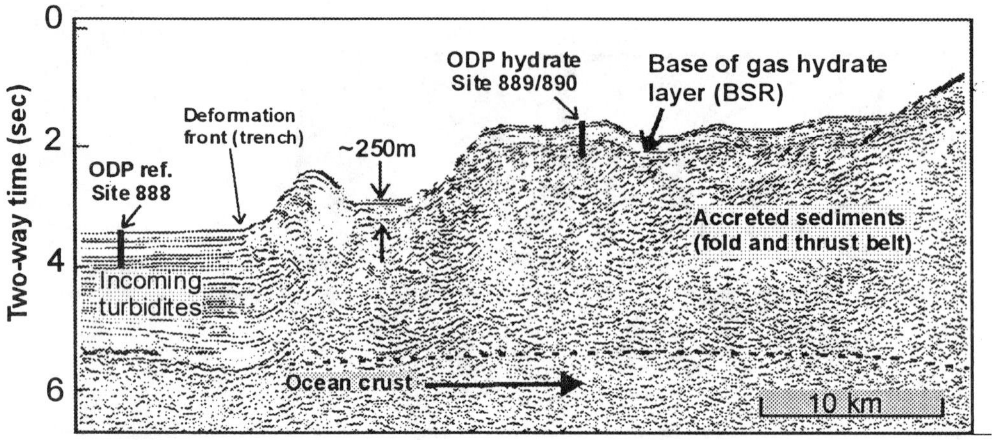

Figure 4. Seismic section across N. Cascadia margin through ODP Site 889/890 showing clear BSR (Line 85-02).

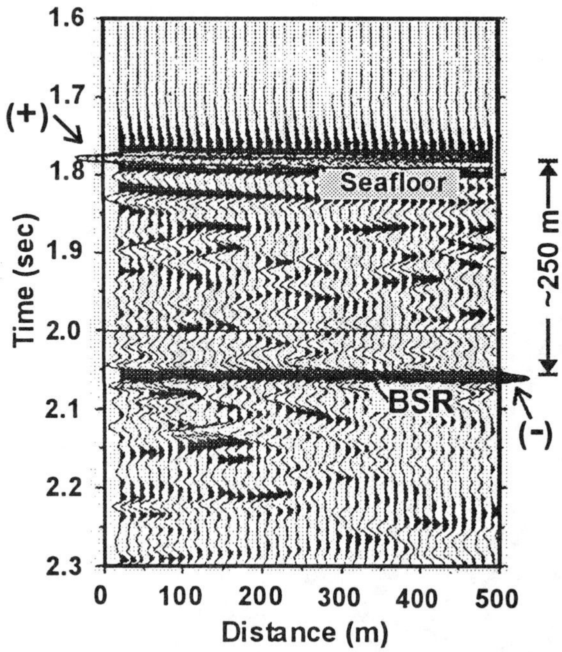

Figure 5. Example of strong BSR near ODP Site 889 showing the simple negative reflection waveform, opposite to that of the seafloor.

Chapman et al., 2001], determining and modeling amplitude-versus-offset (AVO) [Hyndman and Spence, 1992; Yuan et al., 1999; Riedel et al., 2001], full-waveform inversions of the seismic data for velocity-depth profiles [Singh et al., 1993; Singh and Minshull, 1994;Yuan, 1996; Yuan et al., 1999], and geological associations of hydrate BSRs [Hyndman and Davis, 1992; Fink and Spence, 1999; Yuan et al., 1999; Mi, 1999; Ganguly et al, 2000].

Several deployments of OBSs have been carried out near ODP Site 889 in 1993, 1997, and 1999. In 1993, wide-angle data were collected during two deployments of OBSs, recorded with an airgun (120 cu. in.) with a dominant frequency of 75 Hz [Spence et al., 1995]. Simultaneous traveltime inversions were carried out to provide a 2-D velocity model through the drill site [Hobro et al., 1998]. Velocities directly above the BSR ranged from 1.87 to 1.96 km/s. The higher velocities appear to be correlated to an increase in BSR reflection strength which suggests that there may also be a correlation between BSR reflection coefficient and hydrate concentration. A full 3-D velocity inversion has also been completed [Hobro, 1999].

High resolution multichannel seismic data were acquired using the Deep-Tow Acoustics/Geophysics System (DTAGS) [e.g., Rowe and Gettrust, 1993] in the vicinity of ODP Site 889/890 [Gettrust et al., 1999]. The system was towed at a depth of about 1000 m in water depths averaging 1400 m. Because of its proximity to the seafloor, wide-angle data may be recorded on the DTAGS streamer which has a relatively short length of 622 m. The wide angles facilitate velocity estimation and may allow amplitude-versus-offset analysis. However, for velocity analysis and coherent stacking of the data, careful geometry corrections for the varying depth of the streamer were required. The short wavelength associated with the high frequencies necessitates very accurate geometry and inter-hydrophone statics corrections. A method of estimating the depths of the source and hydrophones using sea-surface reflection times was developed by Walia and Hannay [1999]. The frequency band of the DTAGS was 250-650 Hz, providing a vertical resolution of ~2 m (Figure 7). As well, the near-seafloor source and receiver configuration reduces the Fresnel zone size (and horizontal resolution) to about 25 m, compared to ~400 m for 30 Hz conventional seasurface towed seismic systems. In the seismic sections, a high amplitude reflection about 5 ms two-way time below the seafloor often can be resolved (e.g., Figure 7). This near-

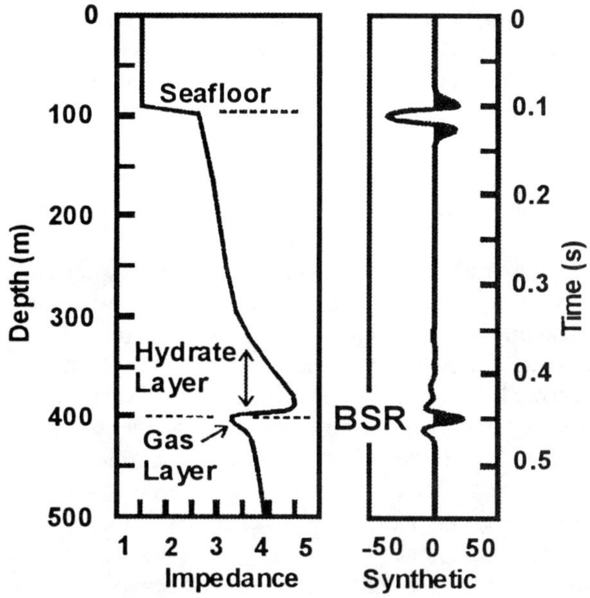

Figure 6. Simple synthetic seismogram that reproduces the main features of the BSRs. The seafloor reflection results mainly from the density contrast and the BSR mainly from the velocity contrast.

seafloor reflection may indicate a carbonate pavement precipitated from upward-advecting fluid [Fink and Spence, 1999]. A number of narrow, near-vertical blank zones are also identified that extend down to the BSR. They may represent fluid expulsion channels as discussed below.

SEISMIC DATA ANALYSES

Distribution of Hydrate BSR

The primary method for mapping the occurrence of gas hydrate is to use the distribution of the BSR. However, evidence for substantial hydrate concentrations where there is no clear BSR has been obtained through detailed seismic interval velocity studies and seafloor electrical resistivity profiling. Thus the region of hydrate distribution based upon the BSR represents a minimum area. The BSR distribution across the continental slope has been defined by data from nine multichannel seismic lines, as shown by Hyndman et al. [1994]; a simplified distribution is shown in Figure 2 . There is a clear BSR over more than half of this region. There are no BSRs in the deep sea Cascadia basin. They first appear on the continental slope 5-10 km landward of the deformation front and BSRs are not evident in water depths of less than 600 m on the upper slope.

Quantitative estimates of the concentration of hydrate are available in only a few locations. However, semi-quantitative estimates are possible by assuming that the BSR reflection coefficient is related to hydrate concentration above the BSR [e.g., Fink and Spence, 1999]. Based on the downhole velocity data at the ODP Site 889, about 2/3 of the BSR velocity contrast appears to result from the high-velocity hydrate and about 1/3 from the low-velocity gas, relative to a no-hydrate no-gas reference velocity-depth [e.g., Yuan et al., 1996]. This ratio is undoubtedly highly variable. For example, on the Blake Ridge the BSR reflection appears to be primarily due to the low-velocity free gas [e.g., Paull et al., 1996]. Also, the reflection coefficient is sensitive to the velocity increase due to the hydrate concentration, only within a wavelength of the main BSR impedance contrast. Thus, although very useful, this approach has a large uncertainty. The estimated hydrate concentrations are significant over a wide area, averaging about half of that at ODP Site 889/890.

A BSR is generally not observed in the well-bedded slope basin sediments (Figure 8), but interpretations vary as to whether or not hydrate is present. Hydrate may not be present because well-bedded sediments reduce permeability compared to the deformed accreted sediments, thus inhibiting vertical fluid and methane flow [e.g., Zuehlsdorff et al., 2000]. Alternatively, hydrate may be present but tectonic subsidence

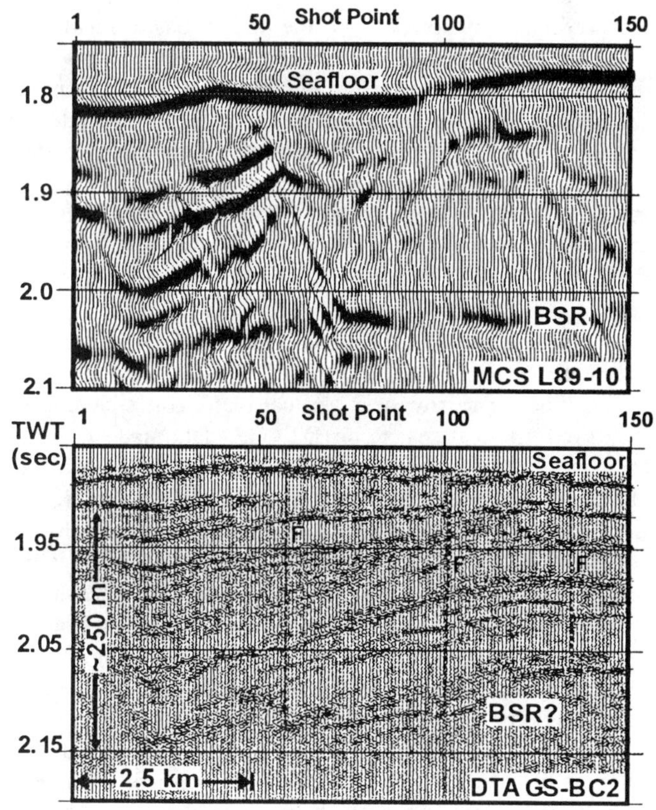

Figure 7. Comparison of conventional surface multichannel data (top) and DTAGS deep towed high resolution seismic section (bottom).

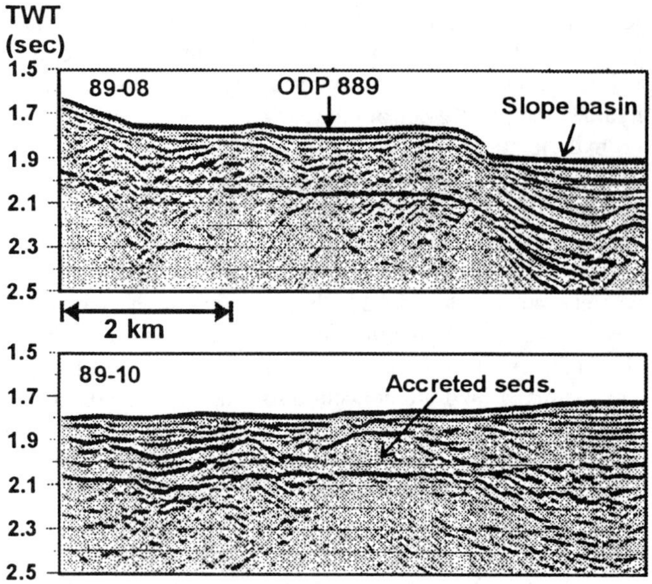

Figure 8. Conventional multichannel examples of clear BSR in deformed accreted sediments and weak or no BSR visible in bedded slope basin sediments (Lines 8 and 10 of Figure 2).

and sediment deposition in the slope basin results in downward movement of the base of the stability field, so that the gas layer is transformed to hydrate and the BSR is much weakened [e.g., von Huene and Pecher, 1999]. The interval velocity data and the recent seafloor electrical resistivity data suggest that hydrate is indeed present above the base of the stability field in these areas. This result supports the explanation of the lack of a significant gas layer for where there is no BSR.

Vertical Fluid Expulsion Channels and Seafloor Venting

A number of narrow near-vertical blank zones have been identified in near-sea floor DTAGS data and in high-frequency sea surface data [Gettrust et al., 1999; Zuelsdorff et al., 2001] that extend down to the BSR, and may represent fluid expulsion channels (Figure 9). A 3-D grid of closely spaced high resolution multichannel data was acquired in the region of ODP Site 889/890 [Riedel et al., 2001]. The survey involved 40 lines at 100 m spacing plus 8 cross lines, and a more detailed small grid at 25 m spacing, focussed on a region of blank zones. One aspect of several blank zones is that they bound regions of anomalously high reflectivity just below the BSR. This high reflectivity may indicate concentrations of free gas below the BSR, which feed the vertical channels. The surface expression of one blank zone, determined by 3.5 kHz sub-bottom profiling (Figure 10), indicated a circular structure with a diameter of about 500 m (Figure 11). Other blank zones are more linear as seen in 3-D time or horizon slices across the seismic reflection grid. The blank zones are thus interpreted as controlled by fractures which act as conduits for the focussed upward expulsion of fluids and methane gas.

BSR Reflection and Reflection Coefficients for Conventional Multichannel and Single Channel Data

The important seismic reflection characteristics of the BSR in northern Cascadia are: (1) The reflection polarity is consistently reversed relative to the seafloor (e.g., Figure 5) indicating, for a simple interface contact, a lower impedance below the boundary, (2) The reflection coefficients are large at the frequencies of conventional multichannel surveys (~30 Hz), up to 50% of that of the seafloor. If the reflection results from an interface, the boundary must occur over a depth interval of less than about a wavelength (~50 m), (3) the BSR reflection is generally a single symmetrical pulse, characteristic of a simple interface. The reversed waveform for the BSR is almost identical to that of the seafloor reflection. There is no systematic pattern giving evidence for either a reflection from the bottom of the free gas layer or from the top of the hydrate layer over a range of seismic frequencies from 15 to 650 Hz. It is concluded that both of these changes occur over depth intervals of greater than several 10's of meters. There is no "flat spot" at the base of the gas layer such as is commonly observed in commercial

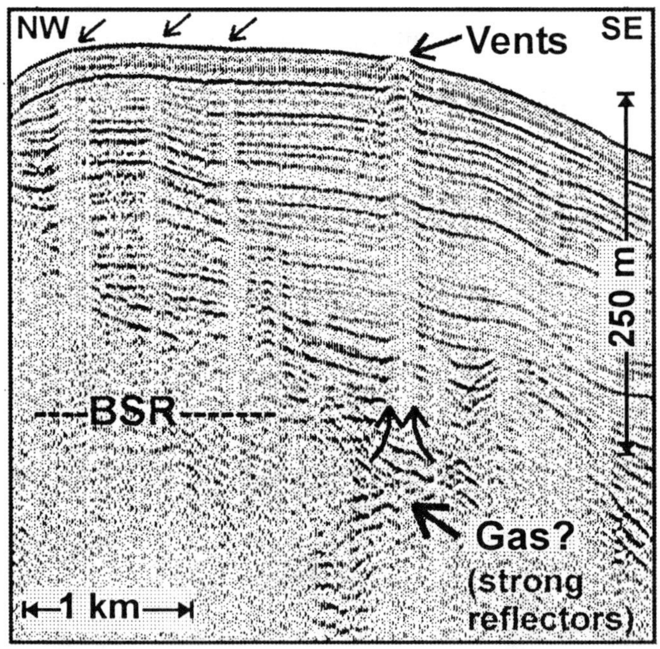

Figure 9. High resolution seismic section showing fluid and gas expulsion vents.

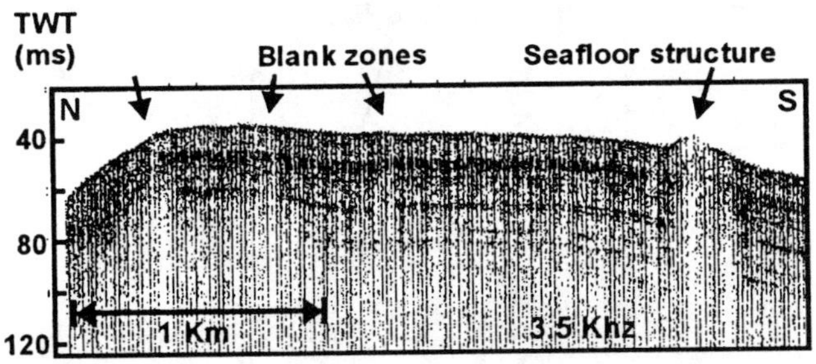

Figure 10. Near-surface expression of vent structures in 3.5 KHz profile.

gas-producing areas. Such flat spots are interpreted to indicate a sharp boundary between overlying high gas concentrations and underlying water saturated sediments.

The BSR reflection coefficients have been estimated by comparison of reflection amplitudes with those of the seafloor. The seafloor reflection coefficient in turn has been estimated from the relative amplitude of the primary seafloor reflection and its multiple [e.g., Warner, 1990], and from the velocity and density of the sediments from piston core data [Mi, 1998]. Hyndman and Spence [1992], Fink and Spence [1999], Yuan et al. [1996], Yuan [1996], and Mi [1998] found BSR reflection coefficients of about 0.1 for conventional multichannel seismic data, compared to typical seafloor reflection coefficients of 0.18-0.24.

Frequency Dependence of BSR

The BSR reflection coefficient in the area of ODP Site 889/890 has been found to decrease markedly with increasing frequency for airgun data, i.e., 15-30 Hz conventional multichannel data to small airgun data of 75-150 Hz [Spence et al., 1995; Zuehlsdorff et al., 2000]. The decrease is even greater for the high frequency DTAGS data of 250-650 Hz, such as to be only intermittently observed [Gettrust et al., 1999; Chapman et al., 2001]. Calibration with respect to the seafloor reflection and the clear sedimentary reflections at similar depths indicate that the low BSR amplitudes at high frequency are not due to insufficient power in the DTAGS signal nor due to scattering losses. The explanation suggested is that the frequency dependence of the BSR reflection amplitudes is a consequence of the BSR being a gradational velocity contrast over a depth interval of 6-10 m [see also Chapman et al., 2001]. This thickness is less than a wavelength for low frequency waves so the boundary appears as a sharp interface, but the thickness is more than a wavelength at high frequencies and the boundary is seen as gradational. The reflection from the BSR therefore is strong in this region only for low frequency seismic data (<100 Hz).

The reflection data over a wide range of frequencies from 20-650 Hz from 5 different seismic survey systems has allowed the frequency dependence of the BSR to be well determined [Chapman et al., 2001]. The frequency dependence has been modeled using synthetic seismograms, and the thickness of the velocity gradient was found to be 6-10 m near ODP Site 889/890, with velocity decreasing by 250 m/s (Figure 12).

Interval Velocity-Depth Profiles

The main quantitative constraint on hydrate and gas concentrations is from very detailed multichannel interval velocities. In addition to velocities in the hydrate and gas zones, velocities to greater depth provide important information on the velocity-depth profile for no hydrate and no gas, which is a critical factor for quantitative estimates of hydrate and gas concentration. Fortunately, the Cascadia sediment section is very homogeneous on scales greater than a seismic wavelength, although there is fine scale turbidite layering. Thus, the reference velocity-depth may be estimated by extrapolating upward the deeper trend of velocity-depth. This extrapolation provides velocities that agree with measured multichannel and piston core velocities near the seafloor where negligible hydrate is estimated. The other source for no-hydrate, no-gas reference velocity-depth data is from ODP density and porosity logs in which hydrate behaves similarly to pore fluid, in contrast to the behavior for velocity. Empirical density-velocity and porosity-velocity relations thus may be used to obtain estimates of no-hydrate velocity-depth [Yuan et al., 1996].

Most of the interval velocity data comes from the conventional multichannel seismic data. Additional high

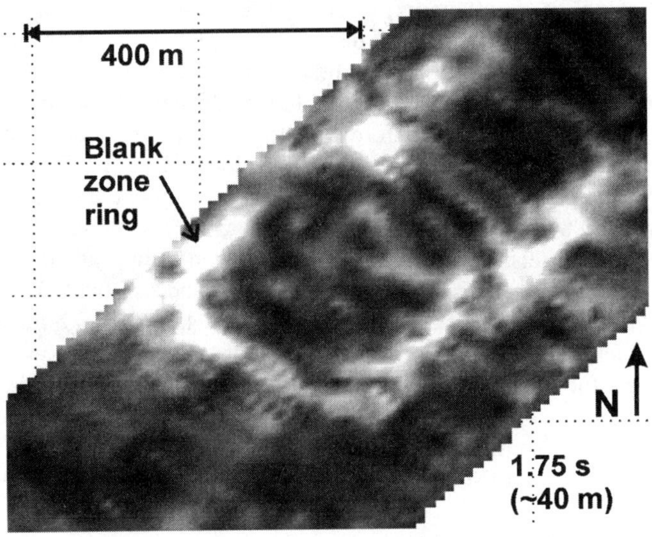

Figure 11. Time slice (1.75 s twt; approx. 40 m depth) over blank zone ("vent" in Figure 9) showing surrounding high-reflectivity ring that may indicate a gas hydrate concentration.

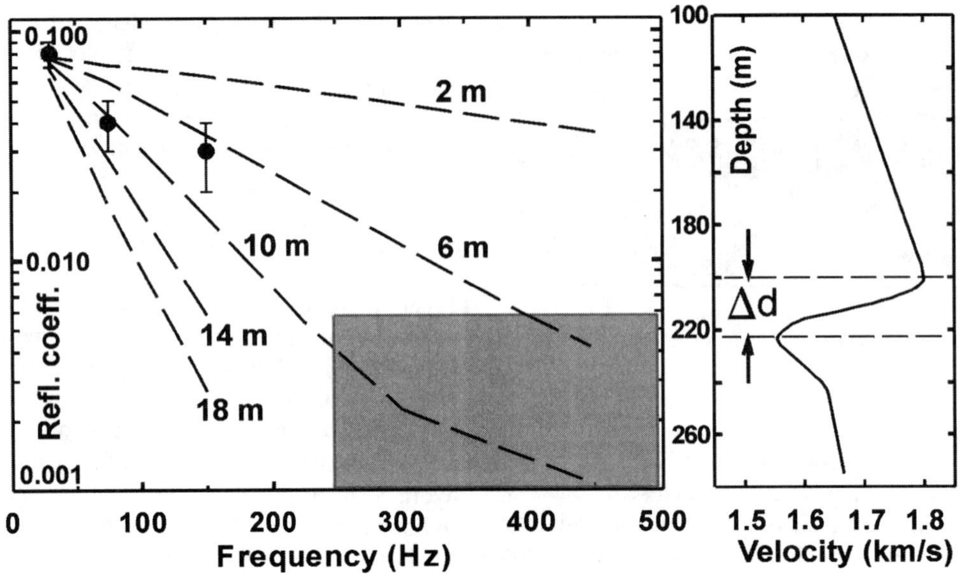

Figure 12. (Left).The observed BSR reflection coefficient with frequency (solid dots). The shaded area is the constraint from DTAGS data. The dashed lines are modeled reflection coefficients with frequency for different transition thicknesses of BSR. (Right) Model BSR velocity-depth.

resolution velocity data giving similar results have been obtained in one area from the deep towed DTAGS multichannel system [Chapman et al., 2001]. The conventional multichannel data were collected using 3500 and 3700 m long hydrophone arrays that are 3 times the water depth in the main study area. This large recording aperture allows unusually high resolution of velocity variations from CDP gathers. Two detailed velocity studies were carried out employing semblance analysis, supplemented by careful visual fit to reflection moveout [Yuan et al., 1994] (Figure 13). The first was to define the regional variations in velocity-depth profiles from the deep sea Cascadia basin to the mid-continental slope. All of the multichannel lines in the basin away from the deformation front (e.g., from bottom left of Figure 3) showed a very similar velocity depth profile. In the second study, detailed velocities were obtained near ODP Site 889/890 for comparison with downhole sonic logs and vertical seismic profiles (VSP), and to constrain estimates of hydrate concentration. In the latter area, the sediment prism has been tectonically shortened and thickened by about a factor of two, and the vertical stretching of the velocity-depth profile results in significantly lower reference velocities compared to the same sub-bottom depths in the deep sea basin. The lower velocities probably reflect lower porosities compared to the basin reference. The lower porosities are in turn interpreted to be maintained by high pore pressures [e.g., Hyndman et al., 1993].

The velocities of the continental slope sediments in the area within 10 km of the ODP Site 889/890 are well constrained and, excluding those velocities in the hydrate above the BSR, they increase smoothly with depth (Figure 13). This velocity-depth profile is the primary no-hydrate no-gas reference that has been used. However, Yuan et al. [1996] considered that the upper 120 m of the section penetrated at ODP Site 889/890 is little-deformed slope sediments. This layer varies in thickness from 0-250 m in the 10 km wide area near the drill site used for the detailed multichannel velocity analyses. For this section, the undeformed deep sea basin sediment velocity-depth profile could be more appropriate. The interval velocities are not sufficiently accurate at shallow depths to resolve this difference. Just above the BSR the observed velocities are about 1800 m/s whereas the reference velocity at that depth is 1650 m/s for the prism reference (Figure 13). For the deep sea reference, the velocity at that depth is about 1730 m/s. If the deep sea velocity-depth reference is used in the areas of largely undeformed slope sediments (0-128 m), the hydrate concentration in those areas is about half of that from using the deformed prism reference.

Amplitude-Versus-Offset

Amplitude-versus-offset (AVO) analyses (Figure 14) that determine the change in reflection amplitude as a function of receiver offset or reflection angle, are an important method used by the petroleum industry for gas detection beneath an interface. AVO interpretations are based upon the theory and observation that free gas substantially reduces the P-wave velocity with little effect on the S-wave velocity [e.g., Gregory, 1976]. As shown by Ostrander [1984], Shuey [1985], Rutherford and Williams [1989] and many others, the

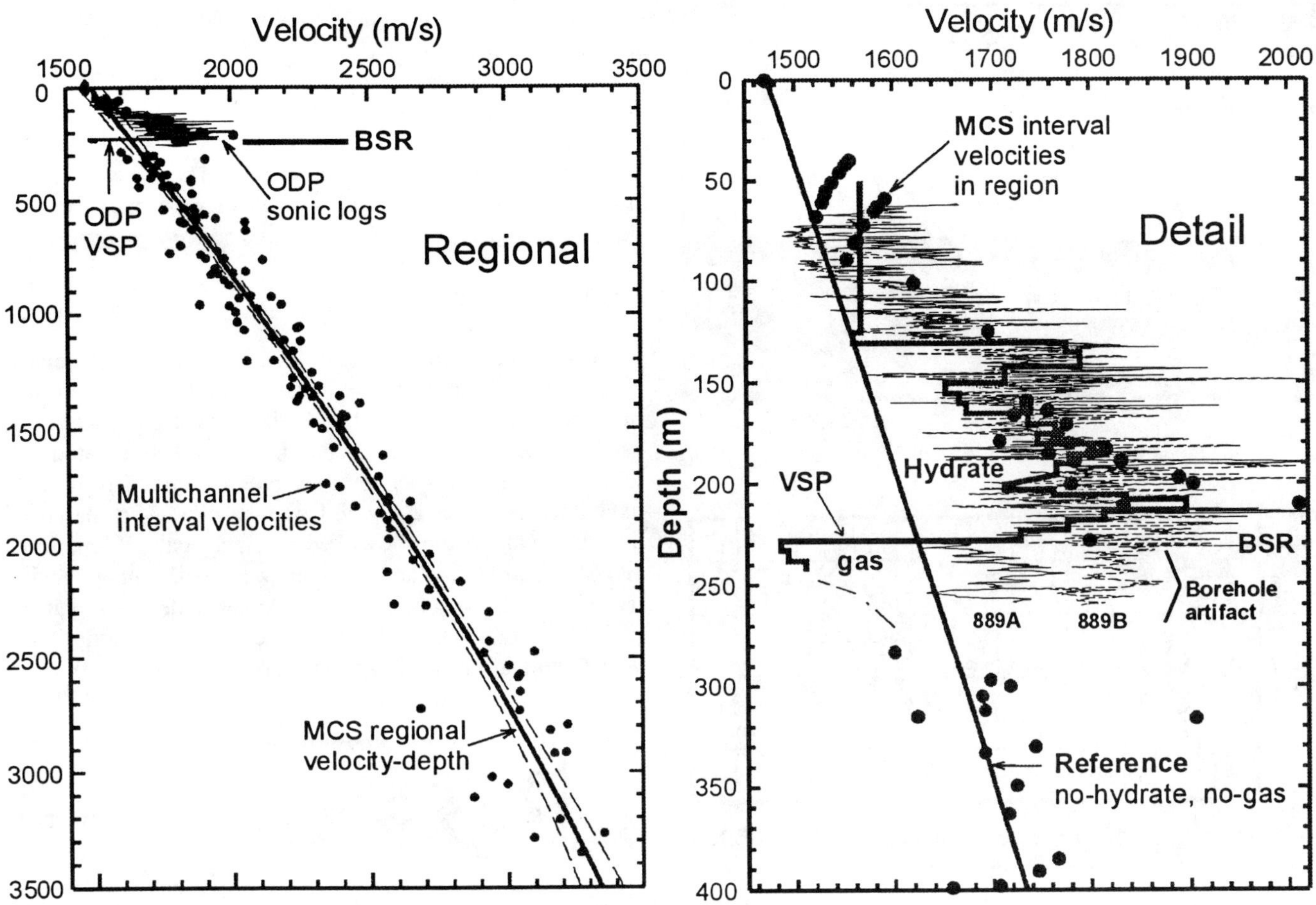

Figure 13. (Left) Deep multichannel interval velocity data from within 10 km of ODP Site 889/890 that provides the no-hydrate, no-gas reference. (Right) Detailed comparison of interval velocity data with ODP sonic logs and downhole VSP data. Only the VSP shows the low-velocity free gas layer.

P-wave reflection amplitude changes with reflection angle depending on the shear wave as well as compressional wave velocities above and below the reflecting interface, especially to the Vp/Vs ratio or to Poisson's Ratio. It was thus hoped that AVO would provide information on whether the BSR is mainly due to underlying free gas or to overlying hydrate, and on the amount of free gas below the BSR [e.g., Hyndman and Spence, 1992]. An example of the change in amplitude and waveform for the hydrate BSR as a function of offset distance is shown in Figure 15 for a location near ODP Site 889/890. At far offsets the amplitude increases rapidly and there is a phase shift.

Accurate AVO requires correction for the directivity of both the airgun array and for the hydrophone array, as well as for wave spreading. For the conventional multichannel systems used on N. Cascadia, the combined hydrophone and airgun array directivity correction is large, about 0.55 for a reflection angle of 45° (an offset of 3000 m for the seafloor and BSR near ODP Site 889/890). Figure 16 shows the seafloor AVO with and without the directivity corrections for locations near ODP Site 889/890. The observed seafloor AVO compared with that predicted by a seafloor impedance model provides a check on these corrections. The expected free gas layer behavior of an approximately linear increase in amplitude with offset is not observed. The amplitude behavior where the BSR is strong is little change or slight decrease at near offsets, but a rapid increase in amplitude for offsets of greater than about 2000 m (incidence angle of about 60°) (Figure 16).

The waveform and amplitude-versus-offset behavior of a number of models has been obtained using the full waveform modeling routine of Fuchs and Muller [1971] for comparison with the observed data [Hyndman and Spence, 1992]. The AVO behavior of the hydrate BSR is found not to be as useful for determining the amount of free gas below the BSR as was earlier hoped [Yuan et al., 1999]. Three models were tested; gas only, gas plus hydrate and hydrate only. The best fits are for the hydrate only and hydrate plus gas models. However, there are only small differences in AVO for the three models

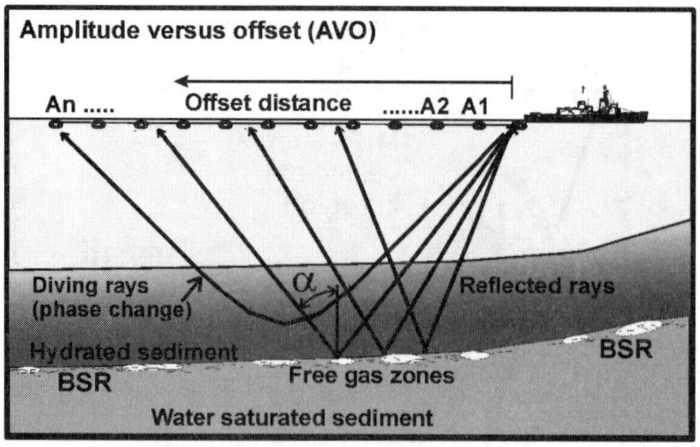

Figure 14. Amplitude versus offset studies.

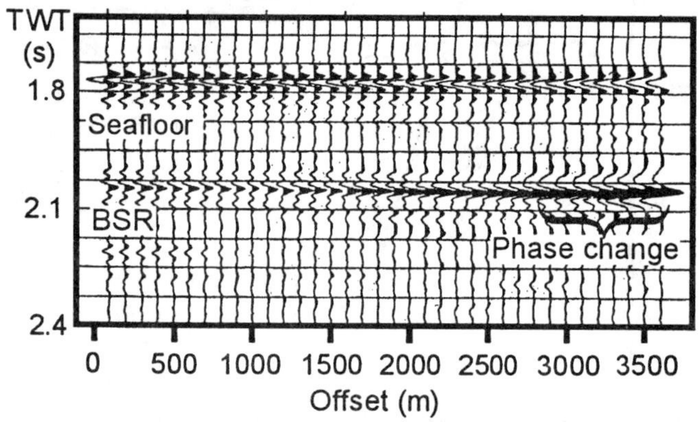

Figure 15. Common depth-point gather illustrating the seafloor and BSR reflection amplitude with offset. There is an amplitude increase and phase change at far offsets for the BSR.

(Figure 17). The lack of a near offset AVO effect appears to be because the high-porosity unconsolidated sediments have high Poisson's ratios (and Vp/Vs ratios). The change due to gas is not sufficient to reduce the Poisson's ratios to the small values required for normal gas AVO behavior. The observed nonlinear AVO behavior, constant or slightly decreasing at near to mid-offsets and large amplitude at far offsets, can be reproduced in the hydrate models if the S-wave velocity increases as expected with hydrate cementation [e.g., Hyndman and Spence, 1992].

The rapid increase in amplitude at far offsets is interpreted to represent turning rays in the high velocity hydrate above the BSR. At far offsets, the downward increasing velocity is sufficient for rays to turn and be refracted upward before they reach the BSR reflection. Since the BSR reflected waves involve a different phase compared to the turning rays, there is a phase shift from near to far offsets (Figure 15). The offset reflection angle where amplitudes increase and the reflection phase changes provide a measure of the velocity gradient above the BSR. This angle thus provides a constraint on hydrate concentration.

As yet the dependence of AVO on a broad range of frequency has not been examined. Only the frequency dependence of vertical incidence reflection as summarized above. A thin gas layer may have a distinctive AVO frequency response, so such studies may be productive in determining gas layer thicknesses and concentrations.

Full Waveform Inversions

Forward modeling analyses of the BSR amplitude-versus-offset used only the maximum wave amplitudes. Additional information is contained in the detailed waveforms that can be used by full waveform inversion. Several automated full waveform inversion schemes for multichannel seismic data have been developed; for N. Cascadia most work has been done using the scheme described by Singh et al. [1993], Singh and Minshull [1994] and Yuan et al. [1999]. This inversion technique tries to find a velocity-depth model such that the synthetic data produced by the model fit both the amplitude, waveform and phase of the data sample by sample, over all

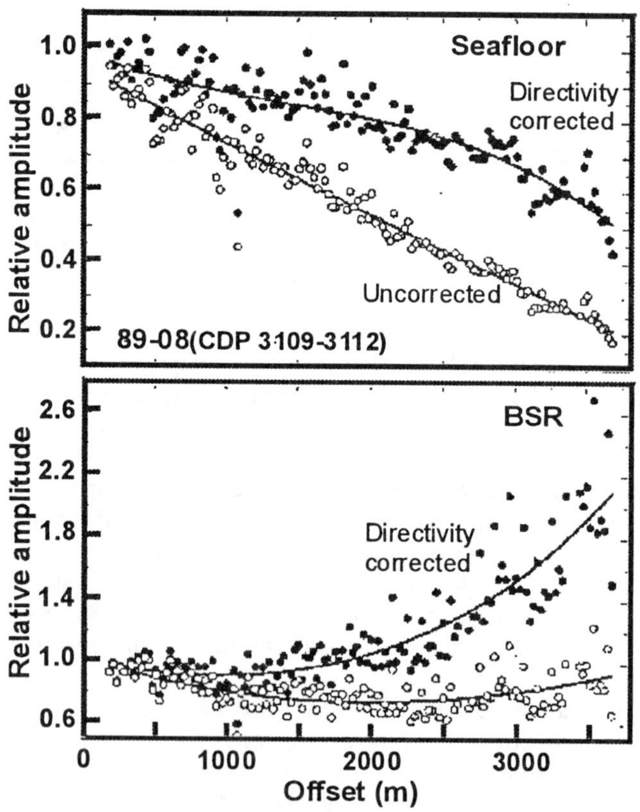

Figure 16. AVO for the seafloor and the BSR observed and directivity corrected, for a composite gather near ODP Site 889/890.

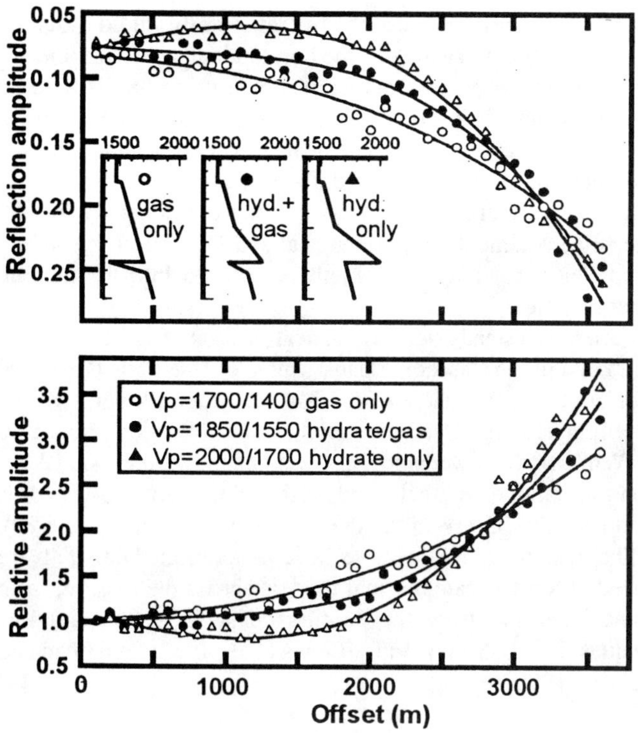

Figure 17. Synthetic reflection amplitudes with receiver offset from three BSR velocity-depth models for hydrate and underlying free gas.

offsets. The inversion was implemented on CDP gathers near ODP Sites 889/890 that are representative of the main different AVO characteristics, and where the seafloor and BSR are nearly horizontal so 1-D models can be used. In the first inversion work the long-wavelength velocity structure was estimated by a Monte Carlo search in the τ-p domain [Singh and Minshull, 1994], but a more accurate starting model was used later based upon the multichannel interval velocities in the area [Yuan et al., 1999]. The inversion was found to constrain the general form of the velocity-depth profile well. However, it provided poor constraint on the long wavelength velocity-depth variation. The long spatial scale velocities were found to be very sensitive to the starting model; equally good fit to the waveforms and amplitudes were obtained for a considerable range of long spatial scale length velocities. Thus the choice of the no-hydrate, no-gas reference has an important effect on the final conclusion as to whether the BSR is mainly due to high-velocity hydrate or to low-velocity gas layers. The best fitting full waveform inversion velocity-depth model, using a smoothed version of the interval velocities as the starting model, is shown in Figure 18. There is good general agreement with both the multichannel interval velocities and with the ODP downhole velocity data.

TEMPERATURE-PRESSURE OF THE BSR AND HYDRATE STABILITY

Temperature measurements in the ODP boreholes provide important information on the pressure-temperature (P-T) relation for the BSR and its relation to laboratory data for the stability field of gas hydrate. Figure 19 shows the temperature estimated for the BSR at ODP Site 889/890 compared to other ODP data, and the stability field for pure water and pure methane and for seawater and methane [Westbrook et al., 1994; Hyndman et al., 1992; Sloan, 1990; Dickens and Quinby-Hunt, 1994]. The P-T for the BSR is used in calculating the heat flow from the BSR depth. The Site 889/890 temperatures are about 1°C lower, but within the measurement uncertainty of the seawater curve. The difference may result from measurement uncertainty, porefluid or gas composition difference, or that the BSR does not exactly correspond to the base of the hydrate stability [e.g., Xu and Ruppel, 1999; Zatsepina and Buffett, 1997].

HEAT FLOW DATA AND DEPTH TO THE BSR

The regional heat flow across the margin shows the effect of Juan de Fuca plate subduction. The heat flow is high in the

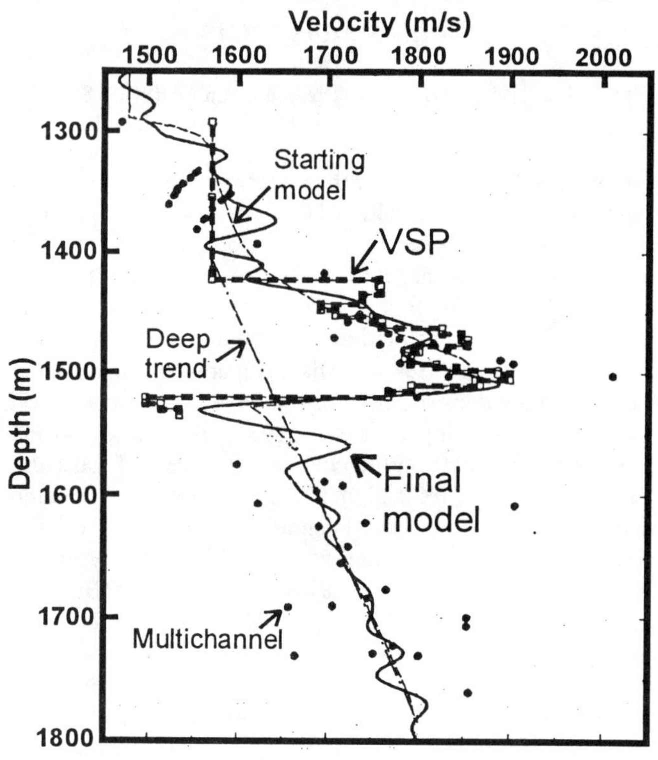

Figure 18. Full waveform inversion (solid curve) compared to interval velocities (open circles) and downhole VSP (solid dashed line). The thin dashed line is the starting model.

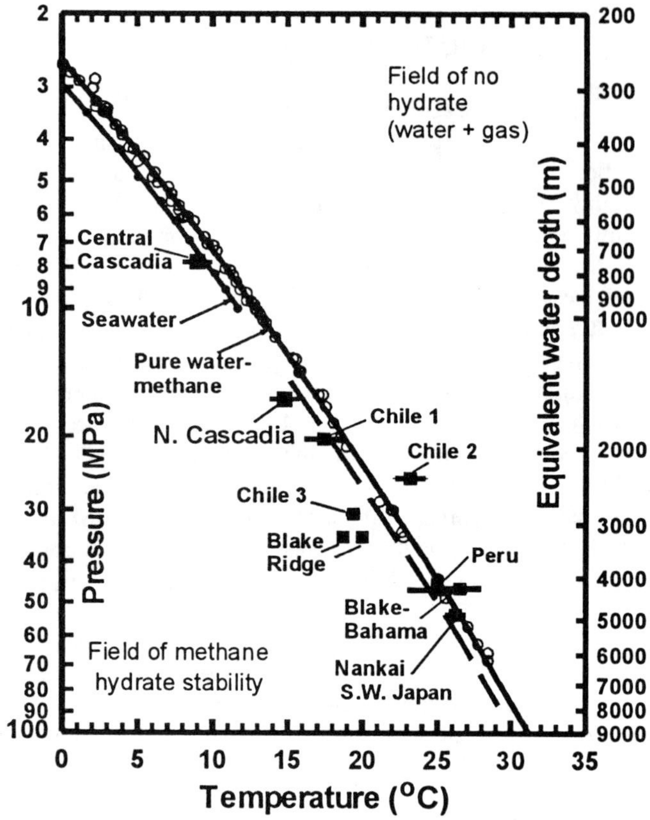

Figure 19. BSR temperature at N. Cascadia ODP Site 889/890 compared to other ODP estimates and laboratory hydrate P-T data.

Cascadia basin associated with the very young oceanic crust. The heat flow decreases inland of the trench due to the heat sink of the cold upper part of the oceanic plate that is underthrusting the margin [e.g., Hyndman and Wang, 1993; Wang et al., 1995]. On a smaller scale, the heat flow is affected by the deformation and fluid expulsion of the accretionary prism. Heat flow data (Figure 20) are available from land boreholes and from petroleum exploration wells on the continental shelf [Lewis et al., 1988; 1991], from marine probe measurements on the continental slope and Cascadia Basin [Davis et al., 1990; Hyndman et al., 1993], and from the depth of the hydrate bottom-simulating reflection on the continental slope [Hyndman et al., 1993; Ganguly et al., 2000]. Good heat flow data are available from ODP Site 889 on the mid-slope. The detailed profiles of "pogo-probe" heat flow across the continental slope and the heat flow from ODP Site 889/890 allow comparison with the estimates from the depth to the hydrate BSR. There is general agreement among the heat probe data, the data in the ODP wells and the BSR heat flow, but in detail there are differences that are not yet explained [Davis et al., 1990; Hyndman et al., 1993]. The temperature data in the ODP wells indicates that the BSR occurs usually slightly cooler than the predicted P-T conditions for the base of the hydrate stability field based on laboratory data. However, the correspondence is sufficiently close to allow measurements of the depth to the BSR to be used to map heat flow on the continental slope, with an accuracy of about 15%. The heat flow map compared to the predictions of numerical subduction zone thermal models shows the thermal effect of the rapid thickening of the accreted sediments [Hyndman et al., 1993], and suggests fluid expulsion that may carry methane upward into the hydrate stability field.

There is a steady decrease in heat flow landward from about 120 mWm^{-2} in Cascadia Basin, which is close to that expected for oceanic lithosphere of 6-7 Ma age, after allowing for the cooling effects of sedimentation. The average is about 90 mWm^{-2} on the lower slope, decreasing landward to 50 mWm^{-2} on the continental shelf, and further decreasing to very low values of 30-35 mWm^{-2} in the continental forearc (Figure 20). There also are radioactive heat generation data, that are needed for accurate thermal models across the margin, from land sites and from the continental shelf wells, with low values from 0.4-0.6 μWm^{-3} [Lewis et al., 1988; Hyndman and Lewis, 1995].

THERMAL MODELS FOR SUBDUCTION AND FOR SEDIMENT ACCRETION

Thermal data provide one indicator of upward fluid expulsion which is inferred to carry methane generated at depth up into the hydrate stability field. To isolate the effect of fluid expulsion on heat flow, the larger-scale thermal variations associated with subduction underthrusting must be subtracted. The large scale thermal regime across the margin has been modeled numerically [Hyndman and Wang, 1993; 1995; Wang et al., 1995]. The model includes the dependence on the age of the subducting plate, the plate converence rate, the thickness of insulating sediments on the incoming oceanic crust, the angle profile of the fault, and the thermal properties of the overlying accretionary prism and forearc crustal material. The model-predicted heat flow (Figure 20) provides a reference for estimating the more localized thermal effect of the sediment prism thickening and regional fluid expulsion, as well as localized fluid expulsion.

A numerical thermal model has been developed that incorporates sediment accretion, consolidation, and fluid expulsion processes [Hyndman et al., 1993; Wang et al., 1993] (Figure 21). The advective thermal effects of the thickening of the sediment prism (vertical stretching of the isotherm spacing) and of fluid expulsion (upward advective heat transport) have been estimated from the reduction in surface heat flow as measured from the BSR depth data compared to the regional model (Figure 21), and from the landward change in porosity-depth as estimated from the

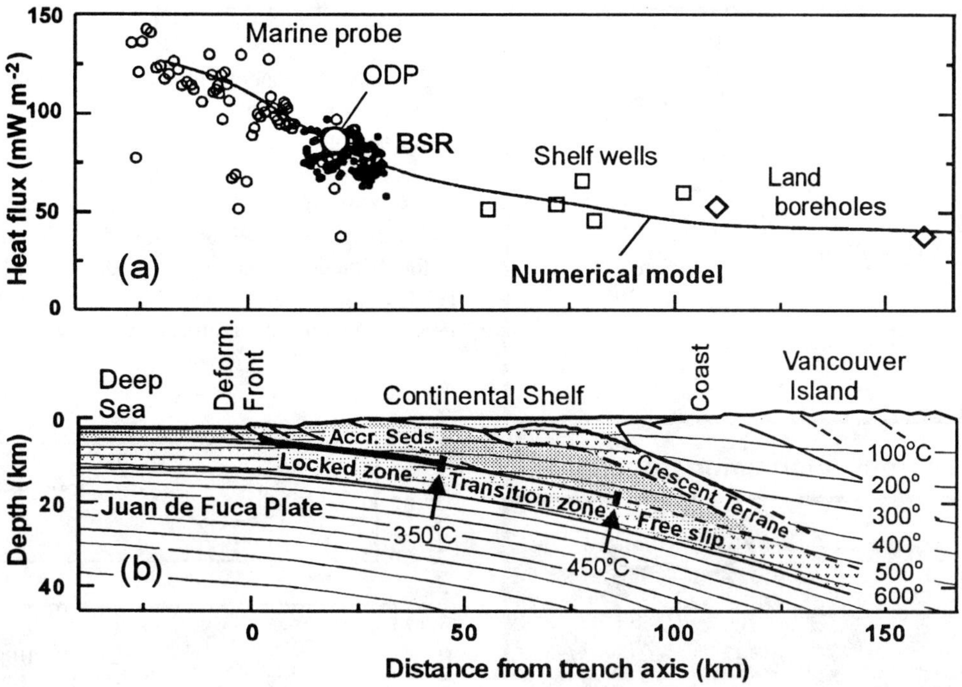

Figure 20. Heat flow across the N. Cascadia margin including from BSR depths, and model temperature cross-section. The BSR temperatures have been corrected for the effect of sediment thickening and fluid expulsion.

multichannel velocity-depth data. The heat flow is reduced to 70-80% of the regional model due to the effect of sediment thickening [Hyndman et al., 1993]. An important inference is that the sediments in the 30 km wide frontal portion of the accretionary prism are underconsolidated and therefore very high pore pressure must exist in the accretionary zone. This pore pressure must be slowly released farther landward by upward fluid expulsion that is distributed across much of the continental slope. If methane is generated over a considerable depth below the BSR, this fluid should produce increasing amounts of gas hydrate landward.

A grid of short-offset multichannel seismic lines was acquired southeast of Site 889 (Area 3 in Figure 2), and the heat flow calculated from the depth of the BSR [Ganguly et al., 2000]. The regional variation of heat flow, decreasing landward, is consistent with the tectonic thickening of the accretionary wedge, but there are significant local variations, notably BSR deflections indicating low heat flow values over topographic highs and high heat flows over the flanks. As shown in Figure 22, heat flow at some localities increased by as much as 50% (from 65 to 100 mW/m^2) over a horizontal distance of 1-2 km. Much of this variation may be due to the focussing and defocussing effects of topography. However, a remaining component may be produced by thrust faulting or by the upward migration of warm fluids that produces localized increases in heat flow. Zwart et al. [1996] observed high temperature gradients over lateral distances of less than 300 m off Oregon in central Cascadia which may be caused by advective heat transport mechanisms. Correlation of anomalously high temperature gradients with out-of-sequence thrusts indicated that these structures are currently acting as conduits for warm fluids. The long-term observations of pressure and temperature in central Cascadia Hole 892B reported by Davis et al. [1995] confirm the occurrence of upward fluid flow in the fault penetrated by that hole.

SEAFLOOR ELECTRICAL PROFILING AND COMPLIANCE ESTIMATES OF HYDRATE

An important new technique for mapping sub-seafloor hydrate concentrations is the electrical profiling systems developed at the University of Toronto [Edwards, 1997]. The dipole-dipole system is towed on the sedimented seafloor. It consists of a signal transmitter and a series of electrical receivers towed at a range of distances. Resistivity-depth information can be obtained to several 100 m depth. As with the seismic surveys, downhole electrical logs provide an important calibration. Sediment core electrical resistivity and pore fluid salinity also are important in the analysis. Excellent agreement has been obtained in the resistivity-depth between the ODP downhole resistivity logs and the electrical profiling system data from the immediate area [Yuan and Edwards, 2001] (Figure 23). Hydrate pore saturations of up to 17-26% are estimated in the area of the ODP Sites 889/890. As with

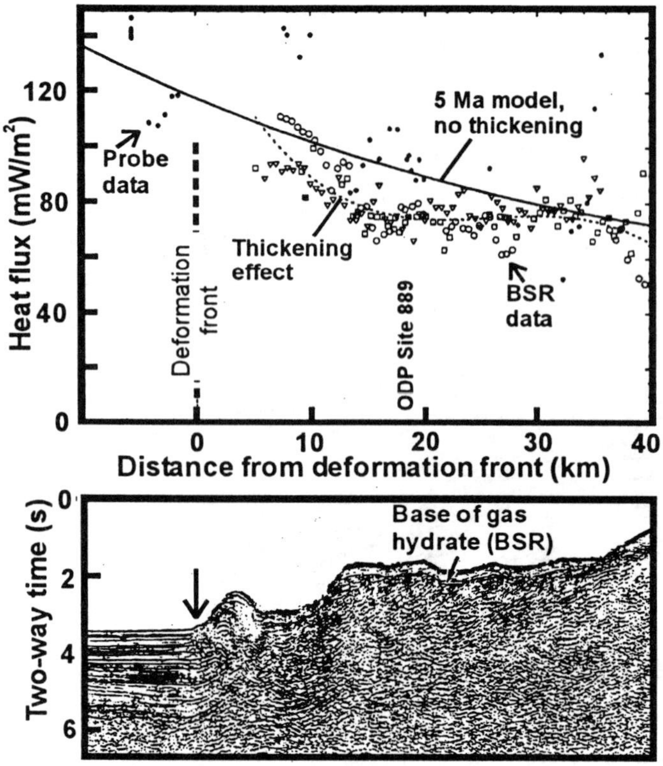

Figure 21. Heat flow across the continental slope from BSR depths along three MCS lines, compared to large scale thermal model prediction. The dashed line is the prediction of the model for sediment thickening and fluid expulsion.

the seismic velocity data, equally high concentrations of hydrate are estimated for areas without a BSR as for those with a strong BSR. It is noted that a comparison of seafloor and downhole data may be affected by anisotropy. The resistivities from the seafloor data apply mainly to the horizontal resistivity whereas the downhole resistivity data are mainly for the vertical resistivity.

Another new technique for studying seafloor hydrate and associated free gas is seafoor compliance. Compliance is the transfer function between the vertical displacement of the seafloor and the seafloor pressure as a function of frequency. It is sensitive mainly to the sediment shear modulus. A differential pressure gauge and a high precision seafloor gravimeter record long period ocean waves and infragravity waves [Willoughby and Edwards, 2001]. Only initial results have as yet been obtained, but for the area of ODP Site 889/890 comparison with models suggest a high shear velocity above the BSR interpreted to be gas hydrate. The presence of free gas will mainly decrease compressional velocities, not shear velocities, so the method is insensitive to free gas.

OCEAN DRILLING PROGRAM DATA

This review concentrates on the geophysical data for gas hydrates on northern Cascadia margin, but important complementary geological and geochemical data were provided by ODP Leg 146 that drilled several adjacent wells (Sites 889 and 890) on the mid-continental slope off Vancouver Island (Figure 2), and a reference hole (Site 888) in the adjacent deep sea Cascadia basin [Westbrook et al., 1994; Carson et al., 1995; MacKay et al., 1994; Jarrard et al., 1995]. Downhole logging and core analyses give important

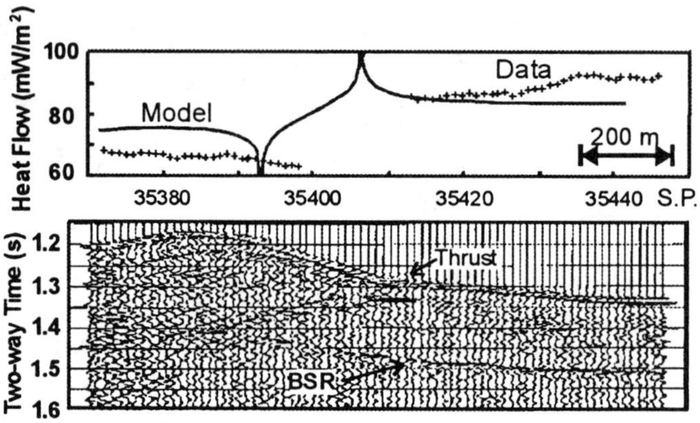

Figure 22. Heat flow estimated from depth to BSR compared to the predicted heat flow for the observed topography. There may be an additional effect from fluid flow along the thrust fault.

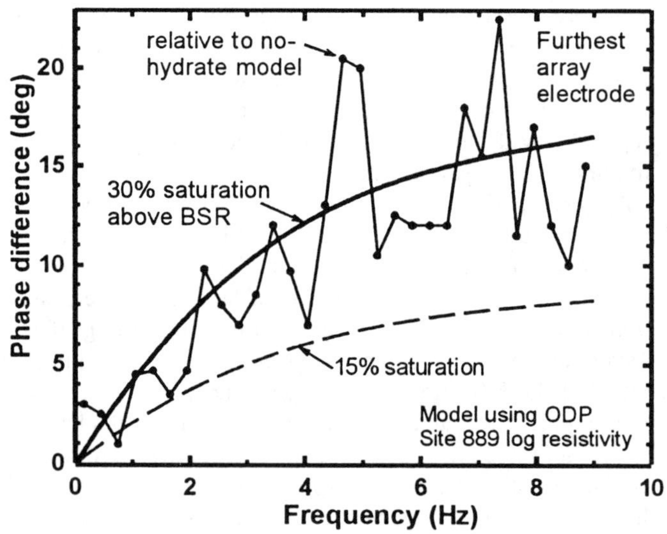

Figure 23. Electrical profiling results in the area of ODP Site 889/890 showing a comparison of observed data compared to model based upon downhole electrical resistivity data (see above).

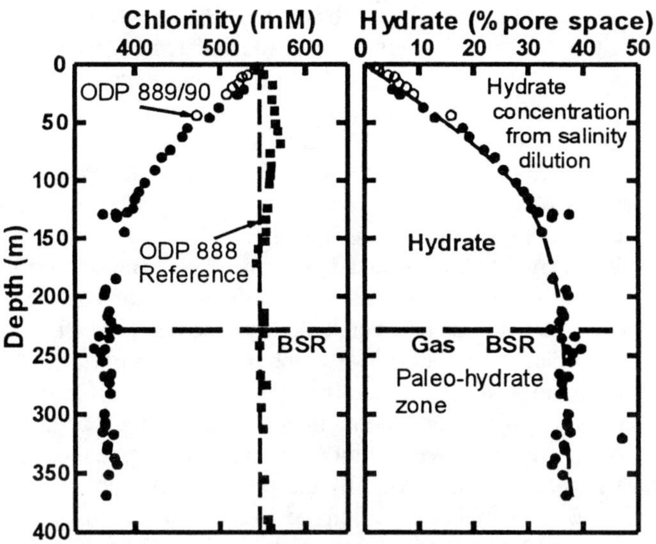

Figure 24. (Left) ODP core pore fluid chlorinity for hydrate Site 889/890 and Cascadia Basin reference Site 888. (Right) The hydrate concentration at Site 889/890 estimated assuming all of the chlorinity difference compared to 888 is due to dissociation of insitu hydrate.

constraints on the concentrations of hydrate and free gas at the drill sites, and provide calibration of the regional geophysical data. The recovered hydrate appeared as disseminated "snow", but much of the hydrate in the core had dissociated before the core recovery. However, cold temperatures in the recovered core due to hydrate dissociation demonstrated that substantial hydrate had been present [Westbrook et al., 1994]. The core pore fluid freshening and other geochemical data also indicated that substantial amounts of hydrate were present [Kastner et al., 1995] (Figure 24). When hydrate is formed, it excludes most of the salt in the seawater pore fluid. If the salt produced when the hydrate formed has diffused or advected away to give normal seawater pore fluid salinities, when the hydrate dissociates it produces freshened pore water.

The downhole geophysical logs provide the most quantitative estimates of hydrate and gas concentrations. The sonic logs and the VSP give excellent agreement with the velocity-depth profiles determined from the multichannel seismic data in the immediate area. The sonic logs (Figure 13) do not resolve the gas layer, probably because hydrate is formed from the gas in an annulus around the hole below the BSR, through cooling by the cold drilling fluid. Alternatively, the drilling circulation washed out the gas immediately adjacent to the borehole wall. However, the gas layer is well resolved by the low velocities in the VSP which has a greater effective depth of penetration. The velocity drops to about 1500 m/s from a reference of ~1650 m/s for 10-20 meters just below the BSR.

The downhole electrical resistivity logs provide an important second constraint on hydrate concentrations. For the resistivity data, the no-hydrate reference has been obtained from the Cascadia basin reference hole and from core measurements after the hydrate has dissociated [Hyndman et al., 1999]. The difference in porosity-depth profile for the basin compared to the accretionary prism has only a small effect on resistivity, in contrast to the larger effect for velocity. The effect of porosity-depth differences between the two areas also is partially cancelled by the difference in the temperature effect on resistivity in the two areas. The heat flow and thus temperature gradient is higher in the basin compared to in the continental slope area.

In the downhole resistivity data from Site 889/890 there is a pronounced increase in resistivity to about 2.0 ohm-m in the 100 m hydrate zone above the BSR, compared to a no-hydrate reference of about 1.0 ohm-m (Figure 25) from Cascadia basin Site 888, and from core measurements after any hydrate has dissociated. It is possible to separate the effect on downhole resistivity due to hydrate from that due to lowered pore fluid salinity if measurements of core resistivity and pore

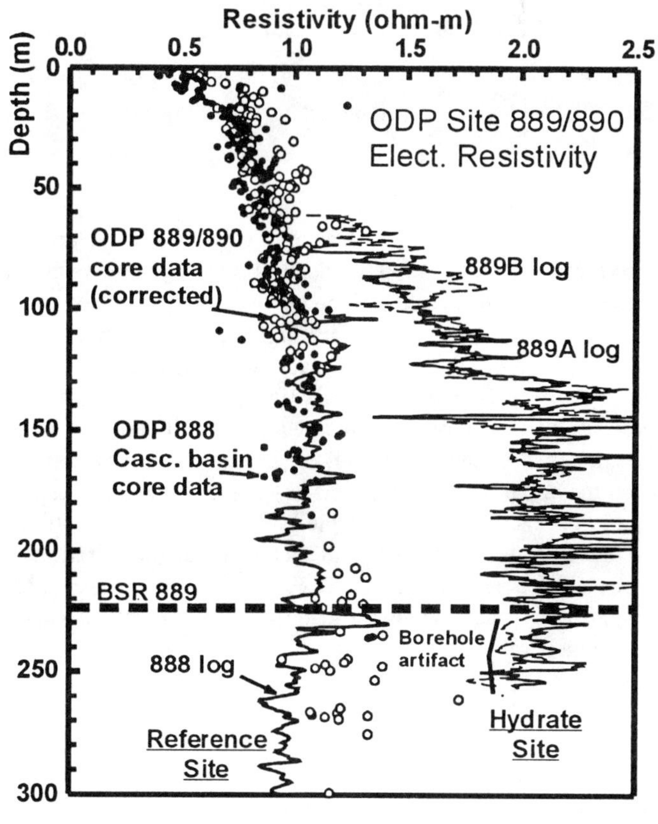

Figure 25. Downhole electrical resistivity at hydrate Site 889/890 compared to reference Site 888 in Cascadia Basin and to laboratory core sample data corrected in insitu conditions.

TABLE 1. Lithology of ODP Site 889/890 [summary from Westbrook et al., 1994].

0-128 m (slope basin sediments)

Upper Quaternary clayey silts, fine sands, and diagenetic carbonates comprising slope and slope basin sediments, deposited as hemipelagites, turbidites, and mass-flow deposits. The interval 0-104 m is dominated by subhorizontal bedding, with little apparent deformation of the slope-basin sediments. The interval 104-127 m is characterized by less common sand/silt beds that dip 40°-70° to the west, and shows evidence of slump folding.

128-345 m (hole bottom) (accreted sediments)

Lower Pleistocene to Pliocene sediments of similar composition to above, but more consolidated and is highly fractured; consists of abyssal-plain silts and clays that have been fractured during accretion. Tectonic stress appears to be accommodated by distributed strain. The dominant fracture fabric dips 45° to 60°. Diagenetic carbonates are observed throughout the section. Abundant authigenic glauconite below 301 m suggests deposition under suboxic conditions.

fluid salinity (after dissociation of any hydrate) are available [Hyndman et al., 1999]. Most of the higher downhole resistivity at Site 889 appears to be due to hydrate. However, above 120 m depth, a part of the high resistivity is calculated to be due to pore fluid salinity that is somewhat lower than for the reference [Hyndman et al., 1999]. The depth interval above 120 m is slope basin sediments (Table 1), in contrast to the deeper accreted Cascadia Basin sediments, so it is possible that the apparent low salinity is an artifact of the calculation assumptions. A simple Archie's Law model of resistivity versus hydrate concentration gives a hydrate concentration of about 30% pore saturation, or 15% of sediment volume. This estimate is somewhat higher than that from the velocity data, but in general agreement.

Hydrate and Free Gas Concentrations

Determining hydrate and free gas concentrations are primary objectives of the N. Cascadia gas hydrate studies. Seismic interval velocities are the main method for hydrate and gas estimates from surface or near-seafloor geophysical data, but seafloor electrical resistivity profiling has provided important additional data. Each of these methods have resolved the hydrate layer above the BSR. However, only seismic full wave inversion has allowed resolution of the underlying free gas layer, and that only semi-quantitatively. For ODP data, the downhole sonic velocity and VSP data, along with the downhole electrical resistivity logs, provide the primary constraints. In the downhole data only the VSP velocity data resolves the low-velocity free gas layer. Both the sonic logs and electrical resistivity logs are strongly affected by borehole effects below the BSR. In addition to the geophysical methods, analyses of the pore fluid chemistry provide strong constraints on hydrate concentrations. In all of the estimation methods, we emphasize the importance of determining the reference profile with depth for each parameter for sediment containing no hydrate and no gas.

As discussed above, the interval velocities above the BSR are significantly increased by the presence of high-velocity hydrate. A number of relations between velocity and hydrate concentration have been suggested. If the models are based on assumed physical parameters, the hydrate concentrations for a particular velocity can be quite different. However, Desmons [1996] showed that if the models are constrained to have the reference velocity for no hydrate, the differences are quite small for up to 30% hydrate (Figure 26). In the figure, the models have been constrained to give the same reference velocity for no hydrate where the porosity is 50%, i.e., just above the BSR for near Site 889/890. All of the models are appropriate only if the hydrate is primarily located in the pore spaces rather than cementing the grain contacts. We note however that extreme models that have hydrate cementing the grain boundary contacts may have a quite different relation [e.g., Dvorkin et al., 1991; Helgerud et al., 1999].

In our studies, we have used two simple model approximations to relate the hydrate concentration to sediment

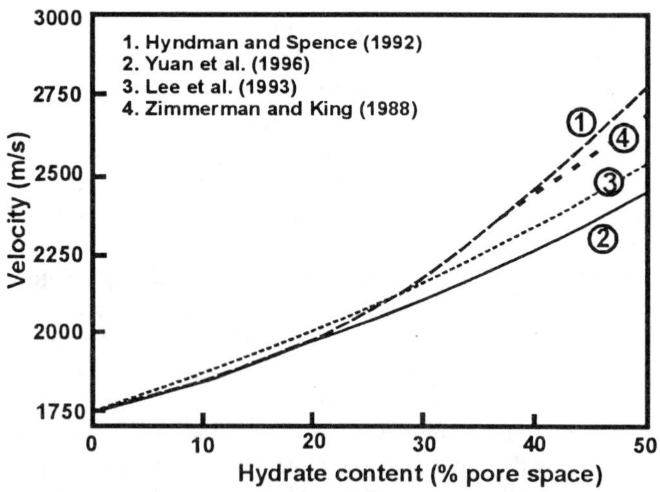

Figure 26. Several model relations between hydrate concentration and velocity. The zero-hydrate velocity has been constrained to the no-hydrate reference for Site 889 just above the BSR.

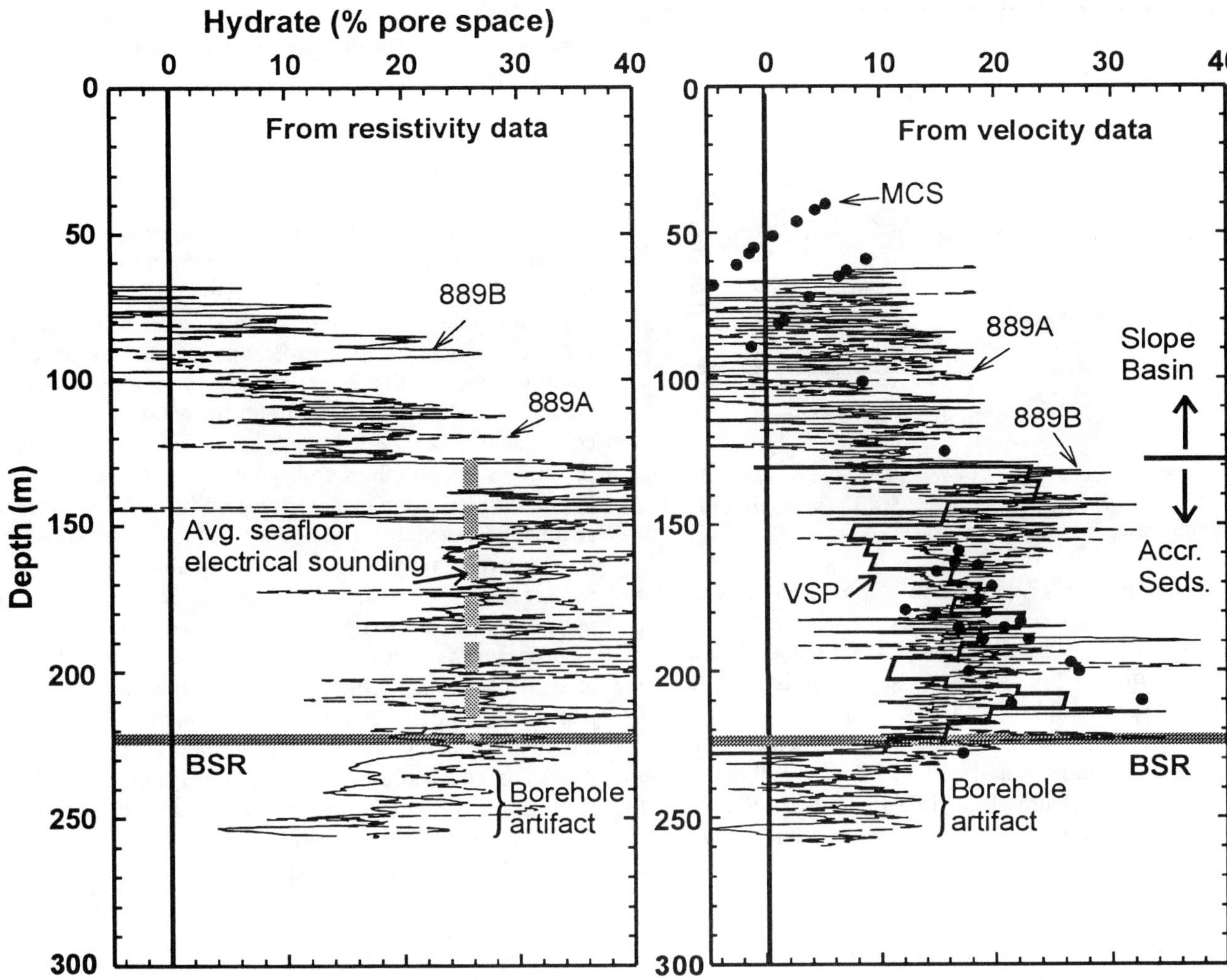

Figure 27. Hydrate concentrations with depth in ODP Boreholes 889/890 from velocity and electrical resistivity data. The broad dashed vertical line indicates the hydrate concentration estimated from the resistivity found in the area from the seafloor electrical profiling.

velocities. In both methods we assume that the hydrate replaces the pore fluid in the pore spaces without significantly changing the pore structure. One method is to assume that the observed velocity increase due to hydrate relative to the reference (see above) can be approximately by a simple reduction in porosity [Hyndman and Spence, 1992; Yuan et al., 1996]. This is justified because hydrate has roughly the same velocity as the sediment matrix. We have used the velocity-porosity for Cascadia sediments [e.g., Yuan et al., 1996] to calculate inferred porosities and hydrate concentrations for three data sets near Site 889/890: (1) downhole velocity logs, (2) VSP velocities, (3) multichannel interval velocities. As shown in Figure 27, concentrations reach a maximum of 20-25% of the pore space above the BSR.

Another method is to obtain a velocity for the combination of pure hydrate and sediment matrix (i.e., no pore fluid), and then to combine that composite matrix with water-saturated sediment to determine an overall model velocity [Lee et al., 1993; Yuan et al., 1996; 1999]. For the observed increase in velocity of about 250 m/s relative to the reference velocity of 1650 m/s, this method suggests that up to 25% of the sediment pore space is occupied by hydrate in the area of Site 889/890.

To determine the amount of free gas below the BSR, the main constraint is the reduction in P-wave velocity from about 1650 to 1400-1500 m/s, as determined from the VSP data [MacKay et al., 1994] and from full waveform inversions in the immediate area [e.g., Yuan et al., 1999]. The amount of free gas has been estimated using the Biot-Gassmann theory by Desmons [1996] for ODP sites at different water depth and

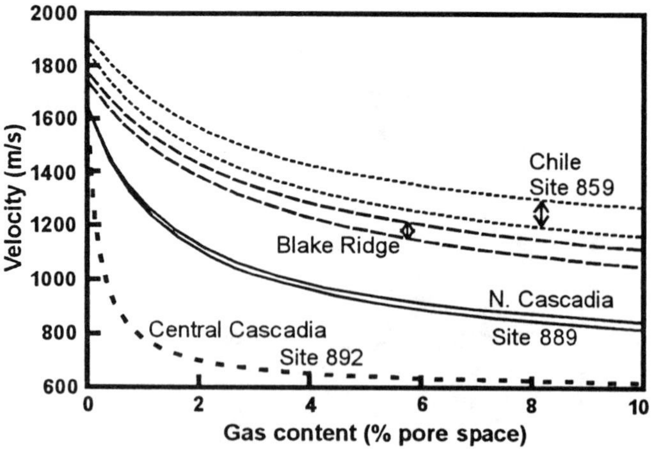

Figure 28. Effect of free gas concentration on velocity for four marine hydrate sites [after Desmons, 1996]. The a strong effect of depth (pressure) is evident.

with different no-gas reference velocities (Figure 28). The plots show the strong dependence on water depth (i.e., pressure). At Site 889 the estimate is only about 0.2% gas saturation (Figure 28). A 1% gas saturation is predicted to result in velocities below 1000 m/s. A limited calibration of this approach is provided by the general agreement between the gas concentration from the Blake Bahama drill hole insitu pressure sampler [Paull et al., 1996] and the predictions of this method. The method has a large uncertainty, but very likely the gas concentration is less than 1% at the N. Cascadia site.

REGIONAL UPWARD FLUID EXPULSION AND HYDRATE FORMATION

The formation of substantial amounts of gas hydrate required large amounts of methane. Geochemical and isotope analyses have shown the methane to be mainly biogenic. Hydrate may be formed from biogenic methane that is generated locally at depths where the P-T conditions are within the stability field. However, there appears to be insufficient organic carbon in the section above the BSR in the clastic Cascadia sediments, and a model has been developed where the hydrate is formed through removal of methane from upward-moving pore fluids as they pass into the hydrate stability field [Hyndman and Davis, 1992] (Figure 29; see also Paull et al. [1994] for a general discussion). Buffett and Zatsepina [2000] provide experimental evidence and review previous work showing that gas hydrate can form directly from aqueous solution. The rising fluid does not need to have methane saturation. In this fluid-expulsion model, which is especially applicable to accretionary sedimentary prisms, the required biogenic methane may be generated throughout a thick sediment section extending well below the BSR. The upward fluid expulsion model predicts the highest concentration of hydrate just above the BSR, decreasing upward, in good agreement with the concentration profile for the Vancouver Island margin.

In the mid-slope region, the rapid thickening of the wedge results in sediment elements being moved to greater depths with little porosity loss and little associated velocity change. The sediments are thus underconsolidated and high pore pressures are generated that are slowly dissipated through upward fluid expulsion [Hyndman et al., 1993]. Fluid expulsion rates that reach about 1 mm/yr appear to be sufficient to carry enough methane upward to the hydrate stability zone to form several 10's % hydrate saturation.

Subsequent to initial hydrate formation on the lower continental slope, the sediments are carried landward into the prism. As a result, the stability field must move upward due to pressure reduction from tectonic uplift, and due to ongoing sediment deposition. The hydrate just above the BSR is thus progressively moved out of the stability field and is dissociated to free gas. The free gas in turn will migrate upward, or be carried upward by the fluid expulsion to be re-incorporated as hydrate above the new re-formed BSR [Paull et al., 1994]. This progressive concentration of methane to just above the BSR results in the maximum concentration of the hydrate being located not far above the BSR [e.g., Zatsepina and Buffett, 1998; Xu and Ruppel, 1999]. The thickness and concentration of the free gas layer is expected to depend upon the permeability of sediments to rising free gas and the rate of fluid expulsion that carries the gas upward. For northern Cascadia, the permeability and fluid expulsion rate appear sufficient that there is only a thin layer a few 10's m thick of free gas with average concentration estimated to be less than 1%. In contrast, the Blake Ridge area off eastern North America, the estimated gas concentrations are much higher, ~5%, and much thicker, ~100 m [e.g., Paull et al., 1996]. In that region, the hydrate is in finer grained sediments and thus probably of lower permeability, and upward fluid expulsion rates are much lower.

COMPARISONS WITH SEVERAL OTHER HYDRATE AREAS

For comparison with the northern Cascadia results we show the hydrate and free gas concentration estimates for two other well-studied areas where quantitative data is available, the Blake ridge off eastern U.S.A., and the Mallik 2L-38 well in the Canadian Arctic (Figure 30). The Blake ridge marine hydrate area has received detailed geophysical study and ODP drilling. This area is on a passive continental margin, in contrast to the Cascadia subduction zone clastic accretionary prisms where sediment accretion produces consolidation and upward fluid expulsion. The Blake area occurrence is in a contourite sediment deposit in which there also may be

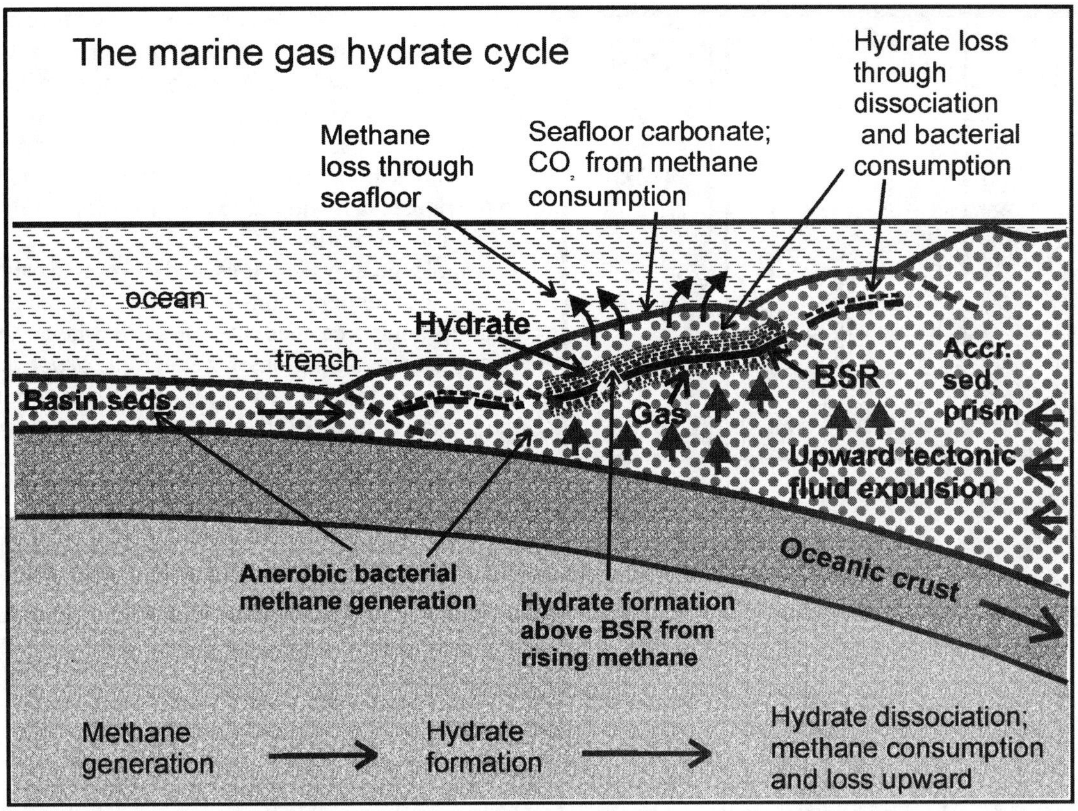

Figure 29. Model for cycle of biogenic methane generation, upward expulsion in accretionary sedimentary prism, hydrate formation and methane loss to seafloor.

consolidation driven upward fluid expulsion [Paull et al., 1996]. The seismic data is summarized by Korenaga et al. [1997] and downhole logging data by Collett and Ladd [2000] and Lee [2000]. The major differences at the Blake Ridge ODP sites compared to Cascadia are: (1) The hydrate concentration is lower, averaging about 3-5% saturation (with thin layers to about 20%) compared to N. Cascadia average of 20-30%. The difference may be primarily because the Blake fluid expulsion rate is lower, (2) The estimated gas concentration is greater, 2-3% compared to less than 1%. The difference may be because the finer grain and lower permeability Blake ridge sediments restrict upward migration of gas to under the BSR. However, a higher hydrate concentration for other areas of the Blake region of ~20% saturation over ~200 m was estimated by Desmons [1996] using seismic interval velocities and inversions by Wood et al. [1994] and Katzman et al. [1994].

The most detailed study of arctic hydrate, for comparison with the marine occurrences, is the recently drilled JAPEX/JNOC/GSC Mallik 2L-38 gas hydrate research well in the Mackenzie Delta in the Canadian Arctic [Dallimore et al., 1999; Lee and Collett, 1999]. The well had exhaustive geological studies and downhole geophysical logs. The well is located near the top of the Mallik anticline structure, where gas appears to have migrated updip. An important difference compared to marine hydrate occurrences is that there is no clear BSR in the seismic data. The hydrate is located beneath thick permafrost. The main hydrate concentrations are just above the base of the hydrate stability field which is at about 1100 m depth. The estimated average hydrate concentrations are very high, ~40% saturation, even higher than those of the N. Cascadia region. However, the hydrate is more irregularly distributed with depth and more geologically controlled, mainly concentrated in the sandier sections. Estimated local concentrations reach as high as 80% saturation. The velocities in the hydrate section from about 900-1100 m depth are up to 3300 m/s [Lee and Collett, 1999; Sakai, 1999; Walia et al., 1999] compared to a reference no-hydrate velocity of about 2500 m/s. The downhole electrical logs give similar estimates of hydrate concentration. Free gas is inferred to occur but it is not clearly defined in the downhole data.

CONCLUSIONS

The wide range of geophysical studies, and downhole and core data from the Ocean Drilling Project (ODP) on the

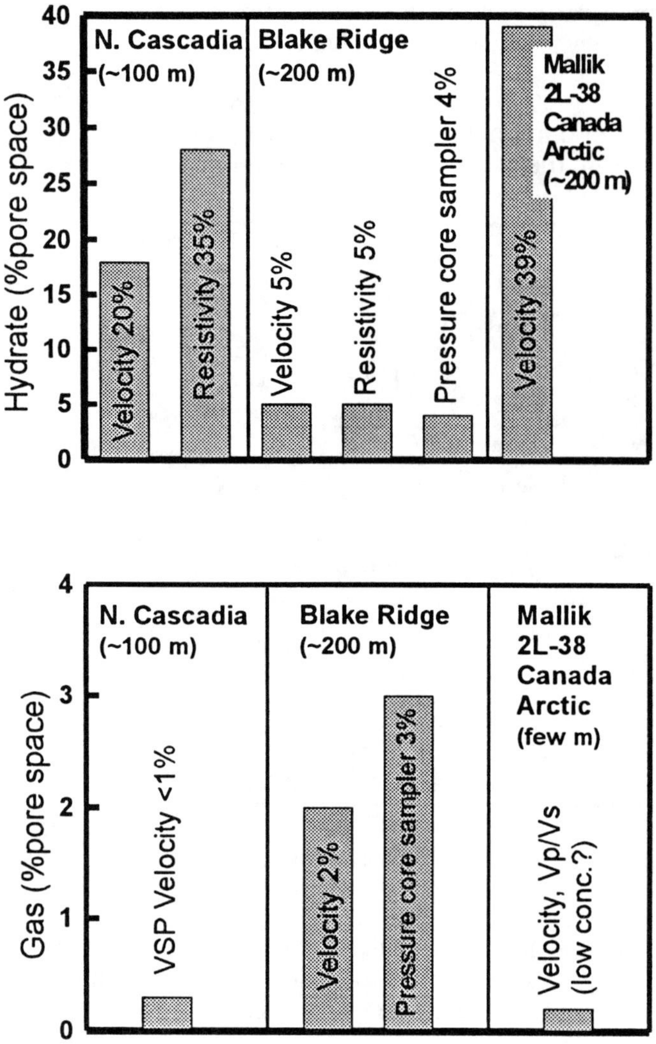

Figure 30. Comparison of estimated gas hydrate and free gas concentrations in N. Cascadia, Blake Ridge, and Mallik Arctic coast.

continental slope of the northern Cascadia subduction zone have provided strong constraints on, (1) the distribution of gas hydrate, (2) the concentration of gas hydrate and underlying free gas, and (3) on the geological controls of hydrate formation. Most information on the gas hydrate has come from seismic surveys that have mapped the area of the characteristic bottom-simulating reflection (BSR), and have determined the depth distribution of hydrate and underlying free gas. BSRs are evident beneath about half of the mid-continental slope. Special seismic studies of BSR reflection coefficients, of the frequency dependence of the BSR amplitude, of BSR amplitude-versus-offset (AVO), and of full wave inversions have constrained the hydrate to be concentrated in ~100 m above the BSR with concentrations reaching about 20% saturation. Important additional information on hydrate concentration and distribution have been obtained from a system that profiles the electrical resistivity of the sub-seafloor to a depth of several 100 m, and measurements of seafloor compliance. Detailed heat flow surveys have been used to estimate the temperature at the BSR and shown general agreement with ODP downhole temperatures and with laboratory data for the P-T conditions of hydrate stability. Ocean Drilling Program Leg 146 downhole geophysical logs and core analyses have given partly independent constraints on the concentrations of hydrate and free gas at the drill sites, and provided confirmation calibration of the regional geophysical data. In the region of the main ODP site on the mid-continental slope, the concentration of hydrate from both geophysical and borehole data is estimated to be 20-30% of the pore space (10-15% of total volume) over a depth interval above the BSR of about 100 m, with an underlying 10-20 m thick layer with less than 1% free gas. The hydrate concentration at Blake Ridge ODP drill sites is lower but the gas concentration higher and thicker than N. Cascadia. In contrast, for the Arctic Mackenzie delta Mallik gas hydrate well the hydrate concentration appears to be somewhat higher, but the gas concentration may be even lower and thinner.

Acknowledgements. We thank the many collaborating scientists and students that have worked on northern Cascadia marine gas hydrate studies, especially J. Gettrust, W. Wood, V. Speiss, L. Zuehlsdorff, T. Lewis, R. Walia, T. Yuan, C. Fink, B. Desmons, Y. Mi, N. Ganguly, E. Willoughby, J. Yuan, and technical support especially by R. Macdonald, W. Hill, and I. Frydecky. We also thank the participants of ODP Leg 146 that drilled the critical gas hydrate wells in the area. Geological Survey of Canada Publication Number 2000022

REFERENCES

Buffett, B.A., and O.Y. Zatsepina, Formation of gas hydrate from dissolved gas in natural porous media, *Marine Geol., 164*, 69-77, 2000.

Carson, B., G.K. Westbrook, R.J. Musgrave, and E. Suess, *Proc. Ocean Drill. Program, Scientific Rep., 146*, Ocean Drill. Program, College Station, Tex., 477p., 1995.

Chapman, R., J. Gettrust, R. Walia, W. Wood, D. Hannay, G. Spence, and R.D. Hyndman, High resolution deep towed multichannel seismic survey of deep sea gas hydrates off western Canada, *Geophysics*, in press, 2001.

Clowes, R.M., C.J. Yorath, and R.D. Hyndman, Reflection mapping across the convergent margin of western Canada, *Geophys. J. R. Astron. Soc, 89*, 79-84, 1987.

Collett, T.S., and J. Ladd, Detection of gas hydrate with downhole logs and assessment of gas hydrate concentrations (saturations) and gas volumes on the Blake Ridge with electrical resistivity log data, *Proc. Ocean Drill Program, Sci. Results, 164*, 179-191, 2000.

Dallimore, S.R., T. Uchida, and T.S. Collett, in *Scientific Results from JAPEX/JNOC/GSC Mallik 2L-38 Gas Hydrate Research Well, Mackenzie Delta, Northwest Territories, Canada*, Geological Survey Canada Bulletin 544, 403p, 1999.

Davis, E.E., and R.D. Hyndman, Accretion and recent deformation of sediments along the northern Cascadia subduction zone, *Geol. Soc. Am. Bull., 101*, 1465-1480, 1989.

Davis, E.E., R.D. Hyndman, and H. Villinger, Rates of fluid expulsion across the northern Cascadia accretionary prism: constraints from new heat flow and multichannel seismic reflection data, *J. Geophys. Res., 95*, 8869-8889, 1990.

Davis, E.E., K. Becker, K. Wang, and B. Carson, Long-term observations of pressure and temperature in Hole 892B, Cascadia accretionary prism, *Proc. Ocean Drill. Prog., Sci. Results, 146*, 299-312, 1995.

Desmons, B., Integrated study of gas hydrates in marine sediments using geophysical and geochemical data, M.Sc. dissertation, Univ. of Victoria, Victoria, B.C., Canada, 1996.

Dickens, G.R., and M.S. Quinby-Hunt, Methane hydrate stability in seawater, *Geophys. Res. Lett., 21*, 2115-2118, 1994.

Dvorkin, J., G. Mavko, and A. Nur, The effect of cementation on the elastic properties of granular material, *Mech. Mater, 12*, 207-217, 1991.

Edwards, R.N., On the resource evaluation of marine gas hydrate deposits using the sea-floor transient electric dipole-dipole method, *Geophysics, 62*, 63-74, 1997.

Fink, C.R., Methane hydrate distribution offshore Vancouver Island from detailed single channel seismic studies, MSc thesis, Univ. Victoria, Victoria, B.C., Canada, 1995.

Fink, C.R., and G.D. Spence, Hydrate distribution off Vancouver Island from multi-frequency single channel seismic reflection data, *J. Geophys. Res., 104*, 2909-2922, 1999.

Fuchs, K., and G. Muller, Computation of synthetic seismograms with the reflectivity method and comparison with observations, *Geophys. J. R. Astr. Soc., 23*, 417-433, 1971.

Ganguly, N., G.D. Spence, N.R. Chapman, and R.D. Hyndman, Heat flow variations from bottom-simulating reflectors on the Cascadia margin, *Marine Geol., 164*, 53-68, 2000.

Gettrust, J., W. Wood, D. Lindwall, R. Chapman, R. Walia, D. Hannay, K. Louden, R. MacDonald, and R.D. Hyndman, New seismic study of deep sea gas hydrates results in greatly improved resolution, *Eos, Trans. Am. Geophys. Un., 38*, 439-440, 1999.

Gregory, A.R., Fluid saturation effects on dynamic elastic properties of sedimentary rocks, *Geophysics, 41*, 895-921, 1976.

Helgerud, M.B., J. Dvorkin, A. Nur, A. Sakai, and T. Collett, Elastic-wave velocity in marine sediments with gas hydrates: Effective medium modeling, *Geophys. Res. Lett., 26*, 2021-2024, 1999.

Hobro, J.W., S. Singh, and T. Minshull, Tomographic seismic studies of the methane hydrate stability zone in the Cascadia margin, *Geol. Soc. Lond. Spec. Publ. 137*, 133-140 1998.

Hobro, J.W., Three-dimensional tomographic inversion of combined reflection and refraction travel-time data, Ph.D. thesis, University of Cambridge, U.K., 1999.

Hyndman, R.D., The Lithoprobe corridor across the Vancouver Island continental margin: The structural and tectonic consequences of subduction, *Can. J. Earth Sci., 32*, 1777-1802, 1995.

Hyndman, R.D., and G.D. Spence, A seismic study of methane hydrate marine bottom-simulating reflectors, *J. Geophys. Res, 97*, 6683-6698, 1992.

Hyndman, R.D., and E.E. Davis, A mechanism for the formation of methane hydrate and seafloor bottom-simulating reflectors by vertical fluid expulsion, *J. Geophys. Res, 97*, 7025-7041, 1992.

Hyndman, R.D., and K. Wang, Thermal constraints on the zone of major thrust earthquake failure: the Cascadia subduction zone, *J. Geophys. Res, 98*, 2039-2060, 1993.

Hyndman, R.D., and T.J., Lewis, Review: The thermal regime along the southern Canadian Cordillera Lithoprobe corridor, *Can. J. Earth Sci., 32*, 1611-1617, 1995.

Hyndman, R.D., J.P. Foucher, M. Yamano, and A. Fisher, Deep sea bottom-simulating reflectors: calibration of the bae of the hydrate stability field as used for heat flow estimates, *Earth Planet. Sci., Lett., 109*, 289-301, 1992.

Hyndman, R.D., K. Wang, T. Yuan, and G.D. Spence, Tectonic sediment thickening, fluid expulsion and the thermal regime of subduction zone accretionary prisms: the Cascadia margin off Vancouver Island, *J. Geophys. Res., 98*, 21, 865-21,876, 1993.

Hyndman, R.D., G.D. Spence, T. Yuan, and E.E. Davis, Regional geophysical and structure framework of the Vancouver Island margin accretionary prism, *Proc. Ocean Drill. Program, Initial Rep., 146*, 399-419, 1994.

Hyndman, R.D., T. Yuan, and K. Moran, The concentration of deep sea gas hydrates from downhole electrical resistivity logs and laboratory data, *Earth Planet. Sci. Lett., 172*, 167-177, 1999.

Jarrard, R.D., M.E. MacKay, G.K. Westbrook, and E.J. Screaton, Log-based porosity of ODP sites on the Cascadia accretionary prism. *Proc. Ocean Drill. Program, Sci. Results., 146*, Ocean Drill. Prog., College Station, Tex., 1995.

Kastner, M., K.A. Kvenvolden, M.J. Whiticar, A. Camerlenghi, and T.D. Lorenson, Relation between pore fluid chemistry and gas hydrate associated with bottom-simulating reflectors at the Cascadia margin, *Proc. Ocean Drill. Program, Sci. Results, 146*, 175-187, 1995.

Katzman, R., W. Holbrook, and C. Paull, Combined vertical-incidence and wide-angle seismic study of a gas hydrate zone, Blake Ridge, *J. Geophys. Res., 99*, 17,975-17,995, 1994.

Korenaga, J., W.S. Holbrook, S.C. Singh, and T.A. Minshull, Natural gas hydrates on the southeast U.S. margin: Constraints from full waveform and travel time inversions of wide-angle data, *J. Geophys. Res., 102*, 15,345-15,365, 1997.

Lee, M.W., Gas hydrates amount estimated from acoustic logs at the Blake Ridge, Sites 994, 995, and 997, *Proc. Ocean Drill. Program, Sci. Results, 164*, 193-198, 2000.

Lee, M.W., D.R. Hutchinson, W.P. Dillon, J.J. Miller, W.F. Agena, and B.A. Swift, Method of estimating the amount of in situ gas hydrate in deep marine sediments, *Mar. Pet. Geol., 10*, 493-506, 1993.

Lee, M.W., and T.S. Collett, Amount of gas hydrate estimated from compressional- and shear-wave velocities at the JAPEX/JNOC/GSC Mallik 2L-38 gas hydrate research well, *in Scientific Results from JAPEX/JNOC/GSC Mallik 2L-38 Gas Hydrate Research Well, Mackenzie Delta, Northwest Territories, Canada*, Eds. Dallimore, S.R., T. Uchida, and T.S. Collett, Geological Survey Canada Bulletin 544, pp. 313-322, 1999.

Lewis, T.J., W.H., Bentkowski, E.E. Davis, R.D. Hyndman, J.G. Souther, and J.A. Wright, Subduction of the Juan de Fuca plate:

thermal consequences, *J. Geophys. Res., 93*, 15,207-15,225, 1988.

Lewis, T.J., W.H., Bentkowski, and R.D. Hyndman, Crustal temperatures near the Lithoprobe Southern Canadian Cordillera transect, *Can. J. Earth Sci., 29*, 1197-1214, 1991.

MacKay, M.E., R.D. Jarrard, G.K. Westbrook, R.D. Hyndman and the Shipboard Scientific Party of ODP Leg 146, Origin of bottom simulating reflectors: Geophysical evidence from the Cascadia accretionary prism, *Geology, 22*, 459-462, 1994.

Mi, Y., Seafloor sediment coring and multichannel seismic studies of gas hydrate, offshore Vancouver Island, M.Sc. thesis, University of Victoria, Victoria, Canada,1998.

Ostrander, W.J., Plane-wave reflection coefficients for gas sands at non-normal angle of incidence, *Geophysics, 49*, 1637-1648, 1984.

Paull, C.K., W. Ussler III, and W.S. Borowski, Sources of biogenic methane to form marine gas hydrates, *Ann. N.Y. Acad. Sci., 715*, 392-411, 1994.

Paull, C.K., R. Matsumoto, P. Wallace et al., *Proc. Ocean Drill. Program, Initial Rep., 164*, 623p., 1996.

Paull, C.K., W. Ussler, and W.S. Borowski, Sources of biogenic methane to form marine gas hydrates: In situ production or upward migration? *New York Acad. Sci. Annal., 715*, 392-409, 1994.

Riedel, M., R. Chapman, G.D. Spence, S. Dosso, and R.D. Hyndman, A pseudo 3-D seismic survey at a gas-hydrate field offshore Vancouver Island, Eos, Trans. Am. Geophys. Un., 80, p. F480, 1999.

Riedel, M., R.D. Hyndman, G.D. Spence, and R. Chapman, Seismic investigations of an apparent active vent field associated with gas hydrate, offshore Vancouver Island, *J. Geophys. Res.*, in press, 2001.

Rowe, M.M., and J.F. Gettrust, Fine structure of methane hydrate-bearing sediments on the Blake Outer Ridge as determined from deep-tow multichannel seismic data, *J. Geophys. Res., 98*, 463-473, 1993.

Rutherford, S.R., and R.H. Williams, Amplitude-versus-offset variations in gas sand, *Geophys., 54*, 6880-688, 1989.

Sakai, A.,Velocity analysis of vertical seismic profile (VSP) survey at JAPEX/JNOC/GSC Mallik 2L-38 gas hydrate research well, and related problems for estimating gas hydrate concentration, in *Scientific Results from JAPEX/JNOC/GSC Mallik 2L-38 Gas Hydrate Research Well, Mackenzie Delta, Northwest Territories, Canada*, editors Dallimore, S.R., T. Uchida, and T.S. Collett, Geological Survey Canada Bulletin 544, pp. 323-340, 1999.

Singh, S.C., and T.A. Minshull, Velocity structure of a gas hydrate reflector at Ocean Drilling Program Site 889 from global seismic waveform inversion, *J. Geophys. Res., 99*, 24,221-24,233, 1994.

Singh, S.C., T.A. Minshull, and G.D. Spence, Velocity structure of a gas hydrate refelctor, *Science, 260*, 204-207, 1993.

Shuey, R.T., A simplification of the Zoeppritz equations, *Geophysics, 50*, 609-614, 1985.

Sloan, E.D., *Clathrate Hydrates of Natural Gas*, Marcel Dekker, New York, 1990.

Spence, G.D., R.D. Hyndman, E.E. Davis, and C.J. Yorath, Multichannel reflection profiles across the Vancouver Island continental shelf and slope, Geol. Surv. Can., Open File Rep., 2391, 1991a.

Spence, G.D., R.D. Hyndman, E.E. Davis, and C.J. Yorath, Seismic structure of the northern Cascadia accretionary prism: Evidence from new multichannel seismic reflection data, in *Continental Lithosphere, Deep Seismic Reflections, Geodyn. Ser., V. 22*, edited by R. Meissner, et al., pp. 257-263, AGU, Washington, D.C., 1991b.

Spence, G.D., T.A. Minshull, and C. Fink, Seismic structure of methane gas hydrate, offshore Vancouver Island, *Proc. Ocean Drill. Program, Sci. Results 146*, 163-174, 1995.

von Huene, R., and I. Pecher, Vertical tectonics and the origins of BSRs along the Peru margin, *Earth Planet. Sci. Lett., 166*, 47-55, 1999.

Walia, R., and D. Hannay, Source and receiver geometry corrections for deep towed multichannel seismic data, *Geophys. Res. Lett., 26*, 1993-1996, 1999.

Walia, R., Y. Mi, R.D. Hyndman, and A. Sakai, Vertical seismic profile (VSP) in the JAPEX/JNOC/GSC Mallik 2L-38 gas hydrate research well, in *Scientific Results from JAPEX/JNOC/GSC Mallik 2L-38 Gas Hydrate Research Well, Mackenzie Delta, Northwest Territories, Canada*, editors Dallimore, S.R., T. Uchida, and T.S. Collett, Geological Survey Canada Bulletin 544, pp. 341-355, 1999.

Wang, K., R.D. Hyndman, and E.E. Davis, Thermal effects of sediment thickening and fluid expulsion in accretionary prisms, *J. Geophys. Res., 98*, 9975-9984, 1993.

Wang, K., T. Mulder, G.C. Rogers, and R.D. Hyndman, Case for very low couupling stress on the Cascadia subduction fault, *J. Geophys. Res., 100*, 12,907-12,918, 1995.

Warner, M., Absolute reflections from deep seismic reflections, *Tectonophys. 173*, 15-23, 1990.

Westbrook, G.K., B. Carson, R.J. Musgrave, and Shipboard Scientists, *Proc. Ocean Drill. Program, Initial Rep., 146*, 611p., 1994.

Willoughby, E.C., and R.N. Edwards, Shear velocities in Cascadia from seafloor compliance measurements, *Geophys. Res. Lett.*, in press, 2001.

Wood, W., P. Stoffa, and T. Shipley, Quantitative detection of methane hydrate through high-resolution seismic velocity analysis, *J. Geophys. Res., 99*, 9681-9695, 1994.

Xu, W. and C. Ruppel, Predicting the occurrence, distribution, and evolution of methane gas hydrate in porous marine sediments, *J. Geophys. Res., 104*, 5081-5096, 1999.

Yuan, T., Seismic studies of the northern Cascadia accretionary prism: Sediment consolidation and gas hydrate, Ph.D. dissertation, Univ. of Victoria, Victoria, B.C., 1996.

Yuan, J., and R.N. Edwards, Towed seafloor electromagnetics and assessment of gas hydrate deposits, *Geophys. Res. Lett.*, in press, 2001.

Yuan, T., G.D. Spence, and R.D. Hyndman, Seismic velocities and inferred porosities in the accretionary wedge sediments at the Cascadia margin, , *J. Geophys. Res., 99*, 4413-4427, 1994.

Yuan, T., R.D. Hyndman, G.D. Spence, and B. Desmons, Seismic velocity increase and deep-sea gas hydrate above a bottom-simulating reflector on the northern Cascadia continental slope, *J. Geophys. Res., 101*, 13,655-13,671, 1996.

Yuan, T., G.D. Spence, R.D. Hyndman, T.A. Minshull, and S.C. Singh, Seismic velocity studies of a gas hydrate bottom-simulating reflector on the northern Cascadia continental margin: Amplitude modeling and full waveform inversion, *J. Geophys. Res., 104*,1179-1191,1999.

Xu, W., and C. Ruppel, Predicting the occurrence, distribution, and evolution of methane gas hydrate in porous sediments, *J. Geophys. Res., 104*, 5081-5095, 1999.

Zatsepina, O.Y., and Buffett, B.A. Thermodynamic conditions for the stability of gas hydrate in the seafloor, *J. Geophys. Res., 103*, 24,127-24,139, 1998.

Zimmerman, R.W. and M.S. King, The effect of the extent of freezing on seismic velocities in unconsolidated permafrost, *Geophysics, 51, 1285-1290*, 1986.

Zuehlsdorff, L., V. Spiess, C. Huebscher, H. Villinger, and A. Rosenberger, Implications for focused fluid transport at the northern Cascadia accretionary prism from a correlation between BSR occurrence and near seafloor reflectivity anomalies imaged in a multi-frequency data set, *Geologische Rundschau, 88*, in press, 2001.

Zwart, G., J.C. Moore, and G.R. Cochrane, Variations in temperature gradients identify active faults in the Oregon accretionary prism, *Earth Planet. Sci. Lett., 139*, 485-495, 1996.

R.D. Hyndman, Pacific Geoscience Centre, Geological Survey of Canada, 9860 W. Saanich Road, Sidney, B.C., V8L4B2, Canada

G.D. Spence, R. Chapman and M. Riedel, School of Earth and Ocean Sciences, University of Victoria, Victoria, B.C., V8W3P6, Canada

R.N. Edwards, Department of Physics, University of Toronto, Toronto, Ontario, M5S1A7, Canada

High-resolution Multibeam Survey of Hydrate Ridge, Offshore Oregon

David A. Clague, Norm Maher, and Charles K. Paull

Monterey Bay Aquarium Research Institute

A new 30 kHz multibeam echo-sounder survey of part of the Oregon margin imaged Hydrate Ridge at a higher resolution than previous surveys conducted in the same area. The bathymetric data show that a northern ridge rises to 588 m and a southern ridge to 773 m. Moderate-to-high backscatter areas, interpreted to indicate patchy, discontinuous chemosynthetic communities, authigenic carbonate, or hydrate accumulations cover about 26 km^2 of the summit area of the northern ridge. Similar moderate-to-high backscatter areas comprise about 4.9 km^2 of the southern ridge. However, the southern ridge does have about 0.095 km^2 of high-backscatter that is believed to be associated with nearly continuous carbonate and hydrate outcrop on the western edge of the summit. A site on the southern ridge where intercalated carbonate and hydrate deposits were previously recovered is within the region with only patchy outcrop.

INTRODUCTION

Hydrate Ridge (Figure 1), located about 85 km offshore of the central Oregon margin, is one of the classic areas where the effects and processes associated with methane hydrate in near-seafloor sediment has been studied. The Ocean Drilling Program drilled Site 892 during Leg 146 on the northwest flank of Hydrate Ridge and numerous submersible dives and geophysical surveys have been conducted to map and collect hydrate, sediment, and biological samples from the site [*Carson and Westbrook*, 1995]. The southern part of Hydrate Ridge has been selected for future ODP drilling after intercalated hydrate and authigenic carbonate deposits were recovered using a giant grab sampler [*Bohrmann et al.*, 1998].

Carson et al. [1994] used GLORIA sidescan imagery to determine the regional coverage of clastic and diagenetic lithologies along the Oregon margin, although their study area included only the western slope of Hydrate Ridge. They attributed high-amplitude backscatter to diagenetic products and the result of biological phenomena associated with fluid discharge sites. These occurrences include methane-derived authigenic carbonate, chemosynthetic communities, and sea floor accumulations of gas hydrates. Areas of low reflectivity are covered with clastic sediment.

Because of interest in this area as a detailed gas hydrate study site, the Monterey Bay Aquarium Research Institute (MBARI) collected high-resolution bathymetric and backscatter data at Hydrate Ridge and a strip of the Oregon margin north and south of the ridge. This report presents the data collected during this survey and provides some interpretation of them in order to place the existing and future drill sites into a more regional perspective.

Geologic Setting

At Hydrate Ridge (Figure 1), the Juan de Fuca plate is obliquely underthrusting beneath the North American plate at about 4.5 cm/yr. Sediment on the subducting plate, consisting largely of turbidite beds, is accreted to the Oregon margin [*MacKay et al.*, 1992, *MacKay*, 1995]. Hydrate Ridge is located near, but south of, a structural transition between seaward vergent thrust packets to the south and landward vergent thrusting to the north [*MacKay*, 1995]. In addition to the change in tectonic style from north to south along the margin, numerous oblique strike-slip faults offset the accretionary prism [*McCaffrey and Goldfinger*, 1995].

Natural Gas Hydrates: Occurrence, Distribution, and Detection
Geophysical Monograph 124

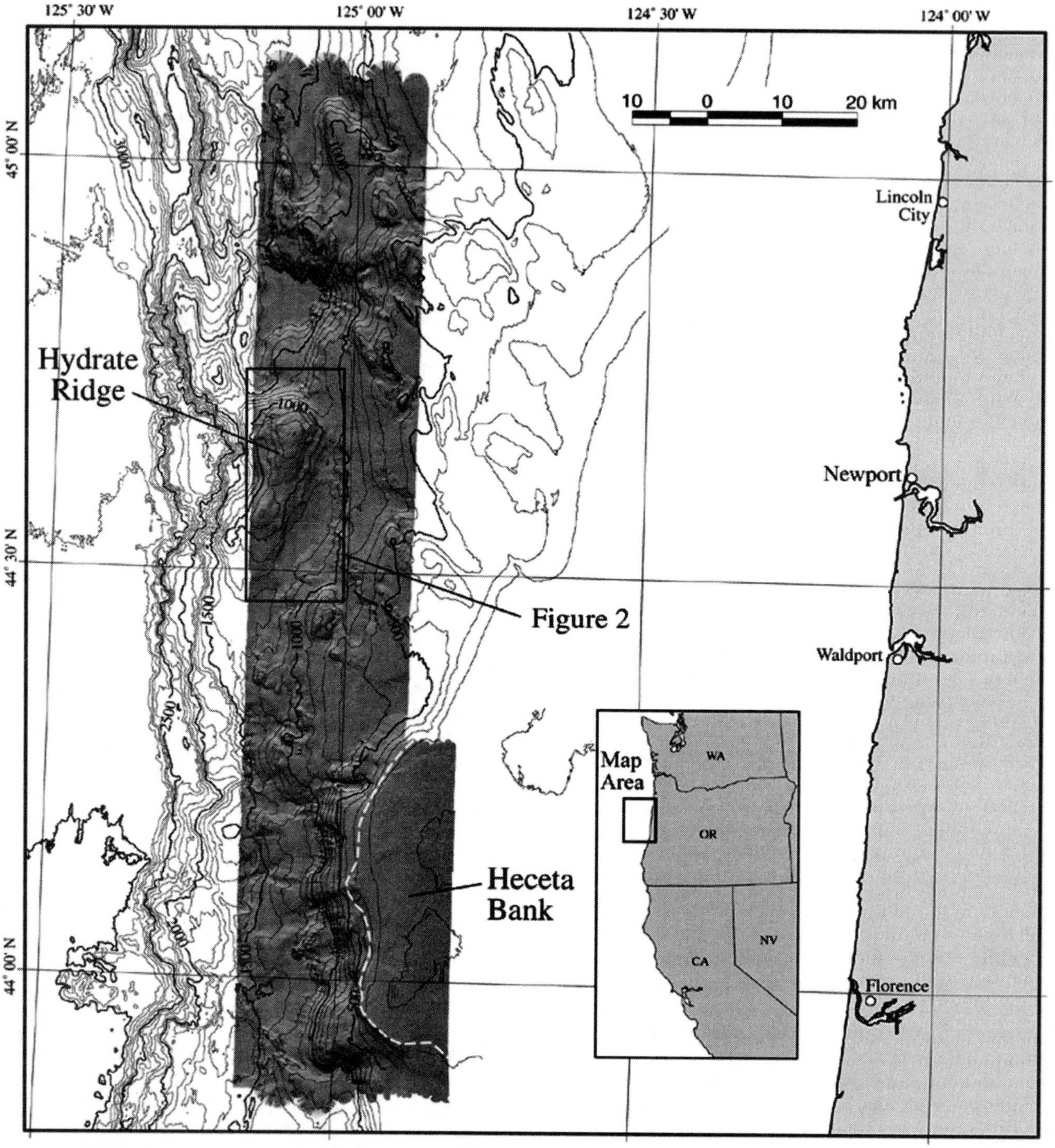

Figure 1. The Oregon margin showing the entire SIMRAD EM300 survey area as illuminated bathymetry. The contours outside the new survey are based on previously collected SeaBeam bathymetric data. The smaller inset map shows regional location of the study area. The box shows the area enlarged in Figure 2 and corresponds with Hydrate Ridge. Dashed line separates data collected by NOAA on Heceta Bank from MBARI's survey to the west.

The Oregon margin is a site of active fluid expulsion that results in communities of chemosynthetic organisms (dominated by clams and tubeworms) [e.g., *Kulm et al.*, 1986] and authigenic carbonate precipitates [Ritger et al., 1987]. A continuous bottom-simulating reflector (BSR) occurs along much of the lower slope of the Oregon margin [e.g. *Trehu et al.*, 1995] and is interpreted to be caused by free methane gas beneath a zone containing some methane hydrate (ODP Leg 146 Scientific Party, 1993). *Trehu et al.* [1995] note that the BSR shallows landward from about 270 m below the sea floor at 1,512 m water depth until it intersects with the seafloor at about 600 m depth, just slightly deeper than the summit of Hydrate Ridge, but still deeper than the stability field for methane hydrate [*Kvenvolden*, 1988]. More recently, focus has broadened to include studies of methane emissions into the water column, and methane hydrates near or at the sea floor [*Bohrmann et al.*, 1998].

DATA ACQUISTION

MBARI contracted C&C Technologies of Layfayette, Louisiana to collect multibeam bathymetric survey data using a Simrad EM300 system on the *R/V Ocean Alert* in May 1998. The EM300 system is a hull-mounted multibeam system that is used to collect detailed bathymetric and acoustic backscatter data. The EM300 system consists of 137 30-kHz acoustic beams that can be divided into three to nine independently beam-steered sectors to correct for vessel pitch, heave, roll, and yaw. Beam angles were adjusted from 1-4 degrees as a function of water depth. The survey was navigated using differential GPS. A CTD cast collected during the survey was used to determine sonic velocity structure in the water column. Bathymetric data has a vertical and horizontal resolution of approximately 2.0% and 0.2% of water depth, respectively. Regional bathymetric data were gridded at 30 m and backscatter data were mosaicked at 10 m (Figure 1). Hydrate Ridge bathymetric data were gridded at 15 m and backscatter data were mosaicked at 5 m (Figure 2). This higher resolution grid shows greater detail for the top of Hydrate Ridge, but introduces more noise on the deeper flanks of the ridge. Illuminated images of the bathymetry were computed with apparent illumination from azimuth 315° and an elevation of 45°. All figures are UTM projections and the illustrated data were processed on shore using MB-system 4.6.10 software developed by David W. Caress and Dale N. Chayes.

NEW BATHYMETRIC AND BACKSCATTER DATA

The MBARI EM300 survey was contiguous with a survey of Heceta Bank collected by NOAA as part of a fisheries study [*R. Embley*, personal communication]. The NOAA data are shown here also (Figure 1). Hydrate Ridge is a relatively small part of the overall survey and is located near the west central boundary of the survey region (Figures 1 and 2).

Hydrate Ridge has two summits (Figure 2); a large northern ridge that rises to 588-m depth and a smaller southern ridge that crests at 773-m depth. ODP site 892 [*Carson and Westbrook*, 1995] is located on the western flank of the northern ridge about 1.9 km west-northwest of the summit. Proposed ODP Site HR-01, HR-02, and HR-3 lie along a transect that crosses the southern ridge about 3 km north of its summit (Figures 2 and 3). Information concerning the planned ODP drilling (Leg 204) can be found at http://www-odp.tamu.edu/publications.

The southern part of Hydrate Ridge (Figure 3) is a relatively smooth feature with little surface roughness at the 10-m pixel scale. The ridge is bounded to the southeast by a steep east-dipping slope striking about N10°E at about 125° 06.5'W. This slope becomes steeper to the south of Figure 4 (see Figure 2) and terminates just north of 44° 31'N. The northern end of the steep slope flattens and merges with the regional slope just south of the proposed drillsites. A small secondary topographic high, located northeast of the summit, but south-southwest of HR-01, rises to 855-m depth. Most of the ridge has low backscatter, but one small patch with high backscatter is located in a rough region just west of the summit. This particular area of high backscatter covers an area of about 0.065 km^2 and is surrounded by a larger region (about 0.550 km^2) with moderate backscatter. Most of the southern ridge has backscatter amplitudes somewhat higher than the adjacent sediment-filled basins. We categorized the backscatter amplitudes roughly into three groups, using the highest and lowest values in Figure 2 as end-member values. For the area shown in Figure 3, only 0.095 and 4.9 km^2 of a total of 54.9 km^2 have high or moderate backscatter, respectively.

The northern part of Hydrate Ridge (Figure 2) is characterized by significantly rougher bottom topography compared to the southern part of Hydrate Ridge (Figures 2 and 3). Numerous small mounds and depressions near the summit and several larger structures on the flanks cause the roughness. The western flank between 800 and 1,000 m depth has several embayments that widen downslope. On the eastern flank at about 800-m depth, a ~300 m in diameter semicircular structure is located at the northeast end of a ridge extending from the summit. Also on the eastern flank, two steep mounds surmount a terrace extending north-northeast at 950-910 m depth

Analysis of the northern ridge (Figure 4), using the same three-backscatter categories outlined above, indicate that 26 and 3.4 km^2 of the total area of 112 km^2 have moderate-to-high backscatter respectively. Thus, they include about 25% of the northern ridge shallower than 1,000 m. Areas with topographic relief are characterized by high-to-moderate backscatter. However, regions of high-to-

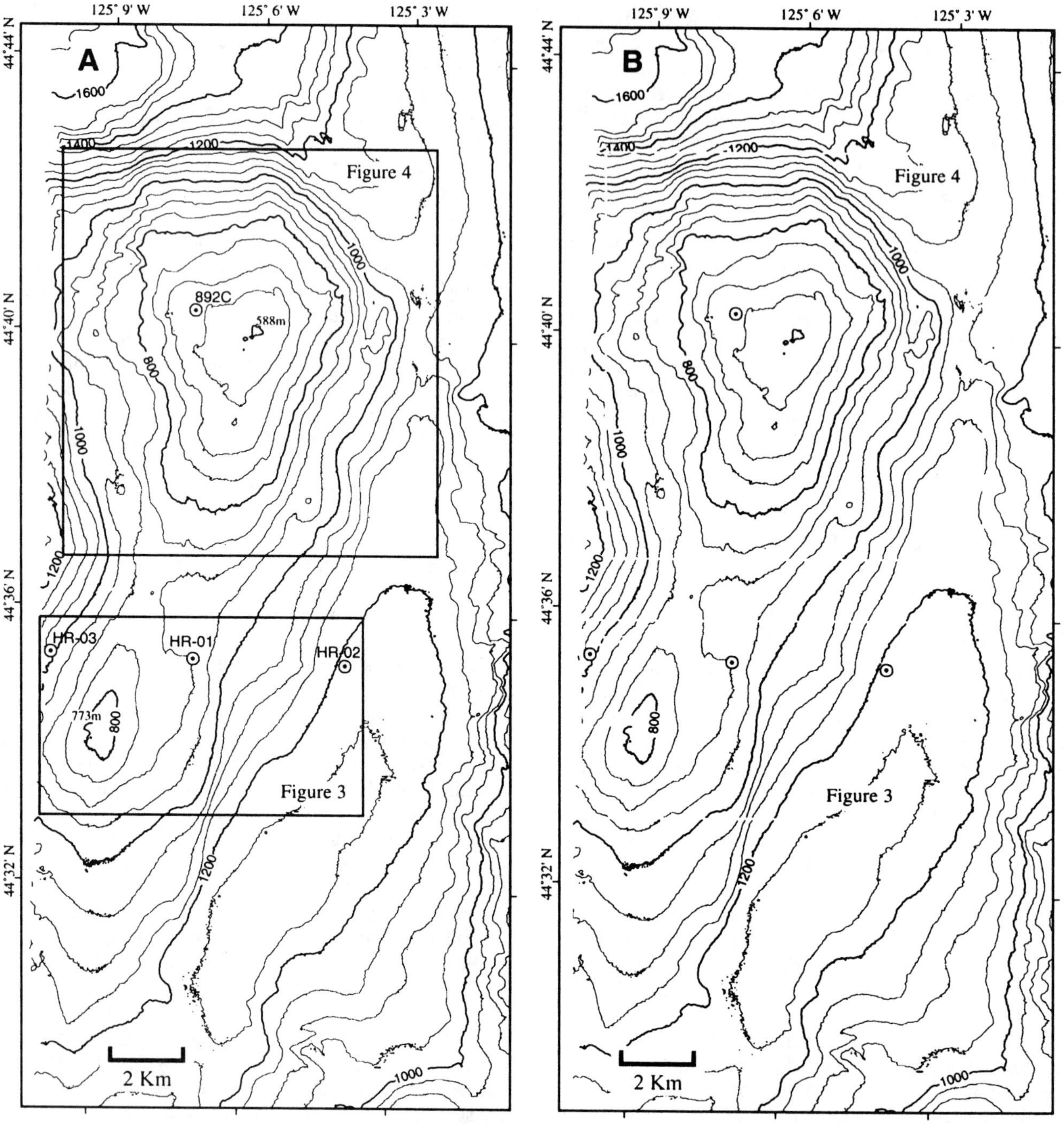

Figure 2. Illuminated bathymetry (a) and backscatter amplitude (b, bright is high backscatter) of Hydrate Ridge with 50-m contours based on the EM-300 data. The 1,000-m contour encircles Hydrate Ridge. ODP site 892C is shown on the northern ridge and proposed ODP sites HR-01 to HR-02 are shown. The enlarged areas shown in Figures 3 and 4 are enclosed in boxes correspond with the Northern and Southern Hydrate Ridge respectively. The north-south trending brighter strips are data collection artifacts associated with the EM-300 nadir (vertically below the vessel).

moderate backscatter are not limited to topographically rough areas. The higher backscatter regions are not evenly distributed nor of constant backscatter value, but occur in large, irregular patches (Figure 2). The largest patch of high backscatter occurs near the summit of the northern ridge.

INTERPRETAION OF THE NEW MULTIBEAM DATA

In our study, we followed the same general interpretation of backscatter data outlined by *Carson et al.* [1994]. We interpret the areas with lowest backscatter as being covered by soft sediment whereas the regions of elevated backscat-

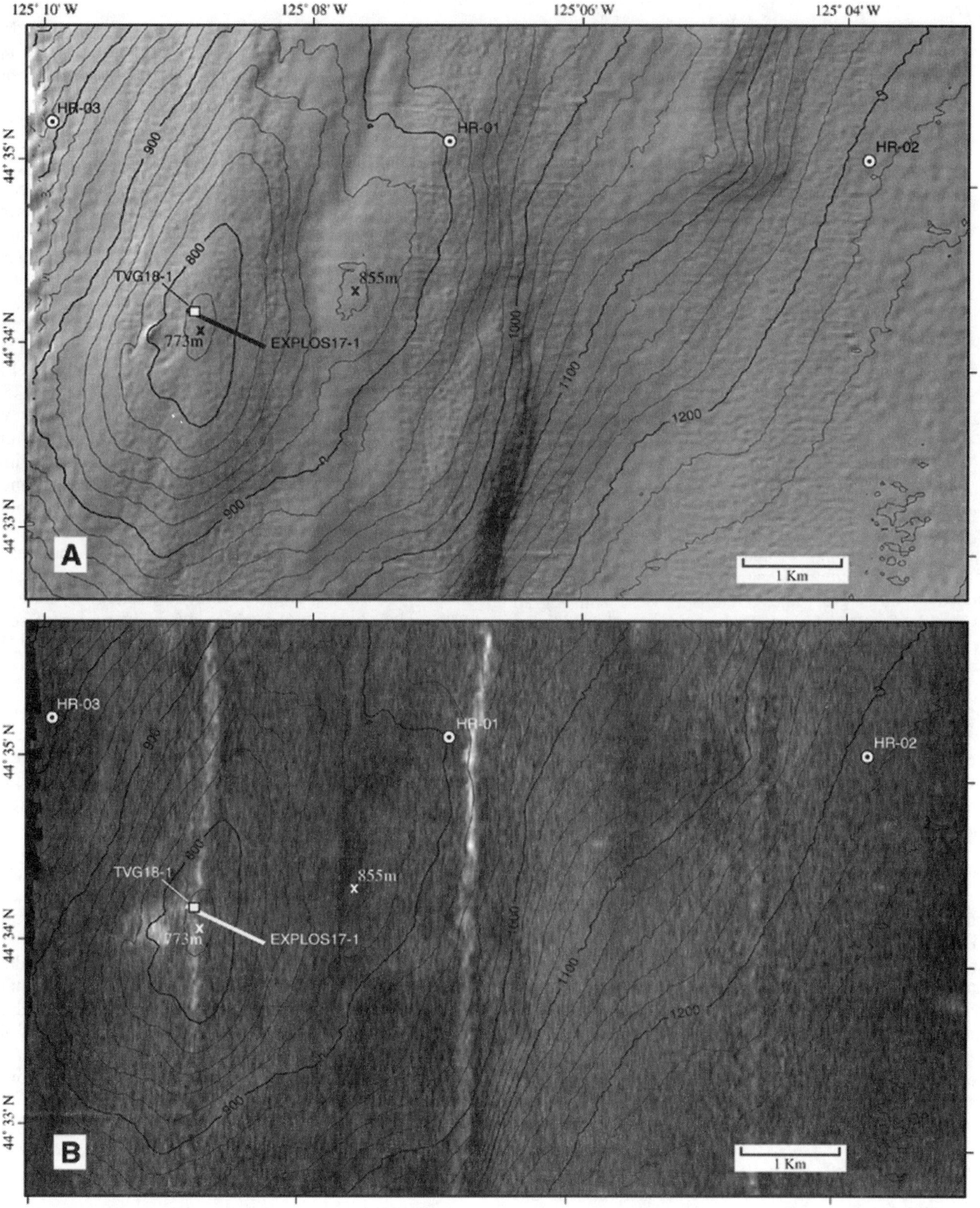

Figure 3. Illuminated bathymetry (a) and backscatter data (b) for the southern part of Hydrate Ridge with 20-m contours. A video camera run [EXPLOS 17-1, *Suess and Bohrmann*, 1998], a grab sample [TVG-18-1, *Bohrmann et al.*, 1999], and three proposed ODP sites [HR-1 to HR-03, *Trehu et al.*, 1998] are labeled and discussed in the text.

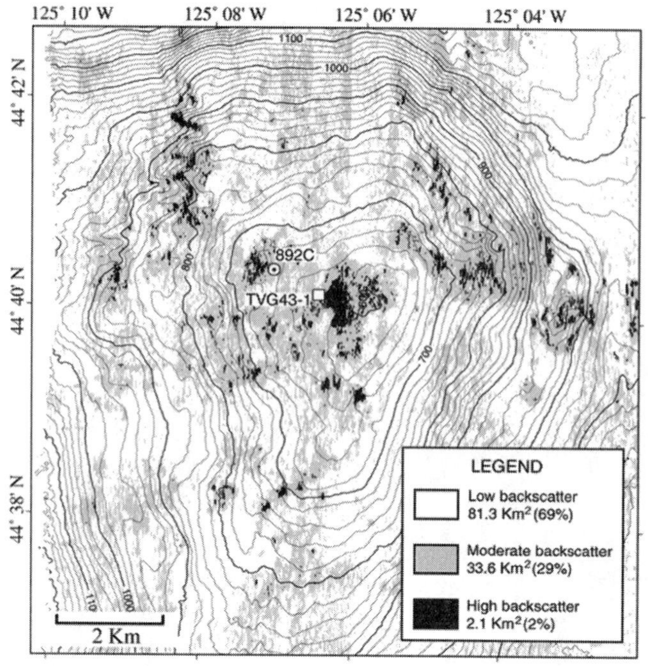

Figure 4. Bathymetry of the northern part of Hydrate Ridge with 20-m contours with the backscatter intensities reduced to three groupings. ODP site 892C and a grab sample [TVG43-1, *Herzig et al.*, 1997] are labeled and discussed in the text.

ter indicate the presence of hard reflective bottom, presumably produced by the carbonate deposits and biological communities that are typically found at fluid seep sites. A video transect (EXPLOS 17-1, shown on Figure 4) near the summit of the southern ridge showed patchy partial bottom coverage of hydrate masses, bacterial mats, and authigenic carbonate deposits [*Suess and Bohrmann*, 1997] and crosses the site where *Bohrmann et al.* [1998] recovered intercalated authigenic carbonate and massive gas hydrate deposits using a grab sampler (station TVG 18-1, shown on Figure 4). The bottom observational data from Sonne cruise SO110 confirm that the regions of moderate backscatter contain scattered seep-related carbonate and hydrate accumulations. Based on these observations, we presume that the highest backscatter regions correspond with nearly continuous hydrate and authigenic carbonate masses. Unfortunately, carbonate, chemosynthetic biological communities, and hydrate accumulations have similar backscatter characteristics and cannot be differentiated with backscatter data.

We categorized the backscatter amplitudes into three roughly equal groups. Moderate and high backscatter areas comprise about 4.9 and 0.095 km^2, respectively, of the southern part of Hydrate Ridge, or about 9% of the surface shown in Figure 3b. The average backscatter value is fairly low, suggesting that most of this area has only very sparse or patchy coverings of authigenic carbonate and hydrate deposits, and chemosynthetic communities. The total bottom-cover of these seep-related materials is probably less than 1 km^2.

The Simrad 30 kHz sonar penetrates surficial bottom sediment, so at least some of the moderate backscatter regions could indicate that authigenic carbonate and hydrate accumulations are buried just beneath the surface (probably <1-m sub-bottom). Distinguishing between a bottom that is covered by patchy carbonate and hydrate masses and a bottom that includes shallowly buried carbonate and hydrate deposits requires additional observational or deep-tow high-resolution seismic reflection data.

Herzig et al. [1997] report carbonate deposits and seep fauna are common in this region and also recovered hydrate masses in a grab sample (TVG 43-1, shown on Figure 3) just west of the summit. We used the same backscatter categorization procedure for the northern ridge and found that 26 km^2 had moderate backscatter and 3.4 km^2 had high backscatter. The 29.4 km^2 of moderate to high backscatter patches (26% of the surface) on the northern ridge suggests that patchy authigenic carbonate accumulations, with some hydrate, cover this entire region. The average backscatter suggests that perhaps 1/3 of the area of these patches, or perhaps 10 km^2, consists of surficial or shallowly buried authigenic carbonate and/or hydrate deposits. The presence of such an extensive region of seep-related deposits certainly justifies naming this feature Hydrate Ridge.

Areas with high backscatter are not distributed in linear patterns, suggesting that fluid venting is diffuse rather than focused along faults. In fact, the distribution of carbonate and hydrate deposits on the northern ridge is broadly concentric and follows bathymetric contours for depths shallower than about 1,000 m. This concentric pattern centered on the summit suggests that fluid flow is focused through the topographic highs. A similar pattern is observed on the Blake Plateau where the shallowest sulfate-methane interface depth, indicative of higher methane fluxes, occurs consistently at or near the ridge crest [*Borowski et al.*, 1999].

Several of the embayments on the west slope of Hydrate Ridge appear to be small-scale slumps, although the material removed is not apparent downslope from the embayments. A semicircular feature on the east flank of the northern ridge appears to be a breached and exhumed anticline, as do a pair of hills aligned on the north-northwest lineament.

Regional Distribution of High Backscatter Caused by Flow Phenomena

Every ridge or hill with closed contours within the entire area that was surveyed displays high backscatter. In fact, Hydrate Ridge does not exhibit either the highest backscatter values or the largest area of high backscatter in comparison with other ridges within this survey. Backscatter amplitude on the ridges northeast of Hydrate Ridge is typically higher and less patchy than elsewhere on Hydrate Ridge. The ridges with the highest amplitude of backscatter and the most extensive and continuous area of high backscatter are located at 44°43.5'N, 124°54.5'W; 44°46'N, 124°56.5'W; 44°49'N, 125°54.5'W; 44°53.5'N, 124°55'W; and 44°57'N, 124°56.5'W. A bathymetric high that also displays high backscatter is located southeast of Hydrate Ridge at 44°26.5'N, 125°02'W. The summit depths of two of these ridges are shallower than 500 m depth and therefore above the stability field for methane hydrate. Surficial authigenic carbonate pavements or active chemosynthetic communities may cause the high backscatter in these regions. However, the high backscatter observed on the crest of the other ridges is probably caused by a combination of surficial or shallowly buried methane hydrate and authigenic carbonate accumulations. At both of these sites, the areas of high backscatter are patchier than at the shallower sites.

CONCLUSIONS

A new high-resolution multibeam bathymetric survey of Hydrate Ridge records backscatter that is interpreted to reveal the distribution of authigenic carbonate and gas hydrate accumulations and chemosynthetic biologic communities. These data indicate that abundant and extensive outcrops of authigenic carbonate and hydrate deposits and chemosynthetic communities characterize the northern part of Hydrate Ridge. These surface accumulations are patchy in their distribution but probably cover nearly 15 km^2 near the summit. In contrast, a much smaller proportion of the southern hill or Hydrate site (about 0.065 km^2 that is within an area slightly larger than 0.5 km^2 that has moderate backscatter values), is associated with the high backscatter interpreted to represent areas of carbonate and hydrate accumulations and chemosynthetic communities. Hydrate Ridge is not the only extensive site of fluid seepage cause deposits in the central Oregon margin. Four other areas, all at the tops of hills that reach above 500-m depth, exhibit more continuous regions of high backscatter. Continuous or near-continuous pavements of carbonate deposits are likely to cause the high backscatter at these sites. Two additional areas, the tops of hills that are deeper than the stability field for methane hydrate, also have high backscatter that is probably caused by the presence of abundant hydrate and carbonate patches. The occurrence of extensive carbonate and hydrate deposits associated with the summits of hills suggests that the fluid seepage responsible for forming the carbonate and hydrate accumulations is focused diffusively through topographic highs rather than along faults.

Acknowledgements. We thank Jennifer Reynolds and David Caress for assistance with the processing of the SIMRAD EM300 data. The paper also benefited from reviews by Dave Scholl and Bill Dillon.

REFERENCES

Borowski, W.S., Paull, C.K., and Ussler, W. III., Global and local variations of interstitial sulfate gradients in deep-water, continental margin sediments: Sensitivity to underlying methane and gas hydrates, *Marine Geology*, 159, 131-154, 1999.

Bohrmann, G., Greinert, J., Suess, E., and Torres, M., Authigenic carbonates from the Cascadia subduction zone and their relation to gas hydrate stability, *Geology*, 26, 647-650, 1998.

Carson, B., Seke, E., Paskevich, V., and Holmes, M.L., Fluid expulsion sites on the Cascadia accretionary prism: Mapping diagenetic deposits with processed GLORIA imagery, *J. Geophys. Res.*, 99, 11,959-11,969, 1994.

Carson, B., and Westbrook, G.K., Modern fluid flow in the Cascadia accretionary wedge: A synthesis, *Proceedings of the Ocean Drilling Program, Scientific Results*, vol. 146 (Pt. 1), Carson, B., Westbrook, G.K., Musgrave, R.J., And Suess, E. (Eds), 413-421, 1995.

Herzig, P., Suess, E., and Linke, P., FS SONNE, Cruise report SO109:HYDROTRACE, Astoria-Victoria-Astoria-Victoria, May 23-July 8, 1996, *GEOMAR report*, 58, 250 p, 1997.

Kulm, L.D., Suess, E., Moore, J.C., Carson, B., Lewis, B.T., Ritger, S.D., Kadko, D.C., Thornburg, T.M., Embley, R.W., Rugh, W.D., Massoth, G.J., Langseth, M.G., Cochrane, G.R., and Scamman, R.L., Oregon subduction zone: Venting, fauna, and carbonates, *Science*, 231, 561-566, 1986.

Kvenvolden, K. A., Gas hydrate—Geological perspective and global change, *Rev. Geophys.*, 31, 173-187, 1988.

MacKay, M.E., Structural variation and landward vergence at the toe of the Oregon accretionary prism, *Tectonics*, 14, 1,309-1,320, 1995.

MacKay, M.E., Moore, G.F., Cochrane, G.R., Moore, J.C., and Kulm, L.D., Landward vergence and oblique structural trends in the Oregon margin accretionary prism: Implications and effects on fluid flow, *Earth Planet. Sci. Lett.*, 109, 477-491, 1992.

ODP Leg 146 Scientiffic Party, ODP Leg 146 examines fluid flow in Cascadia margin, *EOS Trans. Am. Geophys. Union*, 74, 345-347, 1993.

Ritger, S., Carson, B., and Suess, E., Methane-derived authigenic carbonates formed by subduction-induced pore-water expulsion along the Oregon/Washington margin, *Geol. Soc. Am. Bull.*, 98, 147-156, 1987.

Suess, E., and Bohrmann, G., FS SONNE, Cruise report SO110:SO-RO (SONNE-ROPOS), Victoria-Kodiak-Victoria, July 9-August 19, 1996, *GEOMAR report*, 59, 181 p, 1997.

McCaffrey, R., and Goldfinger, C., Forearc deformation and great subduction earthquakes: Implications for Cascadia offshore earthquake potential, *Science*, 267, 856-859, 1995.

Trehu, A.M., Lin, G., Maxwell, E., and Goldfinger, C., A seismic reflection profile across the Cascadia subduction zone offshore central Oregon: New constraints on methane distribution and crustal structure, *J. Geophys. Res.*, 100, 15,101-15,116, 1995.

David A. Clague, Norm Maher, and Charles K. Paull, Monterey Bay Aquarium Research Institute, 7700 Sandholdt Road, Moss Landing, CA 95039-0628

Section 4

IMPLICATIONS

Potential Influence of Gas Hydrates on Seabed Installations

Martin Hovland and Ove Tobias Gudmestad

Statoil, N-4035 Stavanger, Norway

This paper is concerned with the engineering significance of gas hydrates and the possible engineering consequences on seabed installations. A review of physical behaviour of gas hydrate-charged marine soil is presented and the potential influence on the most relevant installations is discussed. Of particular concern are seafloor stability, drilling, sea floor pipelines as well as heavy structures and anchors. A few anecdotal examples of detrimental gas hydrate experienced by the hydrocarbon industry are mentioned. We point to numerous unknown aspects of gas hydrates and call for all means of caution and long term monitoring of engineering structures in hydrate prone areas.

1. INTRODUCTION

Extensive petroleum exploration and development efforts are presently occurring in deep waters where gas hydrates can form. The chances of encountering gas hydrates at the sea floor and within the upper sedimentary layers will only increase as the petroleum industry moves further into deeper waters: off Brasil, in the Gulf of Mexico, off NE Europe, West Africa, and Japan. Petroleum exploration and production involves operations that significantly alter ambient sub-surface conditions. Of particular concern in relation to the occurrence of gas hydrates are disruptions caused by vibrations and pressure gradients as well as temperature increases. Because of this, the presence of gas hydrates may be considered to represent a geohazard for offshore hydrocarbon exploration and exploitation. Also the concern for initiating the development of gas hydrates on or around wellheads and other structures is a major concern.

Although the offshore hydrocarbon industry has experienced numerous setbacks and mishaps because of gas hydrates forming inside production wells and flowlines and topside piping (Gjertsen et al., 1997; Austvik et al., 1997), there are very few reports of incidents due to in-situ gas hydrates. Even so, this paper aims at providing a review of the current knowledge and particularly focusing the aspects of concern to the petroleum industry concerning in-situ gas hydrates.

At water depths and sediment depths within the gas hydrate stability zone (GHSZ), gas hydrates form when light hydrocarbons (methane, ethane, propane, butane), or other non-hydrocarbon gases (including CO_2 and H_2S) common in hydrocarbon-bearing regions are present in adequate concentrations. Of concern to the industry is that gas hydrates may form rapidly when appropriate gases leak into the sediments or water column from below. Furthermore, gas hydrates may dissociate rapidly when disturbed by heating or de-pressurization

The ideal assessment of a potential drilling site or deep-water construction site, including potential pipeline routes, therefore, has to include an evaluation of the possible presence and formation of gas hydrates. However, as many oil companies and academic institutions have already experienced, such an assessment is very difficult to perform, even after having employed dedicated drilling and sediment sampling (Hovland et al., 1999). In such an assessment it is clear that all the available indicators of gas hydrate formation and dissociation in the environment needs to be used, including indirect means. These include interpretation of seismic, sonar, and topographic features, as well as direct ground-truthing means, such as visual observation and seabed sampling (MacDonald et al., 1994; Borowski and

Natural Gas Hydrates: Occurrence, Distribution, and Detection
Geophysical Monograph 124

Paull, 1997; Paull, 1997; Hovland et al., 1997; Dillon et al., 1998; Bouriak et al., 1999; Clennell et al., 1999; Orange et al., 1999; Prior and Hooper, 1999; Suess et al., 1999; Torres et al., 1999; Vogt et al., 1999; Xu and Ruppel, 1999). An assessment of potential gas hydrates should also go hand-in-hand with laboratory work and a theoretical consideration and analysis (Clennell et al., 1999; Austvik et al., 2000).

2. GAS HYDRATE FORMATION AND DISSOCIATION, CASE STUDY

Examples of how gas hydrates may grow and dissociate is best observed during controlled laboratory experiments. A vividly described experiment has thus been provided by Austvik and co-workers (2000): "An initial attempt at forming hydrate plugs in the hydrate reactor was carried out with a gas mixture of C1 to iC4, and tap water in the absence of hydrocarbon liquids. Hydrates were initially formed by injecting water droplets into the gas atmosphere at a pressure of 60 bar. A total of 13.5 liters of water was injected during 1.5 hours at a subcooling of 13°C. Gas was periodically circulated through the plug from the bottom of the reactor and the plug was periodically allowed to mature for several days without disturbances. After a stagnant period 4 more liters of water were injected at the top of the plug. Most of the water was sucked into the plug, working as a sponge, and remained in the plug for the rest of the experiment. Even though the entire experiment lasted for 20 days, the plug remained porous. The plug length was 40 cm. When a pressure difference was introduced over the plug the video recordings demonstrated that gas bubbled easily through the internal channels in the plug. Hence, the nature of the channels determined the permeability of the plug. Gas bubbled through the channels until more hydrates were formed and blocked further migration. The gas then found new channels until these were blocked as well. Thus, the plug gradually became less permeable. By measuring the gas released during melting and the water volume after complete dissociation the plug content was calculated. The volume percentage of hydrates, free water and free gas in the plug was 12%, 38% and 50%, respectively. Hence, the hydrate content in the plug was low and the porosity of the plug was 88%." (Austvik et al., 2000).

3. PHYSICAL BEHAVIOUR OF GAS HYDRATE-CHARGED MARINE SOIL

There is a wide range of features within the sediments and also on the seafloor surface that can be ascribed to the formation and dissociation of gas hydrates. Because we are only in a very early state of knowledge when it comes to understanding, let alone modelling the processes responsible for these features, it is very difficult to predict the exact physical behaviour of gas hydrate-charged marine soil.

However, a few general statements can be made: Because in-situ gas hydrate development is dependent on the availability of excess gases and relatively high sediment water content, it is a practical and fair statement to say that in regions of no hydrocarbon flux, there are no BSRs (bottom simulating reflectors) or significant volumes of gas hydrates (Bouriak et al., 2000; Hovland et al., 1999). Experience from several ODP (Ocean Drilling Program) legs and other academic research proves that the corollary to this statement is also valid: namely that in regions of low to moderate hydrocarbon flux and fine-grained marine sediments, there could be BSRs and a high probability of thin, discrete layers of gas hydrates. The thermodynamic and physico-chemical reasons for this is further discussed and described in Clennell et al. (1999).

But, if there are high-porosity sediments present, such as buried sands and gravels in regions with hydrocarbon flux, these sediment layers are likely to have developed into massive gas hydrate reservoirs, as demonstrated by the Mallik gas hydrate research well in the MacKenzie Delta in 1998 (Dallimore et al., 1999). Also, with increasing hydrocarbon flux and increasing tectonic activity, the amount of gas hydrates within the sediments and near the surface will increase. This has been demonstrated by amongst others, the Geochemical Exploration and Research Group (GERG), in the Gulf of Mexico (Sassen et al., 1993; MacDonald et al., 1994). In addition, the seafloor topography will be increasingly affected, and the presence of gas hydrates may even lead to very large slope failures and seafloor collapse events (Prior et al., 1989; Prior and Hooper, 1999; Dillon et al., 1998; Bouriak et al., 2000).

In regions where there is a very high fluid flux through the sea floor, masses of gas hydrates may form within the sea floor sediments, as described by Suess and co-workers (1999), from the Hydrate Ridge area: "Pure gas hydrate occurred in layers or joints several millimeters to decimeters thick, generally oriented parallel to bedding, but sometimes cutting the bedding planes obliquely. There was not one instance observed where the hydrate occupied the original sediment pore space and cemented the sediment framework on a microscale. Rather, it filled large secondary pore space either as fractures or joints, or seemed to create its own space during growth by fracturing or pushing apart the sediment framework, most often along bedding planes. The internal fabric of the pure hydrate showed a peculiar structure with large filled pores not unlike bubble wrap material." (Suess et al., 1999).

Up till now, very few incidents due to in-situ gas hydrates have been documented by the hydrocarbon industry. Indications of what may happen when gas hydrate-bearing sediments are disturbed have, however, been documented by scientific drilling.

When gas hydrates dissociate they return to a mixture of free gas and water. This may totally change the physical character of sediment conditions and even of the seabed topography. For example, "soupy layers" were repeatedly encountered when drilling through marine sediments above a prominent BSR during Leg 146 of ODP (Westbrook et al., 1995). These sampled layers were at first interpreted as having been severely disturbed by the drilling operations, but later turned out to have been gas hydrate bearing. On dissociation due to de-pressurization, the water released affected the sediments and turned them into a totally structureless "soup", or rather, melange. It proved that the gas hydrates had been concentrated as abundant, but minute crystals dispersed in the clayey sediment matrix of only distinct layers (of slightly higher permeability?). The model presented by Clennell et al. (1999) explains this distribution in fine-grained sediments, and compares it with permafrost formation in arctic soil.

(The authors regret that several of the observations and experiences stated below cannot be scientifically documented, as their time, location, and other particulars come under industrial confidentiality).

Sea floor engineering constructions need to be located at certain distances from fluid escape features, such as pockmarks, acoustically high reflective patches on the surface and at depth, mud diapirs, mud mounds, and gas hydrate mounds (Borowski and Paull., 1997; Prior and Hooper, 1999). The reason is that any drilling on or near such structures may disturb the natural sub-surface "plumbing" system, which may be very delicate and sensitive to pressure and fluid disturbances. However, if it is necessary to perform engineering and drilling on and close by such natural structures, then it calls for great caution and the conduction of various surveys and assessment programs, including modelling and laboratory simulations.

Also the introduction of gas hydrate forming gases into the GHSZ may cause the formation of "unwanted" gas hydrates. Such formation may cause damaging local increases in sediment bulk volumes, and local pressure increases. Gas hydrate formation on and around mechanical structures placed on the seafloor (i.e., wellheads and blowout preventers, BOPs) have been known to become sealed off by rapidly forming gas hydrates caused by unwanted, but apparently insignificant gas seepages, such that mechanical systems are rendered inoperable.

4. ENGINEERING SIGNIFICANCE AND CONSEQUENCES

Extensive data collection efforts on the Louisiana continental slope (Roberts et al., 1999) have demonstrated that "temperature increase in the water column of 1.0 - 1.5°C caused distinct dissociation of the gas hydrate leading to out-gassing. Smaller fluctuations in out-gassing were also recorded at the tidal frequency", implying that very small perturbations in the ambient conditions caused by influx of warmer water or by pressure fluctuations could lead to gas release. Therefore recognising the occurrence of gas hydrates and understanding the significance gas hydrates may have on engineering structures is important to ensure the safe design, installation and operation of such structures.

5. POTENTIAL ENGINEERING CONSEQUENCES

5.1 Sea floor stability

Sediment instability can be caused by gas hydrates dissociating from lower layers, resulting in decreased sediment strength. A similar phenomenon is experienced in permafrost regions onshore in association with the heating of permafrost. Dissociation of gas hydrates can be caused by temperature increases or pressure fluctuations. Of particular concern is the potential reduction in sediment strength of sloping sediments, which could trigger underwater slides and movement of large sediment volumes.

Near a deep-water Gulf of Mexico construction site Prior and Hooper (1999) investigated signs of sediment instability. Their case study demonstrates the effort needed to assess and secure a potentially dangerous geohazard associated with gas and gas hydrates: "A deep water oil and gas production Tension Leg Platform (TLP) system has been successfully installed in 872 m water depth, in an area where there has been a history of point-source repetitive debris flows (Prior and Doyle, 1993). Detailed mapping using deep tow survey techniques revealed that two large fluid expulsion mounds and craters are the sources of debris flows, with chutes and shear-bounded channels leading downslope to a sequence of stacked debris lobes. Analysis of lobe geometries and sediment properties revealed episodes of repeated debris flows that began on slopes of 10° to 15°, and were associated with periodic venting of large volumes of overpressured gas, with debris runout distances of up to 11 km. Development of the TLP site model, particularly using sediment properties and age-dating, yielded the following information:

- the site is 1,800 m downslope from the most recent flows;

-the most recent flows occurred within the last 1,000 years but are low magnitude, very thin flows not exceeding 5 m thick;

-there are indications of about 4 events in the past 12,000 years;

-a deeply buried flow dates from about 25 to 30,000 years BP; and

-older more deeply bedded flows are of Pleistocene age, greater than 30,000 years.

From an engineering perspective, it is evident that the site has not experienced debris flow activity for at least 12,000 years because all the later debris flows stopped short of the site by at least 1,800 m. Furthermore, if debris flows recur, the overall frequency pattern suggests the TLP will not be affected during its lifetime. Also, the general trend in reduction of debris flow magnitude (thickness and runout distance) with time suggests that the maximum flow thickness that could affect the site would be 4 to 5 m. There are geotechnical data for the flow deposits to provide a conservative approach tofoundation design." (Prior and Hooper, 1999, p. 434-435).

The consequences of gas hydrate associated sediment instability to engineering structures would of course be very damaging. The potentially most disruptive engineering structures are considered to be those carrying warm fluids, like wells with hydrocarbons and hot pipelines. An extensive evaluation of the sea floor stability should precede any sea floor installation operation in order to prevent any unwanted instability development. The consequences could be both serious environmental damage by hydrocarbon release through free flowing wells, or by sea floor slumping and total loss of engineering structures. Thus, the economic consequences associated with losses and environmental damage must be considered as well. Often, it is only when such an economic consequence assessment has been performed that the magnitude of an unwanted event is properly understood and action is taken to investigate the true natural conditions.

5.2 Drilling

Drilling into the sea bottom represents a considerable disturbance of the upper sediment layers. Gas hydrates in the sediment could dissociate releasing gas during drilling (Fig. 3). Free gas released during drilling may form gas hydrates elsewhere. If for example gas hydrates form around the wellhead and BOP, it could cause the mechanical systems to fail, even at low rates of formation (Fig. 5). Larger gas releases could have consequences for the stability of the drilling rig, possibly causing it to lose buoyancy, and which would seriously affect the safety of the rig and its crew. Alternatively, the forces from the gas plume could represent a skew load causing the rig to heel (Hovland and Judd, 1988).

The greatest drilling concern, however, is associated with the production of warm hydrocarbons (up to over 100°C) through a casing, heating the surrounding formations. Examples of casing collapse due to excess local pressure caused by dissociating in-situ gas hydrates is known to have occurred (Fig. 4). Although insulation of the casing may be attempted, it is doubted whether the present state of knowledge and technology guarantees sufficient measures of safe operations. Another aspect of concern is the possibility of introducing gas hydrates in the drilling mud with possible loss of drilling mud control and free gas development as the hydrates dissociate.

If there is the slightest probability that wells could be severed due to sediment movement, subsea safety barriers must be installed to avoid free-flowing wells.

5.3 Sea floor pipelines

A warm pipeline on the sea bottom, or buried in sediments will warm the surroundings, and will create a temperature gradient which eventually could cause in-situ gas hydrates to dissociate. The released methane and water could cause sediment instability and slumping. Under such conditions the pipeline could rupture if exposed to an excessive free span, or if the stresses in the buried pipeline become excessive (Figs. 1 and 2). In order to avoid such scenarios, efficient pipeline insulation would be required. In the case of sea floor instability a buried pipeline is more vulnerable to large stresses than a pipeline lying directly on the sea floor. The installation of sea floor safety valves will reduce the potential risk for large oil spills from sea floor pipelines.

Flowlines transporting hot fluids are also prone to upheaval buckling due to thermal expansion and failure of the stability design (gravel or concrete mattress cover). Any in-situ near surface gas hydrates could dissociate and render the seafloor "soupy", and thus cause loss of stability and initiate such unwanted buckling.

5.4 Heavy structures and anchors

Heavy structures placed on the sea floor will exert additional pressure on the sea floor sediments, thus changing the ambient conditions. Such structures are typically linked to hot hydrocarbons in casings, manifolds or pipelines. Their stability on the sea floor would be threatened if gas release causes considerable soil movements or if gas pressure is allowed to build up under compartements of the structure when they protrude into the

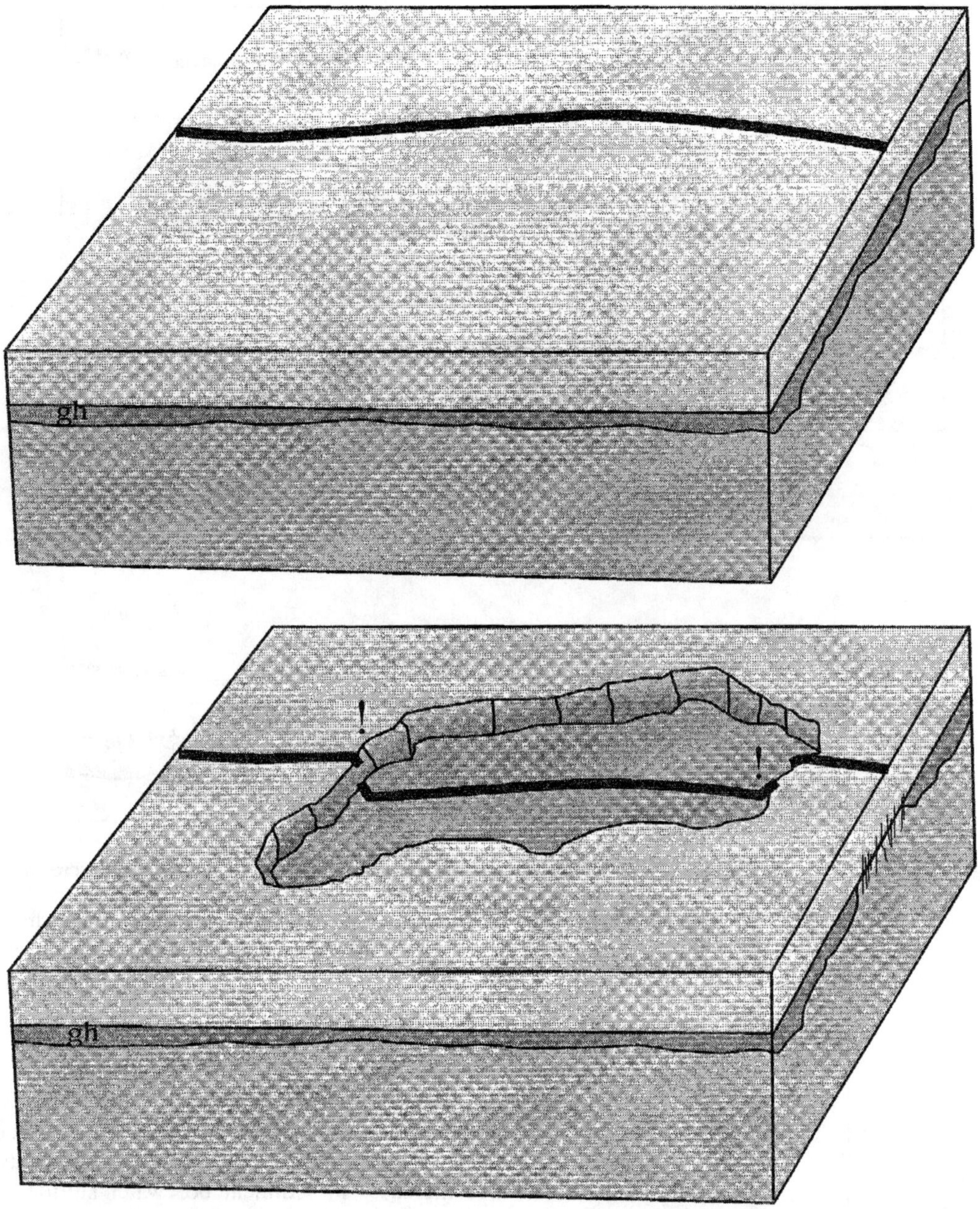

Figure 1. Heat from the pipeline on the seabed may cause gas hydrate dissociation, fluidization of sediments and cratering. The pipeline may become severed as a result (drawing modified from Borowski and Paull, 1997).

sea floor. Such compartements should therefore be equipped with ventilation mechanisms for gas pressure release.

Suction anchors are becoming increasingly popular in deep waters to keep floating structures anchored to the sea floor. As for the heavy structures discussed above, gas could accumulate inside the anchors threatening the safe operation. If the achors exert a suction force, the pressure reduction could cause local gas hydrate dissociation with possible gas build up inside the anchor buckets and soil movements disturbing the friction forces along the walls of the anchor (Fig. 3). In order to avoid hazardous situations, the suction in the anchors should be kept to a minimum for locations where there is a danger of disturbing the stability

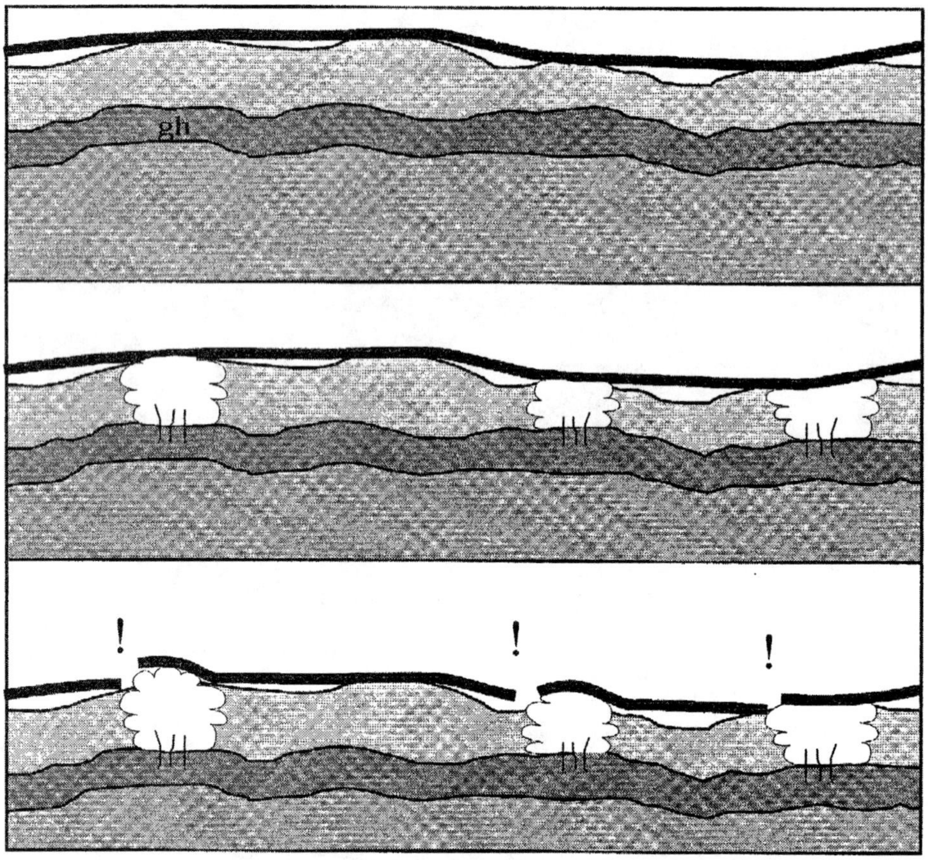

Figure 2. Similar case to Fig. 1, but here the seabed sediments swell as gas hydrates dissociate and re-form.

of in-situ gas hydrates. In such areas the forces on the anchor should whenever possible be taken up horizontally rather than vertically, employing long mooring lines on the sea floor.

5.5 Corrosion

Because gas hydrate dissociation may lead to methane release and anoxic conditions in the sediments, it also leads to increases in sediment porewater sulfide content, which could represent an agressive corrosive environment for steel structures (Sahling et al., 1999). This could call for extra corrosion protection.

6. CONCLUSIONS

In spite of many years of research, observation, and drilling in deep ocean sediments, the true dynamic nature of

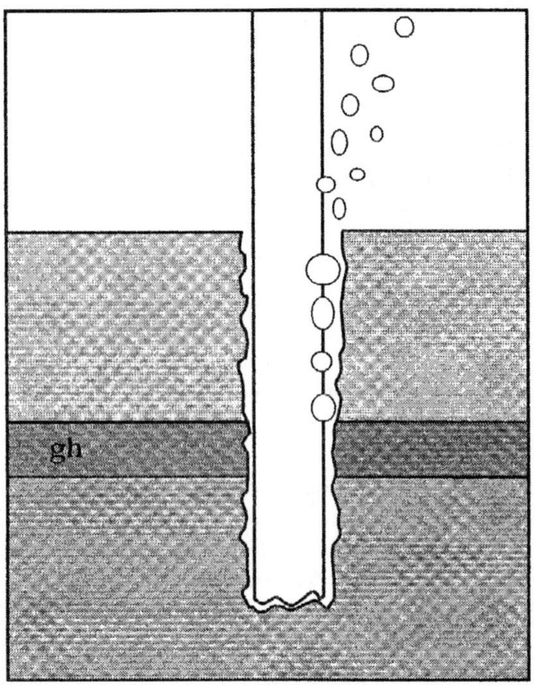

Figure 3. Penetration of gas hydrate bearing layer may lead to gas seepage.

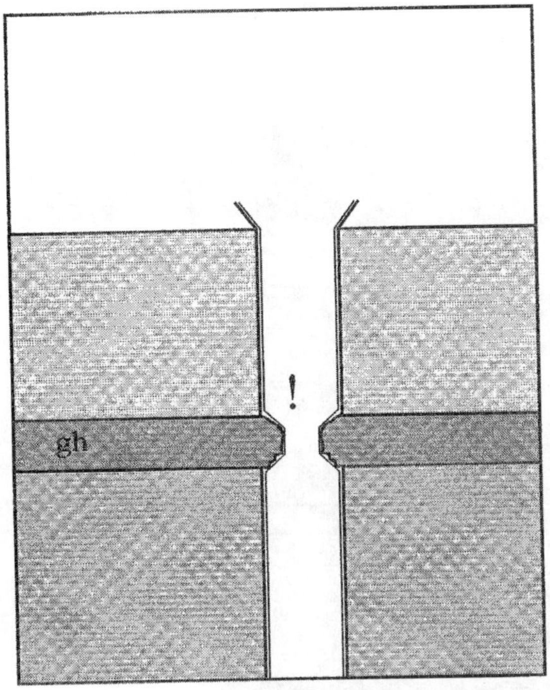

in-situ gas hydrate formation and dissociation is still being discovered. Even though scientific research in areas with abundant natural gas hydrates, such as Hydrate Ridge on the Cascadia accretionary wedge, off Oregon (Torres et al, 1999; Zhou et al., 1999) and parts of the Gulf of Mexico (MacDonald et al., 1994) is currently on the increase, there are still numerous unknown aspects of gas hydrates in marine sediments. The delicate thermodynamic balance and nature of gas hydrates in natural environments, calls for all means of caution.

For any potential deep-water interaction site, the need for mapping, inspection, geochemical, geotechnical and biological sampling and theoretical assessment is even greater than for shallow water sites. Because the hydrates may develop and dissociate slowly over time, long term monitoring of any engineering structures in relevant areas is also called for.

Figure 4. When casing through a gas hydrate bearing layer, the casing may collapse due to high pressure in the zone. A similar incident has occurred at the Messakoya gas field in Siberia.

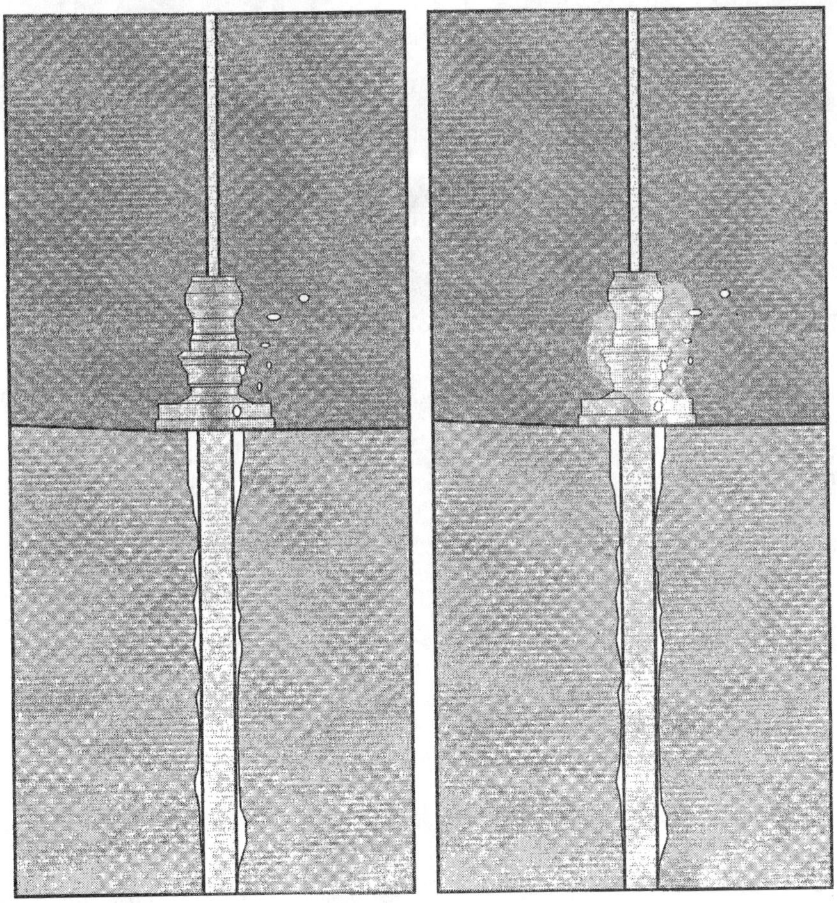

Figure 5. Leaking gas from a well may form gas hydrate ice on the BOP and wellhead, rendering sensitive mechanisms inoperable.

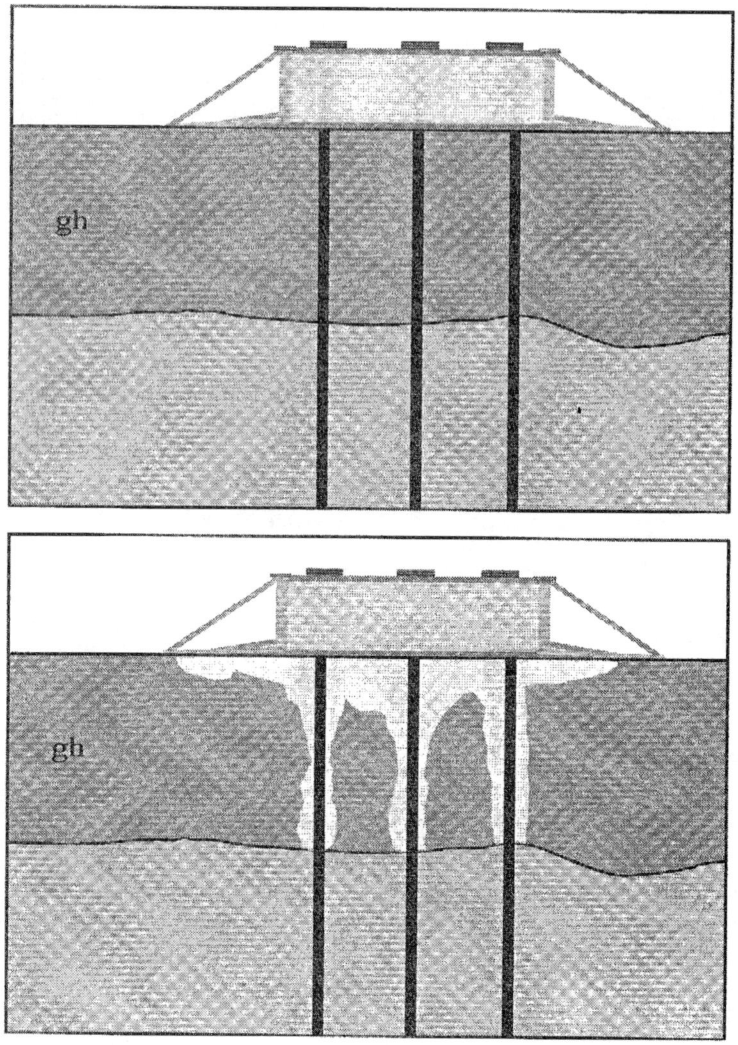

Figure 6. Heat from wells and template may cause fluidization of mud and foundation destabilization.

Acknowledgements. We would like to thank Charles Paull at Monterey Bay Research Institute for comments.

REFERENCES

Austvik, T., E.Hustvedt, L.H. Gjertsen, and O.Urdahl, Formation and Removal of Hydrate Plugs: Field Trial at Tommeliten. *Proceedings of GPA Annual Conference*. San Antonio, USA, 1997.

Austvik, T., X. Li, and L.H. Gjertsen, Hydrate plug properties - formation and removal of plugs. *Conf. on natural gas hydrates*. Salt Lake City, 2000.

Borowski, W.S., and C.K. Paull, The gas hydrate detection problem: Recognition of shallow-subbottom gas hazards in deep-water areas. *Offshore Technology Conference*. Paper no. OTC8297, 211-216, 1997.

Bouriak, S., M. Vanneste, and A. Saoutkine, Inferred gas hydrates, shallow gas accumulations and clay diapirs on the southern edge of the Vøring Plateau, offshore Norway. *Mar. Geol.*, 163, 125-148, 2000.

Clennell, M.B., M. Hovland, J.S. Booth, P. Henry, and W.J. Winters, Formation of natural gas hydrates in marine sediments. 1. Conceptual model of gas hydrate growth conditioned by host sediment properties. *Journal of Geophysical Research,* 104, no. B10, 22985-23003, 1999.

Dallimore, S.R., T. Uchida and T.S. Collett, Scientific Results from JAPEX/JNOC/GSC Mallik 2L-38 gas hydrate research well, Mackenzie Delta,
Northwest Territories, *Canada. Geol. Soc. Canada Bulletin* No. 544, 1999.

Dillon, W.P., W.W. Danforth, D.R. Hutchinson, R.M. Drury, M.H. Taylor, and J.S. Booth, Evidence for faulting related to dissociation of gas hydrate and release of methane off southeastern United States. In: Henriet, J.P. and J. Mienert (Eds.), Gas hydrates: Relevance to world margin stability and climate change. *Geol. Soc. Spec. Publ.*, No. 137, 293-302, 1998.

Ergun, M., and G. Cifci, Gas-saturated sediments in the eastern Black Sea and geohazard effects. *Offshore Technology Conf.*, OTC paper no. 10924, 621-630, 1999.

Fisher, C.R., I.R. MacDonald, S. Macko, K. Nelson, R. Sassen, M. Joy, M., R. Carney, and E. McMullin, Colonization of gas hydrates by metazoans (abstr.). Conference proceedings. Gas hydrates: Resource? Hazard? Origins? *Geoscience 98*, Keele, U.K., April 1998.

Gjertsen L.H., T. Austvik, O.Urdahl, and R.Duus, Removal of a Gas Hydrate Plug from a Subsea Multiphase Pipeline in the North Sea. *Multiphase '97*. 515-526, 1997.

Hovland, M., T.J.G. Francis, G.E. Claypool, and M.M. Ball, M.M., Strategy for scientific drilling of marine gas hydrates. *Joides Journal*, 25, no. 1., 20-24, 1999.

Hovland, M. and A. Judd (Eds.), *Seabed pockmarks and seepages. Impact on geology, biology and the marine environment*, 293 pp., Graham and Trotman, Ltd, London, 1988.

Hovland, M., J.W. Gallagher, M.B. Clennell, and K. Lekvam, K., Gas hydrate and free gas volumes in maarine sediments: Example from the Niger Delta front. *Marine and Petroleum Geology*. 14, No. 3, 245-255, 1997.

MacDonald, I.R., N.L. Guinasso, R. Sassen, J.M. Brooks, L. Lee, and K.T. Scott, Gas hydrates that breaches the sea floor on the continental slope of the Gulf of Mexico. *Geology*, 22, 699-702, 1994.

Orange, D.L., M.M. Angell, and D. Lapp, D., Using seafloor mapping (bathymetry and backscatter) and high resolution sub-bottom profiling for both exploration and production: Detecting seeps, mapping geohazards, and managing data overload with GIS. *Offshore Technology Conf.*, OTC Paper no. 10870, 469 - 487, 1999.

Paull, C.K., 1997. Drilling for gas hydrates: Offshore drilling program Leg 164. Offshore Technology Conference. Paper no. OTC8294, 185-192.

Prior, D.B., E.H. Doyle, and M.J. Kaluza,.,. Evidence for sediment eruption on deep sea floor, Gulf of Mexico. Science 243, 517-519, 1989.

Prior, D.B., and J.R. Hooper, Sea floor engineering geomorphology: recent achievements and future directions. *Geomorphology*, 31, 411-439, 1999.

Sahling, H., D. Rickert, and E. Suess, Faunal community structure along a sulphide gradient: interrelationship between porewater chemistry and organisms associated with gas hydrates, Oregon subduction zone. (Abstr.). *Eos Trans. AGU*, 80, No. 46, F510, 1999.

Sassen, R., H.H. Roberts, P. Ahron, J. Larkin, E.W. Chinn, and R. Carney, Chemosynthetic bacterial mats at cold hydrocarbon seeps, Gulf of Mexico continental slope. *Organic Geochemistry*, 20, 77-89, 1993.

Suess, E., M.E. Torres, G. Bohrmann, R.W. Collier, J. Greinert, P. Linke, G. Rehder, A. Trehu, K. Wallmann, G. Winckler, and E. Zuleger, Gas hydrate destabilization: enhanced dewatering, benthic material turnover and large methane plumes at the Cascadia convergent margin. *Earth and Planetary Science Letters*, 170, 1-15, 1999.

Torres, M.E., P. Linke, A. Trehu, K.M. Brown, and K. Heescen, K., Gas hydrate dynamics at Hydrate Ridge, Cascadia. (Abstr.). *Eos Trans. AGU*, 80, No. 46, F527-F528, 1999.

Vogt, P.R., J. Gardner, and K. Crane, The Norwegian-Barents-Svalbard (NBS) continental margin: Introducing a natural laboratory of mass wasting, hydrates, and ascent of sediment, pore water, and methane. *Geo-Marine Letters*, 19, 2-21, 1999.

Xu, W., and C. Ruppel, Predicting the occurrence, distribution, and evolution of methane gas hydrate in porous marine sediments. *Journal of Geophysical Research*, 104, no. B3, 5081-5095, 1999.

Zhou, Y., C. Goldfinger, J.E. Johnson, M. Torres, A. Trehu, D. Clague, C. Paull, C., and G. Bohrmann, Distribution and morphology of venting-related carboantes near Hydrate Ridge, Oregon Margin, based on sidescan sonar and multibeam imagery. (Abstr.). *Eos Trans. AGU*, 80, No. 46, F510, 1999.

Gudmestad, Ove Tobias and Hovland, Martin, Statoil, N-4035 Stavanger, Norway.